19,21 / 125,-

THE KLEENE SYMPOSIUM

STUDIES IN LOGIC

AND

THE FOUNDATIONS OF MATHEMATICS

VOLUME 101

Editors

NORTH-HOLLAND PUBLISHING COMPANY
AMSTERDAM · NEW YORK · OXFORD

THE KLEENE SYMPOSIUM

Proceedings of the Symposium held June 18–24, 1978 at Madison, Wisconsin, U.S.A.

Edited by

Jon BARWISE
Stanford University, Stanford, CA, U.S.A.

H. Jerome KEISLER
University of Wisconsin-Madison, WI, U.S.A.

Kenneth KUNEN
University of Texas, Austin, TX, U.S.A.

1980

NORTH-HOLLAND PUBLISHING COMPANY
AMSTERDAM · NEW YORK · OXFORD

ISBN: 0 444 55719 9

Published by:

North-Holland Publishing Company—Amsterdam · New York · Oxford

Sole distributors for the U.S.A. and Canada:

Elsevier North-Holland, Inc.
52 Vanderbilt Avenue
New York, N.Y. 10017

Library of Congress Cataloging in Publication Data

Kleene Symposium, University of Wisconsin-Madison, 1978.
The Kleene Symposium.

(Studies in logic and the foundations of mathematics; v. 101)
Bibliography: p. xii
1. Recursion theory—Congresses. 2. Intuitionistic mathematics—Congresses. 3. Kleene, Stephen Cole, 1909– I. Kleene, Stephen Cole, 1909– II. Barwise, Jon. III. Keisler, H. Jerome. IV. Kunen, Kenneth. V. Series.
QA9.6.K53 1978 511'.3 79-20792
ISBN 0-444-55719-9

PRINTED IN THE NETHERLANDS

Contents

Stephen C. Kleene

A Biographical Sketch of Stephen C. Kleene

Stephen Cole Kleene was born in Hartford, Connecticut, on January 5, 1909. His father, Gustav Adolph Kleene, was a professor of economics and his mother, Alice Cole Kleene, was a published poet. He grew up in Hartford, but spent his summers on a family farm in Maine. This farm has been in his family since 1824 and has been visited by many logicians over the years. He still returns there frequently.

Kleene received his Bachelor of Arts degree from Amherst College in 1930. From 1930 to 1935 he was a graduate student and research assistant in mathematics at Princeton University, studying under Professor Alonzo Church. This was an exciting period, when the completeness and incompleteness theorems of Kurt Gödel were announced. Kleene had contact with Gödel at the Institute for Advanced Study, and with J. Barkley Rosser, his friend and contemporary as a graduate student at Princeton and later as a member of the Mathematics Department at Wisconsin. Kleene's major lifelong research interest, recursive function theory, was created during his years at Princeton as a synthesis of the work of Kleene, Church, Gödel, Rosser and a few others, notably Emil L. Post in New York and Alan R. Turing in England.

He took his doctorate under Church in 1934, at the height of the depression. His first full-time academic position in 1935 was an Instructorship at the University of Wisconsin, where he was to spend most of his career. He was promoted to Assistant Professor in 1937 and visited the Institute for Advanced Study in 1939–40. During this time he laid the foundations for recursive function theory and developed a graduate course which would later become his famous book, *Introduction to Metamathematics* (published jointly by North-Holland Publishing Company, P. Noordhoff and D. Van Nostrand in 1952). In 1941 he returned to his alma mater, Amherst, as an Associate Professor.

From 1942 to 1945, in the Second World War, he served in the United States Navy, rising from the rank of Lieutenant (j.g.) to Lieutenant Commander. He was a navigation instructor at the USNR Midshipmen's School in New York,[1] and then a Project Director at the Naval Research

[1]Kleene drafted the *Atlantis Chart*, of which thousands of copies were used by Midshipmen in doing navigation exercises. The island on the chart was originally labeled "Fig I.", meaning Fig Island, as seemed appropriate to the latitude. But the Chief objected that this could be confused with "Figure I", and Kleene changed it by a stroke of the pen to "Pig I."

Laboratory in Washington, D.C. He returned to Wisconsin as an Associate Professor in 1946. He was promoted to Full Professor in 1948 and Cyrus C. MacDuffee Professor in 1964. In the 1940's he turned to a second area of research, intuitionistic logic, and introduced the fundamental method of recursive realizability discussed in the article by Joan Moschovakis in this volume. He visited Amsterdam in 1949–50 as a Guggenheim Fellow, Princeton in 1956–57 as a Visiting Professor, Marburg in 1958–59 as a National Science Foundation Grantee, and the Institute for Advanced Study again in 1965–66. For the remainder of his career he frequently alternated between the two areas of recursive function theory and intuitionism. In the summer of 1951, however, he worked at RAND Corporation and produced a major breakthrough in a third area of research, the theory of finite automata, discussed by Robert Constable in this volume. His research on recursive functions in the 1950s and early 1960s largely shaped the future course of the subject. His work with the late Emil Post was the beginning of the study of degrees of unsolvability, and his work on recursive functionals formed the basis of higher recursion theory. His most recent research, reported at the 1977 Oslo meeting and in the present volume, continues his work on higher recursion theory. In the late 1960s he concentrated on intuitionism in the area discussed in Richard Vesley's article. Kleene's book with Vesley is a product of that period.

Throughout his career, Steve Kleene has attracted graduate students. Many of his former students have contributed to this volume. Kleene was the only faculty member in logic at Wisconsin until the arrival of H. Jerome Keisler in 1962 and J. Barkley Rosser and Michael Morley in 1963. He built a strong tradition in logic which is found at only a handful of other universities, and which has been of great value to the field. The logic picnics at Devil's Lake, featuring a rock climbing expedition led by Steve Kleene, were well established by 1962. The picnic at the Kleene Symposium brought back memories for many of the participants.

He has been a leader in professional organizations as well as at his university. He served as President of the Association of Symbolic Logic in 1956 through 1958. He was President of the International Union of the History and the Philosophy of Science in 1961, and of its Division of Logic, Methodology and Philosophy of Science in 1960–62, and he was one of the founders of both. He was an editor of the Journal of Symbolic Logic for 12 years. In 1969 he was elected to the National Academy of Sciences, and is one of the few mathematicians to have received this honor. He was Chairman of the Mathematics Department at Wisconsin in 1957–58 and 1960–62 and of Numerical Analysis (now renamed Computer Sciences) in 1962–63. In 1966–67, while Rosser was on leave of absence, Kleene was Acting Director of the Mathematics Research Center. He was the driving force behind the building of Van Vleck Hall, dedicated in 1963, and argued successfully that the top floor lounge would be as valuable to mathematical research as a laboratory is to research in the experimental sciences. It is especially appropriate that the Kleene Symposium has taken place in Van Vleck Hall.

During the years 1969–1974 he made the supreme sacrifice to the University, serving as Dean of the College of Letters and Sciences in a period of retrenchment and campus unrest.[2] As a dean he was tough, decisive and fair. While finding time to teach one course each year, he had to put his research aside. After returning to the Mathematics Department in 1974 and spending two years catching up with the literature in recursive function theory, he has successfully made the transition back to research.

Kleene married Nancy Elliott in 1942 and has four children. His wife died in 1970. He married Jeanne Steinmetz in 1978 and became the stepfather of three children.

Kleene is an active supporter of conservation and environmental causes. To lepidopterists he is known as the discoverer of the variety Beloria toddi ammiralis ab. kleenei (Watson) 1921.

Steve Kleene is a vigorous 70 and will become an Emeritus Professor in the summer of 1979. There is every indication that he will continue his research and also continue to lead the rock climbing at future logic picnics at Wisconsin.

[2]Because of the Deanship, Kleene resigned from the top position in faculty governance at Madison, the Chairmanship of the University Committee, and declined to advance from Chariman Designate to Chairman of the Division of Mathematical Sciences of the National Research Council.

A Summary of Kleene's Work in Recursion Theory

In dedicating his book on degrees of unsolvability, J. R. Shoenfield wrote "to S. C. Kleene, who first made recursion theory into a theory." It is indeed hard to imagine any one field of mathematics that owes so much to one person's work. In this summary we touch briefly on the high points of this work.

Kleene's thesis developed a theory of λ-definable functions of natural numbers, and showed how extensive the class of such functions is. In 1934, developing a suggestion of Herbrand, Gödel introduced the class of general recursive functions via the equation calculus. Based on Kleene's thesis, Kleene and Church proved that the classes of general recursive and the λ-definable functions are the same. In 1935, Church enunciated his now famous thesis, that the class of general recursive functions comprises all the intuitively computable functions, and paved the way for showing that various mathematical problems are undecidable.

Kleene's 1936 Normal Form Theorem states that each general recursive function f is expressible in the form

$$f(y) = U(\mu x T(e,y,x))$$

where T is a particular "primitive recursive" predicate (now called the Kleene T-predicate), U is a particular "primitive recursive" function and "μx" means "the least x such that." (The "primitive recursive" functions are just those definable using only explicit definition and definition by induction.) It follows that the predicate of e given by

$$\exists x\, T(e,e,x)$$

is recursively undecidable. (This formula decorated the special Kleene Symposium T-shirts.)

The early work in recursion theory restricted itself to total functions. In 1938, Kleene introduced the class of partial recursive functions and proved the Recursion Theorem. A number e is called an *index* of a (partial) function of natural numbers f if, for all y:

$$f(y) = \begin{cases} U(\mu x T(e,y,x)) & \text{if } \exists x T(e,y,x), \\ \text{undefined} & \text{otherwise.} \end{cases}$$

As $e = 0, 1, 2, \ldots,$ we get indices for all partial recursive functions. Kleene's

Recursion Theorem states that for each binary partial recursive function $g(\cdot,\cdot)$ there is an integer e such that the partial recursive f defined by

$$f(y)=g(e,y)$$

has index e. Intuitively, each such g has a "fixed point" e. This result, with its surprising two line proof, is one of the recursion theorist's most versatile tools.

In 1940, Kleene showed how the predicates of elementary number theory fall into a natural hierarchy, the arithmetical hierarchy, based on alternations of quantifiers. Each arithmetical predicate $P(x)$ is expressible in one of the forms

$$R(x),\quad \exists y_1 R(x,y_1),\quad \forall y_1 R(x,y_1),\quad \exists y_1 \forall y_2 R(x,y_1,y_2),$$
$$\forall y_1 \exists y_2 R(x,y_1,y_2),\quad \exists y_1 \forall y_2 \exists y_3 R(x,y_2,y_3),\ldots$$

where R is recursive. Furthermore, there are predicates of each form that cannot be expressed in any form without more quantifiers. This was independently discovered, somewhat later, by A. Mostowski.

This heirarchy suggested a number of interesting possibilities, which Kleene developed and which now go under the title of generalized recursion theory. One idea is to iterate the hierarchy into the transfinite, along recursive ordinals. This leads to Kleene's famous hyperarithmetical hierarchy.

Another direction suggested by the arithmetic hierarchy is to study a hierarchy based on function quantification (the "analytic hierarchy"). To do so, Kleene extended the theory of recursive predicates to allow variables over total number-theoretic functions, as well as over numbers, in the 1950s. This culminated in several characterizations of the hyperarithmetic predicates, the simplest being that they are those expressible in both one-function quantifier forms. The work on the analytical hierarchy continues to be one of the frontiers of research today.

Another pioneering paper, this one with E. L. Post in 1954, inaugurated the study of the degrees of undecidability.

Most of Kleene's research in recursion theory from the late 1950s to this volume concerns recursive predicates of objects of arbitrary finite type: numbers, functions of numbers, functions of these, and so on. In particular, Kleene established analogues of the hierarchy theorem and recursion theorem in this new setting. This "higher type" recursion theory is another of the frontiers of logic, as the papers in this volume show.

In addition to his pioneering work in recursion theory, Kleene has studied Brouwer's intuitionism with the tools provided by recursion theory. His work in this area is surveyed in the articles by J. R. Moschovakis and R. Vesley in this volume. His work on finite automata and regular events is discussed in the article by R. Constable. His books and his 1952 memoir include contributions to proof theory.

The Bibliography of Stephen C. Kleene

1. Proof by cases in formal logic. Annals of Math., 2s., Vol. 35 (1934), pp. 529–544.
2. A theory of positive integers in formal logic. Ph.D. thesis, Princeton University, September, 1933; revised June, 1934. American Journal of Math., Vol. 57 (1935), pp, 153–173, 219–244.
3. The inconsistency of certain formal logics (with J. B. Rosser). Annals of Math., 2s., Vol. 36 (1935), pp. 630–636.
4. General recursive functions of natural numbers. Mathematische Annalen, Vol. 112 (1936), pp. 727–742. Reprinted, with three notes by the author, in the Undecidable, Basic Papers on Undecidable Propositions, Unsolvable Problems and Computable Functions, ed. by Martin Davis, Raven Press (Hewlett, N.Y.), 1965, p. 236–253.
5. λ-definability and recursiveness. Duke Math. J., Vol. 2 (1936), pp. 340–353.
6. A note on recursive functions. Bull. Amer. Math. Soc., Vol. 42 (1936), pp. 544–546.
7. Formal definitions in the theory of ordinary numbers (with Alonzo Church). Fundamenta Mathematicae, Vol. 28 (1936), pp. 11–21.
8. Construction of formal definitions of functions of ordinal numbers (with Alonzo Church). Abstract, Bulletin of the Amer. Math. Soc., Vol. 42 (1936), p. 639, and Journal of Symbolic Logic, Vol. 1 (1936), pp. 59–60.
9. On notation for ordinal numbers. J. Symbolic Logic, Vol. 3 (1938), pp. 150–155.
10. A postulational basis for probability (with H. P. Evans). Amer. Math. Monthly, Vol. 46 (1939), pp. 141–148.
11. Review of Carnap's "The logical syntax of language." J. Symbolic Logic, Vol. 4 (1939), pp. 82–87.
12. On the term 'analytic' in logical syntax. Preprinted for the members of the Fifth International Congress for the Unity of Science, Cambridge, Mass., Sept. 1939, from the Journal of Unified Science (Erkenntnis), Vol. 9, pp. 189–192.
13. Review of Hilbert and Bernays' "Grundlagen der Mathematik II." Journal of Symbolic Logic, Vol. 5 (1940), pp. 16–20.
14. Review of Gödel's "The consistency of the continuum hypothesis." Mathematical Reviews, Vol. 2 (1941), pp. 66–67.
15. Recursive predicates and quantifiers. Trans. Amer. Math. Soc., Vol. 53 (1943), pp. 41–73. Abstract of Preliminary Report, Bull. Amer. Math. Soc., Vol. 46 (1940), p. 885. Reprinted, with a correction and an addendum by the author, in The Undecidable (cf. 4), 1965, pp. 254–287.
16. On the forms of the predicates in the theory of constructive ordinals. Amer. J. Math., Vol. 66 (1944), pp. 41–58.
17. On the interpretation of intuitionistic number theory. J. Symbolic Logic, Vol. 10 (1945), pp. 109–124.
18. Analysis of lengthening of modulated repetitive pulses. Proc. of the Institute of Radio Engineers, Vol. 35 (1947), pp. 1049–1053.
19. On the intuitionistic logic. Preprinted for the members of the Congress as pages 185–187, from the Proceedings of the Tenth International Congress of Philosophy (Amsterdam, Aug. 11–18, 1948), ed. by E. W. Beth, H. J. Pos and J. H. A. Hollak, North-Holland Publ. Co. (Amsterdam), 1959, Fasc. 2, pp. 741–743.

20. A symmetric form of Gödel's theorem. Koninklijke Nederlandse Akademie van Wetenschappen, Proceedings of the section of sciences, Vol. 53 (1950), pp. 800–802; also Indagationes Mathematicae, Vol. 12, pp. 244–246.
21. Recursive functions and intuitionistic mathematics. Proceedings of the International Congress of Mathematicians, Cambridge, Mass., U.S.A., Aug. 30–Sept. 6, 1950, Amer. Math. Soc. (Providence, R.I.), 1952, Vol. 1, pp. 679–685.
22. Permutability of inferences in Gentzen's calculi **LK** and **LJ**. Two papers on the perdicate calculus, by S. C. Kleene, Memoirs of the Amer. Math. Soc., No. 10 (1952), pp. 1–26 with bibliography on pp. 67–68. Russian translation by V. P. Orevkov and A. V. Sočilina, with added footnotes by the translators, Matematičeskaja teorija logičeskogo vyvoda, ed. by A. V. Idel'son and G. E. Minc, Izdatel'stvo "Nauka," Moscow 1967, pp. 208–236 with bibliography on pp. 283–284. Second printing, with revisions, Amer. Math. Soc. (Providence, R.I.), 1967.
23. Finite axiomatizability of theories in the predicate calculus using additional predicate symbols. Ibid, 1952, pp. 27–66 with bibliography on pp. 67–68. Russian translation by G. V. Davidov, with added footnotes by the translator, ibid. 1967, pp. 237–283 with 283–284. Second printing, with revisions, ibid. 1967.
24. Introduction to metamathematics. North-Holland Publ. Co. (Amsterdam), Noordhoff (Groningen), Van Nostrand (Princeton, New York and Toronto), 1952, X+550 pp. Reprinted in Taiwan. Russian translation, by A. S. Esenin-Vol'pin, ed. by V. A. Uspenskij, with preface and seven appendices by the translator, numerous footnotes by the translator and by the editor, and a greatly expanded bibliography, Izdatel'stvo Inostrannoj Literatury, Moscow, 1957; reprinted about 1972. Translated into Chinese by Moh Shaw-Kwei in 1964–65. Roumanian translation of Chapter III, with added footnotes by the translator, and from the Russian edition, Logică și filozofie, orientări in logica modernă și fundamentele matematicii, ed. by M. Tirnoveanu and Gh. Enescu, Editura politică, Bucharest 1966, pp. 414–462. Tokyo University International Edition 42, in English from the fifth (1967) reprint, Univ. of Tokyo Press. Spanish translation by Manuel Garrido with the collaboration of Rafael Beneyto, Jose Sanmartin and Enrique Casaban, Editorial Tecnos, S. A., Madrid 1974. Sixth (1971) and Seventh (1974) reprints, with two short notes by the author, Wolters-Noordhoff (Groningen), North-Holland Publ. Co. (Amsterdam, Oxford) and American Elsevier (New York).
25. Review of Péter's "Rekursive Funktionen." Bull. Amer. Math. Soc., Vol. 58 (1952), pp. 270–272.
26. The upper semi-lattice of degrees of recursive unsolvability (with Emil L. Post). Annals of Math., Ser. 2, Vol. 59 (1954), pp. 379–407.
27. Arithmetical predicates and function quantifiers. Trans. Amer. Math. Soc., Vol. 79 (1955), pp. 312–340.
28. On the forms of the predicates in the theory of constructive ordinals (second paper). Amer. Journal Math., Vol. 77 (1955), pp. 405–428.
29. Hierarchies of number-theoretic predicates. Bull. Amer. Math. Soc., Vol. 61 (1955), pp. 193–213.
30. Representation of events in nerve nets and finite automata. Automata Studies, ed. by C. E. Shannon and J. McCarthy, Annals of Mathematics Studies, no. 34, Princeton Univ. Press (Princeton, N. J.) 1956, pp. 3–41. Appeared in preliminary form as Project RAND Research Memorandum RM-704, 15 December 1941, 101 pp. Russian translation, Avtomaty, Izdatel'stvo Inostrannoj Literatury, Moscow 1956, pp. 15–67. German translation by Frans Kaltenbeck and Peter Weibel, with a preface (to the volume) by the translators, two added footnotes by the author (pp. 19, 29), and a biographical note (pp. 423–424), Studien zur Theorie der Automaten, C. E. Shannon/J. McCarthy (Hrsg.), Rogner und Bernhard (Munich), 1974, pp. 3–55.

31. A note on computable functionals. Koninklijke Nederlandse Akademie van Wetenschappen te Amsterdam, Proceedings, Series A, Vol. 59, also Indagationes Mathematicae, Vol. 18, 1956, pp. 275–280.
32. Sets, logic, and mathematical foundations. Mimeographed notes by H. William Oliver on lectures at a National Science Foundation Summer Institute for Teachers of Secondary and College Mathematics, Williams College, Williamstown, Mass., 1956, v + 169 pp.
33. Mathematics, foundations of. 1957 and subsequent printings of Encyclopaedia Britannica (800 words). Beginning with the 1974 printing, the article was rearranged to replace Kleene's section on intuitionism by a fuller treatment by Solomon Feferman.
34. Decision procedure. Ibid. up through the 1973 printing (75 words).
35. Realizability. Mimeographed summaries of talks presented at the Summer Institute of Symbolic Logic, 1957, Cornell University, Vol. 1, pp. 100–104. Second edition, Communications Research Division, Institute for Defense Analyses, 1960, pp. 100–104. Published in Constructivity in Mathematics, ed. by A. Heyting, North-Holland Publ. Co. (Amsterdam), 1959, pp. 285–289.
36. Recursive functionals of higher finite types. Ibid (1957), pp. 148–154, with errata Vol. 3, p. 429. Second edition, ibid. (1960), pp. 148–154.
37. A note on function quantification (with J. W. Addison). Proc. Amer. Math. Soc., Vol. 8 (1957), pp. 1002–1006.
38. Extension of an effectively generated class of functions by enumeration. Colloquium Mathematicum, Vol. 6 (1958), pp. 67–78.
39. Recursive functionals and quantifiers of finite types I. Trans. Amer. Math. Soc., Vol. 91 (1959), pp. 1–52.
40. Countable functionals. Constructivity in Mathematics, Proceedings of the Colloquium held in Amsterdam 1957, ed. by A. Heyting, North-Holland Publ. Co. (Amsterdam), 1959, pp. 81–100.
41. Quantification of number-theoretic functions. Compositio Mathematica, Vol. 14 (1959), pp. 23–40.
42. Mathematical logic: constructive and non-constructive operations. Proceedings of the International Congress of Mathematicians, [Edinburgh] 14–21 August 1958, ed. by J. A. Todd, Cambridge at the University Press, 1960, pp. 137–153.
43. Realizability and Shanin's algorithm for the constructive deciphering of mathematical sentences. Logique et Analyse, n. s., 3ᵉ annee (1960), 11–12, pp. 154–165.
44. Mathematical logic. Mimeographed notes by Edward Pols on lectures at a National Science Foundation Summer Institute, Bowdoin College, Brunswick, Me., 1961, xi + 217 pp.
45. Disjunction and existence under implication in elementary intuitionistic formalisms. J. Symbolic Logic, Vol. 27 (1962), pp. 11–18.
46. Turing-machine computable functionals of finite types I. Logic, Methodology and Philosophy of Science: Proceedings of the 1960 International Congress (at Stanford University), ed. by Ernest Nagel, Patrick Suppes and Alfred Tarski, Stanford Univ. Press. (Stanford, California) 1962, pp. 38–45.
47. Turing-machine computable functionals of finite types II. Proc. London Math. Soc., 3s., Vol. 12 (1962), pp. 245–258.
48. Lambda-definable functionals of finite types. Fundamenta Mathematicae, Vol. 50 (1962), pp. 281–303.
49. Herbrand-Gödel-style recursive functionals of finite types. Proceedings of the Symposia in Pure Mathematics, Vol. 5, Recursive Function Theory (New York, April 6–7, 1961), Amer. Math. Soc., Providence, R.I., 1962, pp. 49–75.
50. Recursive functionals and quantifiers of finite types II. Trans. Amer. Math. Soc., Vol. 108 (1963), pp. 106–142.

51. An addendum [to "Disjunction and existence under implication in elementary intuitionistic formalisms"]. J. Symbolic Logic, Vol. 28 (1963), pp. 154–156.
52. Computability. The Voice of America Forum Lectures, Phil. of Science Series, 1964, No. 6, 8 pp. Reprinted in Philosophy of Science Today, ed. by Sidney Morgenbesser, Basic Books (New York, London), 1967, pp. 36–45.
53. The foundations of intuitionistic mathematics, especially in relation to recursive functions (with Richard Eugene Vesley). The preface, Chapter I A formal system of intuitionistic analysis, Chapter II Various notions of realizability, Chapter IV On order in the continuum, and the Bibliography, are by Kleene. North-Holland Publ. Co. (Amsterdam), 1965, VIII+206 pp. Russian translation by F. A. Kabakov and B. A. Kušner, with a preface, and some additional footnotes and additions to the bibliography, by the translators, Izdatel'stvo "Nauka," Moscow 1978.
54. Classical extensions of intuitionistic mathematics. Logic, Methodology and Philosophy of Science, Proceedings of the 1964 International Congress (at Jerusalem), ed. by Yehoshua Bar-Hillel, North-Holland Publ. Co. (Amsterdam), 1965, pp. 31–44.
55. Logical calculus and realizability. Acta Philosophica Fennica, fasc. 18 (1965), pp. 71–80.
56. Empirical mathematics? In a discussion of L, Kalmár "Foundations of mathematics: whither now?", Problems in the philosophy of mathematics, Proceedings of the International Colloquium in the Philosophy of Science, London, 1965, Vol. I, ed. by Imre Lakatos, North-Holland Publ. Co. (Amsterdam), 1967, pp. 195–196.
57. Mathematical Logic. Wiley (New York, London and Sidney), 1967, xiii+398 pp. First corrected printing 1968. French translation by Jean Largeault, Librairie Armand Colin (Paris), 1971. Japanese translation by Ken-ichi Ozawa, with short acknowledgement by the translator (1971), Meiji Tosho Ltd. (Tokyo), in two volumes 1971 and 1973. Russian translation by Ju. A. Gastev, edited by G. E. Minc, with a preface by the translator and editor, two appendices by the editor, and some added footnotes and additions to the bibliography by the translator, Izdatel'stvo "Mir," Moscow 1973.
58. Constructive functions in "The foundations of intuitionistic mathematics." Logic, methodology and philosophy of science III, Proceedings of the Third International Congress for Logic, Methodology and Philosophy of Science, Amsterdam 1967, ed. by B. van Rootselaar and J. F. Staal, North-Holland Publ. Co. (Amsterdam), 1968, pp. 137–144.
59. On the normal form theorem. Les fonctions recursives et leurs applications, Colloque International organizé par l'Association Mathematique Bolyai János avec le soutien du Conseil International des Unions Scientifiques, Tihany, Hongrie, 4–7 September 1967, Texte des Conferences publié par l'Institut Blaise Pascal du Centre national de la Recherche Scientifique [Paris] et l'Association Mathematique Bolyai János (Budapest), 1969, pp. 71–83.
60. Formalized recursive functionals and formalized realizability. Memoirs of the Amer. Math. Soc., No. 89 (1969), 106 pp.
61. The new logic. (A Sigma-Xi-RESA National Lecture, Southeast Tour, Spring 1969.) American Scientist, Vol. 57 (1969), pp. 333–347.
62. Realizability: a retrospective survey. Cambridge Summer School in Mathematical Logic, held in Cambridge, England, August 1–21, 1971, ed. by A. R. D. Mathias and H. Rogers, Lecture Notes in Mathematics 337, Springer-Verlag (Berlin, Heidelberg, New York), 1973, pp. 95–112.
63. Kleene, Stephen Cole. Brief mathematical autobiography, edited and translated into Italian. Scienziati e Tecnologi Contemporanei, Arnoldo Mondadori Editore, Milan, 1974, Vol. 2, pp. 109–111.
64. Reminscences of logicians (with C. C. Chang, John Crossley, Jerry Keisler, Mike and Vivienne Morley, Andrzej Mostowski, Anil Nerode and Gerald Sacks). Algebra and

Logic, Papers from the 1974 Summer Research Institute of the Australian Mathematical Society, Monash University, Australia, ed. by J. N. Crossley, Lecture Notes in Mathematics 450, Springer-Verlag (Berlin, Heidelberg, New York), 1975, pp. 1–41 (continued with other participants, pp. 41–57).

65. The work of Kurt Gödel. Journal of Symbolic Logic, Vol. 41 (1976), pp. 761–778.
66. Recursive functionals and quantifiers of finite types revisited I. Generalized Recursion Theory II, Proceedings of the 1977 Oslo Symposium, ed. by J. E. Fenstad, R. O. Gandy and G. E. Sacks, North-Holland Publ. Co. (Amsterdam, New York, Oxford), 1978, pp. 185–222.
67. An addendum to "The work of Kurt Gödel." Journal of Symbolic Logic, Vol. 43 (1978), p. 613.
68. Recursive functionals and quantifiers of finite types revisited II. This volume, pp. 1–29.

Besides the four reviews included above, Kleene between 1936 and 1958 wrote thirty-three other reviews. The thirty-seven reviews are indexed in the Journal of Symbolic Logic, Vol. 26 (1961), pp. 78, 145.

Ph.D. Students of Stephen C. Kleene

All degrees were from the University of Wisconsin.

1. Nels David Nelson, George Washington University, Ph.D. 1946
   Recursive functions and intuitionistic number theory.
2. Gene Fuerst Rose, California State University at Fullerton, Ph.D. 1952
   Jaśkowski's truth tables and realizability
3. John West Addison, Jr., University of California-Berkeley, Ph.D. 1955
   On some points in the theory of recursive functions
4. Clifford Spector, deceased, Ph.D. 1955
   On degrees of recursive unsolvability and recursive well-orderings
5. Paul Axt, Pennsylvania State University, Ph.D. 1958
   On a subrecursive hierarchy and primitive recursive degrees
6. Richard Eugene Vesley, State University of New York at Buffalo, Ph.D. 1962
   The intuitionistic continuum
7. John (Yiannis) Nicholas Moschovakis, University of California at Los Angeles, Ph.D. 1963
   Recursive analysis
8. Douglas Albert Clarke, University of Toronto, Ph.D. 1964
   Hierarchies of predicates of arbitrary finite types
9. Shih-Chao Liu, Academia Sinica (Taipei), Ph.D. 1965
   On many–one degrees
10. Joan Elizabeth (Rand) Moschovakis, Occidental College, Ph.D. 1965
    Disjunction, existence and λ-eliminability in formalized intuitionistic analysis
11. Robert Lee Constable, Cornell University, Ph.D. 1968
    Extending and refining hierarchies of computable functions
12. Dick Herman Jacobus De Jongh, University of Amsterdam, Ph.D. 1968
    Investigations on the intuitionistic propositional calculus
13. David Philip Kierstead, Ph.D.

The Program of the Kleene Symposium

A symposium in honor of Professor S. C. Kleene's seventieth birthday was held from June 19 through June 23, 1978, in Madison, Wisconsin. The speakers were chosen from among the leading workers in recursion theory and in intuitionism, two of the fields to which Professor Kleene has made enormous contributions.

The location of the meetings on the campus of the University of Wisconsin was a fitting tribute to Professor Kleene's long association with that University, both as a faculty member and as Dean. The lectures were held in the mathematics building, Van Vleck Hall, which was built under Professor Kleene's supervision as Chairman of the Building Committee.

Besides the scientific program, the Symposium included a picnic at Devil's Lake State Park. This continued a long tradition of such semi-annual Logic Picnics established by Professor Kleene, who highlighted the picnic by leading a climb up the cliffs surrounding the lake.

A detailed program of the meeting follows:

Invited lectures for the Kleene Symposium

Monday, June 19, 1978

9:30 Opening Ceremonies
Professor H. Jerome Keisler, Chairman of Organizing Committee
Professor Joshua Chover, Chairman of U.W. Mathematics Department
President Edwin Young, President of the University of Wisconsin
10:00 Alonzo Church
Topic: The development of recursion theory
11:00 Dana Scott
Title: Lambda calculus: Some models, some philosophy
2:00 Yiannis N. Moschovakis
Topic: Recursion in Higher Types
2:45 S. C. Kleene
Title: Recursion in higher types
8:00 Reception in the Alumni Lounge of the Wisconsin Center

Tuesday, June 20, 1978

9:00 J. W. Addison
Topic: Hierarchy theory

9:40 R. O. Gandy
Title: Church's Thesis
11:00 Gerald Sacks
Title: Recursion in higher types–an absolute approach
2:00 Joseph Shoenfield
Title: Degrees of Unsolvability
3:00 Richard Shore
Title: First-order theory of reducibility orderings

Wednesday, June 21, 1978

9:00 Richard Vesley
Topic: Intuitionistic analysis
9:40 Peter Aczel
Title: Propositions, truths and sets
11:00 Stephen Simpson
Title: The hierarchy based on the jump operator
1:30–9:00 Picnic at Devil's Lake

Thursday, June 22, 1978

9:00 Joan Rand Moschovakis
Title: Kleene's notions of recursive realizability
9:40 Harvey Friedman
Title: Classically and intuitionistically provable recursive functions
11:00 A. S. Troelstra
Title: Intension vs. Extension and choice sequences
2:00 D. H. J. deJong
Title: Intuitionistic propositional connectives

Friday, June 23, 1978

9:00 Robert Constable
Title: The role of finite automata in the development of modern computing theory
9:40 A. S. Kechris
Title: The theory of Π^1_3 sets of reals
11:00 Leo Harrington
Title: An admissible set with a negative solution to Post's Problem

Contributed lectures for the Kleene Symposium

Tuesday, June 20, 1978

1:00 Robert P. Daley
Title: The busy beaver method
4:00 Peter Winkler
Title: Straight-line computation in abstract algebra
5:00 David Kierstead
Title: A semantics for Kleene's *j*-expressions

Thursday, June 22, 1978

1:00 D. N. Hoover
Title: First-order logic with an oracle

3:00 Phokion Kolaitis
Title: Recursion and non-monotone induction in a quantifier

Thursday, June 22, 1978

4:00 Arnold W. Miller
Title: Covering 2^ω with ω_1 disjoint closed sets.

5:00 S. C. Liu
Title: Non-standard constructive analysis

Following the Kleene Symposium, there was a meeting of the Association of Symbolic Logic. Hour addresses were presented by Donald Martin and Michael Morley. There were 24 contributed papers.

The organizing committee for the Symposium consisted of John W. Addison, Jon Barwise, H. Jerome Keisler, Kenneth Kunen and Yiannis N. Moschovakis. Financial support for the symposium was received from the National Science Foundation, the University of Wisconsin and from the registration fees of the participants.

J. Barwise, H. J. Keisler and K. Kunen, eds., *The Kleene Symposium*

Recursive Functionals and Quantifiers of Finite Types Revisited II*

S.C. Kleene
University of Wisconsin, Madison, WI 53706, U.S.A.**

Abstract: In my June 13, 1977 Oslo address (Generalized recursion theory II, 1978) I overcame the limitations in my 1959, 1963 theory on substitution of λ-functionals and on the use of the first recursion theorem by introducing a new list of schemata, with computation rules that do not require values of parts to be computed unless and until they are needed. Computation thereby became starkly formal. The present paper takes steps toward developing a certain semantics to flesh out the formal bones of those computations. The old types 0, 1, 2 are represented within new types $\dot{0}, \dot{1}, \dot{2}$, where type $\dot{0} = \{u, 0, 1, 2, \ldots\}$ (u = "undefined" or "unknown"), and the higher types are composed of the "unimonotone" partial one-place functions from the preceding type into $\{0, 1, 2, \ldots\}$. These are the functions which are monotone and can each be embodied in an oracle who, e.g. at type $\dot{2}$, has been programmed by Apollo to respond (if she does), to any question "$\dot{\alpha}^2(\dot{\alpha}^1)$?" we ask, on the basis of information about $\dot{\alpha}^1$ which she obtains by asking in turn questions of an oracle embodying $\dot{\alpha}^1$ and observing her responses.

I propose first to touch briefly on my research of the remote past and of the distant future.

My first significant theorem was that the predecessor function is λ-definable. (It had not been clear in advance whether to expect this to be true or not.) My proof came to me late in January, or early in February, 1932 while I was in a dentist's office in Trenton, NJ (shall I say, on an "excursion" from Princeton?) having four wisdom teeth extracted.

I still have my other 28 teeth, all in good biting condition. So I do not plan to give them up for many years. When I do, can I not expect seven more theorems?

Now I turn to my research of the immediate present. Between the remote past and this present, I worked on several things, which I leave to the other speakers.

*My June 19, 1978 lecture, of which this is the written version, was listed in the program of the Kleene Symposium under the title "Recursion in higher types". I dedicated it to AAHOOO.

**Continuously since September 1935, except for July 1941–December 1945 and some leaves.

4. Review and introduction

4.1. "Recursive functionals and quantifiers of finite types revisited I" is the published version, 1978, of the address I gave on June 13, 1977 in Oslo at the Second Symposium on Generalized Recursion Theory.[1] My objective was to give an alternative to the development in my 1959 and 1963 papers. I mentioned this alternative in KLEENE (1963, p. 111), but did not pursue it then. Instead of the schemata S1–S9 of 1959, this alternative uses a new list of schemata. At this point in the oral presentation of the present paper on June 19, 1978, I projected onto the screen the transparency giving the new list of schemata S0, S1.0, …, S11 which I used at Oslo. I shall assume that persons who would study the present paper in detail will have at their elbows the published text of my Oslo address. The new schemata are on p. 190 (p. 6 of the preprints). However, in the oral presentation and in this written version, I aim to recapitulate enough so that listeners or readers not familiar with that paper will get the gist of the present one.

4.2. In my revisitation, the first recursion theorem (KLEENE IM, 1952a, p. 348) is given a central role; it is expressed by the schema

S11 $\qquad \phi(\Theta;\mathfrak{A})\simeq\psi(\lambda\mathfrak{A}\phi(\Theta;\mathfrak{A}),\Theta;\mathfrak{A}).$

Implicit in the first recursion theorem is the employment, in the computation of the function ϕ being defined, of that function in stages of definition short of its full definition. After all, this is the way recursion works: knowing some values of a function ϕ, by preassigned applications of known operations we obtain the knowledge of another value. Thus, the present S11 throws the determination of the value of a function ϕ for given arguments $\mathfrak{A}$ back on $\mathfrak{A}$ and the same function ϕ, by means of a previously defined functional ψ. In this and the other schemata, Θ is a list $\theta_1,\dots,\theta_l$ (perhaps empty) of assumed functions. In using S11, ψ has to have been introduced earlier with one more such assumed function η than ϕ has, which in the process of using S11 comes to be identified with ϕ. For fixed Θ, $\phi(\Theta;\mathfrak{A})$ is a function of $\mathfrak{A}$, where $\mathfrak{A}$ is a list of variables of the types 0, 1, 2, 3 which I employed in 1959. In 1978, I concentrated on the first four of the 1959 types, where type 0 = the natural numbers $\{0,1,2,\dots\}$, and type $j+1$ = the total one-place functions from type j into the natural numbers. As Θ varies, ϕ is a functional; $\psi(\eta,\Theta;\mathfrak{A})$ is a functional as η,Θ vary. At the end of a sequence of applications of the schemata, the Θ's used in earlier applications may all have been identified with functions fed back in by S11, or some may remain as function variables of the final functional.

Readers familiar with my 1959 treatment will miss the reflection schema S9, the role of which is of course taken over now by S11.

[1]Not everything I said there would be here is. Some things I didn't say are.

Also it is to be noted that I now boldly admit the schema

S4.j ($j=1,2,3$) $\qquad \phi(\Theta;\mathfrak{A})\simeq\psi(\Theta;\lambda\beta^{j-1}\chi(\Theta;\beta^{j-1},\mathfrak{A}),\mathfrak{A})$

which allows any substitution for a function variable, say α^j, of an expression $\lambda\beta^{j-1}\chi(\Theta;\beta^{j-1},\mathfrak{A})$ formed by applying the λ-operator to a previously defined function. In 1959, such substitutions were only provided by S8.2, S8.3 for the argument of a type-2 or -3 variable, and in other cases under caveats by a rather laboriously proved theorem ((XXIII), p. 21). This is closely connected with a new view of computation.

4.3. I repeat a simple illustration from 1978 (p. 196). Suppose a function ϕ is defined by the following list of schema applications with the list Θ empty throughout (constituting a "partial recursive description of ϕ"):

$$\begin{array}{ll} \rho(a,b)\simeq a & \text{by S3,} \\ \psi(b,a)\simeq\rho(a,b) & \text{by S6.0,} \\ \vdots & \\ \chi(a)\simeq\cdots & \cdots, \\ \phi(a)\simeq\psi(\chi(a),a) & \text{by S4.0.} \end{array}$$

Let us undertake to compute the value of the final function ϕ for a given value r of its variable "a". The obvious first step takes us to "$\psi(\chi(a),a)$". In 1959, I said we must next compute the part "$\chi(a)$". Only if a computation of it could be completed, say with result s, did I proceed to "$\psi(b,a)$" with s as the value of "b" and then think about how ψ is defined. Now, I have overcome the inhibitions of my youth (1959), and plow ahead with the impetuosity of middle age (1978) from "$\psi(\chi(a),a)$" to whatever the definition of ψ gives, namely to "$\rho(a,\chi(a))$", without worrying about what "$\chi(a)$" is.

In general, my present computation rules call for carrying parts forward in analyzing wholes, not worrying about what values, if any, the parts may have unless and until I come to steps for which I "need to know". Why worry about what you don't yet "need to know"? In this example, I have made things trivial: ρ is the identity function of its first variable, so "$\chi(a)$" drops out in the obvious next step from "$\rho(a,\chi(a))$", and the computation is completed by writing the numeral R for the value r assigned to the variable "a". Thus I now obtain as the completed computation:

$$\phi(a)-\psi(\chi(a),a)-\rho(a,\chi(a))-a-\mathrm{R}\dagger.$$

A flag "$\dagger$" on the R advertizes that it completes the computation. In general, it may be no finite matter to see whether we can get through to a value of "$\psi(\chi(a),a)$" without having a value of "$\chi(a)$". In 1959, if "$\chi(a)$" has no value (is "undefined"), by fiat "$\psi(\chi(a),a)$" was undefined.

I do the like now in an application of S4.j for $j=1$, 2 or 3 (above) with the part "$\lambda\beta^{j-1}\chi(\Theta;\beta^{j-1},\mathfrak{A})$"; that is, I carry it forward till I need to know something about it.

4.4. It must strike anyone who has worked with the 1959 theory that my present way of operating flouts the convention I had there (KLEENE (1978, Sections 3.7, 3.14)) that a function is undefined if its argument positions are not occupied only by expressions evaluable by type-0 quantities (natural numbers) or (completely defined) type-j functions ($j>0$).

Herewith computation necessarily takes on a starkly formal aspect, in which we manipulate expressions by formal rules, hoping in the end to reach a numeral that will wave a flag "†" saying "Look at me!". Thus in the illustration in Section 4.3 I put "a", "$\psi(\chi(a),a)$", etc. in quotes to emphasize that we are dealing with them as expressions.

I thus perform computations on expressions regarded simply formally, but with an *assignment* Ω of values to their free variables, and to the function symbols θ_t of the list Θ (if it is not empty), in my side pocket. So now I require a formal syntactical definition of the classes of expressions ("j-expressions" for $j=0,1,2,3$) that I will be working with. This definition (1978, p. 198) is given relative to a fixed "partial recursive derivation" $\phi_1,\ldots,\phi_p$ of a function $\phi=\phi_p$ from functions $\theta_1,\ldots,\theta_l$ (briefly, Θ). The Θ_i for each $\phi_i(\Theta_i;\mathfrak{A}_i)$ ($i=1,\ldots,p$) are determined by the derivation; so (relative to it) I can simplify the notation by writing "$\phi_i(\mathfrak{A}_i)$" instead of "$\phi_i(\Theta_i;\mathfrak{A}_i)$". So the j-expressions are simply the expressions, syntactically constructed to represent type-j objects (when the function symbols represent total functions), using only "θ_1",…,"θ_l","ϕ_1",…,"ϕ_p", the symbols "$'$", "$\dot{-}$" and "cs" for three particular functions used in the schemata, "0", "λ", variables of our four types, commas and parentheses. To illustrate by giving just one of the clauses, E4 says that, if A is a $j+1$-*expression* and B is a j-*expression*, then $\{A\}(B)$ or A(B) is a 0-*expression* ($j=0,1,2$). It is 0-expressions that we want to compute.

The computation rules (1978, pp. 200–203) are correlated to the clauses in the definition of "0-expression". Thus E4 includes E4.1, which says that, if α^1 is a type-1 variable and B is a 0-*expression*, α^1(B) is a 0-*expression*. The corresponding clause E4.1 of the inductive definition of computation presupposes, for there to be a computation tree for a 0-expression of the form α^1(B), that there be a subordinate computation tree or computation subtree

$$\mathrm{B}\cdots\mathrm{R}\dagger$$

(R† the numeral for some natural number r) for the sub-(0-expression) B with each free variable and θ_t assigned the same value as under the assignment Ω in question for α^1(B). Using this subtree, the computation tree for α^1(B) is assembled thus,

$$\begin{array}{l}\alpha^1(\mathrm{B})\text{——}\mathrm{N}\dagger,\\ \quad\diagdown\\ \qquad\quad \mathrm{B}\cdots\mathrm{R}\dagger\end{array}$$

where N is the numeral for $n=\alpha^1(r)$ for α^1 the value of the variable "α^1" in the assignment Ω.

As E4.2, we have similarly,

$$\alpha^2(\mathrm{B}) \text{——} \mathrm{M}\dagger,$$

$$\vdots$$

$$\mathrm{B}(c)\cdots\Omega_1\cdots\mathrm{N}_1^\dagger$$

$$\mathrm{B}(c)\cdots\Omega_0\cdots\mathrm{N}_0^\dagger$$

where c is a type-0 variable not occurring free in B, and, for each natural number r, Ω_r is the assignment to B(c) in which each free variable of B has the value it has under Ω and the new variable c has the value r. For the computation tree for α^2(B) to exist as shown, a computation subtree of the form

$$\mathrm{B}(c)\cdots\Omega_r\cdots\mathrm{N}_r^\dagger$$

must exist for each of $r=0,1,2,\ldots$. These subtrees in their aggregate determine a type-1 function α^1 (where, for all r, $\alpha^1(r)=n_r$, the natural number for which N_r is the numeral), and M is the numeral for $m=\alpha^2(\alpha^1)$ for α^2 the value of the variable "α^2" under the assignment Ω. (Hereafter, I usually don't write the quotes.)

The remainder of the published version 1978 of my 1977 Oslo address is devoted to developing some points in the theory of this basis, more or less paralleling 1959.

4.5. I claimed (on p. 204) that the present computation procedure, defined inductively, is just as well-defined as a computation procedure as was the 1959 one that followed the interpretations. As a formal procedure, it stands on its own inductive feet!

But the problem does arise: Can we flesh things out so that we do have an interpretation on the formal bones throughout all the intermediate stages in a computation?

This I now propose to do by extending, or more accurately modifying, the types $0,1,2,3$. The types $0,1,2,3$ of 1959 will be represented within the new types $\dot{0},\dot{1},\dot{2},\dot{3}$. I am describing research currently in progress. One way of fleshing out the 1978 formal apparatus with a semantics is the subject of David P. Kierstead's lecture in this Symposium. I have been working on another, which was introduced by me first, but still needs some development and checking. I think it is of interest to entertain at least these two differing interpretations.

5. The new types $\dot{0},\dot{1},\dot{2},\dot{3}$: general features

5.1. To be able always to interpret $\psi(\chi(a),a)$ as a value of a function ψ for arguments of our types, our lowest type $\dot{0}$ must include u = "undefined" or "unknown".

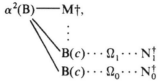

But this is not all. (The cursory reader should be able to understand the following example without checking the references in it to 1978.) Suppose that $\phi_i(a) \simeq \psi(\lambda b \chi_i(b,a),a)$ $(i=1,2)$ by S4.1, where $\psi(\alpha,a) \simeq \sigma(\alpha,\rho(\alpha,a),a)$ by S4.0, $\sigma(\alpha,c,a) \simeq \alpha(c)$ by S7.1, $\rho(\alpha,a)=\rho(a)$ by (IV) with $\rho(a)$ totally undefined by (XIV) (in effect, by IM, Example 1, p. 350), $\chi_1(b,a) \simeq 0$ by S2.0 and $\chi_2(b,a) \simeq b \dot{-} b$ for $\dot{-}$ as in PÉTER (1934) or IM. Indeed, $\dot{-}$ is definable by the successive primitive recursions ($'$ being $+1$ and pd the predecessor function)

$$\begin{cases} \text{pd}(0)=0, \\ \text{pd}(a')=a, \end{cases} \qquad \begin{cases} a \dot{-} 0 = a, \\ a \dot{-} b' = \text{pd}(a \dot{-} b). \end{cases}$$

For all a and b of type 0, $\chi_1(b,a) \simeq \chi_2(b,a) \simeq 0$. And $\phi_1(a) \simeq \psi(\lambda b \chi_1(b,a),a) \simeq 0$. But $\phi_2(a) \simeq \psi(\lambda b \chi_2(b,a),a) \simeq \rho(a) \dot{-} \rho(a) \simeq \text{u}$, since to compute $\rho(a) \dot{-} \rho(a)$ (via the rules for computing $a \dot{-} b$, as rendered in 1978 via (XII) with the help of the function cs of S5.1) we need to know $\rho(a)$. This example shows that the evaluation of an expression $\psi(\lambda b \chi(b,a),a)$ for a given value of a may not depend just on what partial function $\lambda b \chi(b,a)$ is for b ranging on type 0. We can be required to consider it for b ranging on type $\dot{0} = \{\text{u},0,1,2,\ldots\}$.

These indications determine in part the path I shall follow.

A further idea will be introduced in my particular semantics. I shall wish our computation procedure to be determinate.

This will be a departure from the theory of relative partial recursiveness in IM (pp. 326, 334, 337–338), though not from 1959. In IM, for example to compute the representing function $\phi(x)$ of the strong disjunction $Q(x) \vee R(x)$ (say ψ, χ are the representing functions of Q, R), sufficient inputs were $\psi(x)=0$ or $\chi(x)=0$ or $\psi(x)=\chi(x)=1$ with no order specified for seeking them. This was remarked upon by PLATEK (1966, p. 128).[2]

[2]The conception in IM arose rather easily on p. 326 out of the Herbrand–Gödel notion when relativized and partialized. But it was not pursued in any depth, except in one principal case, where it served well. That is the case when, for our illustration, $\psi(x)$ and $\chi(x)$ are in turn either partial recursive functions or functions partial recursive in total functions and predicates Ψ (cf. IM, p. 332). In that case, the normal form theorem (absolute IM XIX (a), p. 330, or relativized to total Ψ XIX (b) with IX*, p. 292) provides determinateness. For, a determinate procedure consists in searching for a least y in the normal forms of ψ and χ, alternating search steps in a prescribed sequence between ψ and χ.

The relativized normal form theorem in IM holds in general only for total Ψ, as I have known for some time.

PLATEK also comments (1966, p. 129) on the treatment of strong definition by cases in IM, pp. 337–338. There not only is determinateness lacking in the general case, but (e.g. for $n=1$, $m=2$) an extension of the domains of definition of the representing functions $\psi_1(x), \psi_2(x)$ of the case hypotheses $Q_1(x), Q_2(x)$ could give an "inconsistency". In IM, such an extension was intended to be excluded by the hypothesis of mutual exclusiveness of $Q_1(x), Q_2(x)$ in the theorem (XX (c), p. 337). In the context of IM, I would now prefer to handle the case hypotheses in the manner of (XI) of KLEENE (1978, p. 211).

5.2. To begin with, our new lowest type $\dot{0}$ shall be $\{u, 0, 1, 2, \ldots\}$ where u = "undefined" or "unknown". (The variables $a, b, \ldots, \alpha^0, \beta^0, \ldots$ used for type 0 may now be dotted, and similarly at higher types.) I shall speak of u, 0, 1, 2, ... as possible *worths* or *interpretations* of a type-$\dot{0}$ variable, since I do not consider u to be a *value*; only 0, 1, 2, ... can be *values*.

The general form of our definition of type $(j+1)^{\cdot}$ for each $j = 0, 1, 2, \ldots$ is that *type* $(j+1)^{\cdot}$ is the "unimonotone" partial one-place functions from *type* $j^{\cdot}$ into the natural numbers $\{0, 1, 2, \ldots\}$. "Unimonotone" will be explained as I proceed, type by type. First, I shall give some definitions that make sense with or without the unimonotonicity.

"Partial" functions (e.g. IM, Chapter XII) are functions whose domain is a subset (proper or improper) of the ostensible domain of the variable (for functions of several variables, of the Cartesian product of the ostensible domains of the separate variables). This usage was introduced by the author in a paper presented at Chapel Hill, NC in December 1936 (published 1938) (see KLEENE (1978, footnote 1)).

Before proceeding, note the notations

$$u^0 = u,$$
$$u^1 = \lambda \dot{\alpha}^0 u,$$
$$u^2 = \lambda \dot{\alpha}^1 u,$$
etc.

$$N^0 = \text{type } \dot{0} - \{u\} = \{0, 1, 2, \ldots\},$$
$$N^1 = \text{type } \dot{1} - \{u^1\},$$
$$N^2 = \text{type } \dot{2} - \{u^2\},$$
etc.

I shall not consider $u^1, u^2, \ldots$ as *values* of type-$\dot{1}$,-$\dot{2}$,... variables, only as *worths* or *interpretations*.

I write $\dot{\alpha}^j \supset \dot{\beta}^j$ (or $\dot{\beta}^j \subset \dot{\alpha}^j$) to express the idea that $\dot{\alpha}^j$ contains the information $\dot{\beta}^j$ contains and maybe more. For $j = 0$, I consider u as no information and any $r \in N^0$ as one bit. So indeed, $\dot{\alpha}^0 \supset \dot{\beta}^0$ iff $\dot{\alpha}^0 = \dot{\beta}^0 \vee (\dot{\alpha}^0 \in N^0 \ \& \ \dot{\beta}^0 = u)$. For $j > 0$, I define $\dot{\alpha}^j \supset \dot{\beta}^j$ to have its usual set-theoretic meaning that $\dot{\beta}^j$ considered as a set of ordered pairs is a subset (proper or improper) of $\dot{\alpha}^j$ likewise considered. This corresponds to considering each ordered pair in a function as one bit of information, and works the way I want because I took $\dot{\alpha}^j, \dot{\beta}^j$ to be partial functions into $\{0, 1, 2, \ldots\}$ rather than total functions into $\{u, 0, 1, 2, \ldots\}$. Thus, if $\dot{\alpha}^j$ comes from $\dot{\beta}^j$ by making $\dot{\alpha}^j$ defined for some arguments $\dot{\alpha}^{j-1}$ for which $\dot{\beta}^j$ is undefined, I am adding some pairs of the form $\langle \dot{\alpha}^{j-1}, n \rangle$ $(n \in N^0)$ to $\dot{\beta}^j$ to get $\dot{\alpha}^j$, whereas if the functions were total into $\{u, 0, 1, 2, \ldots\}$ I would be changing some pairs $\langle \dot{\alpha}^{j-1}, u \rangle$ to pairs $\langle \dot{\alpha}^{j-1}, n \rangle$.

$\dot\alpha^{j+1}$ is *monotone* iff, for all $\dot\alpha^j$ and $\dot\beta^j$, $\dot\alpha^j \supset \dot\beta^j \to$[if $\dot\alpha^{j+1}(\dot\beta^j)$ is defined, so is $\dot\alpha^{j+1}(\dot\alpha^j)$ with the same value].

With the trivial definition of $\dot\alpha^0 \supset \dot\beta^0$ given above, monotonicity of a type-$\dot 1$ function $\dot\alpha^1$ simply means that, if $\dot\alpha^1(\mathrm{u}) = n \in N^0$, then also $\dot\alpha^1(r) = n$ for all $r \in N^0$, so then $\dot\alpha^1$ is the constant function $\lambda\dot\alpha^0 n$. Thus, with monotonicity, we can have $\dot\beta^1(\mathrm{u}) = \mathrm{u}$, $\dot\alpha^1(\mathrm{u}) = n \in N^0$ and $\dot\alpha^1 \supset \dot\beta^1$ only if $\dot\beta^1$ is the constant n on its domain.

For type-$\dot 1$ functions, "unimonotone" shall be synonymous with "monotone".

5.3. It is trivially the fact that (for $\dot\alpha^1$ of type $\dot 1$), if $\dot\alpha^1(\dot\alpha^0)$ is defined, there is a minimum (for $\subset$ as "$\leqslant$") $\dot\beta^0 \subset \dot\alpha^0$ such that $\dot\alpha^1(\dot\beta^0)$ is defined (and by monotonicity, $\dot\alpha^1(\dot\beta^0) = \dot\alpha^1(\dot\alpha^0)$). That is, $\dot\beta^0 \subset \dot\alpha^0$ and $\dot\alpha^1(\dot\beta^0)$ is defined; and for each $\dot{\bar\beta}^0 \subset \dot\alpha^0$ with $\dot\alpha^1(\dot{\bar\beta}^0)$ defined, $\dot\beta^0 \subset \dot{\bar\beta}^0$. I call this (unique) $\dot\beta^0$ the *basis* for $\dot\alpha^0$ with respect to $\dot\alpha^1$.

What are the possible bases for $\dot\alpha^0$ w.r. to a given type-$\dot 1$ object $\dot\alpha^1$ as $\dot\alpha^0$ ranges over type $\dot 0$?

Case 1^1: $\dot\alpha^1(\dot\alpha^0)$ *is undefined for every* $\dot\alpha^0$; i.e. $\dot\alpha^1 = \lambda\dot\alpha^0\,\mathrm{u} = \mathrm{u}^1$. *There are no bases* (none are needed).

Case 2^1: $\dot\alpha^1(\mathrm{u})$ *is defined*, $= n$ say; so by monotonicity $\dot\alpha^1 = \lambda\dot\alpha^0 n$. *Then* u *is the basis for every* $\dot\alpha^0$.

Case 3^1: *Otherwise*; *i.e.* $\dot\alpha^1(\mathrm{u})$ *is undefined* (by not Case 2^1), *but* $\dot\alpha^1(r)$ *is defined for some* r's$\in N^0$ (thence by not Case 1^1). *These* r's *are their own bases*.

The idea of the basis $\dot\beta^0$ w.r. to $\dot\alpha^1$ for an $\dot\alpha^0$ for which $\dot\alpha^1(\dot\alpha^0)$ is defined is that $\dot\beta^0$ contains just the information about $\dot\alpha^0$ that is required and used in determining that $\dot\alpha^1(\dot\alpha^0)$ is defined and its value. In Case 2^1, no information is used; in Case 3^1, one bit (the whole).

Any type-$\dot 1$ function $\dot\alpha^1$ can be represented as "the union of its sections", $\dot\alpha^1 = \bigcup_{\dot\beta^0 \in A^1} S_{\dot\beta^0}$, where A^1 is the set of all the bases $\dot\beta^0$ for $\dot\alpha^0$ w.r. to $\dot\alpha^1$ as $\dot\alpha^0$ varies over type $\dot 0$, and the *section* $S_{\dot\beta^0}$ is the constant function $\dot\alpha^1 \restriction \{\dot\alpha^0 | \dot\alpha^0 \supset \dot\beta^0\}$.

In Case 1^1, there are no sections (A^1 = the empty subset $\varnothing$ of type $\dot 0$; $\dot\alpha^1 = \mathrm{u}^1$ = the empty set $\varnothing$ of ordered pairs). In Case 2^1, $\dot\alpha^1$ consists of the single section $\{\langle \mathrm{u}, n\rangle, \langle 0, n\rangle, \langle 1, n\rangle, \ldots\}$ where $n = \dot\alpha^1(\mathrm{u})$ ($A^1 = \{\mathrm{u}\}$). In Case 3^1, $\dot\alpha^1$ is a union of one or more unit sections $\{\langle r, \dot\alpha^1(r)\rangle\}$, one for each $r \in N^0$ for which $\dot\alpha^1(r)$ is defined (A^1 is the set of these r's).

The following three propositions are immediate. (We are preparing the way for what will be less trivial at higher types.)

(XX.1) $\dot\beta^0 \in A^1$ *iff* $\dot\beta^0$ *is its own basis w.r. to* $\dot\alpha^1$.

(XXI.1) (Basic sublemma, $j=1$.) *For each $\dot{\beta}^0 \in A^1$ and each $\bar{\dot{\alpha}}^0 \supset \dot{\beta}^0$, if $\bar{\dot{\beta}}^0$ be the basis for $\bar{\dot{\alpha}}^0$ w.r. to $\dot{\alpha}^1$, then $\bar{\dot{\beta}}^0 = \dot{\beta}^0$.*

(XXII.1) (Basic lemma, $j=1$.) *Each subunion $\dot{\beta}^1$ of $\dot{\alpha}^1 = \bigcup_{\dot{\beta}^0 \in A^1} S_{\dot{\beta}^0}$, i.e. each $\dot{\beta}^1 = \bigcup_{\dot{\beta}^0 \in B^1} S_{\dot{\beta}^0}$ where $B^1 \subset A^1$ (so $\dot{\beta}^1 \subset \dot{\alpha}^1$), is a type-$\dot{1}$ object; and each $\dot{\alpha}^0$ for which $\dot{\beta}^1(\dot{\alpha}^0)$ is defined has the same basis $\dot{\beta}^0$ w.r. to $\dot{\beta}^1$ as w.r. to $\dot{\alpha}^1$.* (So each section of $\dot{\beta}^1$ is a section of $\dot{\alpha}^1$; and B^1 is all the bases for $\dot{\alpha}^0$ w.r. to $\dot{\beta}^1$ as $\dot{\alpha}^0$ varies over type $\dot{0}$.)

5.4. A partial one-place function $\dot{\alpha}^2$ from type $\dot{1}$ into N^0 shall be *unimonotone* iff it is monotone (as defined in Section 5.2) and has the following further property (to be formulated in two stages). For each $\dot{\alpha}^1$ of type $\dot{1}$ for which $\dot{\alpha}^2(\dot{\alpha}^1)$ is defined, there is a minimum subunion $\dot{\beta}^1$ of the sections $S_{\dot{\beta}^0}$ of $\dot{\alpha}^1$ such that $\dot{\alpha}^2(\dot{\beta}^1)$ is defined. That is: $\dot{\beta}^1 = \bigcup_{\dot{\beta}^0 \in B^1} S_{\dot{\beta}^0}$ with $B^1 \subset A^1$ (so $\dot{\beta}^1 \subset \dot{\alpha}^1$, and by (XXII.1) $\dot{\beta}^1 \in$ type $\dot{1}$) and $\dot{\alpha}^2(\dot{\beta}^1)$ is defined (so by monotonicity $\dot{\alpha}^2(\dot{\beta}^1) = \dot{\alpha}^2(\dot{\alpha}^1)$); and, for each $\bar{\dot{\beta}}^1 = \bigcup_{\dot{\beta}^0 \in \bar{B}^1} S_{\dot{\beta}^0}$ with $\bar{B}^1 \subset A^1$ and $\dot{\alpha}^2(\bar{\dot{\beta}}^1)$ defined, $\dot{\beta}^1 \subset \bar{\dot{\beta}}^1$. I call this (unique) $\dot{\beta}^1$ the *basis* for $\dot{\alpha}^1$ w.r. to $\dot{\alpha}^2$. Further, the basis $\dot{\beta}^1$ (for each such $\dot{\alpha}^1$) is "intrinsically determined", as will be explained in Section 6. (*Unimonotone*$\equiv$*monotone* with *uni*que *i*ntrinsically determined basis.)

Leaving the curtain drawn on the "intrinsic determination" (which will be the crowning jewel in this semantics, if it can be proved to be genuine), let us consider what are the possible bases for $\dot{\alpha}^1$ w.r. to a given type-$\dot{2}$ object $\dot{\alpha}^2$, as $\dot{\alpha}^1$ varies over type $\dot{1}$.

Case 1^2: *$\dot{\alpha}^2(\dot{\alpha}^1)$ is undefined for every $\dot{\alpha}^1$*; i.e. $\dot{\alpha}^2 = \lambda\dot{\alpha}^1 \mathrm{u} = \mathrm{u}^2$. *No bases.*

Case 2^2: *$\dot{\alpha}^2(\mathrm{u}^1)$ is defined,* $=m$ say; so by monotonicity (since every $\dot{\alpha}^1 \supset \mathrm{u}^1 = \varnothing$) $\dot{\alpha}^2(\dot{\alpha}^1) = m$ for every $\dot{\alpha}^1$, i.e. $\dot{\alpha}^2 = \lambda\dot{\alpha}^1 m$. *Then* u^1 *is the basis for every* $\dot{\alpha}^1$ (u^1 is the empty union of sections of $\dot{\alpha}^1$).

Case 3^2: *Otherwise.* Then (by not Case 1^2) $\dot{\alpha}^2(\dot{\alpha}^1)$ is defined for some $\dot{\alpha}^1$'s, and (by not Case 2^2) each such $\dot{\alpha}^1$ must be defined for some $\dot{\alpha}^0$. We say for each $\dot{\alpha}^1$ what its basis can be by subcases. *Subcase* 3.1^2: $\dot{\alpha}^1 = \mathrm{u}^1$ (Case 1^1). Then $\dot{\alpha}^2(\dot{\alpha}^1) = \mathrm{u}$. *No basis. Subcase* 3.2^2: *$\dot{\alpha}^1(\mathrm{u})$ is defined* (Case 2^1), $=n$ say; so $\dot{\alpha}^1 = \lambda\dot{\alpha}^0 n$. Then $\dot{\alpha}^1$ has only one section, itself. *If $\dot{\alpha}^2(\dot{\alpha}^1)$ is defined, $\dot{\alpha}^1$ is its own basis.* (The empty union u^1 would not suffice, by not Case 2^2.) *Subcase* 3.3^2: *$\dot{\alpha}^1(\mathrm{u})$ is undefined, but $\dot{\alpha}^1(r)$ is defined for some r's* $\in N^0$ (Case 3^1). *If $\dot{\alpha}^2(\dot{\alpha}^1)$ is defined, then* the basis $\dot{\beta}^1$ is some non-empty subunion of the set of the unit sections $\{\langle r, \dot{\alpha}^1(r)\rangle\}$ (non-empty by not Case 2^2). Thus *the basis $\dot{\beta}^1$ for $\dot{\alpha}^1$ is a $\dot{\beta}^1 \subset \dot{\alpha}^1$ defined on some non-empty subset of those r's* (i.e. a non-empty partial function on N^0), *the minimum one for which $\dot{\alpha}^2(\dot{\beta}^1)$ is defined.*

In this $j=2$ case, the basis existence for an $\dot{\alpha}^1$ for which $\dot{\alpha}^2(\dot{\alpha}^1)$ is defined is not given automatically by just monotonicity. For example, let $\dot{\alpha}^2(\dot{\alpha}^1)=0$ if $\dot{\alpha}^1\supset\{\langle 0,0\rangle\}\vee\dot{\alpha}^1\supset\{\langle 1,1\rangle\}$ and $\dot{\alpha}^2(\dot{\alpha}^1)=\mathrm{u}$ otherwise. Then $\dot{\alpha}^2$ is monotone, but $\{\langle 0,0\rangle\}$ and $\{\langle 1,1\rangle\}$ are two different minimal subunions $\dot{\beta}^1$ of $\{\langle 0,0\rangle,\langle 1,1\rangle\}$ for which $\dot{\alpha}^2(\dot{\beta}^1)$ is defined, and there is no minimum one.

Also, under our definition of "basis", the basis $\dot{\beta}^1$ for an $\dot{\alpha}^1$ for which $\dot{\alpha}^2(\dot{\alpha}^1)$ is defined is not necessarily the minimum $\dot{\beta}^1\subset\dot{\alpha}^1$ with $\dot{\alpha}^2(\dot{\beta}^1)$ defined. For example, let $\dot{\alpha}^2(\dot{\alpha}^1)=0$ if $\dot{\alpha}^1\supset\{\langle 0,0\rangle\}$ and $\dot{\alpha}^2(\dot{\alpha}^1)=\mathrm{u}$ otherwise. Now $\dot{\alpha}^2(\lambda\dot{\alpha}^0 0)$ is defined, with $\lambda\dot{\alpha}^0 0$ as the basis; but the minimum $\dot{\beta}^1\subset\lambda\dot{\alpha}^0 0$ for which $\dot{\alpha}^2(\dot{\beta}^1)$ is defined is $\{\langle 0,0\rangle\}$.

Recalling the motivation for our definition of $\dot{\beta}^j\subset\dot{\alpha}^j$, the monotonicity of an $\dot{\alpha}^{j+1}$ means that, if $\dot{\alpha}^{j+1}(\dot{\beta}^j)$ is determined (using only information in $\dot{\beta}^j$), then replacing $\dot{\beta}^j$ by an $\dot{\alpha}^j$ with the same or more information in it will not alter the result. The basis $\dot{\beta}^j$ for an $\dot{\alpha}^j$ for which $\dot{\alpha}^{j+1}(\dot{\alpha}^j)$ is defined is thought of as the minimum amount of information about $\dot{\alpha}^j$ which is needed and used in the determination of $\dot{\alpha}^{j+1}(\dot{\alpha}^j)$. In rendering this idea precisely, there is a wrinkle: if $\dot{\alpha}^2(\dot{\alpha}^1)$ is determined to be m for an $\dot{\alpha}^1$ whose value $\dot{\alpha}^1(\dot{\alpha}^0)$ is determined to be n without using any information about $\dot{\alpha}^0$, we regard the information that $\langle \mathrm{u},n\rangle\in\dot{\alpha}^1$ as dragging with it all the other argument-value pairs $\langle r,n\rangle$. In other words, for $\dot{\alpha}^1=\lambda\dot{\alpha}^0 n$, the pairs $\langle r,n\rangle$ are not allowed to be separated out as bits of information independent of $\langle \mathrm{u},n\rangle$.

Although the following results will not be needed till in the sequel to this paper and were not given in my June 19, 1978 lecture, their treatment fits conveniently here. The cursory reader may skip to Section 6.

Again, each type-$\dot{2}$ object $\dot{\alpha}^2$ is a union of sections, $\dot{\alpha}^2=\bigcup_{\dot{\beta}^1\in A^2} S_{\dot{\beta}^1}$, where A^2 is all the bases $\dot{\beta}^1$ for $\dot{\alpha}^1$ w.r.t. to $\dot{\alpha}^2$ as $\dot{\alpha}^1$ varies over type $\dot{1}$, and $S_{\dot{\beta}^1}=\dot{\alpha}^2\upharpoonright\{\dot{\alpha}^1|\dot{\alpha}^1\supset\dot{\beta}^1\}$.

(XX.2) *$\dot{\beta}^1\in A^2$ iff $\dot{\beta}^1$ is its own basis w.r. to $\dot{\alpha}^2$.*

Proof. The "if" is immediate. For the "only if", remember that, if $\dot{\beta}^1$ is the basis for $\dot{\alpha}^1$, each section of $\dot{\beta}^1$ is a section of $\dot{\alpha}^1$, by (XXII.1). So if $\dot{\beta}^1$ were not its own basis, $\dot{\beta}^1$ could not be the minimum subunion of $\dot{\alpha}^1$ for which $\dot{\alpha}^2(\dot{\beta}^1)$ is defined.

In contrast to the situation for bases w.r. to a type-$\dot{1}$ function, it is not always the case that $[\dot{\alpha}^2(\dot{\alpha}^1)$ is defined$]$ & $\bar{\dot{\alpha}}^1\supset\dot{\alpha}^1\rightarrow[\bar{\dot{\alpha}}^1$ and $\dot{\alpha}^1$ have the same basis w.r. to $\dot{\alpha}^2]$.[3] The exceptions are rather special, however. In

[3]Thus, with $\dot{\alpha}^2$ fixed, $\lambda\dot{\alpha}^1$ [the basis for $\dot{\alpha}^1$ w.r. to $\dot{\alpha}^2$] may not be a monotone partial function from type $\dot{1}$ into type $\dot{1}$.

Also $\Gamma=\lambda\dot{\alpha}^1$ [the basis for 0 w.r. to $\dot{\alpha}^1$] is not a monotone partial function from type-$\dot{1}$ into type-$\dot{0}$. For, take $\dot{\alpha}^1_1=\lambda\dot{\alpha}^0 n\supset\{\langle r,n\rangle|r\in N^0\}=\dot{\alpha}^1_2$. Then $\Gamma(\dot{\alpha}^1_1)=\mathrm{u}$ and $\Gamma(\dot{\alpha}^1_2)=0$.

Cases 1^2 and 2^2 (whose hypotheses depend only on $\dot{\alpha}^2$), there are respectively no bases or only one basis for all $\dot{\alpha}^1$. So consider Case 3^2 with $\dot{\alpha}^2(\dot{\alpha}^1)$ defined. If both $\bar{\dot{\alpha}}^1$ and $\dot{\alpha}^1$ come under Subcase 3.2^2, then $\bar{\dot{\alpha}}^1 \supset \dot{\alpha}^1 \rightarrow \bar{\dot{\alpha}}^1 = \dot{\alpha}^1$. If both come under Subcase 3.3^2 with $\bar{\dot{\alpha}}^1 \supset \dot{\alpha}^1$, then $\bar{\dot{\alpha}}^1$ and $\dot{\alpha}^1$ having different bases would contradict the uniqueness of the minimum subunion which constitutes the basis for $\bar{\dot{\alpha}}^1$, as the basis for $\dot{\alpha}^1$ would be a subunion of $\bar{\dot{\alpha}}^1$ also (since each of $\bar{\dot{\alpha}}^1$ and $\dot{\alpha}^1$ is a union of unit sections). However, if $\bar{\dot{\alpha}}^1 = \lambda \dot{\alpha}^0 n$ under Subcase 3.2^2, and $\dot{\alpha}^1$ under Subcase 3.3^2 is the constant n over a non-empty subset of N^0, then $\bar{\dot{\alpha}}^1 \supset \dot{\alpha}^1$ holds with $\bar{\dot{\alpha}}^1$ and $\dot{\alpha}^1$ having different bases, namely $\bar{\dot{\alpha}}^1$ itself and a non-empty subfunction $\dot{\beta}^1$ of $\dot{\alpha}^1$, respectively. In this situation, $\{\langle \lambda \dot{\alpha}^0 n, \dot{\alpha}^2(\lambda \dot{\alpha}^0 n) \rangle\}$ is a unit section of $\dot{\alpha}^2$ that is a proper subset of $S_{\dot{\beta}^1}$.

(XXI.2) (Basic sublemma, $j = 2$.) *For each $\dot{\beta}^1 \in A^2$ and each $\bar{\dot{\alpha}}^1 \supset \dot{\beta}^1$, if $\bar{\dot{\beta}}^1$ be the basis for $\bar{\dot{\alpha}}^1$ w.r. to $\dot{\alpha}^2$, then $\bar{\dot{\beta}}^1 \supset \dot{\beta}^1$.*

Proof. We have just seen that, if $\dot{\beta}^1 \in A^2$ (so $\dot{\alpha}^2(\dot{\beta}^1)$ is defined) and $\bar{\dot{\alpha}}^1 \supset \dot{\beta}^1$, then, except in a special situation, $\bar{\dot{\alpha}}^1$ and $\dot{\beta}^1$ have the same basis w.r. to $\dot{\alpha}^2$, so $\bar{\dot{\beta}}^1 = \dot{\beta}^1$ since by (XX.2) $\dot{\beta}^1$ is its own basis. In the special situation, $\bar{\dot{\alpha}}^1 = \lambda \dot{\alpha}^0 n = \bar{\dot{\beta}}^1 \supset \dot{\beta}^1$ where $\dot{\beta}^1$ is a constant non-empty partial function on N^0 with value n.

(XXII.2) (Basic lemma, $j = 2$.) *Each subunion $\dot{\beta}^2 = \bigcup_{\dot{\beta}^1 \in B^2} S_{\dot{\beta}^1}$ of $\dot{\alpha}^2 = \bigcup_{\dot{\beta}^1 \in A^2} S_{\dot{\beta}^1}$, where $B^2 \subset A^2$ (so $\dot{\beta}^2 \subset \dot{\alpha}^2$), is a type-$\dot{2}$ object; and each $\dot{\alpha}^1$ for which $\dot{\beta}^2(\dot{\alpha}^1)$ is defined has the same basis $\dot{\beta}^1$ w.r. to $\dot{\beta}^2$ as w.r. to $\dot{\alpha}^2$. (So each section of $\dot{\beta}^2$ is a section of $\dot{\alpha}^2$; but unlike the $j = 1$ case, B^2 may not be all the bases for $\dot{\alpha}^1$ w.r. to $\dot{\beta}^2$ as $\dot{\alpha}^1$ varies over type $\dot{1}$.)*

Proof. As the intrinsic-determination part of the definition of *type $\dot{2}$* has not yet been given, I can prove now only that $\dot{\beta}^2$ satisfies the part of the definition already given (assuming the like for $\dot{\alpha}^2$). At the end of Section 6.3, I shall supply what has to be missing now.

First, $\dot{\beta}^2$ is monotone, since $\dot{\alpha}^2$ is monotone and the domain of each section $S_{\dot{\beta}^1}$ is closed under $\supset$.

Consider any $\bar{\dot{\alpha}}^1$ such that $\dot{\beta}^2(\bar{\dot{\alpha}}^1)$ is defined; so $\bar{\dot{\alpha}}^1 \supset \dot{\beta}^1$ for some $\dot{\beta}^1 \in B^2 \subset A^2$, and $\dot{\beta}^2(\bar{\dot{\alpha}}^1) = \dot{\alpha}^2(\bar{\dot{\alpha}}^1)$. For $\bar{\dot{\beta}}^1$ as in (XXI.2), $\bar{\dot{\beta}}^1 \supset \dot{\beta}^1$, so $\dot{\beta}^2(\bar{\dot{\beta}}^1)$ is defined $(= \dot{\alpha}^2(\bar{\dot{\beta}}^1))$.

Now I show that $\bar{\dot{\beta}}^1$ is the basis for $\bar{\dot{\alpha}}^1$ also w.r. to $\dot{\beta}^2$. By definition, $\bar{\dot{\beta}}^1$ is the minimum union of sections of $\bar{\dot{\alpha}}^1$ with $\dot{\alpha}^2(\bar{\dot{\beta}}^1)$ defined; and $\dot{\beta}^2(\bar{\dot{\beta}}^1)$ is defined (as we just saw). Consider any union $\dot{\gamma}^1$ of sections of $\bar{\dot{\alpha}}^1$ with $\dot{\beta}^2(\dot{\gamma}^1)$ defined. By $\dot{\beta}^2 \subset \dot{\alpha}^2$, $\dot{\alpha}^2(\dot{\gamma}^1)$ is defined. So $\bar{\dot{\beta}}^1 \subset \dot{\gamma}^1$ by $\bar{\dot{\beta}}^1$ being minimum w.r.t. to $\dot{\alpha}^2$. Thus $\bar{\dot{\beta}}^1$ is minimum w.r.t. to $\dot{\beta}^2$.

6. Intrinsic determination of bases w.r.t. to type-$\dot{1}$ and -$\dot{2}$ objects

6.1. Now I lift the curtain. In saying that the basis $\dot{\beta}^j \subset \dot{\alpha}^j$ w.r.t. to $\dot{\alpha}^{j+1}$ for an $\dot{\alpha}^j$ for which $\dot{\alpha}^{j+1}(\dot{\alpha}^j)$ is defined is "intrinsically determined", my idea is roughly that the $\dot{\beta}^j$ is obtained by working from within $\dot{\beta}^j$ itself, without looking at $\dot{\alpha}^j$ outside of its subfunction $\dot{\beta}^j$.

I shall render this explicit by elaborating on the idea of an oracle, which Turing introduced in 1939 for the special case of computing one fixed number-theoretic predicate from another, and which I extended to type-1 variables in 1952 and to variables of the higher finite types 2,3,... in 1959 (see KLEENE (1978, footnote 2)).

Now I extend the idea to type-$\dot{1}$ and -$\dot{2}$ variables (and to certain θ_t's), and at the end of this paper I shall give some preliminary indications for type $\dot{3}$.

I do not now simply postulate that an oracle produces the value of the function when presented an argument for which the function is defined. Instead, I shall give a detailed hyperanthropomorphic picture of how the oracle behaves in doing so. I might say "anthropomorphic" simply, but are not oracles hyperhuman?

6.2. The type-$\dot{1}$ case is rather simple, but it lays the foundation for the type-$\dot{2}$ case.

I conceive of there being an oracle who may give us values of $\dot{\alpha}^1$ on request. How do we get an answer to a question, "What is $\dot{\alpha}^1(\dot{\alpha}^0)$?", if indeed we do get an answer? An answer must be a natural number, since $\dot{\alpha}^1$ is a partial function from type $\dot{0}$ into N^0.

To ask the oracle for $\dot{\alpha}^1$ a question, "What is $\dot{\alpha}^1(\dot{\alpha}^0)$?", we present him/her with a closed envelope containing the natural number r if $\dot{\alpha}^0 = r \in N^0$, empty if $\dot{\alpha}^0 = \mathrm{u}$.

Let us pause to settle the matter of sex. The Selli of Dodona were male oracles. More recently, the Delphic oracles and the Cumaean Sibyl were female. Hereafter, I fix our pronoun to agree with the more recent practice.

She—the oracle—behaves deterministically; she always does the same thing under the same circumstances. All unopened envelopes look alike to her. Upon our presenting her with the envelope, there are three possibilities.

Case $\bar{1}^1$: She stands (or sits) mute (or perhaps disappears into her temple, not to reappear again). So $\dot{\alpha}^1(\dot{\alpha}^0) = \text{u}$ for every $\dot{\alpha}^0$; i.e. $\dot{\alpha}^1 = \lambda\dot{\alpha}^0\,\text{u} = \text{u}^1$.

Case $\bar{2}^1$: Without opening the envelope, she pronounces that $\dot{\alpha}^1(\dot{\alpha}^0) = n$, for some $n \in N^0$. So $\dot{\alpha}^1(\dot{\alpha}^0) = n$ for every $\dot{\alpha}^0$ (in particular, $\dot{\alpha}^1(\text{u}) = n$); $\dot{\alpha}^1 = \lambda\dot{\alpha}^0 n$.

Case $\bar{3}^1$: She opens the envelope, revealing that she will not answer our question "What is $\dot{\alpha}^1(\dot{\alpha}^0)$?" without first receiving an answer to her question "What is $\dot{\alpha}^0$?". If she finds the envelope empty ($\dot{\alpha}^0 = \text{u}$), she stands mute. (Were she willing to answer "$\dot{\alpha}^1(\dot{\alpha}^0)$?" without information about $\dot{\alpha}^0$, she would do so under Case $\bar{2}^1$.) If she finds in the envelope a natural number r, then, depending in general on what number r is, she may stand mute or pronounce that $\dot{\alpha}^1(\dot{\alpha}^0) = n$.

Any partial one-place function from type $\dot{0}$ into N^0 whose values are revealed by an oracle in the above manner is monotone, and thus of type $\dot{1}$, with the bases being as catalogued under Cases 2^1 and 3^1 in Section 5.3. For, if $\dot{\alpha}^1(\dot{\alpha}^0)$ is defined for some $\dot{\alpha}^0$ (i.e. if Case 1^1 does not apply), then Cases $\bar{2}^1$ and $\bar{3}^1$ here coincide with the Cases 2^1 and 3^1 of Section 5.3.

Conversely, if $\dot{\alpha}^1$ is a monotone partial one-place function from type $\dot{0}$ into N^0, there is an oracle whose answers to the questions "$\dot{\alpha}^1(\dot{\alpha}^0)$?" give us the values of $\dot{\alpha}^1(\dot{\alpha}^0)$. So, when this is what we mean by the bases being determined intrinsically, their being so determined (as well as their existence) is already assured at type $\dot{1}$ by the monotonicity. This justifies the last sentence of Section 5.2.

If Case 1^1 in Section 5.3 applies ($\dot{\alpha}^1 = \text{u}^1$), an $\dot{\alpha}^1$-oracle might operate under Case $\bar{1}^1$. Or she might operate under Case $\bar{3}^1$, opening the envelope but then always standing mute, there being no $r \in N^0$ we can put in it which would please her.[4]

Other than thus in Case 1^1, the oracle for any $\dot{\alpha}^1$ of type $\dot{1}$ is unique (if not in her person, at least in her performance).

6.3. For a type-$\dot{2}$ object $\dot{\alpha}^2$, we need an "archoracle" or "$\dot{\alpha}^2$-oracle".

To ask her, "What is $\dot{\alpha}^2(\dot{\alpha}^1)$?", we present her with a closed envelope containing an $\dot{\alpha}^1$-oracle. Let us not worry ourselves whether it is unkind to imagine an $\dot{\alpha}^1$-oracle as two-dimensional so that she would be comfortable in an envelope. If confronted with this protest, we could provide her with a padded chest.

[4]Our concept is that an oracle has been programmed by Apollo to respond to local situations. An oracle operating under Case $\bar{3}^1$ might not have the global self-knowledge that she will turn up her nose at every $r \in N^0$, so that she could more efficiently operate under Case $\bar{1}^1$.

Upon our presenting the envelope (or chest) to the $\dot{\alpha}^2$-oracle, there are again three possibilities.

Case $\bar{1}^2$: The $\dot{\alpha}^2$-oracle stands mute; $\dot{\alpha}^2$ is the totally undefined function $\lambda\dot{\alpha}^1\ \mathrm{u}=\mathrm{u}^2$.

Case $\bar{2}^2$: Without opening the envelope, the $\dot{\alpha}^2$-oracle pronounces that $\dot{\alpha}^2(\dot{\alpha}^1)=m$; so $\dot{\alpha}^2(\mathrm{u}^1)=m$ and $\dot{\alpha}^2=\lambda\dot{\alpha}^1 m$.

Case $\bar{3}^2$: The $\dot{\alpha}^2$-oracle opens the envelope, revealing that she wants information about $\dot{\alpha}^1$ before deciding whether and how to respond to our question "$\dot{\alpha}^2(\dot{\alpha}^1)$?". Indeed, she will not answer our question without learning some values of $\dot{\alpha}^1$. (Were she willing to answer "$\dot{\alpha}^2(\dot{\alpha}^1)$?" without learning some values of $\dot{\alpha}^1$, she would do so under Case $\bar{2}^2$.) To obtain such information, she proceeds to question the $\dot{\alpha}^1$-oracle who was inside the envelope. There is nothing lost by our supposing that the $\dot{\alpha}^2$-oracle starts with the preliminary question "$\dot{\alpha}^1(\dot{\alpha}^0)$?" using an empty envelope ($\dot{\alpha}^0=\mathrm{u}$). *Subcase* $\overline{3.1}^2$: The $\dot{\alpha}^1$-oracle does not respond at all (Case $\bar{1}^1$ in Section 6.2). Then the $\dot{\alpha}^2$-oracle stands mute; $\dot{\alpha}^2(\dot{\alpha}^1)=\mathrm{u}$. *Subcase* $\overline{3.2}^2$: The $\dot{\alpha}^1$-oracle, without opening the envelope, pronounces that $\dot{\alpha}^1(\dot{\alpha}^0)=n$ (Case $\bar{2}^1$). Observing this, the $\dot{\alpha}^2$-oracle learns everything about $\dot{\alpha}^1$, namely that $\dot{\alpha}^1=\lambda\dot{\alpha}^0 n$. The $\dot{\alpha}^2$-oracle, depending in general on the n, then either stands mute or declares that $\dot{\alpha}^2(\dot{\alpha}^1)=m$. *Subcase* $\overline{3.3}^2$: The $\dot{\alpha}^1$-oracle opens the envelope (Case $\bar{3}^1$). The $\dot{\alpha}^2$-oracle, observing this, may either stand mute, or pose a first non-preliminary question "$\dot{\alpha}^1(r_0)$?" (passing over the fact that $\dot{\alpha}^1$, finding the envelope empty, does not answer the preliminary question). For this, she will need to pick an $r_0 \in N^0$. Since she acts deterministically, she must pick the same r_0 for all $\dot{\alpha}^1$'s who open envelopes. As $\dot{\alpha}^2$ has learned, $\dot{\alpha}^1$ opens all envelopes. Depending in general on the number r_0 the $\dot{\alpha}^1$-oracle now finds inside, the $\dot{\alpha}^1$-oracle either stands mute or pronounces that $\dot{\alpha}^1(r_0)=n_0$. In the first case, the $\dot{\alpha}^2$-oracle stands mute. In the second case, depending in general on the n_0, the $\dot{\alpha}^2$-oracle may either stand mute (in effect declaring that [$\dot{\alpha}^1$ opens envelopes] & $\dot{\alpha}^1(r_0)=n_0$ is sufficient information to rule out that $\dot{\alpha}^2(\dot{\alpha}^1)$ be defined), or declare that $\dot{\alpha}^2(\dot{\alpha}^1)=m$, or query $\dot{\alpha}^1$ with an envelope containing another integer r_1; etc. In general, in this Subcase $\overline{3.3}^2$, a series of questions (possibly extending into the transfinite) will be asked of $\dot{\alpha}^1$ by $\dot{\alpha}^2$ with distinct numbers

$$r_0, r_1, \ldots, r_\zeta, \ldots,$$

and will be answered by $\dot{\alpha}^1$ with numbers

$$n_0, n_1, \ldots, n_\zeta, \ldots$$

($n_\zeta=\dot{\alpha}^1(r_\zeta)$, where what r_ζ is is determined by $\dot{\alpha}^2$ from only the information that $\dot{\alpha}^1$ opens envelopes and $\dot{\alpha}^1(r_\eta)=n_\eta$ for $\eta<\zeta$. That is, the question "$\dot{\alpha}^1(r_\zeta)$?" asked by $\dot{\alpha}^2$ is the same of all $\dot{\alpha}^1$'s who open envelopes and agree in their values on $\{r_\eta \mid \eta<\zeta\}$. This continues until either $\dot{\alpha}^2$ asks "$\dot{\alpha}^1(r_\zeta)$?"

for an r_ζ such that $\dot\alpha^1$ does not answer, which makes $\dot\alpha^2(\dot\alpha^1)$ undefined; or through all $\zeta<\xi$ for some $\xi<\omega_1$, after which $\dot\alpha^2$ first decides either that [$\dot\alpha^1$ opens envelopes] & $\dot\alpha^1 \supset \{\langle r_\zeta, n_\zeta\rangle | \zeta<\xi\}$ renders an answer to "$\dot\alpha^2(\dot\alpha^1)$?" out of the question ($\dot\alpha^2(\dot\alpha^1)=\mathrm{u}$), or (only after at least one question has been answered) that it is a sufficient basis for pronouncing that "$\dot\alpha^2(\dot\alpha^1)=m$".

We question the $\dot\alpha^2$-oracle with an oracle for $\dot\alpha^1$ in the envelope. But each $\dot\alpha^1$ has a unique oracle, except in the case of $\dot\alpha^1=\mathrm{u}^1$ (end Section 6.2). Then the $\dot\alpha^2$-oracle can only answer our question "$\dot\alpha^2(\dot\alpha^1)$?" via Case $\bar{2}^2$. So indeed the answer to a question "$\dot\alpha^2(\dot\alpha^1)$?" (whatever the case) depends only on $\dot\alpha^1$ as a function.

If $\dot\alpha^2$ (a partial one-place function from type $\dot{1}$ into N^0) be the function whose values are revealed by an $\dot\alpha^2$-oracle in the manner described above, then, for each $\dot\alpha^1$ for which the oracle gives $\dot\alpha^2(\dot\alpha^1)$ a value, there is a basis $\dot\beta^1$ as called for in Section 5.4 (intrinsically determined now). Indeed, in the last event under Subcase $\overline{3.3}^2$ (i.e. a successful conclusion), the basis for $\dot\alpha^1$ w.r. to $\dot\alpha^2$ is $\dot\beta^1=\{\langle r_\zeta,n_\zeta\rangle|\zeta<\xi\}$, which is of the form cataloged under Subcase 3.3^2 of Section 5.4. For, this $\dot\beta^1$ is a union of sections of $\dot\alpha^1$, since for $\dot\alpha^1$ under Case $\bar{3}^1$ (=Case 3^1 of Section 5.3, for $\dot\alpha^1\neq\mathrm{u}^1$) the sections of $\dot\alpha^1$ are the unit sets $\{\langle r,n\rangle\}$ where $\dot\alpha^1(r)=n$. And $\dot\beta^1$ is the minimum union of these sections with $\dot\alpha^2(\dot\beta^1)$ defined, since $\dot\alpha^2$'s questioning requires for a successful conclusion answers from $\dot\alpha^1$ to exactly the questions r_ζ ($\zeta<\xi$). Under Subcase $\overline{3.2}^2$, if $\dot\alpha^2(\dot\alpha^1)=m$, clearly the basis is $\dot\alpha^1$ itself, as in Subcase 3.2^2. Under Case $\bar{2}^2$ (=Case 2^2), the basis is u^1.

Next I ask: Will the partial function $\dot\alpha^2$ an $\dot\alpha^2$-oracle determines as above necessarily be monotone; i.e. whenever $\dot\alpha^2(\dot\alpha^1)$ is defined and $\bar{\dot\alpha}^1\supset\dot\alpha^1$, will $\dot\alpha^2(\bar{\dot\alpha}^1)$ be defined and $=\dot\alpha^2(\dot\alpha^1)$? This holds vacuously under Case $\bar{1}^2$ and trivially under Case $\bar{2}^2$. So suppose she operates under Case $\bar{3}^2$. If both $\bar{\dot\alpha}^1$ and $\dot\alpha^1$ come under Case $\bar{2}^1$, then $\bar{\dot\alpha}^1=\dot\alpha^1$. If both $\bar{\dot\alpha}^1$ and $\dot\alpha^1$ come under Case $\bar{3}^1$, then the $\dot\alpha^2$-oracle will ask the same questions $r_0,r_1,\ldots,r_\zeta,\ldots$ ($\zeta<\xi$) and receive the same answers $n_0,n_1,\ldots,n_\zeta,\ldots$ ($\zeta<\xi$) from $\bar{\dot\alpha}^1$ as from $\dot\alpha^1$, so her answer that $\dot\alpha^2(\dot\alpha^1)=m$ will also say that $\dot\alpha^2(\bar{\dot\alpha}^1)=m$. But for $\bar{\dot\alpha}^1$ under Case $\bar{2}^1$ and $\dot\alpha^1$ under Case $\bar{3}^1$, it shall be a requirement that I impose on $\dot\alpha^2$-oracles to assure monotonicity that, if under Subcase $\overline{3.3}^2$ an $\dot\alpha^2$-oracle pronounces that $\dot\alpha^2(\dot\alpha^1)=m$ on the basis of knowing that $\dot\alpha^1(r_\zeta)=n$ for all $\zeta<\xi$, then under Subcase $\overline{3.2}^2$ she will pronounce that $\dot\alpha^2(\dot\alpha^1)=m$ for an $\dot\alpha^1$ who declares without opening the envelope that $\dot\alpha^1(\dot\alpha^0)=n$.

So we have reached as our goal (capsulated in Sections 5.2 and 5.4 in the adjective "unimonotone") that a type-$\dot{2}$ object $\dot\alpha^2$ is a partial one-place function from type $\dot{1}$ into N^0 which can be embodied in a (monotonicity-assuring) $\dot\alpha^2$-oracle.

Can a given partial one-place function $\dot\alpha^2$ from type $\dot{1}$ into N^0 be embodied in more than one oracle?

Yes. There is the same possibility as before under Case 1^1 (end Section 6.2) of there being two oracles under Case 1^2, one of which operates under Case $\bar{1}^2$ and the other operates under Case $\bar{3}^2$ without ever giving an answer.

Similarly, there is now also the possibility that one oracle for a given $\dot{\alpha}^2$ might under Subcase $\overline{3.3}^2$, for some $\dot{\alpha}^1$, stand mute after asking $r_0, r_1, \ldots, r_\eta, \ldots$ and being answered $n_0, n_1, \ldots, n_\eta, \ldots$ $(\eta < \zeta)$ where ζ is the least ordinal for which $\dot{\alpha}^1 \supset \{\langle r_\eta, n_\eta \rangle | \eta < \zeta\} \rightarrow \dot{\alpha}^2(\dot{\alpha}^1) = \mathrm{u}$, while other oracles for $\dot{\alpha}^2$ would prolong the questioning (maybe to various lengths depending on the next answers) but eventually stand mute no matter what. I do not impose on our $\dot{\alpha}^2$-oracles the requirement that they stand mute at the least such ζ.

In contrast, an $\dot{\alpha}^2$-oracle must declare the value $\dot{\alpha}^2(\dot{\alpha}^1) = m$ of $\dot{\alpha}^2(\dot{\alpha}^1)$ at the least ordinal ζ (called ξ above) at which $\dot{\alpha}^1 \supset \{\langle r_\eta, n_\eta \rangle | \eta < \zeta\} \rightarrow [\dot{\alpha}^2(\dot{\alpha}^1)$ is defined, $= m$ say]. For otherwise, i.e. if she then went mute or asked another question r_ζ (which $\dot{\beta}^1 = \{\langle r_\eta, n_\eta \rangle | \eta < \zeta\}$ would not answer!), she would be incorrectly making $\dot{\alpha}^2(\dot{\beta}^1)$ undefined. Indeed, I am only recapitulating the observation that a $\dot{\beta}^1 \subset \dot{\alpha}^1$ at which the $\dot{\alpha}^2$-oracle concludes her questioning by declaring that $\dot{\alpha}^2(\dot{\alpha}^1) = m$ is the basis for $\dot{\alpha}^1$ w.r. to $\dot{\alpha}^2$.

Now we have another possibility of a plurality of oracles for one and the same function $\dot{\alpha}^2$, when $\dot{\alpha}^2$ is defined for some $\dot{\alpha}^1$'s under Subcase $\overline{3.3}^2$ with $\xi > 1$. Different oracles for $\dot{\alpha}^2$ could then, for a given such $\dot{\alpha}^1$, ask their questions r_ζ $(\zeta < \xi)$ in different orders. But two such oracles must indeed use in the end the same sets of questions. For these two sets determine the basis as respectively $\dot{\beta}^1 = \{\langle r_\zeta, n_\zeta \rangle | \zeta < \xi\}$ and $\bar{\dot{\beta}}^1 = \{\langle \bar{r}_\zeta, \bar{n}_\zeta \rangle | \zeta < \bar{\xi}\}$; and the basis is unique, so $\dot{\beta}^1 = \bar{\dot{\beta}}^1$.

The requirement that the basis be intrinsically determined, as made concrete by our notion of an $\dot{\alpha}^2$-oracle, goes beyond monotonicity plus the existence of minimum bases. Here is an example of a partial function $\dot{\alpha}^2$ from type $\dot{1}$ into N^0 which is monotone, and such that for each $\dot{\alpha}^1$ for which $\dot{\alpha}^2(\dot{\alpha}^1)$ is defined there is a minimum basis w.r. to $\dot{\alpha}^2$, which basis, however, is not intrinsically determinable. Let

$$\dot{\beta}_1 = \{\langle 0,0 \rangle, \langle 1,1 \rangle \qquad \},$$

$$\dot{\beta}_2 = \{ \qquad \langle 1,0 \rangle, \langle 2,1 \rangle \},$$

$$\dot{\beta}_3 = \{\langle 0,1 \rangle, \qquad \langle 2,0 \rangle \}.$$

Now let $\dot{\alpha}^2(\dot{\alpha}^1) = 0$ if $\dot{\alpha}^1 \supset \dot{\beta}_1 \vee \dot{\alpha}^1 \supset \dot{\beta}_2 \vee \dot{\alpha}^1 \supset \dot{\beta}_3$, and be undefined otherwise. Subcase 3.3^2 in Section 5.4 applies if $\dot{\alpha}^2(\dot{\alpha}^1)$ is defined. An oracle for $\dot{\alpha}^2$ would have to operate under Subcase $\overline{3.3}^2$ in determining the value of $\dot{\alpha}^2(\dot{\alpha}^1)$ for such an $\dot{\alpha}^1$. But there is no choice of r_0 which the oracle can make in advance of any knowledge about $\dot{\alpha}^1$ (except that its oracle opens envelopes) so that she can be sure of receiving an answer to "$\dot{\alpha}^1(r_0)$?" for every such $\dot{\alpha}^1$.

To include intrinsicality in the proof of (XXII.2) end Section 5, suppose we have an oracle for $\dot{\alpha}^2$. We know that, for an $\dot{\alpha}^1$ for which $\dot{\alpha}^2(\dot{\alpha}^1)$ is defined, the basis is u^1 or $\dot{\alpha}^1$ $(=\lambda\dot{\alpha}^0 n)$ itself or $\{\langle r_\zeta, n_\zeta\rangle | \zeta < \xi\}$, according as the α^2-oracle produces the value $\dot{\alpha}^2(\dot{\alpha}^1)$ under Case $\overline{2}^2$, Subcase $\overline{3.2}^2$ or Subcase $\overline{3.3}^2$, respectively. To get an oracle for $\dot{\beta}^2$, we simply reprogram the oracle for $\dot{\alpha}^2$ by muzzling her in Case $\overline{2}^2$ if $u^1 \notin B^2$, in Subcase $\overline{3.2}^2$ if neither $\dot{\alpha}^1$ $(=\lambda\dot{\alpha}^0 n)$ nor a constant function n on a non-empty subset of $N^0 \in B^2$, in Subcase $\overline{3.3}^2$ if $\{\langle r_\zeta, n_\zeta\rangle | \zeta < \xi\} \notin B^2$.

6.4. In this paper I stop short of a full treatment of type $\dot{3}$.

Consider how we should interpret our assumed functions θ_t, now usually written $\dot{\theta}_t$.

I shall use the same illustration as in KLEENE (1978, Section 2.4) for E7, except that I omit the type-2 and -3 variables; thus I consider simply $\dot{\theta}_t(\dot{a}, \dot{\alpha}_1, \dot{\alpha}_2)$. An assumed function with just one variable and that of type $\dot{2}$, say $\dot{\theta}_t(\dot{\alpha}^2)$, would indeed be a type-$\dot{3}$ object.

I confine our attention to assumed functions $\dot{\theta}_t(\dot{a}, \dot{\alpha}_1, \dot{\alpha}_2)$, called *unimonotone*, such that the following obtains. $\dot{\theta}_t$ is a partial function from (type $\dot{0}$)$\times$(type $\dot{1}$)^2 (briefly, type $(\dot{0}, \dot{1}, \dot{1})$) into N^0 which is monotone in each of its variables holding the others fixed. And such that, to each triple $(\dot{a}, \dot{\alpha}_1, \dot{\alpha}_2)$ of worths of the variables for which $\dot{\theta}_t(\dot{a}, \dot{\alpha}_1, \dot{\alpha}_2)$ is defined, there is a (unique) *basis* $(\dot{b}, \dot{\beta}_1, \dot{\beta}_2)$ of the following sort: $\dot{b} \subset \dot{a}$, and $\dot{\beta}_1, \dot{\beta}_2$ are respectively subunions of $\dot{\alpha}_1, \dot{\alpha}_2$ considered as unions of sections; $\dot{\theta}_t(\dot{b}, \dot{\beta}_1, \dot{\beta}_2)$ is defined (and $= \dot{\theta}_t(\dot{a}, \dot{\alpha}_1, \dot{\alpha}_2)$ by monotonicity); and for $\bar{\dot{\beta}}_1, \bar{\dot{\beta}}_2$ also subunions of $\dot{\alpha}_1, \dot{\alpha}_2$ respectively, if $\dot{\theta}_t(\bar{\dot{b}}, \bar{\dot{\beta}}_1, \bar{\dot{\beta}}_2)$ is defined, then $\dot{b} \subset \bar{\dot{b}}$ & $\dot{\beta}_1 \subset \bar{\dot{\beta}}_1$ & $\dot{\beta}_2 \subset \bar{\dot{\beta}}_2$.[5] Finally, the basis $(\dot{b}, \dot{\beta}_1, \dot{\beta}_2)$ is determined intrinsically, as I shall now describe in terms of oracular behaviour. Such a function $\dot{\theta}_t(\dot{a}, \dot{\alpha}_1, \dot{\alpha}_2)$ I say is of type $(\dot{0}:\dot{0}, \dot{1}, \dot{1})$. An assumed function $\theta_t(a, \alpha_1, \alpha_2)$ as used in KLEENE (1978), where a, α_1, α_2 ranged over types $0, 1, 1$ but the function θ_t could be partial and not total, I say is of type $(\dot{0}:0, 1, 1)$.[6] (In this notation, type $1=$type $(0:0)$, type $\dot{1}=$type $(\dot{0}:\dot{0})$, type $2=$type $(0:1)$, type $\dot{2}=$type $(\dot{0}:\dot{1})$, etc.)

[5]An example of a partial function $\phi(x, \psi, \chi)$ from type $(0, \dot{1}, \dot{1})$ into N^0 which fails the basis requirement for unimonotonicity is the representing function of the strong disjunction $Q(x) \vee R(x)$ of IM, pp. 334, 337 when ψ, χ are the representing functions of Q, R (cf. end Section 5.1).

[6]This adapts the notation of KLEENE (1959a, Section 5.2), for which purpose we think of partial functions into N^0 as total functions into type $\dot{0}$. And when the argument types are dotted (with 0 as the value type), unimonotonicity is assumed.

To ask a $\dot{\theta}_t$-oracle "What is $\dot{\theta}_t(\dot{a},\dot{\alpha}_1,\dot{\alpha}_2)$?", we present her with three closed envelopes, numbered #1, #2 and #3, containing respectively $\dot{a}$ if $\dot{a}\neq$ u (empty if $\dot{a}=$ u) and oracles for $\dot{\alpha}_1$ and $\dot{\alpha}_2$.

Case $\bar{1}^{(0:0,1,1)}$: The $\dot{\theta}_t$-oracle stands mute. Then $\dot{\theta}_t(\dot{a},\dot{\alpha}_1,\dot{\alpha}_2)$ is undefined for every $(\dot{a},\dot{\alpha}_1,\dot{\alpha}_2)$ of type $(\dot{0},\dot{1},\dot{1})$.

Case $\bar{2}^{(0:0,1,1)}$: Without opening any of the three envelopes, the $\dot{\theta}_t$-oracle declares that $\dot{\theta}_t(\dot{a},\dot{\alpha}_1,\dot{\alpha}_2)=w$.

Case $\bar{3}^{(0:0,1,1)}$ (Principal Case): The $\dot{\theta}_t$-oracle performs a sequence of one or more steps, each of which consists of opening one or another of the envelopes #1, #2 and #3, and/or in the case of #2 or #3 of questioning the $\dot{\alpha}_1$- or $\dot{\alpha}_2$-oracle found inside. Of course, the first step must consist of (or at least include) opening an envelope.

In the principal case, by these steps the $\dot{\theta}_t$-oracle progressively accumulates information about $\dot{a}$, $\dot{\alpha}_1$ and $\dot{\alpha}_2$ toward reaching a stage at which she will either rule out that $\dot{\theta}_t(\dot{a},\dot{\alpha}_1,\dot{\alpha}_2)$ be defined or declare that $\dot{\theta}_t(\dot{a},\dot{\alpha}_1,\dot{\alpha}_2)=w$.

She performs no step frivolously. Thus, if she opens an envelope, it is because, in order to give a value w to $\dot{\theta}_t(\dot{a},\dot{\alpha}_1,\dot{\alpha}_2)$, she needs a (natural number) value of $\dot{a}$ or some values of $\dot{\alpha}_1$ or $\dot{\alpha}_2$. So e.g. her opening #1 and finding it empty will cause her to clam up (i.e. stand mute), making $\dot{\theta}_t(\dot{a},\dot{\alpha}_1,\dot{\alpha}_2)=$ u. If she opens e.g. #2, she will next (as part of the same step) query with $\dot{\alpha}^0=$ u the $\dot{\alpha}_1$-oracle who steps out to see whether or not the latter opens envelopes (as the $\dot{\alpha}^2$-oracle questioned the $\dot{\alpha}^1$-oracle in Section 6.3 Case $\bar{3}^2$). If the $\dot{\alpha}_1$-oracle fails to respond (Subcase $\overline{3.1}^2$), the $\dot{\theta}_t$-oracle clams up. If the $\dot{\alpha}_1$-oracle without opening the envelope replies that $\dot{\alpha}_1(\dot{\alpha}^0)=n$ (Subcase $\overline{3.2}^2$), the $\dot{\theta}_t$-oracle learns everything about $\dot{\alpha}_1$, and will not question $\dot{\alpha}_1$ further. If the $\dot{\alpha}_1$-oracle opens the envelope, the $\dot{\theta}_t$-oracle observes this, and will eventually query the $\dot{\alpha}_1$-oracle with a series of one or more numbers $r_{10},r_{11},\ldots$ (as the $\dot{\alpha}^2$-oracle did the $\dot{\alpha}^1$-oracle in Subcase $\overline{3.3}^2$). However, these queries need not all be posed at once; steps involving #1 or #3 may be interspersed and thus influence the choices of $r_{10},r_{11},\ldots$. Each step the $\dot{\theta}_t$-oracle takes is determined by her from only the information accumulated in the preceding steps; i.e. it is the same for all $\dot{a},\dot{\alpha}_1,\dot{\alpha}_2$ which would behave the same in those steps.

I illustrate, using a simple example due to Kierstead.[7] Let $\dot{\theta}_t=\lambda a\alpha_1\alpha_2\alpha_1(a)$. Here is the program for our $\dot{\theta}_t$-oracle. First, she opens envelope #2, and questions the $\dot{\alpha}_1$-oracle with $\dot{\alpha}^0=$ u. If the $\dot{\alpha}_1$-oracle does not respond, the $\dot{\theta}_t$-oracle clams up. If the $\dot{\alpha}_1$-oracle, without opening the envelope, replies that $\dot{\alpha}_1(\dot{\alpha}^0)=n$, the $\dot{\theta}_t$-oracle declares that $\dot{\theta}_t(\dot{a},\dot{\alpha}_1,\dot{\alpha}_2)=n$. If the $\dot{\alpha}^1$-oracle opens the envelope, the $\dot{\theta}_t$-oracle then opens #1; and if she finds an $r\in N^0$ inside, she next queries the $\dot{\alpha}_1$-oracle with that r (otherwise

[7]This example corrected an earlier treatment by me of $\dot{\theta}_t$-oracles.

she clams up). If the $\dot{\alpha}_1$-oracle thereupon replies that $\dot{\alpha}_1(r)=n$, the $\dot{\theta}_t$-oracle finally declares that $\dot{\theta}_t(\dot{a},\dot{\alpha}_1,\dot{\alpha}_2)=n$ (otherwise she clams up). The reader should find it easy to convince (him/her)-self that this gives the right results; i.e. that I have indeed described an oracle for the function $\lambda a\alpha_1\alpha_2\alpha_1(a)$. The key to dealing with this example is letting the response of $\dot{\alpha}_1$ to "$\dot{\alpha}_1$(u)?" (without its stopping the process if $\dot{\alpha}_1$(u) is undefined) determine whether $\dot{\theta}_t$ needs to open #1. If $\dot{\alpha}_1$(u) is defined, opening #1 could be the wrong thing for the $\dot{\theta}_t$-oracle to do, as in case $\dot{a}=$u it would incorrectly render $\dot{\theta}_t(\dot{a},\dot{\alpha}_1,\dot{\alpha}_2)$ undefined.

The $\dot{\theta}_t$-oracle must observe monotonicity. For example, if $\dot{\alpha}_2(r)=n$ for all r's for which she questions an envelope-opening $\dot{\alpha}_2$, along with the information she has acquired about the other arguments, leads to her declaring that $\dot{\theta}_t(\dot{a},\dot{\alpha}_1,\dot{\alpha}_2)=w$, she must make the same declaration for an $\dot{\alpha}_2$ who gives the value n without opening envelopes, along with the same information about the other arguments.

A given partial function $\dot{\theta}_t$ from type $(\dot{0},\dot{1},\dot{1})$ into N^0 may have different oracles, in the same manner as a given partial function $\dot{\alpha}^2$ from type $\dot{1}$ into N^0.

7. Computation using the types $\dot{0},\dot{1},\dot{2}$

7.1. I must now reformulate the computation rules (KLEENE (1978, Section 2.4), reviewed in Section 4.4 here) to use an assignment Ω of worths of the types $\dot{0},\dot{1},\dot{2}$ (instead of the types 0, 1, 2) to all the free variables, and of the like to the symbols for assumed functions.

The definition of j-expression does not need to be altered, though now I will usually write the variables dotted.[8]

Only the rules E1, E4.1, E4.2 and E7 which involve the assignment Ω need to be changed.

E1: E is a type-$\dot{0}$ variable $\dot{a}$ assigned the worth $\dot{\alpha}^0$.

Case 1^0: $\dot{\alpha}^0=$u. There is no computation tree for E under Ω.

Case 2^0: $\dot{\alpha}^0=r\in N^0$. As before, the computation tree is (with R the numeral for r)

$$\dot{a} \text{———} \mathrm{R}\dagger.$$

E4.1: E is $\dot{\alpha}^1$(B), with $\dot{\alpha}^1$ assigned the worth $\dot{\alpha}^1$, for which I suppose given a specific oracle. (I trust this audience will not be confused by my

[8]When we are thinking primarily of the formal j-expressions, it seems natural to use just the one kind of variable for each j, which when introduced in KLEENE (1978, Section 2.2) were written undotted. When we have the interpretation on our mind, now using the types $\dot{0},\dot{1},\dot{2}$ instead of 0, 1, 2, the dotted variables seem natural.

using the same notation "$\dot{\alpha}^1$" to name the variable and to name its worth under the assignment Ω.)

Case $\bar{1}^1$: The $\dot{\alpha}^1$-oracle, presented with any envelope, stands mute ($\dot{\alpha}^1 = \mathrm{u}^1$). There is no computation tree for E under Ω.

Case $\bar{2}^1$: The $\dot{\alpha}^1$-oracle, without opening the envelope, declares that $\dot{\alpha}^1(\dot{\alpha}^0) = n$ ($\dot{\alpha}^1 = \lambda\dot{\alpha}^0 n$). The computation tree for E is then

$$\dot{\alpha}^1(\mathrm{B}) \text{——} \mathrm{N}\dagger$$

where N is the numeral for n.

Case $\bar{3}^1$: The $\dot{\alpha}^1$-oracle opens the envelope. For there to be a computation tree for E, there must be a subordinate computation tree for B, under Ω restricted to the free variables of B, leading to a value r (of B) for which $\dot{\alpha}^1(r)$ is defined, $= n$ say. Then the computation tree for E is

$$\begin{array}{l} \dot{\alpha}^1(\mathrm{B}) \text{——} \mathrm{N}\dagger. \\ \qquad\quad \diagdown \\ \qquad\qquad \mathrm{B}\cdots\mathrm{R}\dagger \end{array}$$

This treatment, with three cases, parallels our account of $\dot{\alpha}^1$-oracles in Section 6.2, with the envelope containing an $r \in N^0$ or being empty transposed here into there being or not being a computation tree for B under Ω-restricted-to-B.

7.2. Before dealing with E4.2 and E7, let us reflect on the structure of our definition. In KLEENE (1978, Section 2.3, paragraph 4 and Section 2.4, paragraph 1), I billed it as an inductive definition of completed computation trees. At the end of KLEENE (1978, Section 2.4), I also viewed computation as a reduction procedure, starting with a given E and Ω and proceeding from left to right.

Implicit in the latter view is an inductive definition of computations (of 0-expressions E under assignments Ω with specific choices of the oracles) not necessarily completed. Thus if E is any 0-expression, it is itself a computation. If we already have a computation with E at the right end of a branch, then under the circumstances of E2 that branch can be extended to get another computation by adding "——F"; of E4.λ, by adding "——$S_{\mathrm{B}}^{\alpha^j}$A|"; of E4.1 Case $\bar{3}^1$ (the holding of which depends in Case 1^1 on the choice of the oracle for $\dot{\alpha}^1$), by adding

$$\begin{array}{l} \diagdown \\ \quad \mathrm{B}, \end{array}$$

etc. If under E4.1 Case $\bar{3}^1$, we have a computation of the form

$$\begin{array}{ll} \cdots\dot{\alpha}^1(\mathrm{B}) & \\ \qquad\diagdown & \\ & \mathrm{B}\cdots\mathrm{R}\dagger \end{array}$$

where "$\mathrm{B}\cdots\mathrm{R}\dagger$" is a completed computation of B with the flagged numeral R† for r at the right end of its principal (uppermost) branch, and if $\dot{\alpha}^1(r)=n$, then also

$$\begin{array}{ll} \cdots\dot{\alpha}^1(\mathrm{B})\text{——}\mathrm{N}\dagger & \\ \qquad\diagdown & \\ & \mathrm{B}\cdots\mathrm{R}\dagger \end{array}$$

is a computation. In brief, computations are the stages in our constructions as we seek to build a completed computation tree. The latter is a computation with a flagged numeral at the right end of its principal branch, which entails every lower branch also ending in a flagged numeral.

7.3. E4.2: E is $\alpha^2(\mathrm{B})$ with α^2 assigned the worth $\dot{\alpha}^2$, for which I suppose given a specific oracle. For the second view in Section 7.2, I am supposing we have a computation before us with this E at the right end of a branch. I shall parallel the account of $\dot{\alpha}^2$-oracles in Section 6.3. The responses there of an $\dot{\alpha}^1$-oracle which the $\dot{\alpha}^2$-oracle is questioning (answerings of questions without opening envelopes, openings of envelopes, answerings of questions after opening envelopes) are transposed here (as we shall see in detail) into certain computational events.

Case $\bar{1}^2$: The $\dot{\alpha}^2$-oracle, presented with any envelope, stands mute ($\dot{\alpha}^2=\mathrm{u}^2$). There is no completed computation of E under Ω; or viewed the other way, there is no extension at all of a given computation with a branch ending in E. Having reached this E, we are blocked.

Case $\bar{2}^2$: The $\dot{\alpha}^2$-oracle, without opening the envelope, declares that $\dot{\alpha}^2(\dot{\alpha}^1)=m$ ($\dot{\alpha}^2=\lambda\dot{\alpha}^1 m$). Our computation with E (i.e. $\alpha^2(\mathrm{B})$) at the end of a branch can be extended thus:

$$\cdots\dot{\alpha}^2(\mathrm{B})\text{——}\mathrm{M}\dagger.$$

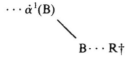

Case $\bar{3}^2$: The $\dot{\alpha}^2$-oracle opens envelopes. Let c be a type-0 variable not occurring free in B. Let Ω_r (or Ω_{u}) be the assignment obtained by using the same worths as in Ω for the free variables of B and $r\in N^0$ (or u) as the worth of c. We can extend our computation, which had reached E at the end of a branch, by starting a subordinate branch (hopefully to become a completed subcomputation) with $B(c)$ under Ω_u as the assignment:

$$\cdots \dot{\alpha}^2(B)$$
$$\diagdown$$
$$B(c)\cdots \Omega_u \cdots .$$

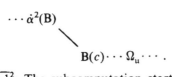

Subcase $\overline{3.1}^2$: The subcomputation started with $B(c)$ under Ω_u cannot be completed, or even carried to a vertex where a branch ends with c (i.e. to a "surfacing" of c), where to continue a value $r \in N^0$ of c is called for. Then no step can be taken horizontally rightward from $\dot{\alpha}^2(B)$. *Subcase* $\overline{3.2}^2$: The subcomputation of $B(c)$ under Ω_u can be completed with $B(c)$ receiving the value n, and $\dot{\alpha}^2(\lambda\dot{\alpha}^0 n) = m$. Then the following is a computation:

$$\cdots \dot{\alpha}^2(B) \text{---} M\dagger.$$
$$\diagdown$$
$$B(c)\cdots \Omega_u \cdots N\dagger$$

Subcase $\overline{3.3}^2$: The subcomputation of $B(c)$ under Ω_u can be carried to a vertex where c is the 0-expression (c surfaces), whereupon a value r ($\in N^0$) of c is called for if the computation is to be continued. (For any r, the computation starting from $B(c)$ up to this surfacing, which was under Ω_u, will be good under Ω_r also.) This corresponds to the $\dot{\alpha}^1$-oracle in Section 6.3 opening the envelope. If the $\dot{\alpha}^2$-oracle vis-a-vis an $\dot{\alpha}^1$-oracle who opens envelopes does not supply an r_0 for a question "$\dot{\alpha}^1(r_0)$?", the computation cannot be continued further. If the $\dot{\alpha}^2$-oracle does supply an r_0, we assign this r_0 to c, and continue the computation of $B(c)$ as under Ω_{r_0} (with the part already constructed under Ω_u being reconstruable as under Ω_{r_0}). If we can thus reach a value n_0 for $B(c)$ in this as the lowest subcomputation, then we may be able to start a next subcomputation under Ω_{r_1} or perhaps take a step "—M†" rightward from E. Following the pattern in Subcase $\overline{3.3}^2$ of Section 6.3, we may have an uncompletable computation by reason of some subcomputation being uncompletable or of $\dot{\alpha}^2$ clamming up, or we may be led in the end to a completed computation of $\dot{\alpha}^2(B)$, extending any given computation "$\cdots \dot{\alpha}^2(B)$" thus

$$\cdots \dot{\alpha}^2(B) \text{---} M\dagger$$
$$\vdots$$
$$B(c)\cdots \Omega_{r_\zeta} \cdots N_\zeta\dagger \quad (\zeta < \xi)$$
$$\vdots$$
$$B(c)\cdots \Omega_{r_1} \cdots N_1\dagger$$
$$B(c)\cdots \Omega_{r_0} \cdots N_0\dagger$$

where $\dot{\alpha}^2(\dot{\beta}_1)$ is defined for $\dot{\beta}^1 = \{\langle r_\zeta, n_\zeta \rangle | \zeta < \xi\}$ with the value m.

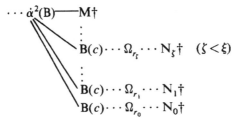

7.4. E7: E is $\dot{\theta}_t(\mathrm{A}, \mathrm{B}_1, \mathrm{B}_2)$. Now I am guided by a specific $\dot{\theta}_t$-oracle, paralleling Section 6.4. Cases $\bar{1}^{(0:0,1,1)}$ and $\bar{2}^{(0:0,1,1)}$ are obvious: no computation step proceeding rightward from E under Ω, or the single step "——W†", respectively.

So consider the principal case (Case $\bar{3}^{(0:0,1,1)}$) that the $\dot{\theta}_t$-oracle performs steps. Steps affecting envelopes #1, #2 or #3 will be represented here by subcomputations starting with $\mathrm{A}, \mathrm{B}_1(c_1)$ or $\mathrm{B}_2(c_2)$ under Ω^{A}, $\Omega^{\mathrm{B}_1}_{r_1}$ or $\Omega^{\mathrm{B}_1}_{\mathrm{u}}$, or $\Omega^{\mathrm{B}_2}_{r_2}$ or $\Omega^{\mathrm{B}_2}_{\mathrm{u}}$, as we shall illustrate. These subcomputations are well-ordered upward in the order in which the steps that call for them occur. For example, if the $\dot{\theta}_t$-oracle opens #1, there will be a subcomputation begun with A under Ω-restricted-to-A (write it Ω^{A}), which must terminate in a value a if the computation is to be continuable. When the $\dot{\theta}_t$-oracle first opens #2 (if she does), a subcomputation of $\mathrm{B}_1(c_1)$ under Ω-restricted-to-B_1 plus u as assignment to a variable c_1 not occurring free in B_1 (write this assignment $\Omega^{\mathrm{B}_1}_{\mathrm{u}}$) will be begun. If it fails either to be completable with a value n or to come to a surfacing of c_1 (Subcase $\overline{3.1}^2$ for an $\dot{\alpha}^2$-oracle), the whole computation is uncompletable. If the subcomputation is completable with a value n (Subcase $\overline{3.2}^2$), its completion

$$\mathrm{B}_1(c_1) \cdots \Omega^{\mathrm{B}_1}_{\mathrm{u}} \cdots \mathrm{N}\dagger$$

becomes the only subcomputation begun with $\mathrm{B}_1(c_1)$. If c_1 surfaces (Subcase $\overline{3.3}^2$), which blocks the completion of the subcomputation of $\mathrm{B}_1(c_1)$ under $\Omega^{\mathrm{B}_1}_{\mathrm{u}}$, the $\dot{\theta}_t$-oracle must supply one or more values $r_{10}, r_{11}, \ldots$ for c_1 for respective subcomputations of $\mathrm{B}_1(c_1)$ under $\Omega^{\mathrm{B}_1}_{r_{10}}, \Omega^{\mathrm{B}_1}_{r_{11}}, \ldots$, but maybe only after pursuing other subcomputations, as e.g. of A if $\dot{\theta}_t = \lambda a\alpha_1\alpha_2\alpha_1(a)$ as in the example of Section 6.4. If and when enough subcomputations are first completed to represent a triple $(\dot{b}, \dot{\beta}_1, \dot{\beta}_2)$ for which $\dot{\theta}_t(\dot{b}, \dot{\beta}_1, \dot{\beta}_2) = w$, then from the vertex in question (occupied by $\dot{\theta}_t(\mathrm{A}, \mathrm{B}_1, \mathrm{B}_2)$) we have the step "——W†" horizontally rightward.

7.5. Here, in contrast to KLEENE (1978, Section 2.4), the definition of a computation beginning with E involves not just an assignment Ω of worths to the free variables of E and the $\dot{\theta}_t$'s but, in the case of free variables of type $\dot{1}$ or $\dot{2}$ and of $\dot{\theta}_t$'s with non-empty lists of variables, a choice of a set of oracles to embody the functions assigned.

Now I would like to see that, if a value w is obtained for E under Ω with one choice of the set of oracles, a value being obtained and the value w itself are independent of the choice. That is, if there be a completed computation tree with the principal branch

$$\mathrm{E} \cdots \mathrm{W}\dagger$$

under one choice of the set of oracles, there will be a completed tree under every other choice, and, although the trees may be different for different choices, the numerals W at the right ends of their principal branches will be the same. Thus the predicate $E \simeq_{\Omega} w$ is defined so as to be true, with a given E and Ω (sans a choice of the oracles), for at most one $w \in N^0$, called (when it exists) the *value* of E under Ω.

To this end, I show, by induction over a completed computation tree for E under Ω with a given choice of the set of oracles required, that under any other choice there is also a completed computation tree, and the trees are in 1–1 correspondence with the leftmost vertices corresponding, preserving the relationships of next-right-horizontally and of next-right-but-lower, and with the same 0-expressions and assignments at corresponding vertices.

As an oracle gets into the act only at applications of E4.1, E4.2 or E7, we need for this induction to examine only the cases of them. The other basis and induction steps are immediate.

E4.1. An $\dot{\alpha}^1$-oracle is unique except when $\dot{\alpha}^1 = \mathrm{u}^1$ (end Section 6.2), and in that case there is no completed computation tree from a vertex bearing $\dot{\alpha}^1(\mathrm{B})$. So there is no problem.

E4.2. The possibility of two oracles for $\dot{\alpha}^2$ under Case 1^2 ($\dot{\alpha}^2 = \mathrm{u}^2$) with Cases $\overline{1}^2$ and $\overline{3}^2$ respectively presents no problem (no completed tree), nor similarly of two oracles under Subcase $\overline{3.3}^2$ when they end a sequence of questions $r_0, r_1, \ldots, r_\eta, \ldots$ by standing mute after less or more questions (Section 6.3 near the end). There remains the possibility of two oracles for $\dot{\alpha}^2$ under Subcase $\overline{3.3}^2$ asking the same set of questions

$$\{r_\zeta | \zeta < \xi\} = \{\bar{r}_\zeta | \zeta < \bar{\xi}\}$$

in different orders. By hyp. ind. over the given computation tree, at each subtree next-right in that computation tree, begun with B(c) under Ω_r, there is a computation tree under the other choice (including perhaps a different oracle for $\dot{\alpha}^2$), begun with the same B(c) under the same assignment Ω_r, in a 1–1 correspondence with the stated features to that subtree. Now, to get a completed tree, begun with $\dot{\alpha}^2(\mathrm{B})$ under Ω with the other choice, in such a 1–1 correspondence to the given tree, I let the 1–1 correspondence match the pairs of vertices (beginning subtrees) occupied by B(c) under the same Ω_r, and use the aforesaid 1–1 correspondences given by the hyp. ind. in the respective subtrees issuing from the two vertices of each pair. This is legitimate, since by the aforesaid application of the hyp. ind. the same function

$$\dot{\beta}^1 = \{\langle r_\zeta, n_\zeta \rangle | \zeta < \xi\} = \{\langle \bar{r}_\zeta, \bar{n}_\zeta \rangle | \zeta < \bar{\xi}\}$$

is represented by the two sets of subtrees, so the E4.2 step is completed in each by "——M†" in the principal branch with the same $m = \dot{\alpha}^2(\dot{\beta}^1)$.

E7. Quite clearly similar considerations apply.

This can be summarized by saying that, with two different choices of the oracles, the computation trees are the same except perhaps for the orders of the subtrees (as identified each by the 0-expression and assignment at its initial vertex) in the applications of E4.2 and E7.

8. The perceived shape of things to come

8.1. In the computation rules of Section 7 we were guided by the interpretation using the modified types which were introduced in Sections 5 and 6. How does this provide a semantics to flesh out the computations of KLEENE (1978, Section 2.4), as we proposed in Section 4.5? For the present, I hold the types to $0, 1, 2$, $(0:j_1,\dots,j_k)$, and $\dot{0}, \dot{1}, \dot{2}$, $(\dot{0}:j_1,\dots,j_k)$, with each $j_i \leq 1$.

It has been simplest, and sufficient thus far, to work only with assignments Ω ("minimal assignments") to just the free variables in the expression E under consideration at the moment (besides to the function symbols θ_t of the list Θ for the presupposed partial recursive derivation $\phi_1,\dots,\phi_p$). But in considering several expressions at once (as in #B below), we may operate with an assignment Ω to the variables occurring free in any one of the expressions (so Ω may not be minimal for each expression); and indeed, we may sometimes have some variables to which we assign worths to be kept throughout an extended discussion—what I called a reserved list V of variables in KLEENE (1978 p. 200). Obviously, worths assigned by Ω to variables not occurring free in E have no effect on the process of computing E under Ω. But in operating with assignments that are not minimal, in order to avoid conflict in assigning worths to c under E4.2 Case $\bar{3}^2$ (and similarly under E7), we will wish to choose as the c for an application of E4.2 a variable which is distinct both from the free variables of B and from any other variables previously assigned worths that we are not ready to erase.

We write "$\simeq_\Omega$" between two expressions, each either a 0-expression or denoting a member of type $\dot{0}$ (e.g. "$E \simeq_\Omega \dot{\alpha}^0$", "$E \simeq_\Omega \dot{\alpha}^j(\dot{\alpha}^{j-1})$", "$E \simeq_\Omega F$") to say that the two members have the same worth (equivalently, both are defined with the same value, or both are undefined), under the assignment Ω in the case of 0-expressions (see KLEENE (1978, footnote 4)). (This use of "$\simeq$" was begun in KLEENE (1938).)

I anticipate proving the following as theorems.[9] In #A and #B we see that the computation rules of Section 7 do conform to the interpretations

[9]Officially, it can stand for the moment that I am proposing the problems of either proving, or finding a counterexample, to each of #A, #B, #C, #D. But I have little doubt that proofs I have in manuscript, when finally checked, will establish them for the stated values of j. For $j=3$, the situation is more problematical.

In the six months following the June 1977 Oslo Symposium I undertook an overambitious project in which I wrote out several times as much material (85 pages of typescript) as I presented at Oslo. This included what I thought were proofs of #A, #B, #C and #D, but not restricted to $j<2$. In February 1978, Kierstead gave me first a memorandum containing his correction to my treatment therein of θ_t-oracles,[7] and then a memorandum containing a counterexample to my treatment therein of $\dot{\alpha}^3$-oracles (cf. Section 8.2). In May 1978, I found time to study the latter memorandum, and I worked out what I believe will prove to be an adequate revision. Subsequently, I examined his correction for θ_t-oracles.

using the types of Sections 5 and 6; and in #C and #D we see how the computations of KLEENE (1978, Section 2.4) can be represented within those of Section 7.

#A. Trivially ($j=0$): *To each 0-expression* E, *and each assignment* Ω *to its free variables and the* θ_t's, *there is a unique type-*$\dot{0}$ *object* $\dot{\alpha}^0$ *such that*

$$E \simeq_{\Omega} \dot{\alpha}^0.$$

Non-trivially: *For* $j=1$ *or* 2, *to each 0-expression* $E_{\gamma^{j-1}}$ *and each assignment* Ω *to its free variables and the* θ_t's *except to the type-*$j-1$ *variable* γ^{j-1}, *there is a unique type-*$j^{\cdot}$ *object* $\dot{\alpha}^j$ *such that, for each* $\dot{\alpha}^{j-1}$ *of type* $(j-1)^{\cdot}$,

$$E_{\gamma^{j-1}} \simeq_{\Omega_{\dot{\alpha}^{j-1}}} \dot{\alpha}^j(\dot{\alpha}^{j-1})$$

where $\Omega_{\dot{\alpha}^{j-1}}$ *is the assignment coming from* Ω *by assigning* $\dot{\alpha}^{j-1}$ *to* γ^{j-1}. Applying this to $\{\lambda\gamma^{j-1}E\}(\gamma^{j-1})$ as the $E_{\gamma^{j-1}}$, each type-j λ-functor $\lambda\gamma^{j-1}E$ is interpretable, under any assignment to its free variables and θ_t's, as a type-$j^{\cdot}$ object.

Similarly, e.g.: *To each 0-expression* E_{c,γ_1,γ_2} *and each assignment* Ω *to its free variables and the* θ_t's *except to the variables* c,α_1,α_2 *of types* 0, 1, 1, *there is a unique type-*$(\dot{0}:\dot{0},\dot{1},\dot{1})$ *object* $\dot{\theta}$ *such that, for each* $\dot{a},\dot{\alpha}_1,\dot{\alpha}_2$ *of types* $\dot{0},\dot{1},\dot{1}$,

$$E_{c,\gamma_1,\gamma_2} \simeq_{\Omega_{\dot{a},\dot{\alpha}_1,\dot{\alpha}_2}} \dot{\theta}(a,\dot{\alpha}_1,\dot{\alpha}_2)$$

where $\Omega_{\dot{a},\dot{\alpha}_1,\dot{\alpha}_2}$ *is the assignment coming from* Ω *by assigning* $\dot{a},\dot{\alpha}_1,\dot{\alpha}_2$ *to* c,γ_1,γ_2.

#B. Non-trivially: *Let* E_D *be a 0-expression containing a tribe of specified free occurrences of a j-expression* D ($j=0$, 1 *or* 2); *and let* $\bar{D}$ *be a j-expression such that, replacing in* E_D *each of the occurrences of the tribe by* $\bar{D}$ *to obtain* $E_{\bar{D}}$, *the resulting occurrences of* $\bar{D}$ *in* $E_{\bar{D}}$ *are free. If, under an assignment* Ω *to the free variables of* D, $\bar{D}$ *and* E_D *and the* θ_t's, D *and* $\bar{D}$ *have the same worth as j-expressions (using* #A *if* D *or* $\bar{D}$ *is not just a variable), then*

$$E_D \simeq_{\Omega} E_{\bar{D}}.$$

#C. Trivially: *For* $j=0$ *or* 1, *the type-j objects are a subset of the type-*$j^{\cdot}$ *objects, and each* α^0 ($\in$ *type* 0) *is its own basis w.r. to any* α^1 ($\in$ *type* 1) *as a type-*$\dot{1}$ *object.*

Easily: *There is a* 1–1 *mapping* $\alpha^2 \leftrightarrow \mathring{\alpha}^2$ *of the type-2 objects into the type-*$\dot{2}$ *objects such that, for each* α^2,α^1 *(of types* 2, 1), $\mathring{\alpha}^2(\alpha^1)=\alpha^2(\alpha^1)$ *and* α^1 *is its own basis w.r. to* $\mathring{\alpha}^2$.

Similarly, e.g.: *There is a* 1–1 *mapping* $\theta \leftrightarrow \mathring{\theta}$ *of the type-*(0:0, 1, 1) *objects into the type-*$(\dot{0}:\dot{0},\dot{1},\dot{1})$ *objects such that, for each* a,α_1,α_2 *(of types* 0, 1, 1), $\mathring{\theta}(a,\alpha_1,\alpha_2)=\theta(a,\alpha_1,\alpha_2)$ *and* (a,α_1,α_2) *is its own basis w.r. to* $\mathring{\theta}$.

#D. Easily: $E \simeq_{\Omega} w$ *under the computation rules of Section* 2.4 *if and only if* $E \simeq_{\mathring{\Omega}} w$ *under the computation rules of Section* 7 *where (using* #C) $\mathring{\Omega}$ *comes from* Ω *by replacing each* α^2 *by* $\mathring{\alpha}^2$ *and each* θ *by* $\mathring{\theta}$.

8.2. That a semantics based on these conceptions works up through type $\dot{2}$ (as I believe, subject to final checking) I consider to be of interest in any case.

I now give the counterexample of Kierstead to a formulation I had entertained for $j=\dot{3}$. I had supposed that, just as we can require an $\dot{\alpha}^2$-oracle in Subcase $\overline{3.3}^2$ of Section 6.3 to come up sui generis with a first question "$\dot{\alpha}^1(r_0)$?" with $r_0 \in N^0$, the same to any $\dot{\alpha}^1$-oracle who opens envelopes, etc., we could ask an $\dot{\alpha}^3$-oracle in Subcase $\overline{3.3}^3$ to come up with a first question "$\dot{\alpha}^2(\dot{\chi}_0)$?" with $\dot{\chi}_0 \in N^1$, the same to any envelope-opening $\dot{\alpha}^2$-oracle, etc.

Kierstead observed that, if E_{γ^2} is $\gamma^2(\lambda a[\gamma^2(\lambda b[a])])$, then $\lambda\gamma^2 E_{\gamma^2}$ is not a type-$\dot{3}$ object under that formulation (contrary to #A for $j=3$). We can see this at once under the obvious interpretation (without explicitly invoking the rules for computing E_{γ^2} for the various type-$\dot{2}$ interpretations of γ^2, which rules are certainly consistent with that interpretation).

For each natural number i, following Kierstead I let

$$\dot{\alpha}_i^2(\dot{\beta}) \simeq \begin{cases} n & \text{if } \dot{\beta} \supset \{\langle i, i+n\rangle, \langle i+1, i+n\rangle, \langle i+2, i+n\rangle, \dots\}, \\ i & \text{if } \dot{\beta} \supset \{\langle i, 0\rangle, \langle i+1, 1\rangle, \langle i+2, 2\rangle, \dots\}, \\ \text{u} & \text{otherwise.} \end{cases}$$

Each $\dot{\alpha}_i^2$ is a type-$\dot{2}$ object. An oracle for $\dot{\alpha}_i^2$ can be programmed to ask successively "$\dot{\beta}(i)$?", "$\dot{\beta}(i+1)$?", "$\dot{\beta}(i+2)$?", ... If the result is the constant $i+n$ for some $n \geq 0$, she declares that $\dot{\alpha}_i^2(\dot{\beta}) = n$. If the result is the sequence $0, 1, 2, \dots$, she says that $\dot{\alpha}_i^2(\dot{\beta}) = i$. If anything else happens (in particular, if $\dot{\beta}$ is the constant n for some $n < i$), she stands mute. Thus

$$\begin{aligned} \dot{\alpha}_i^2(\lambda b[0]) &\simeq \text{u}, \\ &\vdots \\ \dot{\alpha}_i^2(\lambda b[i-1]) &\simeq \text{u}, \\ \dot{\alpha}_i^2(\lambda b[i]) &\simeq 0, \\ \dot{\alpha}_i^2(\lambda b[i+1]) &\simeq 1, \\ &\vdots \end{aligned}$$

Therefore $\dot{\alpha}_i^2(\lambda a[\dot{\alpha}_i^2(\lambda b[a])]) \simeq i$.

How could an $\dot{\alpha}^3$-oracle under the proposed formulation behave, if $E_{\gamma^2} \simeq_{\Omega_{\alpha^2}} \dot{\alpha}^3(\dot{\alpha}^2)$ for each $\dot{\alpha}^2$? Since then $\dot{\alpha}^3 \neq \text{u}^3$ but $\dot{\alpha}^3(\text{u}^2) = \text{u}$, she must ask a first question "$\dot{\alpha}^2(\dot{\chi}_0)$?" of any envelope-opening $\dot{\alpha}^2$. For $\dot{\alpha}_0^2(\dot{\chi}_0)$ be defined, $\dot{\chi}_0 \upharpoonright N^0$ must be either the constant n for some n or the identity function. For $\dot{\alpha}_i^2(\dot{\chi}_0)$ with $i > 0$ to be defined, $\dot{\chi}_0 \upharpoonright N^0$ must be neither the constant n for any $n < i$ nor the identity. Thus, for no $\dot{\chi}_0$ is $\dot{\alpha}_i^2(\dot{\chi}_0)$ defined for all i, though $\dot{\alpha}^3(\dot{\alpha}_i^2)$ is defined. Forget KLEENE's "A manual for $\dot{\alpha}^3$-oracles", Edition of 1977.

8.3. Reflecting on this example (in May 1978), I concluded that the faulted edition of "A manual" did not allow the full and flexible use of all the possibilities, within Kleene's basic conception, for an $\dot{\alpha}^3$-oracle to accumulate information about an $\dot{\alpha}^2$ for which an oracle is available for her to question.

Consider how an unfettered $\dot{\alpha}^3$-oracle could represent

$$\dot{\alpha}^3 = \lambda\gamma^2\gamma^2(\lambda a[\gamma^2(\lambda b[a])]).$$

For each $\dot{\alpha}^2 \in$ type $\dot{2}$, she must give the correct value, if defined, of $\dot{\alpha}^2(\lambda a[\dot{\alpha}^2(\lambda b[a])])$, and no value otherwise. Let "D" say that this expression is defined. Clearly Case $\overline{3}^3$ (corresponding to Case $\overline{3}^2$ for $\dot{\alpha}^2$-oracles) must apply. First, our $\dot{\alpha}^3$-oracle asks "$\dot{\alpha}^2(\dot{\alpha}^1)$?" for $\dot{\alpha}^1 = \mathrm{u}^1$ embodied in an $\dot{\alpha}^1$-oracle who opens envelopes but is impossible to please (end Section 6.2). If, without opening the envelope, the $\dot{\alpha}^2$-oracle says that $\dot{\alpha}^2(\dot{\alpha}^1) = m$ (Subcase $\overline{3.2}^3$), then also $\dot{\alpha}^2(\lambda a[\dot{\alpha}^2(\lambda b[a])]) = m$, and the $\dot{\alpha}^3$-oracle will be correct in saying that $\dot{\alpha}^3(\dot{\alpha}^2) = m$. Now suppose (Subcase $\overline{3.3}^3$) the $\dot{\alpha}^2$-oracle opens the envelope. The $\dot{\alpha}^3$-oracle observes this, and concludes that $\dot{\alpha}^2(\lambda b[\dot{a}])$ is undefined for $\dot{a} = \mathrm{u}$. Thus, in our subcase, the argument $\lambda a \cdots$ of the outer $\dot{\alpha}^2$ is undefined for u as its argument. So if $\dot{\alpha}^2(\lambda a \cdots)$ is defined, i.e. if D, then $\dot{\alpha}^2(\dot{\rho})$ is defined for some non-empty partial function $\dot{\rho}$ ($\subset \lambda a[\dot{\alpha}^2(\lambda b[a])]$) on N^0. The $\dot{\alpha}^3$-oracle lets the question $\dot{\alpha}^1 = \mathrm{u}^1$ stand. The $\dot{\alpha}^2$-oracle will first question with u this $\dot{\alpha}^1$-oracle, find that the $\dot{\alpha}^1$-oracle opens envelopes, and (if D) then question her with the r_0 which she was ready to ask next of any envelope-opening $\dot{\alpha}^1$-oracle. The $\dot{\alpha}^3$-oracle observes all this (in particular, what the r_0 is), and concludes that (if D) the argument $\lambda a \cdots$ of the outer $\dot{\alpha}^2$ must have r_0 in its domain, so the inner $\dot{\alpha}^2$ must be defined for the argument $\lambda b[r_0]$. So the $\dot{\alpha}^3$-oracle next questions the $\dot{\alpha}^2$-oracle with $\lambda b[r_0]$ embodied in an $\dot{\alpha}^1$-oracle who gives the answer r_0 without opening envelopes. If D, the $\dot{\alpha}^2$-oracle answers that $\dot{\alpha}^2(\lambda b[r_0]) = n_0$, say. So the $\dot{\alpha}^3$-oracle learns one argument-value pair $\langle r_0, n_0 \rangle$ that the function $\lambda a[\dot{\alpha}^2(\lambda b[a])]$ must have if D. Now the $\dot{\alpha}^3$-lady questions the $\dot{\alpha}^2$-dame with $\dot{\alpha}^1 = \{\langle r_0, n_0 \rangle\}$. If D, the dame will via questions learn that this $\dot{\alpha}^1$ opens envelopes and answers "$\dot{\alpha}^1(r_0)$?" by n_0. Then, if D, the $\dot{\alpha}^2$-dame must either respond that $\dot{\alpha}^2(\dot{\alpha}^1) = m$ say, whereupon the $\dot{\alpha}^3$-Pythoness declares correctly that $\dot{\alpha}^3(\dot{\alpha}^2) = m$; or ask "$\dot{\alpha}^1(r_1)$?". If the latter, the Pythoness knows that, if D, $\lambda a[\dot{\alpha}^2(\lambda b[a])]$ must be defined for the argument r_1. So she queries $\dot{\alpha}^2$ with $\lambda b[r_1]$, receiving (if D) the answer n_1. Then she questions the $\dot{\alpha}^2$-dame with $\dot{\alpha}^1 = \{\langle r_0, n_0 \rangle, \langle r_1, n_1 \rangle\}$. And so on. There can be transfinite stages, the first when $\dot{\alpha}^1$ consists of ω pairs. If D, eventually

$\dot{\alpha}^3$ will find $\dot{\alpha}^2$ giving a value m to $\dot{\alpha}^2(\dot{\alpha}^1)$, which she will then give correctly as the value of $\dot{\alpha}^3(\dot{\alpha}^2)$. For, if D, $\dot{\alpha}^2$ questioned successively with u^1, $\{\langle r_0, n_0\rangle\}, \{\langle r_0, n_0\rangle, \langle r_1, n_1\rangle\}, \ldots$ will keep dredging up arguments $r_0, r_1, r_2, \ldots$ of a function $\dot{\rho}$ (a subfunction of $\lambda a[\dot{\alpha}^2(\lambda b[a])]$) for which $\dot{\alpha}^2(\dot{\rho})$ is defined.

Under this procedure, the $\dot{\alpha}^3$-oracle has (if D) received answers "$\dot{\alpha}^2(\dot{\chi}) = m$" to "$\dot{\alpha}^2(\dot{\chi})$?" for various $\dot{\chi}$'s (specifically, for $\lambda b[r_0], \lambda b[r_1], \ldots$ and for a subfunction $\dot{\rho}$ of $\lambda a[\dot{\alpha}^2(\lambda b[a])]$). Her answer to our question "$\dot{\alpha}^3(\dot{\alpha}^2)$?" is determined by these answers. But she did not simply demand first an answer "$\dot{\alpha}^2(\dot{\chi}_0) = m$" to "$\dot{\alpha}^2(\dot{\chi}_0)$?" for a $\dot{\chi}_0$ she picks without any acquaintance with $\dot{\alpha}^2$, then to "$\dot{\alpha}^2(\dot{\chi}_1)$?" for a $\dot{\chi}_1$ she picks using only that answer, etc. Instead, she employs questions "$\dot{\alpha}^2(\dot{\alpha}^1)$?" "tentatively", being ready to accept as sufficient responses either an answer or the opening of the envelope or the asking of a question "$\dot{\alpha}^1(r)$?". In this way she explores $\dot{\alpha}^2$'s mind toward finding functions $\dot{\rho}$ such that $\dot{\alpha}^2$ will answer "$\dot{\alpha}^2(\dot{\rho})$?".

This example illustrates the manner in which I propose to undertake to surmount Kierstead's roadblock to extending my oracular semantics to type 3.[10]

References

Other references are at the end of KLEENE (1978).

KIERSTEAD, D. P.

[1979] A semantics for Kleene's *j*-expressions, in: *The Kleene Symposium,* edited by J. Barwise, H. J. Keisler and K. Kunen (North-Holland, Amsterdam) [this volume].

KLEENE, S. C.

[1959a] Countable functionals, in: *Constructivity in Mathematics*, Proceedings of the Colloquium held at Amsterdam, 1957 (North-Holland, Amsterdam) pp. 81–100.

[1978] Recursive functionals and quantifiers of finite types revisited I, in: *Generalized Recursion Theory II*, Proceedings of the 1977 Oslo Symposium, edited by J. E. Fenstad, R. O. Gandy and G. E. Sacks (North-Holland, Amsterdam) pp. 185–222.

PÉTER, R.

[1934] Über den Zusammenhang der verschiedenen Begriffe der rekursiven Funktion, *Math. Ann.*, **110**, 612–632.

[10]A detailed formulation and treatment are planned for a RFQFTR III; the present II is long enough. My dedication of my lecture to AAHOOO? It is to the Ancient and Honorable Order of Oracles.

J. Barwise, H. J. Keisler and K. Kunen, eds., *The Kleene Symposium*

Frege Structures and the Notions of Proposition, Truth and Set*

Peter Aczel
Department of Mathematics, Manchester University, Manchester 13, England

Dedicated to Professor S. C. Kleene on the occasion of his 70th birthday

Abstract: The notion of Frege structure is introduced and shown to give a coherent context for the rigorous development of Frege's logical notion of set and an explanation of Russell's paradox.

1. Introduction

The so called naive or logical notion of set was first explicitly formulated by FREGE in his Grundgesetze der Arithmetik (1893 & 1903). The notion, as formulated by Frege, was seen to be incoherent with the discovery of Russell's paradox. There have been many attempts to give a coherent reformulation of the original idea, but with little general acceptance. Instead, starting from Russell's theory of types and Zermelo's axioms there has been a development of ideas leading to modern axiomatic set theory. Modern mathematicians generally tend to accept the axioms (of ZFC say) without undue reflection on the intuitive justification for them. But set theorists are well aware of the elegant explanation of those axioms in terms of what has been called the iterative or cumulative hierarchy notion of set. (See SCOTT (1974) or SHOENFIELD (1977).) This notion has sets arranged in a hierarchy with each set coming after all its elements, and is quite distinct from Frege's notion.

The logical notion of set may be described in the following way. Sets are "extensions" $\{x|f(x)\}$ of propositional functions f, such that for any object a ($a\in\{x|f(x)\}$) is a proposition which is true iff $f(a)$ is true. Each extension of a function is an object, where Frege insists on a rigid distinction between functions and objects. As an example there is the universal set $\{x|x=x\}$, a set that is not available in the cumulative hierarchy.

*The present paper is a greatly revised version of a talk presented at the symposium.

The aim of this paper is to give a suitable general context for the rigorous consistent formulation of the logical notion of set as described above. In my presentation I have preferred to avoid the use of formal systems or any formal use of syntax in the belief that a structural approach will be more transparent. The central notion is that of a Frege structure. Roughly, such structures are models of the lambda calculus, the elements of the model being called "objects", together with a collection of objects called "propositions" and a subcollection of propositions called "truths". In addition there is a list of the usual logical constants, each satisfying a logical schema that expresses how a proposition is built up using the logical constant and further specifies truth conditions for the resulting proposition. The extension $\lambda x f(x)$ of a function f in a Frege structure is defined using the internal lambda abstraction operation that is available on any model of the lambda calculus. So $\{x|f(x)\}$ is taken to be just another notation for $\lambda x f(x)$, particularly when f is a propositional function, i.e. a function of the Frege structure all of whose values are propositions.

The underlying structure of the formal language in Frege's Grundgesetze is essentially that of a Frege structure satisfying the following additional condition. There are exactly two propositions, "the true" and "the false", with the former being of course the only truth. Russell's paradox shows us that no Frege structure can satisfy this extra condition. I locate the flaw in Frege's condition by formulating the notion of "internal" definability for collections of objects on a Frege structure. Frege's condition implies that there is an internal definition of the collection of truths, and the argument of Russell's paradox shows that there can be no such definition. This is clearly related to the classic result of Tarski on truth. But I go on to show that neither the collection of propositions nor the collection of sets can have an internal definition. Even more extreme results can be obtained. Thus, for any set of a Frege structure, the collection of its subsets is not internally definable. The point is that extensional equality between sets must not be confused with the equality relation between sets as objects. So, for example, the collection of subsets of say $\{x|\neg(x=x)\}$ is the collection of empty sets. Although there is extensionally only one empty set, the collection of empty sets is not internally definable. These results suggest that the traditional distinction between the logical and semantical paradoxes is a false one as both forms have their origin in the same confusion between internal and external definability.

At this point the reader may well be seeking some assurance that the notion of Frege structure is consistent. I provide such assurance by presenting a general construction which can be made to yield a large variety of Frege structures. Roughly, I start with any model of the lambda calculus. I construct a monotone operator on an associated partial ordering so that any fixed point of the operator will yield a Frege structure. The partial ordering has the property that every chain has a least upper bound and it is a standard result of classical set theory that any monotone operator on such a partial ordering has a least fixed point. Hence Frege structures exist.

Frege wrote the Grundgesetze in order to justify his conviction that arithmetic and analysis are parts of logic. Russell's paradox destroyed his plan. Can we carry through Frege's construction of arithmetic on a Frege structure? A positive answer would finally vindicate Frege's work. Unfortunately, as the reader can check for himself, Frege's construction does not go through. (For a start the natural number 0 is supposed to be the "set" of empty sets, but, as mentioned earlier, the collection of empty sets is not internally definable and hence there cannot be a set of empty sets. Similar difficulties occur at other stages of Frege's construction.) In fact Frege structures in general are not sufficiently powerful even for the internal development of arithmetic. What is always possible is a representation of the natural numbers as objects of any Frege structure in such a way that the standard functions of arithmetic are represented as functions of the Frege structure. But in general the resulting collection of natural numbers may not be internally definable. If it is I say that the Frege structure is N-standard. There is no difficulty in constructing N-standard Frege structures, and on them it is possible to give an internal development of arithmetic and elementary analysis. In order to give internal developments of other branches of mathematics it may be necessary that additional collections of objects are internally definable on the Frege structure. Examples of such collections are the second and higher number classes of ordinals. But there are other examples of inductively defined collections of objects that may be internally definable on suitably constructed Frege structures, and I plan to consider them in a later publication.

I have been at pains to present the ideas in this paper in such a way that they can be understood from both a constructive and a classical point of view. For that reason I have taken all the intuitionistic logical constants as primitive, although there is some redundancy from the classical point of view. Note that the logical schemata do not specify the internal logic of the Frege structure, but only reflect the external logic.

The notion of Frege structure was originally conceived as the appropriate notion for lambda-calculus interpretations of MARTIN-LÖF's theory of types (1975). The present paper was written in the context of correspondence and discussion with Per Martin-Löf on the possibility of developing a foundational language for constructive mathematics whose underlying structure would be something like a Frege structure. This language would be type free and would treat the notion of proposition as primitive rather than reduce it to that of type as in the theory of types. But there appear to be weaknesses in attempts to formulate an adequate theory of meaning for such a language. So the foundational significance of the notion of Frege structure is not yet clear. My paper makes no claim to present a foundational framework, but I hope that it may contribute to the future development of such a framework.

Some indication of the interpretation in a Frege structure of Martin-Löf's theory of types may be found in Section 7 of this paper. For details of the interpretation in a restricted context see JAN SMITH's (1978).

In ACZEL (1977) I introduced a first order formal system of combinatory formal arithmetic with reflection axioms, called CFA_1. There I showed that Martin-Löf's theory of types with one universe could be interpreted in CFA_1. This was a preliminary step in the determination of the exact proof theoretic strength of that theory of types. The N-standard Frege structures are essentially almost the same as the ω-standard models of CFA_1. It is not difficult to formulate a first order system whose ω-standard models are just the N-standard Frege structures.

There are several points of contact between the ideas reported here and the work of other logicians. A conspicuous example is the work of Feferman and others on systems of explicit mathematics. This work originated with FEFERMAN's (1975). For more recent publications see FEFERMAN (1978) and BEESON (1977). Feferman's approach differs from that presented here in two main respects. First, Feferman has a partially defined binary application function instead of the totally defined function available in the lambda calculus. Hence where I use models of the lambda calculus Feferman has used applicative models such as those obtained from ordinary and generalised recursion theory. Second, Feferman takes the notions of classification (his notion of set) and membership as primitive, while I prefer to treat the notions of proposition and truth as primitive and define the notions of set and membership in terms of them. It appears that from the technical point of view the two approaches run parallel to each other in the sense that any technical result for one approach can be reconstructed for the other. Thus where Feferman would use the applicative model obtained from ordinary recursion theory to model Church's thesis I would use the term model of the lambda calculus or alternatively the submodel of Scott's graph model consisting of the recursively enumerable sets. Scott's work on the lambda calculus must also be mentioned. His discovery of the set theoretical models of the lambda calculus have shown us that there is as much freedom for constructing such models as there is for constructing applicative models. See SCOTT (1972) for his lattice theory models and SCOTT (1975a) for the graph model. See also SCOTT (1975b) for his treatment of class abstraction using lambda abstraction.

The inductive definability construction that I use in the construction of Frege structures is a formulation of an idea that seems to have become popular only in recent years, although it has its origins in early work of Fitch. The same idea is also used in FEFERMAN (1975), SCOTT (1975b) and ACZEL (1977).

I cannot end this sketchy survey of related work without mentioning the earlier work of Church, Curry and others on the development of systems combining the lambda calculus or combinators with logic. An early attempt is CHURCH's (1932 and 1933). Curry has developed systems of illiative combinatory logic (see CURRY ET AL., 1972). In relation to Curry's work it is worth pointing out something that I have only recently noticed. The non-standard notion of implication that I use in this paper has been considered long ago by Curry (see CURRY ET AL., 1972, 15B5; see also STENLUND, 1975). My own original motivation for introducing it was the need to carry through the interpretation of type theory in a Frege structure.

In this paper I claim to give a structural explanation for Russell's paradox in the context of Frege's Grundgesetze. It is worth emphasising that this explanation does not depend on any particular approach to Frege's philosophy of language. My terminology of "objects" and "functions" in relation to Frege structures may suggest that I am making use of Frege's threefold distinction between "expression", "sense" and "reference" and placing Frege structures in the role of reference. But this is not my intention. I have not attempted to present any foundational language or any theory of meaning here. This task still remains and I believe that it will be pursued by Per Martin-Löf in future publications.

In this paper I will refrain as much as possible from using the terminology of "set" and "$\in$" in the metalanguage. Instead their use will be restricted to their technical use in Frege structures and the terminology "collection" and "in" will be used in the metalanguage. Also the lambda abstraction notation will only be used technically in Frege structures or, more generally, lambda structures. Instead I will use the following notation in the metalanguage. Let $e[x_1,\ldots,x_n]$ be an expression of the metalanguage involving free occurrences of variables from the list $x_1,\ldots,x_n$, where each variable x_i has a specified range. Then $\langle e[x_1,\ldots,x_n] \| x_1,\ldots,x_n \rangle$ denotes the n-place function f such that $f(a_1,\ldots,a_n)$ denotes the value of $e[a_1,\ldots,a_n]$ whenever a_i is in the range of x_i for $i=1,\ldots,n$. Here $e[a_1,\ldots,a_n]$ is the result of simultaneously replacing each occurrence of the free variable x_i in $e[x_1,\ldots,x_n]$ by a_i for $i=1,\ldots,n$. Functions will always be treated extensionally, so that two functions that have the same domain and always have the same value are equal.

The usual infix notation (afb) instead of $f(a,b)$ will often be used for two place functions. Also if F is a one place function having functions as arguments and $\langle e[x] \| x\rangle$ is in the domain of F, then $(Fxe[x])$ will usually be written instead of $F(\langle e[x]\|x\rangle)$. This agrees with the usual notation for the quantifiers and lambda abstraction.

2. Logical systems

We will be concerned with families $\mathfrak{F} = \mathfrak{F}_0, \mathfrak{F}_1, \ldots$ where $\mathfrak{F}_0$ is a collection of entities to be called the *objects on* $\mathfrak{F}$ and for $n = 1, 2, \ldots$ $\mathfrak{F}_n$ is a collection of n-place functions of $\mathfrak{F}_0$ called the n-place *$\mathfrak{F}$-functions*. We shall always require that $\mathfrak{F}$ is *explicitly closed*, i.e. all constant functions and projection functions on $\mathfrak{F}_0$ are $\mathfrak{F}$-functions and the $\mathfrak{F}$-functions are closed under composition. A more applicable characterisation of the explicit closure condition is obtained by considering expressions built up, in the obvious way, from variables with range $\mathfrak{F}_0$ and constants for the objects of $\mathfrak{F}$ and the $\mathfrak{F}$-functions. $\mathfrak{F}$ is explicitly closed iff for every such expression $e[x_1, \ldots, x_n]$, all of whose variables are among $x_1, \ldots, x_n$, the n-place function $\langle e[x_1, \ldots, x_n] \| x_1, \ldots, x_n\rangle$ on $\mathfrak{F}_0$ is an $\mathfrak{F}$-function. We say that $\mathfrak{F}$ is *trivial* if $\mathfrak{F}_0$, and hence each $\mathfrak{F}_n$, has exactly one element.

Definition 2.1. Given an explicitly closed family $\mathfrak{F}$ a function $F : \mathfrak{F}_{n_1} \times \cdots \times \mathfrak{F}_{n_k} \to \mathfrak{F}_0$ is an *$\mathfrak{F}$-functional* if, for every $m > 0$ and functions f_1 in $\mathfrak{F}_{m+n_1}, \ldots, f_k$ in $\mathfrak{F}_{m+n_k}$, the function

$$\langle F(\langle f_1(\bar{y}, \bar{x}_1) | \bar{x}_1\rangle, \ldots, \langle f_k(\bar{y}, \bar{x}_k) | \bar{x}_k\rangle) | \bar{y}\rangle$$

is in $\mathfrak{F}_m$. Here $\bar{y}$ is a list of m variables and x_i is a list of n_i variables for $i = 1, \ldots, k$, each of these variables having range $\mathfrak{F}_0$.

When $n_1 = \cdots = n_k = 0$ this notion coincides with that of $\mathfrak{F}$-function.

We may strengthen our explicit closure properties for $\mathfrak{F}$ by allowing expressions to contain variables for $\mathfrak{F}$-functions and constants for $\mathfrak{F}$-functionals. So if "F" is a constant for an $\mathfrak{F}$-functional: $\mathfrak{F}_{n_1} \times \cdots \times \mathfrak{F}_{n_k} \to \mathfrak{F}_0$, $e_1, \ldots, e_k$ are previously formed expressions and $\bar{x}_i$ is a list of n_i variables for $i = 1, \ldots, k$, then we may form the expression

$$F(\langle e_1 | \bar{x}_1\rangle, \ldots, \langle e_k | \bar{x}_k\rangle).$$

Of course we need to distinguish between free and bound variables in our expressions, but I assume that this is taken care of in a standard way.

It is now straightforward but tedious to prove the following result.

Theorem 2.2. *Let $\mathfrak{F}$ be an explicitly closed family. Let $e[\xi_1, \ldots, \xi_m]$ be an expression built up as above using constants for objects of $\mathfrak{F}$, $\mathfrak{F}$-functions and*

$\mathfrak{F}$-functionals, and variables, each of range $\mathfrak{F}_n$ for some n, such that the free variables of the expression are among $\xi_1,\ldots,\xi_m$. Then $\langle e[\xi_1,\ldots,\xi_m]\|\xi_1,\ldots,\xi_m\rangle$ is an $\mathfrak{F}$-functional.

We shall formulate the central notion of "logical system" relative to an explicitly closed family $\mathfrak{F}$ and a list of distinguished $\mathfrak{F}$-functionals called the *logical constants*. These logical constants are $\neg : \mathfrak{F}_0 \rightarrow \mathfrak{F}_0$, $\&, \vee, \supset$, $\doteq : \mathfrak{F}_0 \times \mathfrak{F}_0 \rightarrow \mathfrak{F}_0$, and $\forall, \exists : \mathfrak{F}_1 \rightarrow \mathfrak{F}_0$.

Definition 2.3. Relative to a list, as above, of logical constants on an explicitly closed family $\mathfrak{F}$, a *logical system* consists of a pair of collections. The first is a collection of objects called *propositions*, and the second is a collection of propositions called *truths*. We will use the natural terminology "*a* is true" instead of "*a* is a truth". These collections must satisfy the following *logical schemata*, one for each logical constant.

The logical schemata

Negation. If a is a proposition, then $\neg a$ is a proposition such that

$$\neg a \text{ is true} \quad \text{iff } a \text{ is not true.}$$

Conjunction. If a,b are propositions, then (a & b) is a proposition such that

$$(a \,\&\, b) \text{ is true} \quad \text{iff } a \text{ is true and } b \text{ is true.}$$

Disjunction. If a,b are propositions, then $(a \vee b)$ is a proposition such that

$$(a \vee b) \text{ is true} \quad \text{iff } a \text{ is true or } b \text{ is true.}$$

Implication. If a is a proposition and the object b is a proposition provided that a is true, then $(a \supset b)$ is a proposition such that

$$(a \supset b) \text{ is true} \quad \text{iff } a \text{ is true implies } b \text{ is true.}$$

In the next two schemata we need the following notion. A *propositional function* is a function all of whose values are propositions.

Universal quantification. If f is a propositional function in $\mathfrak{F}_1$, then $\forall x f(x)$ is a proposition such that

$$\forall x f(x) \text{ is true} \quad \text{iff } f(a) \text{ is true for all objects } a.$$

Existential quantification. If f is a propositional function in $\mathfrak{F}_1$, then $\exists x f(x)$ is a proposition such that

$$\exists x f(x) \text{ is true} \quad \text{iff } f(a) \text{ is true for some object } a.$$

Equality. If a,b are objects, then $(a \doteq b)$ *is a proposition such that*

$$(a \doteq b) \text{ is true} \quad \text{iff } a = b.$$

Our formulation of the implication schema is perhaps non-standard, but will be important. Note that if a is a proposition that is not true, then $(a \supset b)$ is a proposition (in fact a true one) whatever the object b.

We use $(a \ \& \ \supset b)$ as an abbreviation for $(a \ \& \ (a \supset b))$. It is an extended form of conjunction for which there is the derived scheme.

Extended conjunction. If a is a proposition and the object b is a proposition provided that a is true, then $(a \ \& \ \supset b)$ *is a proposition such that*

$$(a \ \& \ \supset b) \text{ is true} \quad \text{iff } a \text{ is true and } b \text{ is true}.$$

We use $(a \equiv b)$ as an abbreviation for $((a \supset b) \ \& \ (b \supset a))$. It has the derived schema.

Bi-implication. If a,b are propositions, then $(a \equiv b)$ *is a proposition such that*

$$(a \equiv b) \text{ is true} \quad \text{iff } (a \text{ is true iff } b \text{ is true}).$$

3. Frege structures and Russell's paradox

The notion of Frege structure that we introduce in this section is intended to isolate the structure of that part of Frege's Grundgesetze that we consider to be correct. Frege's notion of the "course of values" of a function f will here be rendered as $\lambda x f(x)$ where λ is given as part of the following notion.

Definition 3.1. A *lambda system* for an explicitly closed family $\mathfrak{F}$ consists of two $\mathfrak{F}$-functionals $\lambda : \mathfrak{F}_1 \to \mathfrak{F}_0$ and $\mathrm{APP} : \mathfrak{F}_0 \times \mathfrak{F}_0 \to \mathfrak{F}_0$ such that

$$\mathrm{APP}(\lambda x f(x), a) = f(a)$$

for all f in $\mathfrak{F}_1$ and a in $\mathfrak{F}_0$.

Definition 3.2. A *Frege structure* is a logical system, relative to a list of logical constants on an explicitly closed family $\mathfrak{F}$, together with a lambda system for $\mathfrak{F}$.

In the rest of this section we shall work with a fixed Frege structure. Note that we have the truth of Axiom V of Frege's Grundgesetze. Thus

Proposition 3.3. *For all functions f,g in* $\mathfrak{F}_1$ *the following proposition is true.*

$$((\lambda x f(x) \doteq \lambda x g(x)) \equiv \forall x (f(x) \doteq g(x))).$$

Proof. If $(\lambda x f(x) \doteq \lambda x g(x))$ is true, then $\lambda x f(x) = \lambda x g(x)$ so that by Definition 3.1

$$\begin{aligned} f(a) &= \text{APP}(\lambda x f(x), a) \\ &= \text{APP}(\lambda x g(x), a) \\ &= g(a) \end{aligned}$$

for all objects a. Hence $(f(a) \doteq g(a))$ is true for all objects a, so that $\forall x (f(x) \doteq g(x))$ is true.

Conversely, if $\forall x (f(x) \doteq g(x))$ is true, then $f(a) = g(a)$ for all objects a so that $f = g$. Hence $\lambda x f(x) = \lambda x g(x)$ so that $(\lambda x f(x) \doteq \lambda x g(x))$ is true.

Definition 3.4. An object is a *set* if it has the form $\lambda x f(x)$ for some propositional function f in $\mathfrak{F}_1$.

We will use the suggestive notation $\{x | e[x]\}$ for $\lambda x e[x]$, particularly when $\langle e[x] \| x \rangle$ is a propositional function in $\mathfrak{F}_1$. Also we will write $(a \in b)$ for $\text{APP}(b, a)$, particularly when b is a set.

Note that using the new notation the equality in Definition 3.1 becomes

$$(a \in \{x | f(x)\}) = f(a).$$

Hence we can derive the following schemata.

Predication. If b is a set, then $(a \in b)$ is a proposition for any object a.

Comprehension. If f is a propositional function in $\mathfrak{F}_1$, then $\{x | f(x)\}$ is a set b such that for any object a

$$(a \in b) \text{ is true} \quad \text{iff } f(a) \text{ is true.}$$

I claim that these schemata express intuitively correct and therefore consistent principles. In Section 6 we will construct mathematical examples of Frege structures, so there can be no doubt concerning consistency.

It will be instructive, at this point, to see what happens if we try to derive Russell's paradox on a Frege structure. Let R denote the object $\{x | \neg(x \in x)|$. If we could show that $\langle (x \in x) | x \rangle$ were a propositional function, then R would be a set such that for any object a

$$\begin{aligned} (a \in R) \text{ is true} \quad & \text{iff } \neg(a \in a) \text{ is true} \\ & \text{iff } (a \in a) \text{ is not true.} \end{aligned}$$

Now if we take a to be R itself we get a contradiction. For $\langle x \in x | x \rangle$ to be a propositional function we would need to know that $(a \in a)$ is a proposition for every object a. But our predication schema is not strong enough to yield this. All we get is that $(a \in a)$ is a proposition for all sets a. So our attempt to derive a paradox has failed. Russell's paradox can be derived in

Frege's Grundgesetze and we now list the essential ways in which the structure of the Grundgesetze differs from a Frege structure.

(1) The Grundgesetze has exactly two propositions, namely "the true" and "the false". "The true" is the only true proposition.

(2) The Grundgesetze has a propositional function in $\mathfrak{F}_1$ called the *horizontal stroke*. When applied to an object a it yields a proposition $—a$ which is the true if a is the true and is the false if a is not the true. So it satisfies the schema.

Horizontal stroke. For any object a, $—a$ *is a proposition such that*

$$—a \text{ is true} \quad \text{iff } a \text{ is true}.$$

(3) The Grundgesetze has $\neg$, $\supset$, $\forall$ and $\doteq$ as the basic logical constants the remaining logical constants $\vee$, & and $\exists$ are defined from the basic ones as is standard in classical logic. The logical constants $\neg$, $\supset$ and $\forall$ are made to incorporate the horizontal stroke. Thus Frege uses $\top$ instead of $\neg$ and thinks of it as incorporating a horizontal stroke on either side, i.e. as "—"+"|"+"—". Frege uses similar notation for $\supset$ and $\forall$.

This leads to the following strengthened forms of the negation, implication and Universal quantification schemata.

Strong negation. For any object a $\neg a$ *is a proposition such that*

$$\neg a \text{ is true} \quad \text{iff } a \text{ is not true}.$$

Strong implication. For any objects a, b $(a \supset b)$ *is a proposition such that*

$$(a \supset b) \text{ is true} \quad \text{iff } (a \text{ is true implies } b \text{ is true}).$$

Strong universal quantification. For any function f *in* $\mathfrak{F}_1$ $\forall x f(x)$ *is a proposition such that*

$$\forall x f(x) \text{ is true} \quad \text{iff } f(a) \text{ is true for all objects } a.$$

(4) The Grundgesetze does not treat APP as given. Instead it is explicitly postulated in Axiom V that λ is to be one–one. I.e. for all functions f, g in $\mathfrak{F}_1$

$$(\lambda x f(x) \doteq \lambda x g(x)) \equiv \forall x (f(x) \doteq g(x)) \text{ is true.}$$

(Note that because of (1) Frege uses $\doteq$ instead of $\equiv$.)

In addition the Grundgesetze allows function quantification and definite descriptions and then defines APP(b,a) to be $\iota y \exists f(b \doteq \lambda x f(x) \,\&\, f(a) \doteq y)$ (informally, the y such that for some f in $\mathfrak{F}_1$ $b = \lambda x f(x)$ and $f(a) = y$).

Now $(a \in b)$ can be defined as $\text{---}\mathrm{APP}(b,a)$ or else directly as $\exists f(b \doteq \lambda(f)\ \&\ f(a))$ without using definite descriptions. In either case $(a \in b)$ incorporates the horizontal stroke, the comprehension schema holds and the following strengthening of the Predication schema holds.

Strong predication. For any objects a, b $(a \in b)$ is a proposition.

Frege's critical mistake is (1). Given (1) the horizontal stroke can be defined by letting $\text{---}a$ be $a \doteq (a \doteq a)$, and using the horizontal stroke the functions $\neg$, $\supset$, $\forall$ and $\in$ can be redefined so that the strengthened schemata hold. Note that Russell's paradox can be derived using any one of the horizontal stroke, strong negation or strong predication schemata. For example, given strong predication $\langle x \in x | x \rangle$ is a propositional function and hence $\{x | \neg(x \in x)\}$ is a set. Alternatively given strong negation $\langle \neg(x \in x) | x \rangle$ is a propositional function so that $\{x | \neg(x \in x)\}$ is still a set. Finally, given the horizontal stroke $\langle \text{---}(x \in x) | x \rangle$ is a propositional function so that $\{x | \neg\text{---}(x \in x)\}$ is a set and one can still derive Russell's paradox.

I believe that the essence of Frege's mistake lies in the seemingly innocuous horizontal stroke schema. What this schema does is to give an internal definition of truth. Tarski's distinction between object language and metalanguage and his result on the undefinability of truth in the object language should put us on our guard.

It will be helpful to introduce some terminology.

Definition 3.5. If $\mathcal{C}$ is a collection of objects on a Frege structure $\mathfrak{F}$ and C is a propositional function in $\mathfrak{F}_1$ we write that C *internally defines* $\mathcal{C}$ if for all objects a

$$C(a) \text{ is true} \quad \text{iff } a \text{ is in } \mathcal{C}.$$

The horizontal stroke schema expresses that the horizontal stroke internally defines the collection of truths. Our discussion above shows that this is not possible and we have recaptured Tarski's result on the undefinability of truth for our Frege structures.

But on a Frege structure we can go on to obtain further undefinability results. Suppose that P is a propositional function in $\mathfrak{F}_1$ internally defining the collection of propositions. Then it is easy to see that $\langle (P(x)\ \&\ \supset x) | x \rangle$ is a propositional function internally defining the collection of truths. Hence there can be no such definition of the collection of propositions. Similarly if S is a propositional function in $\mathfrak{F}_1$ defining the collection of sets, then $\langle S(\{y | x\}) | x \rangle$ is a propositional function defining the collection of propositions. Note that for any object a, $\{y | a\}$ is $\lambda x f(x)$ where f is the function in $\mathfrak{F}_1$ with constant value a, and is a set iff a is a proposition. It follows that there can be no such definition of the collection of sets.

Summarising the above work we have the following result.

Theorem 3.6. *Each of the collections of propositions, truths and sets on a Frege structure are not internally definable.*

We may obtain yet more undefinability results on a Frege structure. Let a be a set. We define the set b to be a *subset of* a if $\forall x(x \in b \supset x \in a)$ is true. Suppose that A internally defines the collection of subsets of the set a. Let F be $\langle A(\{y|x\})|x\rangle$. For each object b $F(b)$ is a proposition such that

$$F(b) \text{ is true} \quad \text{iff } b \text{ is a proposition such that } \{y|b\} \text{ is a subset of } a, \\ \text{iff } b \text{ is a proposition such that } (b \supset \forall y(y \in a)) \text{ is true.}$$

But the following result tells us that there can be no such F.

Lemma 3.7. *For any proposition c, in a Frege structure, the collection of propositions b such that $(b \supset c)$ is true is not internally definable.*

Proof. Suppose that the collection specified in the lemma is defined by C. Let R be $\{x|C(x \in x)\}$. Clearly R is a set and

$$(R \in R) \text{ is true} \quad \text{iff } C(R \in R) \text{ is true} \\ \text{iff } (R \in R) \supset c \text{ is true.}$$

By standard reasoning it follows that $(R \in R)$ and hence c is true. It follows that $(b \supset c)$ is true for all propositions b so that C defines the collection of propositions, which is impossible.

Letting c be $\forall y(y \in a)$ in the lemma we see that there can be no definition A of the collection of subsets of a. So we have proved the following result.

Theorem 3.8. *For any set a, in a Frege structure, the collection of subsets of a is not internally definable.*

Note that this result applies to every set including the set $\{x|\neg(x \doteq x)\}$. The subsets of this set are just those sets that are empty. It follows that the collection of empty sets is undefinable in the Frege structure. Of course any two empty sets are extensionally equal even though they may not be equal. In the Grundgesetze extensional equality between sets coincided with equality, but as should be clear this cannot hold in a Frege structure.

4. The lambda calculus and lambda structures

Definition 4.1. A *lambda structure* is an explicitly closed family that has a lambda system.

Lambda structures are essentially just models of the lambda calculus. For the standard terminology concerning the lambda calculus we refer the reader to Barendregt (1977). There, the notion corresponding to our lambda structures are called weakly extensional λ-algebras. Our notion has the advantage of avoiding all reference to syntax.

To see in exactly what sense lambda structures are models of the lambda calculus let $\mathfrak{F}$ be a lambda structure and let λ, APP be a lambda system for it. To each term M of the lambda calculus, and each valuation ρ assigning an object of $\mathfrak{F}$ to each variable of the lambda calculus we assign an object $\|M\|_\rho^{\mathfrak{F}}$ of $\mathfrak{F}$ as follows.

$$\|x\|_\rho^{\mathfrak{F}}=\rho(x),$$
$$\|(MN)\|_\rho^{\mathfrak{F}}=\mathrm{APP}(\|M\|_\rho^{\mathfrak{F}},\|N\|_\rho^{\mathfrak{F}}),$$
$$\|(\lambda xM)\|_\rho^{\mathfrak{F}}=\lambda b\|M\|_{\rho(b/x)}^{\mathfrak{F}}$$

where $\rho(b/x)$ is the valuation ρ' that is like ρ except that $\rho'(x)=b$. Here "b" is being used as a metalanguage variable with range $\mathfrak{F}_0$ in contrast to "x" which is a variable of the lambda calculus. In order to carry the definition through we need to know that $\langle\|M\|_{\rho(b/x)}^{\mathfrak{F}}|b\rangle$ is in $\mathfrak{F}_1$. In general one needs to prove the following result simultaneously with the definition of $\|M\|_\rho^{\mathfrak{F}}$.

Lemma 4.2. *For every sequence of pairwise distinct variables $x_1,\ldots,x_n$ of the lambda calculus and every sequence of pairwise distinct variables $b_1,\ldots,b_n$ of the metalanguage $\langle\|M\|_{\rho(b_1/x_1)\cdots(b_n/x_n)}^{\mathfrak{F}}|b_1,\ldots,b_n\rangle$ is in $\mathfrak{F}_n$.*

Proof. This is by induction on the way M is built up making use of the fact that λ and APP are $\mathfrak{F}$-functionals.

Lambda structures model the conversion relation of the lambda calculus in the following sense.

Theorem 4.3. *If $\lambda\vdash M=N$, then*

$$\|M\|_\rho^{\mathfrak{F}}=\|N\|_\rho^{\mathfrak{F}}$$

for all lambda structures $\mathfrak{F}$ and valuations ρ.

Proof By induction on the derivation of $\lambda\vdash M=N$.

There are two methods for constructing models of the lambda calculus. The first method uses the traditional syntactic approach to obtain term models which are seen to be non-trivial via the Church–Rosser theorem. The second method was initiated and developed by Dana Scott and is non-syntactic. The simplest example of this second approach is the graph model $\mathcal{P}\omega$.

Here we shall present the term model and the graph model as lambda structures.

The collection $\mathcal{T}_0$ of objects of the *term model* $\mathcal{T}$ consists of equivalence classes $|M|$ of terms M relative to the conversion relation. So for all terms M and N

$$|M|=|N| \quad \text{iff} \quad \lambda \vdash M=N.$$

The collection $\mathcal{T}_n (n>0)$ of n-place functions of the term model consists of those functions $f: \mathcal{T}_0^n \to \mathcal{T}_0$ such that there is a term M and pairwise distinct variables $x_1, \ldots, x_n$ such that for all terms $N_1, \ldots, N_n$

$$f(|N_1|, \ldots, |N_n|) = |M[N_1, \ldots, N_n / x_1, \ldots, x_n]|.$$

Here $M[N_1, \ldots, N_n / x_1, \ldots, x_n]$ denotes the result of simultaneously substituting N_i for x_i in M for $i=1, \ldots, n$, changing bound variables to avoid clashes. Now define $\lambda: \mathcal{T}_1 \to \mathcal{T}_0$ and $\mathrm{APP}: \mathcal{T}_0 \times \mathcal{T}_0 \to \mathcal{T}_0$ so that if f is in $\mathcal{T}_1$ such that $f(|N|) = |M[N/x]|$ for all terms N then $\lambda(f) = |(\lambda x M)|$ and for all terms M and N $\mathrm{APP}(|M|, |N|) = |(MN)|$. Using standard results of the lambda calculus we can show that λ and APP are well-defined and are $\mathcal{T}$-functionals. Moreover we can show that $\mathcal{T}$ is explicitly closed and that λ, APP form a lambda system. Using the Church–Rosser theorem to get non-triviality we obtain the following result.

Theorem 4.4. *$\mathcal{T}$ is a non-trivial lambda structure.*

We now turn to the definition of the graph model $\mathcal{P}\omega$ as a lambda structure $\mathcal{G}$. The collection $\mathcal{G}_0$ of objects of the *graph model* $\mathcal{G}$ is $\mathcal{P}\omega$, the power set of the set of natural numbers. The collection $\mathcal{G}_n (n>0)$ of n-place functions of the graph model consists of the continuous functions $f: \mathcal{G}_0^n \to \mathcal{G}_0$, where $\mathcal{G}_0^n$ is given the product topology and $\mathcal{G}_0$ is given the topology with basic open sets of the form $\{a \in \mathcal{P}\omega \mid e \subseteq a\}$ for finite sets of natural numbers e. It is obvious that $\mathcal{G}$ is an explicitly closed family. To define the lambda system on $\mathcal{G}$, let $(\cdot, \cdot)$ be a primitive recursive coding of pairs of natural numbers, such as $\langle \frac{1}{2}(n+m)(n+m+1)+m \mid n, m \rangle$, and let $\langle e_n \mid n \rangle$ be the effective one–one enumeration of finite sets of natural numbers given by

$$e_n = \{k_0, \ldots, k_n\} \quad \text{if} \quad n = 2^{k_0} + \cdots + 2^{k_n}$$

with $k_0 < \cdots < k_n$. Now let

$$\lambda(f) = \{(n,m) | m \in f(e_n)\}$$

and

$$\mathrm{APP}(a,b) = \{m | e_n \subseteq b \text{ for some } n \text{ such that } (n,m) \in a\},$$

for all f in $\mathcal{G}_1$ and a, b in $\mathcal{G}_0$.

It is straightforward to show that λ, APP form a lambda system. So we get the result:

Theorem 4.5. *$\mathcal{G}$ is a non-trivial lambda structure.*

For details concerning the graph model we refer to BARENDREGT (1977).

5. Some properties of lambda structures

In this section we assume given a fixed non-trivial lambda structure $\mathcal{F}$ with lambda system λ, APP. Some routine constructions on $\mathcal{F}$ are used to obtain the main result, Theorem 5.15.

As λ is an $\mathcal{F}$-functional we have an operation $\lambda_m^{m+1} : \mathcal{F}_{m+1} \to \mathcal{F}_m$, for each m, given by $\lambda_m^{m+1}(f)(\bar{a}) = \lambda(\langle f(\bar{a},x) | x \rangle)$ for all f in $\mathcal{F}_{m+1}$ and $\bar{a}$ in $\mathcal{F}_0^m$.

Lemma 5.1. *For all m, f in $\mathcal{F}_{m+1}$, $\bar{a}$ in $\mathcal{F}_0^m$ and b in $\mathcal{F}_0$,*

$$\mathrm{APP}(\lambda_m^{m+1}(f)(\bar{a}), b) = f(\bar{a}, \bar{b}).$$

Proof. Trivial.

Starting from λ_m^{m+1} we define

$$\lambda_m^{m+n} : \mathcal{F}_{m+n} \to \mathcal{F}_m$$

by recursion on $n > 0$. So given λ_m^{m+n} for all m we let

$$\lambda_m^{m+n+1}(f) = \lambda_m^{m+1}(\lambda_{m+1}^{m+1+n}(f))$$

for all f in $\mathcal{F}_{m+n+1}$. Starting with $\mathrm{APP}_1 = \mathrm{APP}$ we also define APP_n in $\mathcal{F}_{n+1}$ by recursion on $n > 0$. So given APP_n we let $\mathrm{APP}_{n+1}(e, a, \bar{b}) = \mathrm{APP}_n(\mathrm{APP}(e,a), \bar{b})$ for all e, a in $\mathcal{F}_0$ and $\bar{b}$ in $\mathcal{F}_0^n$.

Lemma 5.2. *For each $n > 0$ and all m*

$$\mathrm{APP}_n(\lambda_m^{m+n}(f)(\bar{a}), \bar{b}) = f(\bar{a}, \bar{b})$$

for all f in $\mathcal{F}_{m+n}$, $\bar{a}$ in $\mathcal{F}_0^m$ and $\bar{b}$ in $\mathcal{F}_0^n$.

Proof. We prove that the equality holds for all m by induction on $n>0$. The case $n=1$ is the previous lemma. Now suppose that the result is true for n. Then if f is in $\mathfrak{F}_{m+n+1}$, $\bar{a}$ is in $\mathfrak{F}_0^m$, b is in $\mathfrak{F}_0$ and $\bar{b}$ is in $\mathfrak{F}_0^n$,

$$\begin{aligned} &\mathrm{APP}_{n+1}(\lambda_m^{m+n+1}(f)(\bar{a})\ b,\bar{b}) \\ &\quad = \mathrm{APP}_n(\mathrm{APP}(\lambda_m^{m+1}(\lambda_{m+1}^{m+1+n}(f))(\bar{a}),b),\bar{b}) \\ &\quad = \mathrm{APP}_n(\lambda_{m+1}^{m+1+n}(f)(\bar{a},b),\bar{b}) \\ &\quad = f(\bar{a},b,\bar{b}), \end{aligned}$$

by the induction hypothesis.

Corollary 5.3. *For each $n>0$*

$$\mathrm{APP}_n(\lambda_0^n(f),\bar{b})=f(\bar{b})$$

for all f in $\mathfrak{F}_n$ and $\bar{b}$ in $\mathfrak{F}_0^n$.

Proof. Put $m=0$ in the lemma.

Note that for each $n>0$ $\lambda_0^n : \mathfrak{F}_n \to \mathfrak{F}_0$ is an $\mathfrak{F}$-functional. For each $n>0$ and object e let $|e|_n$ in $\mathfrak{F}_n$ be $\langle \mathrm{APP}_n(e,\bar{x})|\bar{x}\rangle$. Observe that we have the following equalities.

Theorem 5.4. *For each $n>0$,*

(i) $$|\mathrm{APP}(e,a)|_n(\bar{b})=|e|_{n+1}(a,\bar{b})$$

for all e, a in $\mathfrak{F}_0$ and $\bar{b}$ in $\mathfrak{F}_0^n$.

(ii) $$|\lambda_0^n(f)|_n=f$$

for all f in $\mathfrak{F}_n$.

Fixed point theorem 5.5. (i) *For each h in $\mathfrak{F}_1$ there is an object e such that*

$$e=h(e).$$

(ii) *For each $n>0$ and f in $\mathfrak{F}_{n+1}$ there is an object e such that for all $\bar{a}$ in $\mathfrak{F}_0^n$*

$$|e|_n(\bar{a})=f(e,\bar{a}).$$

(iii) *For each $n>0$ and $\mathfrak{F}$-functional $F: \mathfrak{F}_n \times \mathfrak{F}_0^n \to \mathfrak{F}_0$ there is g in $\mathfrak{F}_n$ such that*

$$g(a)=F(g,\bar{a})$$

for all $\bar{a}$ in $\mathfrak{F}_0^n$.

(iv) *For each* $n>0$ *and* $\mathfrak{F}$*-functional*

$$H:\mathfrak{F}_{n_1}\times\cdots\times\mathfrak{F}_{n_k}\times\mathfrak{F}_n\times\mathfrak{F}_0^n\to\mathfrak{F}_0$$

there is an $\mathfrak{F}$*-functional*

$$G:\mathfrak{F}_{n_1}\times\cdots\times\mathfrak{F}_{n_k}\times\mathfrak{F}_0^n\to\mathfrak{F}_0$$

such that for all $\bar{f}$ *in* $\mathfrak{F}_{n_1}\times\cdots\times\mathfrak{F}_{n_k}$ *and* $\bar{a}$ *in* $\mathfrak{F}_0^n$,

$$G(\bar{f},\bar{a})=H(\bar{f},\langle G(\bar{f},\bar{x})|\bar{x}\rangle,\bar{a}).$$

Proof. (i) Let $e=\mathrm{APP}(c,c)$ where c is $\lambda xh(\mathrm{APP}(x,x))$. Then

$$\begin{aligned}e&=\mathrm{APP}(\lambda xh(\mathrm{APP}(x,x)),c)\\&=h(\mathrm{APP}(c,c))\\&=h(e).\end{aligned}$$

(ii) Let h in $\mathfrak{F}_1$ be $\langle\mathrm{APP}(\lambda_0^{n+1}(f),x)|x\rangle$. By (i) choose an object e such that $e=h(e)$. Then

$$\begin{aligned}|e|_n(\bar{a})&=|h(e)|_n(\bar{a})\\&=|\mathrm{APP}(\lambda_0^{n+1}(f),e)|_n(\bar{a})\\&=|\lambda_0^{n+1}(f)|_{n+1}(e,\bar{a})\\&=f(e,\bar{a}),\quad\text{using Theorem 5.4,}\end{aligned}$$

for all $\bar{a}$ in $\mathfrak{F}_0^n$.

(iii) Let f in $\mathfrak{F}_{n+1}$ be $\langle F(|x|_n,\bar{x})|x,\bar{x}\rangle$. Use (ii) to choose e in $\mathfrak{F}_0$ such that $|e|_n(\bar{a})=f(e,\bar{a})$ for all $\bar{a}$ in $\mathfrak{F}_0^n$. Let g be $|e|_n$. Then for $\bar{a}$ in $\mathfrak{F}_0^n$

$$\begin{aligned}g(\bar{a})&=f(e,\bar{a})\\&=F(|e|_n,\bar{a})\\&=F(g,\bar{a})\end{aligned}$$

for all $\bar{a}$ in $\mathfrak{F}_0^n$.

(iv) Let F be the $\mathfrak{F}$-functional given by

$$F(f,\bar{e},\bar{a})=H(|e_1|_{n_1},\ldots,|e_k|_{n_k},\langle f(\bar{e},\bar{x})|\bar{x}\rangle,\bar{a})$$

for all f in $\mathfrak{F}_{n+k}$, $\bar{e}=e,\ldots,e_k$ in $\mathfrak{F}_0^k$ and $\bar{a}$ in $\mathfrak{F}_0^n$. Now use (iii) to obtain g in $\mathfrak{F}_{n+k}$ such that

$$g(\bar{e},\bar{a})=F(g,\bar{e},\bar{a})$$

for all $\bar{e}$ in $\mathfrak{F}_0^k$ and $\bar{a}$ in $\mathfrak{F}_0^n$. Finally let G be the $\mathfrak{F}$-functional given by

$$G(\bar{f},\bar{a})=g(\lambda_0^{n_1}(f_1),\ldots,\lambda_0^{n_k}(f_k),\bar{a})$$

for all $\bar{f}=f_1,\ldots,f_k$ in $\mathfrak{F}_{n_1}\times\cdots\mathfrak{F}_{n_k}$ and $\bar{a}$ in $\mathfrak{F}_0^n$. Now given $\bar{f}$, $\bar{a}$ as above let

$\bar{e}=e_1,\ldots,e_k$ where $e_i=\lambda_0^{n_i}(f_i)$ for $i=1,\ldots,k$. Then

$$\begin{aligned} G(\bar{f},\bar{a}) &= g(\bar{e},\bar{a}) \\ &= F(g,\bar{e},\bar{a}) \\ &= H(|e_1|_{n_1},\ldots,|e_k|_{n_k},\langle g(\bar{e},\bar{x})|\bar{x}\rangle,\bar{a}) \\ &= H(\bar{f},\langle G(\bar{f},\bar{x})|\bar{x}\rangle,\bar{a}), \end{aligned}$$

where Theorem 5.4 is used to show that $|e_i|_{n_i}=f_i$ for $i=1,\ldots,k$.

The following notation will be useful below. Let

$$\Delta_m^n=\lambda_0^n(\langle x_m|x_1,\ldots,x_n\rangle)$$

for $n>0$ and $m=1,\ldots,n$. Then, for $\bar{a}$ in $\mathcal{F}_0^n$,

$$|\Delta_m^n|_n(\bar{a})=a_m \quad \textit{for } m=1,\ldots,n.$$

We shall write Δ_1 and Δ_2 for Δ_1^2 and Δ_2^2 respectively. Note that because the lambda structure is non-trivial $\Delta_1\neq\Delta_2$, for if $\Delta_1=\Delta_2$, then for any objects $a,b,a=|\Delta_1|_2(a,b)=|\Delta_2|_2(a,b)=b$.

Definition 5.6. A *pairing system* consists of $\mathcal{F}$-functions PAIR : $\mathcal{F}_0\times\mathcal{F}_0\to\mathcal{F}_0$ and $p,q:\mathcal{F}_0\to\mathcal{F}_0$ such that for all objects a,b

$$p(\text{PAIR}(a,b))=a \quad \text{and} \quad q(\text{PAIR}(a,b))=b.$$

Lemma 5.7. *There is a pairing system.*

Proof. $(a,b)=\lambda x|x_2|(a,b)$ for all objects a, b and for each object c let $p(c)=\text{APP}(c,\Delta_1)$ and $q(c)=\text{APP}(c,\Delta_2)$.

Clearly these are $\mathcal{F}$-functions and for all objects a,b

$$\begin{aligned} p(\text{PAIR}(a,b)) &= \text{APP}(\text{PAIR}(a,b),\Delta_1) \\ &= \text{APP}(\lambda x|x|_2(a,b),\Delta_1) \\ &= |\Delta_1|_2(a,b) \\ &= a \end{aligned}$$

and

$$q(\text{PAIR}(a,b))=|\Delta_2|_2(a,b)=b.$$

Corollary 5.8. *For each $n>0$ there are $\mathcal{F}$-functions*

$$J_n:\mathcal{F}_0^n\to\mathcal{F}_0 \quad \textit{and} \quad p_m^n:\mathcal{F}_0\to\mathcal{F}_0 \quad \textit{for } m=1,\ldots,n$$

such that $p_m^n(J_n(\bar{a}))=a_m$ for $\bar{a}=a_1,\ldots,a_n$ in $\mathcal{F}_0^n$.

Proof. Let J_1 be $\langle x|x\rangle$ and given J_n let J_{n+1} be $\langle \mathrm{PAIR}(J_n(\bar{x}),y)|\bar{x},y\rangle$. Further, let p_1^1 be $\langle x|x\rangle$ and for $n>0$ let p_{n+1}^{n+1} be q and given p_m^n let $p_m^{n+1}=\langle p_m^n(p(x))|x\rangle$ for $m=1,\ldots,n$.

Definition 5.9. A *Peano system* consists of an object 0 and $\mathfrak{F}$-functions s, PRED and DEC: $\mathfrak{F}_0\to\mathfrak{F}_0$ such that for all objects a

$$\mathrm{PRED}(s(a))=a,$$

$$\mathrm{DEC}(s(a))=\Delta_2$$

and

$$\mathrm{DEC}(0)=\Delta_1.$$

Lemma 5.10. *There is a Peano system.*

Proof. Let $0=\mathrm{PAIR}(\Delta_1,\Delta_2), s=\langle \mathrm{PAIR}(\Delta_2,x)|x\rangle, \mathrm{PRED}=q$ and $\mathrm{DEC}=p$.

In the following we assume that the pairing and Peano systems are kept fixed. We may define the collection of *natural numbers* to be the smallest collection of objects such that

(i) 0 is natural number, and

(ii) if a is a natural number, then so is $s(a)$.

Clearly this collection satisfies all the Peano axioms. (Note that as $\Delta_1\neq\Delta_2$ it follows that $s(a)\neq 0$ for all objects a.)

Using the fixed point theorem we may find $\mathfrak{F}$-functions satisfying schemes of primitive recursion.

Lemma 5.11. *There is an $\mathfrak{F}$-functional $R:\mathfrak{F}_0\times\mathfrak{F}_0\times\mathfrak{F}_2\to\mathfrak{F}_0$ such that*

$$R(0,b,f)=b$$

and

$$R(s(a),b,f)=f(a,R(a,b,f))$$

for all a,b in $\mathfrak{F}_0$ and f in F_2.

Proof. Let $H:\mathfrak{F}_0\times\mathfrak{F}_2\times\mathfrak{F}_1\times\mathfrak{F}_0\to\mathfrak{F}_0$ be the $\mathfrak{F}$-functional given by

$$H(b,f,g,a)=|\mathrm{DEC}(a)|_2(b,f(\mathrm{PRED}(a),g(\mathrm{PRED}(a))))$$

for a, b in $\mathfrak{F}_0$, f in $\mathfrak{F}_2$ and g in $\mathfrak{F}_1$. By Theorem 5.5 (iv) there is an $\mathfrak{F}$-functional

$$G:\mathfrak{F}_0\times\mathfrak{F}_2\times\mathfrak{F}_0\to\mathfrak{F}_0$$

such that $G(b,f,a)= F(b,f,\langle G(b,f,x)|x\rangle,a)$ for all a, b in $\mathfrak{F}_0$ and f in $\mathfrak{F}_2$. Now let $R(a,b,f)= G(b,f,a)$ for a, b in $\mathfrak{F}_0$ and f in $\mathfrak{F}_2$.

Using R we may define the usual arithmetical functions. For example we may define $+$ in $\mathfrak{F}_2$ by

$$(a+b)=R(b,a,\langle s(y)|x,y\rangle)$$

for all objects a, b. Then $(a+0)=a$ and $(a+s(b))=s(a+b)$ for all objects a, b.

As another application of R we obtain the following result.

Lemma 5.12. *There is an* $\mathfrak{F}$*-function* $\delta_{=}:\mathfrak{F}_0\times\mathfrak{F}_0\to\mathfrak{F}_0$ *such that for all natural numbers* a, b

$$\delta_{=}(a,b)=\begin{cases}\Delta_1 & \text{if } a=b,\\ \Delta_2 & \text{if } a\neq b.\end{cases}$$

Proof. Let f in $\mathfrak{F}_2$ be given by

$$f(a,b)=\lambda x|\mathrm{DEC}(x)|_2(\Delta_2,\mathrm{APP}(b,\mathrm{PRED}(x)))$$

for all objects a, b, and let h in $\mathfrak{F}_1$ be given by

$$h(a)=R(a,\lambda(\mathrm{DEC}),f)$$

for all objects a. Then

$$h(0)=\lambda(\mathrm{DEC}),$$

and

$$h(s(a))=\lambda x|\mathrm{DEC}(x)|_2(\Delta_2,\mathrm{APP}(h(a),\mathrm{PRED}(x)))$$

for all objects a. Now let $\delta_{=}$ in $\mathfrak{F}_2$ be given by $\delta_{=}(a,b)=\mathrm{APP}(h(a),b)$ for all objects a, b. Then

(i) $$\begin{aligned}\delta_{=}(0,0)&=\mathrm{APP}(\lambda(\mathrm{DEC}),0)\\&=\mathrm{DEC}(0)\\&=\Delta_1,\end{aligned}$$

(ii) $$\begin{aligned}\delta_{=}(0,s(b))&=\mathrm{APP}(\lambda(\mathrm{DEC}),s(b))\\&=\mathrm{DEC}(s(b))\\&=\Delta_2,\end{aligned}$$

(iii) $$\begin{aligned}\delta_{=}(s(a),0)&=|\mathrm{DEC}(0)|_2(\Delta_2,\mathrm{APP}(h(a),\mathrm{PRED}(0)))\\&=\Delta_2,\end{aligned}$$

(iv) $$\begin{aligned}\delta_{=}(s(a),s(b))&=|\mathrm{DEC}(s(b))|_2(\Delta_2,\mathrm{APP}(h(a),\mathrm{PRED}(s(b))))\\&=\mathrm{APP}(h(a),b)\\&=\delta_{=}(a,b)\end{aligned}$$

for all objects a, b.

Now using (i)–(iv) and a double induction on the natural numbers we see that $\delta_{=}$ has the desired properties.

The following notions will be important in our construction of Frege structures.

Definition 5.13. An $\mathfrak{F}$-functional $F: \mathfrak{F}_{n_1} \times \cdots \times \mathfrak{F}_{n_k} \to \mathfrak{F}_0$ is *primitive* if F has *projection* functions P_1 in $\mathfrak{F}_{n_1+1}, \ldots, P_k$ in $\mathfrak{F}_{n_k+1}$ such that for $i = 1, \ldots, k$

$$P_i(F(\bar{f}), \bar{a}) = f_i(\bar{a})$$

for all $\bar{f} = f_1, \ldots, f_k$ in $\mathfrak{F}_{n_1} \times \cdots \times \mathfrak{F}_{n_k}$ and $\bar{a}$ in $\mathfrak{F}_0^{n_i}$.

As examples of primitive $\mathfrak{F}$-functionals we have λ (with projection APP) and more generally λ_0^n (with projection APP_n) for each $n > 0$. Also we have PAIR (with projections p, q) and more generally J_n (with projections p_m^n for $m = 1, \ldots, n$) for each $n > 0$. Finally any object is primitive, so in particular 0 is, and s is primitive (with projection PRED).

Definition 5.14. A family $\{F_i\}_i$ of $\mathfrak{F}$-functionals is *independent* if for each i F_i has a *discriminator* δ_i in $\mathfrak{F}_1$ such that if c is a value of F_i, then $\delta_i(c) = \Delta_1$ and if c is a value of F_j for some $j \neq i$, then $\delta_i(c) = \Delta_2$.

An example of an independent family is the family consisting of just 0, s where 0 has DEC as a discriminator.

The main result of this section gives us a free hand in forming countable independent families of primitive $\mathfrak{F}$-functionals.

Theorem 5.15. *For each natural number m let $\nu(m,1), \ldots, \nu(m,k(m))$ be a finite sequence of natural numbers. Then there is an independent family of primitive $\mathfrak{F}$-functionals*

$$F_m : \mathfrak{F}_{\nu(m,1)} \times \cdots \times \mathfrak{F}_{\nu(m,k(m))} \to \mathfrak{F}_0$$

for $m = 0, 1, 2, \ldots$.

Proof. For each natural number m let

$$F_m(\bar{f}) = J_{k(m)+1}\left(m, \lambda_0^{\nu(m,1)}(f_1), \ldots, \lambda_0^{\nu(m,k(m))}(f_{k(m)})\right)$$

for all $\bar{f} = f_1, \ldots, f_{k(m)}$ in $\mathfrak{F}_{\nu(m,1)} \times \cdots \times \mathfrak{F}_{\nu(m,k(m))}$. Clearly each F_m is an $\mathfrak{F}$-functional. Also each F_m is primitive, having projections $P_1, \ldots, P_{k(m)}$ given by

$$P_j(a, \bar{b}) = |P_{j+1}^{k(m)+1}(a)|_{\nu(m,j)}(\bar{b})$$

for $j = 1, \ldots, m, a$ in $\mathfrak{F}_0$ and $\bar{b}$ in $\mathfrak{F}_0^{\nu(m,j)}$. Clearly each P_j is an $\mathfrak{F}$-functional

and

$$P_j(F_m(\bar{f}),\bar{b}) = |\lambda_0^{\nu(m,j)}(f_j)|_{\nu(m,j)}(\bar{b})$$
$$= f_j(\bar{b})$$

for all $\bar{f} = f_1, \ldots, f_k$ in $\mathfrak{F}_{\nu(m,1)} \times \cdots \times \mathfrak{F}_{\nu(m,k(m))}$ and $\bar{b}$ in $\mathfrak{F}_0^{\nu(m,j)}$.

Finally F_m for $m = 0, 1, 2, \ldots$ form an independent family because each F_m has the discriminator δ_m given by

$$\delta_m(c) = \delta_=(m, p_1^{k(m)+1}(c))$$

for all objects c.

6. The construction of Frege structures

In this section we show how to enlarge any lambda structure to a Frege structure. By Theorem 5.15 we see that given a lambda structure $\mathfrak{F}$ we may find an independent family of primitive $\mathfrak{F}$-functionals which include a list of logical constants $\neg : \mathfrak{F}_0 \to \mathfrak{F}_0, \&, \vee, \supset, \doteq : \mathfrak{F}_0 \times \mathfrak{F}_0 \to \mathfrak{F}_0$ and $\forall, \exists : \mathfrak{F}_1 \to \mathfrak{F}_0$. It now only remains to find a logical system relative to these logical constants. To do this it will be helpful to express the logical schemata in a uniform way.

Definition 6.1. By an *$\mathfrak{F}$-system* we mean a pair $\mathfrak{X}$ of collections $\mathfrak{X}_0, \mathfrak{X}_1$ of objects of $\mathfrak{F}$ such that $\mathfrak{X}_1$ is a subcollection of $\mathfrak{X}_0$.

We associate with each logical constant F two predicates Φ_F and Ψ_F. In the following $\mathfrak{X}$ is an $\mathfrak{F}$-system $\mathfrak{X}_0, \mathfrak{X}_1, x, y$ are objects and ξ is a function in $\mathfrak{F}_1$.

Negation.

$\Phi_\neg(\mathfrak{X}, x)$ *is "x is in $\mathfrak{X}_0$",*

$\Psi_\neg(\mathfrak{X}, x)$ *is "x is not in $\mathfrak{X}_1$".*

Conjunction.

$\Phi_\&(\mathfrak{X}, x, y)$ *is "x, y are in $\mathfrak{X}_0$",*

$\Psi_\&(\mathfrak{X}, x, y)$ *is "x, y are in $\mathfrak{X}_1$".*

Disjunction.

$\Phi_\vee(\mathfrak{X}, x, y)$ *is "x, y are in $\mathfrak{X}_0$",*

$\Psi_\vee(\mathfrak{X}, x, y)$ *is "x is in $\mathfrak{X}_1$ or y is in $\mathfrak{X}_1$".*

Implication.

$\Phi_{\supset}(\mathcal{X},x,y)$ *is "x is in* $\mathcal{X}_0$ *and y is in* $\mathcal{X}_0$ *provided x is in* $\mathcal{X}_1$*",*

$\Psi_{\supset}(\mathcal{X},x,y)$ *is "x is in* $\mathcal{X}_1$ *implies y is in* $\mathcal{X}_1$*".*

Universal quantification.

$\Phi_{\forall}(\mathcal{X},\xi)$ *is "*$\xi(x)$ *is in* $\mathcal{X}_0$ *for all x",*

$\Psi_{\forall}(\mathcal{X},\xi)$ *is "*$\xi(x)$ *is in* $\mathcal{X}_1$ *for all x".*

Existential quantification.

$\Phi_{\exists}(\mathcal{X},\xi)$ *is "*$\xi(x)$ *is in* $\mathcal{X}_0$ *for all x",*

$\Psi_{\exists}(\mathcal{X},\xi)$ *is "*$\xi(x)$ *is in* $\mathcal{X}_1$ *for some x".*

Equality.

$\Phi_{\doteq}(\mathcal{X},x,y)$ *is "*$x=x$*",*

$\Psi_{\doteq}(\mathcal{X},x,y)$ *is "*$x=y$*".*

Note that for each logical constant $F, \Psi_F(\mathcal{X},\ )$ does not depend on $\mathcal{X}_0$ while $\Phi_F(\mathcal{X},\ )$ depends on $\mathcal{X}_1$ only in the case when F is $\supset$ and always depends on $\mathcal{X}_0$ positively. If $F: \mathcal{F}_{n_1} \times \cdots \times \mathcal{F}_{n_k} \to \mathcal{F}_0$ is a logical constant and $\mathcal{X}$ is the $\mathcal{F}$-system $\mathcal{X}_0, \mathcal{X}_1$ where $\mathcal{X}_0$ is the collection of propositions and $\mathcal{X}_1$ is the collection of truths, then the logical schema for F may be formulated as follows.

Logical schema for F. If $\bar{f}$ *is in* $\mathcal{F}_{n_1} \times \cdots \times \mathcal{F}_{n_k}$ *such that* $\Phi_F(\mathcal{X},\bar{f})$*, then* $F(\bar{f})$ *is in* $\mathcal{X}_0$ *such that*

$$F(\bar{f}) \text{ is in } \mathcal{X}_1 \quad \text{iff } \Psi_F(\mathcal{X},\bar{f}).$$

Definition 6.2. Given an $\mathcal{F}$-system $\mathcal{X}$ let $\theta(\mathcal{X})$ be the $\mathcal{F}$-system $\theta_0(\mathcal{X})$, $\theta_1(\mathcal{X})$, where $\theta_0(\mathcal{X})$ is the collection of those objects $F(\bar{f})$ where F is a logical constant and $\Phi_F(\mathcal{X},\bar{f})$, and $\theta_1(\mathcal{X})$ is the collection of those objects $F(\bar{f})$ where F is a logical constant and both $\Phi_F(\mathcal{X},\bar{f})$ and $\Psi_F(\mathcal{X},\bar{f})$.

Lemma 6.3. *Any* $\mathcal{F}$*-system* $\mathcal{X}$ *such that* $\theta(\mathcal{X})=\mathcal{X}$ *is a logical system.*

Proof. Let $\mathcal{X}$ be an $\mathcal{F}$-system such that $\theta(\mathcal{X})=\mathcal{X}$ and let $F: \mathcal{F}_{n_1} \times \cdots \times \mathcal{F}_{n_k} \to \mathcal{F}_0$ be a logical constant. We show that the logical schema for F holds. So let $\bar{f}$ be in $\mathcal{F}_{n_1} \times \cdots \times \mathcal{F}_{n_k}$ such that $\Phi_F(\mathcal{X},\bar{f})$. Then by definition $F(\bar{f})$ is in $\theta_0(\mathcal{X})$ and, as $\theta(\mathcal{X})=\mathcal{X}$, it follows that $F(\bar{f})$ is in $\mathcal{X}_0$.

Further, if $\psi_F(\mathcal{X},\bar{f})$, then $F(\bar{f})$ is in $\theta_1(\mathcal{X})$ and hence in $\mathcal{X}_1$. Conversely suppose, as before, that $\Phi_F(\mathcal{X},\bar{f})$ and $F(\bar{f})$ is in $\mathcal{X}_1$. Then, as $\theta(\mathcal{X})=\mathcal{X}, F(\bar{f})$ is in $\theta_1(\mathcal{X})$. Hence there must be a logical constant G and sequence of $\mathcal{F}$-functions $\bar{g}$ such that $F(\bar{f})=G(\bar{g})$, $\Phi_G(\mathcal{X},\bar{g})$ and $\Psi_G(\mathcal{X},\bar{g})$. As the logical constants form an independent family $F=G$, and, as F is primitive, $\bar{f}=\bar{g}$. Hence we have $\Psi_F(\mathcal{X},\bar{f})$, concluding the proof of the logical schema for F. Hence $\mathcal{X}$ is a logical system.

By the above lemma our problem is to show that θ has a fixed point. We now show that.

Definition 6.4. For $\mathcal{F}$-systems $\mathcal{X},\mathcal{Y}$ let $\mathcal{X}\leqslant\mathcal{Y}$ if $\mathcal{X}_0$ is a subcollection of $\mathcal{Y}_0$ such that for a in $\mathcal{X}_0, a$ is in $\mathcal{X}_1$ iff a is in $\mathcal{Y}_1$.

Lemma 6.5. $<$ *is a partial ordering of* $\mathcal{F}$*-systems such that every chain has a least upper bound.*

Proof. Obvious.

Lemma 6.6. θ *is a monotone operator on* $\mathcal{F}$*-systems. I.e. if* $\mathcal{X}\leqslant\mathcal{Y}$, *then* $\theta(\mathcal{X})<\theta(\mathcal{Y})$.

Proof. By a trivial inspection of each logical schema we may show that if $\mathcal{X}<\mathcal{Y}$, then for each logical constant F, if $\Phi_F(\mathcal{X},\bar{f})$, then

$$\Phi_F(\mathcal{Y},\bar{f})$$

and

$$\psi_F(\mathcal{Y},\bar{f}) \quad \text{iff } \Psi_F(\mathcal{X},\bar{f}).$$

The monotonicity of θ now follows easily.

Theorem 6.7. *There is a (unique)* $\mathcal{F}$*-system* θ^∞ *such that* $\theta(\theta^\infty)=\theta^\infty$ *and* $\theta^\infty<\mathcal{X}$ *whenever* $\theta(\mathcal{X})\leqslant\mathcal{X}$.

Proof. The result is a standard consequence of the previous two lemmata when working in classical set theory. We review the construction of θ^∞. By induction on the ordinal α we simultaneously prove that the family of $\mathcal{F}$-systems θ^β for $\beta<\alpha$ is an increasing chain having a least upper bound $\theta^{<\alpha}$ and define θ^α to be $\theta(\theta^{<\alpha})$. Finally we show that the family of $\mathcal{F}$-systems θ^α, for α an ordinal, is an eventually constant increasing chain and define θ^∞ to be its least upper bound. Note that we are interpreting "collection" as "set" and working in classical set theory. For large enough ordinals α $\theta^{\infty}=\theta^{<\alpha}$ so that $\theta(\theta^{\infty})=\theta(\theta^{<\alpha})=\theta^{\alpha}\leqslant\theta^{\infty}$.

Now if $\theta(\mathcal{X})\leqslant\mathcal{X}$, then we may show by induction on α that $\theta^{\alpha}\leqslant\mathcal{X}$, and hence that $\theta^{\infty}\leqslant\mathcal{X}$. As θ is monotone and $\theta(\theta^{\infty})\leqslant\theta^{\infty}$ it follows that $\theta(\theta^{\infty})$ is an $\mathcal{X}$ such that $\theta(\mathcal{X})\leqslant\mathcal{X}$. So, by the above, $\theta^{\infty}\leqslant\theta(\theta^{\infty})$. As we have already obtained $\theta(\theta^{\infty})\leqslant\theta^{\infty}$ we get $\theta(\theta^{\infty})=\theta^{\infty}$.

The above is a standard construction in the theory of inductive definitions but it involves non-constructive reasoning at one or two points. For example the proof that the families of $\mathcal{F}$-systems involved form chains requires that ordinals are comparable, and the proof that the increasing family of $\mathcal{F}$-systems θ^{α} is eventually constant seems to require a nonconstructive indirect argument. Nevertheless I believe the result to be constructively valid. A rigorous elaboration of this point would require an explicit discussion of the role of inductive definitions in constructive mathematics. It will have to suffice here if I simply assert that the logical schemata form the clauses of an inductive definition that generate the propositions and simultaneously give conditions for their truth. The resulting logical system satisfies the conditions of our theorem.

Combining the previous results we obtain the following.

Theorem 6.8. *Any lambda structure can be enlarged to a Frege structure.*

This result may be strengthened considerably. Here I only consider a mild strengthening.

Definition 6.9. A Frege structure is *N-standard* if the collection of natural numbers, relative to some Peano system on the structure, is internally definable.

Theorem 6.10. *Any lambda structure can be enlarged to an N-standard Frege structure.*

Proof. If $\mathcal{F}$ is a Frege structure, then a function N in $\mathcal{F}_1$ is an internal definition of the natural numbers just in case the following schema holds.

Natural numbers. If a is an object, then $N(a)$ is a proposition such that

$N(a)$ is true iff a is a natural number.

N-standard Frege structures may be constructed exactly as in the construction of Frege structures if N is treated as a logical constant and the above schema as the logical schema for it. It is only necessary to suitably define the predicates Φ_N and Ψ_N. But that may be done as follows.

$$\Phi_N(\mathcal{X},x) \textit{ is } \text{“}x=x\text{”},$$

$$\psi_N(\mathcal{X},x) \textit{ is } \text{“}x \textit{ is a natural number”}.$$

The above theorem is a special case of the following more general result which can be proved in exactly the same way.

Theorem 6.12. *Any lambda structure can be enlarged to a Frege structure so that each of any previously specified finite or infinite sequence of collections of objects is internally definable.*

Note that any such collections that are to be given internal definitions must be specified without reference to the logical system itself.

7. Further developments

The original stimulus for the ideas in this paper was the observation that the intuitionistic theory of types presented in MARTIN-LÖF (1975, Section 1) can be modelled in suitably chosen Frege structures. A detailed formal presentation of this result in a restricted context may be found in JAN SMITH (1978). Here I shall briefly describe the modelling of a theory of types in an N-standard Frege structure $\mathfrak{F}$. The theory of types to be modelled will be that used in ACZEL (1979). Types will be interpreted as collections of objects. The starting types N and N_k for $k=0,1,\ldots$ are interpreted as the collection of natural numbers and the collection of objects $\Delta_1^k,\ldots,\Delta_k^k$ for $k=0,1,\ldots$ respectively. The disjoint union $\mathfrak{A}+\mathfrak{B}$ of the two collections $\mathfrak{A},\mathfrak{B}$ is defined to be the collection of objects PAIR(Δ_1,a) for a in $\mathfrak{A}$ or PAIR(Δ_1,b) for b in $\mathfrak{B}$. If $\mathfrak{A}$ is a collection of objects and $\mathfrak{B}(a)$ is a collection of objects for each a in $\mathfrak{A}$, then the collections $(\Pi x\in\mathfrak{A})\mathfrak{B}(x)$ and $(\Sigma x\in\mathfrak{A})\mathfrak{B}(x)$ are defined as follows. $(\Pi x\in\mathfrak{A})\mathfrak{B}(x)$ is defined to be the collection of those objects $\lambda x f(x)$ such that f is in $\mathfrak{F}_1$ and $f(a)$ is in $\mathfrak{B}(a)$ for all a in $\mathfrak{A}$. $(\Sigma x\in\mathfrak{A})\mathfrak{B}(x)$ is defined to be the collection of those objects PAIR(a,b) such that a is in $\mathfrak{A}$ and b is in $\mathfrak{B}(a)$. Finally the “type of small types” V is interpreted as the collection $\mathfrak{V}$ of sets of the Frege structure. Each a in $\mathfrak{V}$ has associated with it the collection $\mathfrak{F}(a)$ of those objects b such that $(b\in a)$ is true. These collections $\mathfrak{F}(a)$ are the “small types” in our interpretation. Note that they are just those collections that are internally definable. This interpretation does not quite model the theory in MARTIN-LÖF (1975) or ACZEL (1979) as there the objects in V *are* the small types whereas they must here be treated only as indices for them as in ACZEL (1977). But this is a minor point which will be ignored here. ACZEL (1979) introduced one further type U, there called the type of sets. This type is interpreted as the smallest collection of objects $\mathfrak{U}$ such that:

If a is in $\mathfrak{V}$ and f is a function in $\mathfrak{F}_1$ such that $f(b)$ is in $\mathfrak{U}$ for all b in $\mathfrak{F}(a)$, then $\{f(x)|x\in a\}$ is in $\mathfrak{U}$.

In the above $\{f(x)|x\in a\}$ is an abbreviation for $\{y|\exists x(x\in a \,\&\, (y=f(x)))\}$.

The main result of ACZEL (1979) is an interpretation of a system CZF of constructive set theory in a type theory. When combined with the above interpretation of type theory we obtain an interpretation of CZF in the N-standard Frege structure $\mathfrak{F}$. This interpretation can be carried out directly and more simply by using, instead of the collection $\mathfrak{U}$, a larger collection of sets called the *iterative sets*. This is defined to be the smallest collection of objects $\mathfrak{X}$ such that

every set of $\mathfrak{X}$'s is in $\mathfrak{X}$,

where the set a is a *set of* $\mathfrak{X}$'s if $\mathfrak{F}(a)$ is a subcollection of $\mathfrak{X}$. Note that every set in $\mathfrak{U}$ is iterative, but for the converse, it turns out only that every iterative set is extensionally equal to a set in $\mathfrak{U}$, where extensional equality on iterative sets is to be defined hereditarily as in ACZEL (1979). I plan to give a detailed account of iterative sets and the above mentioned interpretation of constructive set theory in a future publication.

References

ACZEL, P.

[1977] The strength of Martin-Löf's intuitionistic type theory with one universe, in: *Proceedings of the Symposium on Mathematical Logic (Oulu 1974)*, edited by S. Miettinen and J. Vaananen, Report No. 2 of Dept. Philosophy, University of Helsinki, pp. 1–32.

[1979] The type theoretic interpretation of constructive set theory, in: *Logic Colloquium '77*, edited by A. Macintyre, L. Pacholski and J. Paris (North-Holland, Amsterdam).

BARENDREGT, H. P.

[1977] The type free lambda calculus, in: *Handbook of Mathematical Logic*, edited by J. Barwise (North-Holland, Amsterdam) pp. 1091–1132.

BEESON, M. J.

[1977] Principles of continuous choice and continuity of functions in formal systems for constructive mathematics, *Ann. Math. Logic*, **12**, 249–322.

CHURCH, A.

[1932 & 1933] A set of postulates for the foundations of logic, *Ann. Math.*, **33**, 346–366, and **34**, 839–864.

CURRY, H. B., ET AL.

[1972] *Combinatory Logic, Vol. 2* (North-Holland, Amsterdam).

FEFERMAN, S.

[1975] A language and axioms for explicit mathematics, in: *Algebra and Logic, Lecture Notes in Mathematics*, **450** (Springer, Berlin) pp. 87–139.

[1978] Recursion theory and set theory: a marriage of convenience, in: *Generalised Recursion Theory II* edited by J. E. Fenstad, R. O. Gandy and G. E. Sacks (North-Holland, Amsterdam) pp. 55–98.

FREGE, G.

[1893 & 1903] *Grundgesetze der Arithmetik, begriffsschriftlich abgeleitet* Vols. I and II (Jena), partial english translation in Furth [1964].

FURTH, M.

[1964] *The Basic Laws of Arithmetic* (Univ. of Los Angeles Press, Berkeley and Los Angeles).

MARTIN-LÖF, P.

[1975] An intuitionistic theory of types: predicative part, in: *Logic Colloquim '73*, edited by H. E. Rose and J. C. Shepherdson (North-Holland, Amsterdam) pp. 73–118.

SHOENFIELD, J.R.

[1977] Axioms of set theory, in: *Handbook of Mathematical Logic*, edited by J. Barwise (North-Holland, Amsterdam) pp. 321–344.

SCOTT, D.

[1972] Continuous lattices, in: *Toposes, Algebraic Geometry and Logic*, edited by F. W. Lawvere, *Lecture Notes in Mathematics*, **274** (Springer, Berlin) pp. 97–136.

[1974] Axiomatizing set theory, in: *Proceedings of the Summer Institute on Axiomatic Set Theory*, 1967, Part II, edited by T. Jech, AMS Proceedings of Symposia, volume XIII, Part II, pp. 207–214.

[1975a] Lambda calculus and recursion theory, in: *Proceedings of the Third Scandinavian Logic Symposium*, edited by S. Kanger (North-Holland, Amsterdam) pp. 154–193.

[1975b] Combinators and classes, in: *λ-Calculus and Computer Science*, edited by C. Böhm, *Lecture Notes in Computer Science*, **37** (Springer, Berlin) pp. 1–26.

SMITH, J.

[1978] On the relation between a type theoretic and a logical formulation of the theory of constructions. Dissertation, (Dept. of Mathematics, Göteborg, Sweden).

STENLUND, S.

[1975] Descriptions in intuitionistic logic, in: *Proceedings of the Third Scandinavian Logic Symposium*, edited by S. Kanger (North-Holland, Amsterdam) pp. 197–212.

This page intentionally left blank

J. Barwise, H. J. Keisler and K. Kunen, eds., *The Kleene Symposium*
©North-Holland Publishing Company (1980) 61–83

The Role of Finite Automata in the Development of Modern Computing Theory*

Robert L. Constable
Computer Science Department, Cornell University, Ithaca, NY 14853, U.S.A.

Dedicated to Professor S.C. Kleene on the occasion of his 70th birthday

Abstract: Regular patterns described by finite automata are evident in the behavior of computers, in the structure of programming languages and in the rules for reasoning about programs. The systematic study of these patterns has shaped computing theory, providing theorems, techniques and a paradigm with far reaching and subtle effects on many aspects of modern computing theory. The essence of the development is illustrated in this paper by following the mainstream of work flowing from Kleene's 1951 theorem on regular events to recent (circa 1976) results in computational complexity and algorithmic logic.

1. Introduction

1.1. The terms "finite state machine" and "regular expression" are basic to the vocabulary of. computer science. Even a casual user of computer systems may encounter regular expressions in the command repertoire of the text editor, as in UNIX. A novice programmer is likely to encounter finite state machines, in the form of syntax charts that define the programming language. When the high level language statements are being compiled into machine language, a finite automaton scans the input looking for the atomic tokens. The compiler writer probably learned theorems about the automata that helped her recognize which parts of the language required more than finite state processing.

If several programs are run in a multiprocessing environment, regular expressions may be used in synchronization. At some universities students can formally verify their programs. At Cornell, the verifier's arithmetic routines use a version of the dynamic programming algorithm Kleene first discovered to transform state diagrams into regular expressions.

1.2. In this talk we will examine a few important discoveries about finite automata and the influence of those discoveries on the modern theory of

*Lecture presented at Kleene Symposium, Madison, Wisconsin, June 23, 1978. Research reported in Section 4 was partially supported by NSF grant MCS 76-14293.

computing. The subject is vast; SAMUEL EILENBERG has already written a massive two volume account of it (1974). Moreover, MICHAEL RABIN (1967) wrote an excellent survey article on finite automata for the American Mathematical Society in 1967. So I will concentrate mainly on those aspects of automata derived from Kleene's fundamental theorem and those aspects directly related to logic. Whenever possible I will mention results since 1967, especially those answering questions from Rabin's article.

2. Historical origins, basic definitions and constructions

2.1. The concept of a finite automaton appears to have arisen in the 1943 paper "A logical calculus of the ideas immanent in nervous activity" by WARREN MCCULLOCK and WALTER PITTS (1943). These neurobiologists set out to model the behavior of neural nets, having noticed a relationship between neural activity and logic:

> "The 'all-or-none' law of nervous activity is sufficient to insure that the activity of any neuron may be represented as a proposition. ... To each reaction of any neuron there is a corresponding assertion of a simple proposition."

McCullock and Pitts described neural nets in terms of the formulas of the predicate calculus with a time parameter. At the 1948 Hixon Symposium, JOHN VON NEUMANN (1948) summarized this work as follows:

> "The Main Result of the McCullock–Pitts Theory. McCullock and Pitts' important result is that any functioning in this sense which can be defined at all logically, strictly and unambiguously in a finite number of words can also be realized by such a formal neural network."

Speaking after von Neumann's lecture, McCullock said,

> "... it was not until I saw Turing's paper that I began to get going the right way around, and with Pitts' help formulated the required logical calculus. What we thought we were doing (and I think we succeeded fairly well) was treating the brain as a Turing machine. The important thing was, for us, that we had to take a logic and subscript it for the time of the occurrence of a signal. ... The delightful thing is that the very simplest set of appropriate assumptions is sufficient to show that a nervous system can compute any computable number. It is that kind of device, if you like —a Turing machine."

2.2. In the summer of 1951 Professor Kleene was invited to work at RAND. Merrill Flood, an old graduate school friend at RAND, put Kleene on to reading McCullock and Pitts. He was also influenced to work in this area by reading the Von Neumann lecture quoted from above

("The general and logical theory of automata", the Hixon Symposium, 1948). Part way through his term at RAND and for two months of the fall 1951 semester at Madison, Kleene worked to produce his very influential paper, "Representation of events in nerve nets and finite automata" (RAND RM-704, 101 pages).

In his own words, what Kleene set out to do was discover

> "what kind of events are capable of being represented in the state of an automaton." In general,"... we are investigating McCullock–Pitts nerve nets only partly for their own sake as providing a simplified model of nervous activity, but also as an illustration of the general theory of automata, including robots, computing machines and the like. What a finite automaton can and cannot do is thought to be of some mathematical interest intrinsically, and may also contribute to better understanding of problems which arise on the practical level."

In retrospect it was the clarity of the questions and the beauty of the result which influenced a talented group of mathematicians to explore this subject thoroughly and to set Kleene's theorem in a clean rich context. Let us first examine the modern setting for this theorem, then examine its proofs and finally trace its influence.

2.3. *Basic definitions*

Definition: Let $\Sigma=\{a_1,\dots,a_n\}$ be finite set (of letters). A *word* on Σ is a finite sequence of letters. The empty word is denoted Λ and has no letters. Let Σ^* be the set of all words (including Λ). (Note $\langle\Sigma^*, \text{concatenation}, \Lambda\rangle$ is a semigroup.)

Definition: A *finite automaton* (fa) $\mathcal{A}$ over Σ is an algebraic structure $\mathcal{A}=\langle S,\delta,s_0,F\rangle$ such that

(i) S is a finite set of *states*,

(ii) $\delta : S\times\Sigma\to S$ is the *transition function*,

(iii) $s_0\in S$ is the *initial state*,

(iv) $F\subseteq S$ is the set of *final states*.

When we think of $\mathcal{A}$ as a machine, we sometimes draw it as:

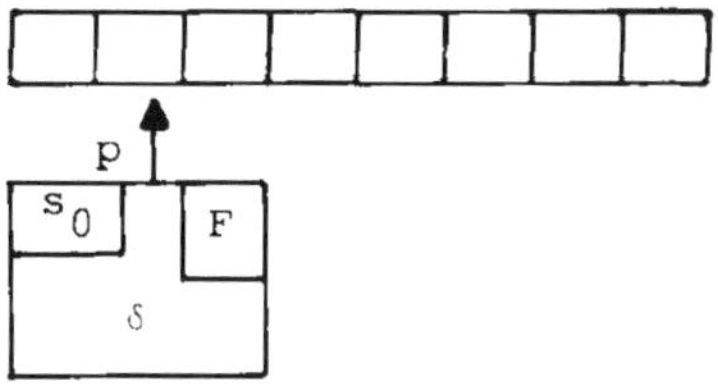

Figure 1.

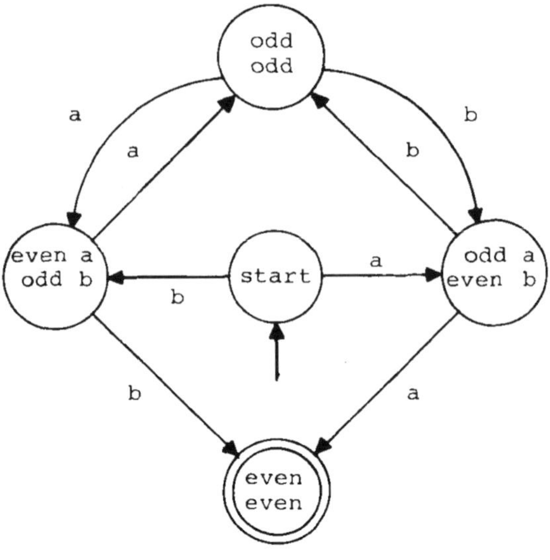

Figure 2.

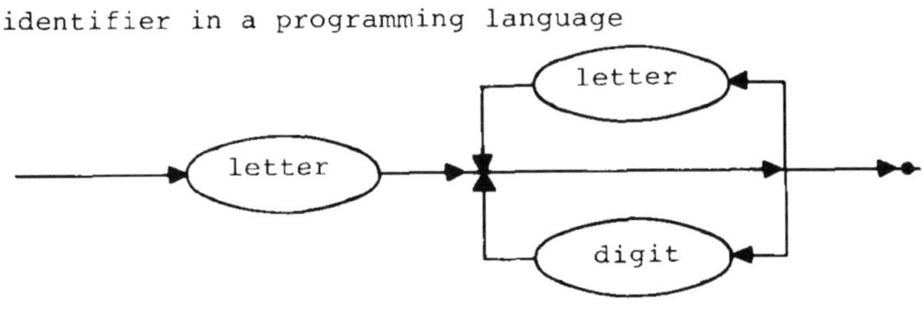

Figure 3.

Definition: $\mathrm{T}(\mathcal{A})=\{x\in\Sigma^*\mid \delta^*(s_0,x)\in F\}$ is called the set (language, or event) *accepted* by $\mathcal{A}$ where δ^* is the natural extension of δ to $S\times\Sigma^*, \delta^*(s,ax)=\delta^*(\delta(s,a),x)$.

A finite automaton can be represented by a transition diagram or state diagram. Figs.2 and 3 give two examples.

2.4. *Regular expressions*

To describe the events accepted by a finite automaton, Kleene introduced the class of regular expressions. Before defining them, let us see what they describe.

Let A, B be subsets of Σ^* (*languages*, events).

Define $A \cup B = \{x \mid x \in A \text{ or } X \in B\}$,

$$A \cdot B = \{xy \mid x \in A, y \in B\},$$

$$A^* = \bigcup_{i=0}^{\infty} A^i, \quad A^0 = \{\Lambda\}, \quad A^i = \underbrace{A \cdot \ldots \cdot A}_{i}$$

Let $\varnothing$ denote the empty set.

Regular expressions are like algebraic terms over constants. Here is an inductive definition.

Definition: (1) $\varnothing$, Λ are regular expressions (Λ denotes $\{\Lambda\}$.)

(2) If $a \in \Sigma$, then a is a regular expression (denoting $\{a\}$).

(3) If E, F are regular expressions, then so are

$(E + F)$

$(E \cdot F)$

$(E)^*$

(4) Nothing else is a regular expression.

We usually omit parentheses, with the precendence $*$, $\cdot$, $\cup$. We also omit the $\cdot$ operator and write EF for $E \cdot F$.

2.5. *Kleene's theorem*

Kleene introduced regular expressions to describe the behavior of finite automata. He proved the beautiful result that they describe this behavior exactly.

Definition: An event (language) is *regular* over Σ if and only if there is a regular expression which denotes it.

Theorem (Kleene, 1951). (1) (Theorem 3) *To each regular event over* Σ, *we can find a finite automaton over*Σ *which accepts it.*

(2) (Theorem 5) *The set accepted by a finite automaton is a regular event (whose defining regular expression can be constructed).*

(Kleene used the term *representable* event for our term *accepted* and used the word *event* instead of language. The theorem numbers are from his paper.)

We will examine the proof of part 1 (Theorem 3) later when we view subsequent developments in the theory of automata.

The most elegant proof I know of part 2 (Theorem 5) of the theorem involves considering transition diagrams whose edges can be labeled by regular expressions as in Fig 4.

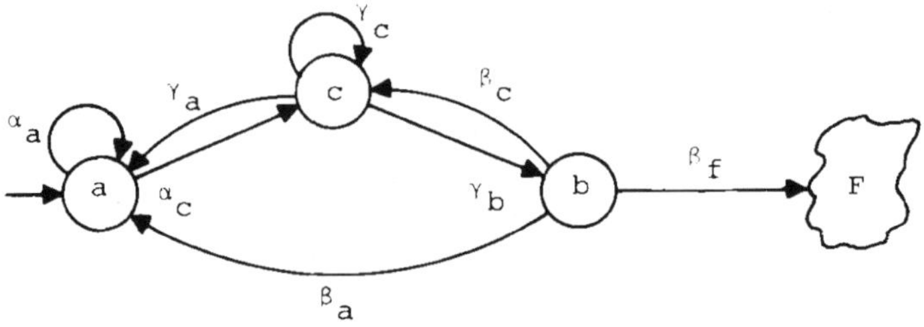

Figure 4.

The meaning of such diagrams should be clear, any string in the labeling events can cause transition along the labeled edge.

Given one such diagram, we can construct an equivalent diagram with fewer states and more complex labeling expressions. By repeating the process starting from the automaton diagram, we generate a sequence $\mathcal{A}_1$, $\mathcal{A}_2, \ldots, \mathcal{A}_n$ and arrive at a diagram with no state and only regular expressions.

By definition, $\mathcal{A}$ accepts every string for which there is a path from s_0 to F. We preserve this property as we change $\mathcal{A}_i$ to $\mathcal{A}_{i+1}$ by the process described above.

To prove this, consider removing any state c, and consider any path from s_0 to F which passes through c, entering from a and departing to b:

$$s_0 \cdots a \quad \underset{\text{enter } c}{c \text{———} c} \quad b \cdots s_f.$$

If we add to the diagram an edge from a to b with a label describing all paths from a to c, from c to itself and from c to b, then any such path will appear in the new diagram, $\mathcal{A}_{i+1}$. Moreover, any path in $\mathcal{A}_{i+1}$ corresponds to a path in $\mathcal{A}_i$.

To describe the construction of the new labels, we use the simple three state example of Fig. 4. The resulting smaller diagram is given in Fig. 5.

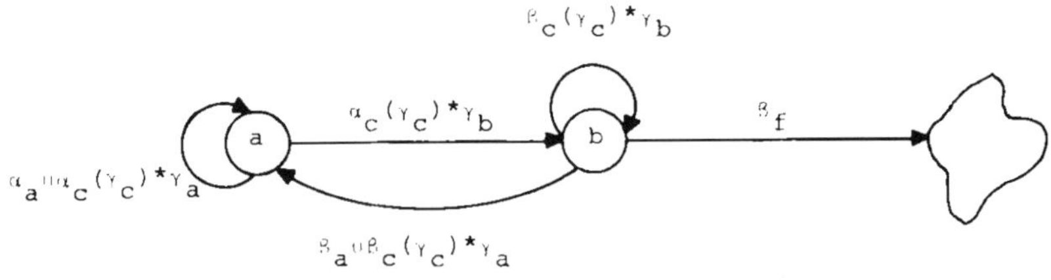

Figure 5.

3. Early automata theory and its influence

3.1. *Minimization and equivalence*

In the early 50's other work on automata theory arose more directly from the phenomena surrounding electronic digital computers and communications equipment. In particular D. A. HUFFMAN (1954) and E. F. MOORE (1956) studied finite state machines.

When finite automata are used to model the construction of hardware (e.g. sequential machine circuits) or software (e.g. lexical scanners), it is important to examine the question of whether a smaller equivalent device can be built. Both Huffman and Moore were concerned about such modeling and both discovered practical algorithms for minimizing a finite automaton. For an n state automaton these algorithms required n^2 steps.

When this state minimization algorithm became significant in practice there was motivation to find a more efficient version. One such was discovered by JOHN HOPCROFT (1971).

Theorem (HOPCROFT). *There is an* $O(n\log(n))$ *algorithm to minimize a finite automaton.*

This minimization procedure also leads to a feasible method of testing automaton equivalence. The minimal automaton is unique (up to isomorphism), so one method is to reduce the automata to their minimal forms and see if they are identical. The minimal form is a cannonical representation of the autômaton. There is an even more efficient variant of this due to J. E. HOPCROFT and R. KARP (personal communication, 1971). (The algorithm requires that the start states are given.)

Theorem (HOPCROFT and KARP) *There is an* $O(nA^{-1}(n))$ *algorithm to test the equivalence of two finite automata, where n is the maximum of the numbers of states and A is Ackermann's function.*

These results immediately suggest the question of whether there is an equally efficient algorithm to test the equivalence of regular expressions. We shall return to the problem later. Moreover these results serve as a model of inquiry for other classes of automata. One is thus led to ask, for example, whether the equivalence problem is solvable for deterministic pushdown automata (dpda), see HOPCROFT and ULLMAN (1969) or (1979). This remains a well-known open problem in automata theory.

3.2. *Algebraic view*

There is an algebraic view of minimization and characterization of finite automata, originating with John Myhill and Anil Nerode. Just as Kleene in attempting to understand McCullock and Pitts abstracted the concept of finite automaton from the nerve nets and the concept of regular events and expressions from the logical calculus of McCullock and Pitts, so John Myhill in attempting to understand Kleene, abstracted finite automata to transition diagrams (deterministic) and events to unions of equivalence classes. Just as Kleene's work attracted the logicians, Myhill and Nerode would attract the algebraists. Nerode was led to the algebraic characterization by attempting to find ways of showing the existence or impossibility of a finite automaton without writing the transition diagrams, which in the cases he was examining were quite complex.

I will state the definitions and theorems in a form due to Nerode, a form instrumental in proving one of the gems of the theory—the equivalence of two-way and one-way finite automata. (This is the way the theorem is stated in RABIN and SCOTT (1959).)

Definition: An equivalence relation R over Σ^* is *right invariant* if and only if whenever xRy, then $xzRyz$ for all $z \in \Sigma^*$. An equivalence relation is of *finite index* if and only if there are only finitely many equivalence classes under R.

Theorem (MYHILL–NERODE). *Let* A *be an event in* Σ^*. *The following conditions are equivalent*:

(i) $A = T(\mathcal{A})$ *for a finite automaton* $\mathcal{A}$.

(ii) A *is the union of some of the equivalence classes of a right invariant equivalence relation of finite index.*

(iii) *the relation* $x \equiv y$ *if and only if for all* $z \in \Sigma^* \, xz \in A \Leftrightarrow yz \in A$ *is of finite index*.

The equivalence relation $\equiv$ defines the unique minimal finite automaton. Any other right invariant equivalence relation defining A is a refinement of this minimal one.

An interesting corollary of this theory is the so called pumping lemma.

Pumping Lemma: *If* $A = T(\mathcal{A})$ *for a finite automaton* $\mathcal{A}$ *with* n *states, then for any string* $x \in A$ *where* $|x| > n$, *there is a substring* w *such that* $x = lwr$ *and* $lw^i r \in A$ *for* $i = 0, 1, 2, \ldots$.

We can use this to show trivially that certain events are not regular. For example $0^n 1 0^n$ is not regular. If it were, it would be accepted by some automaton $\mathcal{A}$ with m states. When $n > m$, one of the strings $0^{n+p} 1 0^n$, $0^n 1 0^{n+p}$ or $0^q (0 \cdots 1 \cdots 0)^p 0^r$ must be accepted which is not allowed.

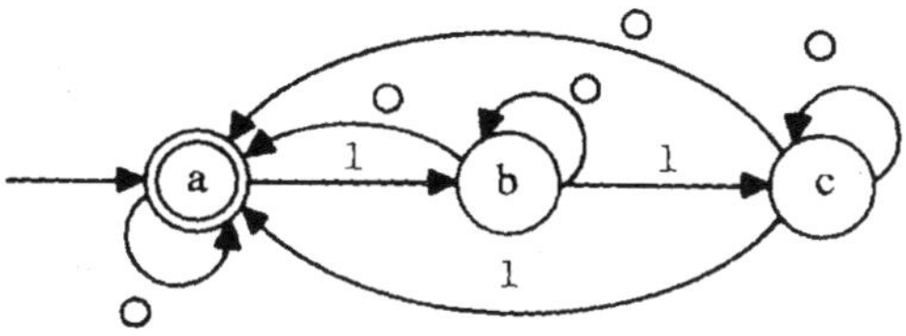

Figure 6.

3.3. *Nondeterminism*

When a finite automaton is viewed as a transition diagram, it is natural to consider several edges with the same label leading from a node, as in Fig. 6

Machines corresponding to these diagrams with multiple edges with the same label are called *nondeterministic finite automata*. They were introduced first by M. O. RABIN and D. SCOTT in their paper "Finite automata and their decision problems" (1959).

Nondeterministic automata are free to make choices at certain steps of their computation. If any of these choices leads to an accepting state, the input is accepted.

Nondeterministic machines can easily express certain operations. For example, given $\mathcal{A}$ accepting A and $\mathcal{B}$ accepting B, we can construct a nondeterministic machine accepting AB and A^*. For example, to accept AB simply connect $\mathcal{A}$ and $\mathcal{B}$ by considering the accept states of $\mathcal{A}$ as if they were start states of $\mathcal{B}$ and adding the appropriate transitions. (To compute A^*, add a new start state to $\mathcal{A}$ which is also an accept state; give it the same outgoing edges as the start state. Also treat the accept states as start states.)

It appears that these nondeterministic machines are more powerful than deterministic machines, but Rabin and Scott proved:

Theorem (RABIN and SCOTT, 1959). *Given any nondeterministic finite automaton $\mathcal{A}$ (with n states), we can construct an equivalent deterministic machine $\mathcal{A}$ (with at most 2^n states).*

The proof is quite straightforward. The states of the deterministic machine are subsets of the states of the nondeterministic machine. Specifically, given $\mathcal{A} = \langle S, \delta, s_0, F\rangle$ the deterministic $\hat{\mathcal{A}}$ is $\langle P(S), \delta, \{s_0\}, \hat{F}\rangle$ where

$$\hat{\delta}(G,a) = \{\delta(s,a) \mid s \in G \subseteq S\}$$

and

$$\hat{F}=\{G \mid G \subseteq S \,\&\, f\in G \text{ for some } f\in F\}.$$

It is easy to prove inductively on the length of accepted strings that $T(\mathfrak{A})=T(\hat{\mathfrak{A}})$.

Remark. In RABIN's 1967 survey paper he raises the question of whether 2^n states are ever required. This question was answered by MEYER and FISCHER (1972). The example I gave in Fig. 6 can be generalized to any n. Any deterministic machine equivalent to it requires 2^n states.

The concept of nondeterministic computation seems somewhat artificial and specific to set acceptance. Nevertheless, there has been practical interest in it. For example, the programming language PLANNER permits nondeterministic programs. They are useful in backtrack searching algorithms. More recently E. Dijkstra has introduced another style of "undetermined" program which he calls nondeterministic. In these programs the order of executing statements is undetermined, but the result must be the same regardless of which choice is taken.

Nondeterminism has played a major role in the theory of automata. Every type of machine comes in two modes, as does every type of complexity class.

	Deterministic	*Nondeterministic*
Finite automaton	fa	nfa
Pushdown automaton	dpda	pda
Linear bounded automaton	dlba	lba
Polynomial time class	P	NP

The natural question to ask for each class is whether the deterministic and nondeterministic machines are equivalent in power. For a long time one of the most interesting of these questions has been "the LBA problem." The question is whether the deterministic and nondeterministic linear bounded automata of John Myhill are equivalent. J. HARTMANIS and H. HUNT wrote a nice survey of this problem (1974), "The LBA problem and its importance in the theory of computing."

Nowadays one of the most interesting questions in computing theory (for reasons I cannot explain here) is whether P=NP. For a thorough treatment of this problem, see HOPCROFT and ULLMAN (1979).

4. Modern themes

Finite automata are central not only to the subjects of automata theory and formal languages, but they are such a fundamental descriptive tool that they appear in every branch of computing theory. Two of the most important areas of modern computing theory are, broadly speaking, the *theory* of *computational complexity*, dealing with the efficiency of algorithms, and the *theory* of *programming logics*, dealing with the reliability of programs. Let us examine briefly the appearance of automata in these two areas. In both areas, especially in logic, we will see interesting connections to other aspects of Professor Kleene's work.

4.1. *Computational complexity*

In the mid 1960's computer scientists actively began studying a theory of asymptotic computational complexity. One of the first areas of investigation related computational complexity to Professor Kleene's and P. Axt's work on subrecursive degrees. The primitive recursive functions were given very interesting interpretations in terms of computational complexity hierarchies (see the textbook of BRAINERD and LANDWEBER (1974) for results).

Definition: Given a Turing machine M the time complexity of M on input $x, T_M(x)$, is the number of steps taken by M. The tape complexity, $\mathrm{TAPE}_M(x)$ is the number of tape cells used if M halts (otherwise it is undefined). The class of languages accepted in time t is called *t-bounded time complexity* class, TIME(t). The class of languages accepted in space t is denoted SPACE(t) and is called the t-space bounded complexity class. If nondeterministic machines are used, the classes are denoted NTIME(t), NSPACE(t). The class $\bigcup_{i=0}^{\infty}\mathrm{NTIME}(n^i)$ is denoted NP and the class $\bigcup_{i=0}^{\infty}\mathrm{TIME}(n^i)$ is denoted P.

In the early days of complexity theory, one of the major concerns was to find "natural" examples of problems which required very large computing resources, e.g. requiring exponential time. It was therefore remarkable when ALBERT MEYER (1975) discovered that RABIN's decision procedure (1969) for SnS was nonelementary and even more remarkable when MEYER and STOCKMEYER (1972) discovered that the equivalence problem for extended regular expressions, those involving $\cup, \cdot, *, -$, though known to be decidable, was also nonelementary. Moreover, they discovered good evidence that even the equivalence problem for $\cup, \cdot, *$ would be very hard; and furthermore, the equivalence problem for $\cup, \cdot$ alone would be hard in the sense that finding a polynomial time algorithm would be a major breakthrough in the theory of computing. Let us examine these last two results and discuss some of the others.

It was well-known that for reasonable recursive complexity bounds, a slight increase in complexity would permit the computation of new functions (see HARTMANIS and STEARNS (1965)). Also given any recursive complexity bound, there were problems requiring more resources than this limit. Building on these results, MEYER and STOCKMEYER proved (1972):

Theorem. *If there is a (deterministic) polynomial time (in the length) algorithm to decide equivalence of regular expressions in* $\cup, \cdot$, *then* P = NP, *i.e. any polynomial time bounded nondeterministic Turing machine can be replaced by a deterministic polynomial time bounded Turing machine.*

Remark. The equivalence of P and NP would have profound consequence for our understanding of computation. The best guess is that P $\neq$ NP.

Proof sketch: Consider the computation of any polynomially time bounded nondeterminisitic *Tm*, M, on input $x \in \Sigma_0^*$ with length $|x| = n$. Let $p(n)$ be the polynomial bound. Look at instantaneous descriptions (id's) of the computation between moves. They represent the finite automaton diagram in Fig. 1. That is, an id is a sequence of letters from the alphabet Σ_0 indicating the contents of the tape, and a symbol q, representing the state, placed in the string at the position immediately to the left of the tape cell which the machine is reading. (The placement of q next to a tape symbol x is written as a single special symbol, $[qx]$.) The length of every id is bounded by $p(n)$. Let us make them all of uniform size $p(n)$ (by so modifying the machine M). Let us also represent the head positions and the state on the tape with the special symbol $[qx]$, $q \in$ States, $x \in \Sigma_0$. Let $\Delta = \Sigma_0 \cup \{[q,x] \mid q \in \text{States}, x \in \Sigma_0\}$. Let us also write the sequence of id's as a single string over $(\Delta \cup \{\#\})$. So a computation has the form

$$\#\text{id}_1\#\text{id}_2\#\cdots\#\text{id}_{p(n)}\#.$$

Let $\Gamma = (\Delta \cup \{\#\})^{p(n)^2}$, then given any string in Γ we can say precisely when it is an *accepting computation* on input x. We can also say when it is neither a valid computation nor an accepting computation. If M accepts x, in $p(n)$ time, then there will be a valid accepting computation (perhaps several). If it does not, then there will be no accepting computation (one ending in a final state).

We will now show that there is a regular expression in $\cup, \cdot$ which describes the strings in Γ which are either not accepting computations or not valid computations at all. Let us agree to call all these strings *invalid computations*. Moreover this expression is of polynomial length in x and can be constructed by a polynomial time algorithm. Therefore, if we can determine in polynomial time whether this expression, call it $R_{(\text{M},x)}$, is equal to Γ, then we could tell in polynomial time whether M accepts x. Namely given M and any x, construct $R_{(\text{M},x)}$ and apply the polynomial time algorithm.

So to prove the theorem, we only need to show how to construct $R_{(M,x)}$. Let us consider a string of Γ, when is it an invalid computation?

E1. If its format is wrong, or
E2. if it starts wrong, or
E3. if it does not have a final state, or
E4. there is an erroneous move.

Let us write a regular expression for each of these and keep track of its size, then $R_{(M,x)}$ will be $E1 \cup E2 \cup E3 \cup E4$. To simplify writing these expressions we let $\Sigma = (\Delta \cup \{\#\})$.

E1 expression The format is wrong if:

(1) the string is too small,
any string of length 1, Σ
any string of length 2, $\Sigma \cdot \Sigma$
any string of length $p(n)-1$, $\underbrace{\Sigma \cdot \cdots \cdot \Sigma}_{p(n)-1}$

This requires at most a $p(n)^2$ length expression.

(2) there is a mistake in the first id representation,
initial # is missing $(\Sigma - \{\#\}) \cdot \Sigma \cdot \cdots \cdot \Sigma$
space between # and # is too small,
space is 1, $\#\Sigma\# \; \Sigma \cdot \cdots \cdot \Sigma$
space is 2, $\#\Sigma \cdot \Sigma \; \#\Sigma \cdot \cdots \cdot \Sigma$
⋮
space is $p(n)-1$ $\# \underbrace{\Sigma \cdot \cdots \cdot \Sigma}_{[p(n)-1]} \; \# \; \Sigma \cdot \cdots \cdot \Sigma$
space between markers is too large, $\#\Sigma^{p(n)+1} \cdot \Sigma \cdot \cdots \cdot \Sigma$
wrong number of heads appear in the id,
no heads, $\#(\Sigma - [qx]) \cdots (\Sigma - [qx])$
for $q \in$ States, $x \in \Sigma_0$
more than one head,

$$p(n)\left\{\begin{array}{l} \text{one head at 1 the other at 2,} \\ \text{one head at 1 the other at 3,} \\ \quad\vdots \\ \text{one head at 2 the other at 3,} \\ \quad\vdots \end{array}\right.$$

This expression is at most $C \cdot p(n)^3$.

(3) there is a mistake in the second id.

⋮

$(p(n))$ there is a mistake in the last id.

So E1 can be described as a union of these possible errors by an expression which is at most $d \cdot \mathrm{p}(n)^4$.

E2 expression. We need only say that the first id is not $[q_0 x]\Sigma \cdot \cdots \cdot \Sigma$.

E3 expression. We can arrange for the computation to end with the head in the right most cell in the final state. So this expression is easy to write.

E4 expression. In describing the erroneous moves, we can assume that the format is correct. We need only say when one id does not follow from another by a legal move. This can be determined by looking at each cell of one id and comparing it to the corresponding cell, $p(n)+1$ symbols away, in the next id. If the cell in question is not under the read write head, then it should not change. To describe the change resulting from the action of the machine, we need only examine the cell being scanned and its neighbors,

$$x_{i_1}\left[qx_{i_2}\right]x_{i_3}.$$

These three cells determine precisely what must occur at the position $p(n)+1$ symbols away from $[qx_{i_2}]$. For example, if the machine writes a y and enters state q', the symbol will be $[q'y]$. If it writes a y, enters q' and moves right, the symbol will be y and the symbol at $p(n)+2$ will be $[q'x_{i_3}]$.

We can describe these errors as follows:

(1) first transition (id_1 to id_2) is wrong
    first cell does not correspond
        triple of symbols is $\# a_1 a_1$
        triple of symbols is $\# a_1 a_2$
        ⋮
    second cell does not correspond
        ⋮
    ⋮
    $p(n)$th cell does not correspond.
(2) second transition (id_2 to id_3) is wrong
    ⋮
⋮
$(p(n))$th transition is wrong.

Putting all this together we have a $O(p(n)^4)$ length expression to describe $R_{(\mathrm{M},x)}$. This proves the result.

We can now consider the sequence of refinements of this technique to obtain progressively stronger theorems. First let us allow the Kleene star,

*, in the regular expressions. Now we can shorten the description of E1. For example, to describe the fact that some id has two or more read heads in it, we can write $\Sigma^* \# \Delta(Sx\Sigma_0)\Delta^*(Sx\Sigma_0)\Delta^* \#$ rather than enumerating all the combinations of errors. We can simplify all of the descriptions, but in the case of describing transition errors we can do significantly more. In the $\cup, \cdot$ expression, to describe an error we must write a long description of where it starts, e.g.

$$\underbrace{\Sigma \cdot \cdots \cdot \Sigma \#}_{\text{initial piece}} \uparrow \Sigma \cdot \Sigma \cdots \Sigma \uparrow \Sigma \cdots \Sigma$$

This limits us to describing polynomially long computations. But if we have *, we can describe the error as

$$\Sigma^* \# \Delta^* \uparrow \Sigma^{p(n)} \uparrow \Sigma^*.$$

This means we can use the same technique plus the * operator to describe polynomial space bounded computations (which may of course require exponential time). We thus have

Theorem (MEYER and STOCKMEYER, 1972). *The problem of deciding equivalence of regular expressions requires as much time as deciding membership in any polynomial space bounded language.*

Another way to view this was discovered by HARTMANIS and HUNT (1974).

Corollary. *If a deterministic lba can decide whether a regular expression R is not equal to Σ^*, then det and nondet lbas are equivalent.*

Proof: A det lba can write down the regular expression in $\cup$, $\cdot$, * which we described above.

Note: there is a polynomial space algorithm to decide $R \neq \Sigma^*$, but it is not linear.

These results can be dramatically improved in the following way. The size of $R_{(M, x)}$ is now determined by the length of the expression used to separate the read/write heads in two adjacent id's. To write $\Sigma^{p(n)}$ required $p(n)$ steps. However, HUNT showed (1972) that if $\cap$ is added as an operator, then this Σ^n can be described by a log (n) expression. This shows that equivalence for $\cup, \cdot, *, \cap$ is nonpolynomial.

MEYER and STOCKMEYER proved (1972) that by adding the complement operator, $\neg$, very short expressions could be written to describe long strings used to separate the reading heads (these are called *yardsticks*). They showed:

Theorem (MEYER and STOCKMEYER, 1972). *The equivalence problem for extended regular expressions,* $\cup, \cdot, ^*, \neg$, *is nonelementary.*

4.2. *The logic of programming and automata*

We are seeing in the late 1970's a fascinating confluence of streams in Professor Kleene's work. In the mid 1950's he had finished his RAND report and was thinking further about realizability and the applications of recursive function theory to logic. Now 20 years later the techniques of automata theory are being applied to logics of programming in somewhat the same way, i.e. to provide a computational semantics. Let us see how this is developing.

The entire computing community is actively interested in problems of "software reliability." How can software engineers guarantee the performance of their products? How can highly critical programs, as in space craft, nuclear reactors, air traffic control and in banking be made extremely reliable? How can costs of program debugging and repair be lowered in ordinary industrial applications?

When the computational task can be described independently of the program solving it, then the program can be regarded as a constructive "proof" of the assertion that the task is accomplished. This method of achieving reliability was proposed by ALAN TURING (1949) and by GOLDSTINE and VON NEUMANN (1947) for numerical programs. In the numerical case, the statement of the computational task is often a familiar mathematical assertion and the program formalizes a familiar informal algorithm. So the method is little more than adding specific algorithmic content to ordinary theorems. These ideas have been interestingly refined by JOHN MCCARTHY (1963), ROBERT FLOYD (1967) and more recently by C. A. R. HOARE (1969) and E. DIJKSTRA (1976). These computer scientists have extended the methods to nonnumerical problems and have proposed specific ways of encorporating modern algorithmic notation into the proofs (and mathematical assertions into programs).

The need to guarantee that practical programs perform as required has led to a class of logics dealing directly with modern high level programming languages. This focus on programs has also led people to consider the reliability problem as one of "program verification" or "program proving." But in fact, from the logician's point of view, it is a special class of proofs that is being verified, namely proofs that use programs. (It is perhaps because the programs are "used" by the computer and the proofs are not that people focus on the programs.) The term "program proving" seems more acceptable because its ambiguity allows one to interpret it as "giving a proof that involves writing a program." Regardless of what the subject is called, let us pursue it by first considering two of the typical operations of programming languages such as ALGOL 60, ALGOL 68, PL/I, PASCAL, etc., the so-called ALGOL-*like languages*. The meaning of the two commands should be self-explanatory. They manipulate a *state*, which is a map from variables to values.

Informal name	*Algol notation*	*Abbreviation*
while loop	*while b do α od*	$b*\alpha$
conditional	*if b then α else β fi*	$b\rightarrow\alpha,\beta$

The b denotes a *boolean expression*, whose meaning is a map from the state of the computation to *true* or *false*. The α, β denote commands, which map states to states. For example a typical program segment is:

while $x>0$ *do if* even (x) *then* $x:=x/2$ else $x:=(x-1)/2$ *fi od*

where $x:=x/2$ means the value of variable x, as given by the state, is changed to $x/2$ (producing thereby a new state).

The earliest practical work dealing with logics for ALGOL-like languages was by C. A. R. HOARE (1969). He based his logic on the primitive $\{P\}\ \alpha\ \{Q\}$ meaning that if P is true of the state before executing command α, and if α terminates, then Q is true of the state after executing α. Early theoretical work such as ENGELER (1968) produced more systematic logics. My own experiments in mechanically assisted reasoning about programs have led to a very simple logic in the spirit of these systematic theories. The system I will describe, introduced in CONSTABLE (1977) is based on a single new concept beyond the everyday ALGOL-like commands and the predicate calculus. Automata theory plays a basic part in the metamathematics of these logics. I can illustrate its role by sketching a theorem about validity in the "propositional calculus of programs."

The model we have in mind is that commands and assertions refer to the *state of the computation*, s. For the predicate calculus, the state is a mapping from variables to values, i.e. it acts just like the variable assignment component of a first order predicate calculus interpretation. If $\mathfrak{M}$ is a model, then we write $\mathfrak{M}, s \models p$ to mean that assertion p is *satisfied by* $\mathfrak{M}$ and s. Commands map states to states, thus $\mathfrak{M}\ [\alpha]$: states$\rightarrow$states where α is a command. At the propositional level of the logic, the nature of the state is unspecified.

Syntax: The basic idea for joining commands and assertions is that if α is a command and F a formula, then $\alpha;F$ means that command α halts and F is true of the resulting state. Commands are formed using these operations: composition $(\alpha;\beta)$, conditional $(b\rightarrow\alpha,\beta)$ and iteration $(b*\alpha)$. The atomic commands are f_i. Formulas are formed from $\&, \vee, \neg, \Rightarrow$ and composition. The atomic formulas are atomic propositions denoted b, p_i, q_i, r_i and atomic commands.

(1) If α, β are commands and f_i are atomic commands, and b is an atomic proposition, then $(\alpha;\beta)$, $(b*\alpha)$, $(b\rightarrow\alpha,\beta)$ and f_i are commands.

(2) IF F, G are formulas and α_i are atomic commands, then α_i, f_i, $(F\&G)$, $(F\vee G)$, $(F\Rightarrow G)$, $(F;G)$ and $\neg F$ are formulas.

Semantics: Every formula has a dual meaning, both *imperative*, $\mathfrak{M}[F]$: states→states, and *declarative*, $\mathfrak{M}, s\models F$. The imperative meaning of a pure proposition is the identity map, while the declarative meaning of a command is that it terminates on the state.

Here are typical valid formulas:

(1) $\alpha;(p\&q)\Leftrightarrow\alpha;p\&\alpha;q,$

(2) $\neg(\alpha;p)\Rightarrow\neg\alpha\vee\alpha;\neg p,$

(3) $\alpha;p\Rightarrow\alpha,$

(4) $(b*\alpha);p\Rightarrow\neg b\&p\vee b;\alpha;(b*\alpha);p,$

(5) $b*\alpha\Rightarrow(b*\alpha);\neg b,$

(6) $(b\rightarrow\alpha,\beta);p\Rightarrow(\neg b\rightarrow\beta,\alpha);p,$

(7) $b\&b*\alpha;F\Rightarrow\alpha;(b*\alpha);F.$

There is an algorithm to determine whether a formula is valid (satisfied by all models and states). The algorithm constructs a finite automaton $\mathcal{Q}$ based on the formula F. F is valid if and only if $\mathcal{Q}$ does not accept any strings.

To understand the algorithm we must relate programs (compound commands) to finite automata. Consider the program α

$$f_1;(b_1*(b_2\rightarrow f_1,f_2));h$$

The following sequence represents a computation of α in an obvious way:

$$f_1;b_1;\neg b_2;f_2;b_1;b_2;f_1;\neg b_1;h.$$

In the computation the loop body is executed twice. If f_i, b_i are regarded as symbols, then this computation is one string of a formal language. The set of all strings corresponding to computations of α is called the *extended value language* of α, denoted $L(\alpha)$. For those α defined by the commands of the logic, $L(\alpha)$ is easily seen to be regular . (See GRIEBACH (1975) for details.)

Now given an arbitrary formula F, it is valid if and only if $\neg F$ is not satisfiable. That is, F is valid if and only if $\exists s\neg F$ is not satisfiable. The formula F can be written in normal form:

$$(F_{11}\& F_{12}\& \cdots)\vee(F_{21}\& F_{22}\& \cdots)\vee \cdots \vee(F_{n1}\& F_{n2}\& \cdots)$$

where each F_{ij} is either: α, $\neg\alpha$, $\alpha;p$ or $\alpha;\neg p$ for p an atomic proposition, α a program. Let the formulas of the form $\neg\alpha_i$ be denoted by γ_i. Then the form of one of these $F_{i1}\& F_{i2}\& \cdots \& F_{in}$ is

$$(*) \qquad \alpha_1;p_1 \;\&\; \alpha_2;p_2 \;\&\; \cdots \;\&\; \alpha_k;p_k \;\&\; \gamma_1 \;\&\; \cdots \;\&\; \gamma_q$$

(we take the α to be α; *true* so some of these p_j might be identically true; we may need to renumber the α_i to write them in this way).

The question of whether this formula (*) is satisfiable is a question of whether the entire collection of programs behaves in a certain way. In fact, we can consider running the $\alpha_1,\ldots,\alpha_k,\gamma_1,\ldots,\gamma_q$ together (in parallel). This corresponds to forming the product automaton, $\alpha_1\times\alpha_2\times\cdots\times\gamma_q$ where α_i denotes the automaton accepting $L(\alpha_i)$. We examine this machine as it processes inputs of the form $(\langle f_{i_1},\ldots,f_{i_{k+q}}\rangle\langle B_1,\ldots,B_{k+q}\rangle)^*$ where each B_i is $\langle[\neg]b_1,[\neg]b_2,\ldots,[\neg]p_1,[\neg]p_2,\ldots\rangle$ where $\neg b_i$, $\neg p_j$ is written if b_i,p_j is false; and where the b_i,p_j are the atomic propositions occurring in the commands and assertions respectively. An input of this form is *consistent* if each predicate b_i,p_j is given the same value on the same sequence of atomic commands. An input of this form represents a set of computations that satisfy (*) if it is consistent, if the ith component of the vectors is the value language of the ith machine, and if the computation sequence does not end so that the γ_j are in a state from which one of them must halt. All of these conditions can be checked by adding a finite number of simple control states to the product automaton $\alpha_1\times\cdots\times\gamma_q$. The formula (*) is satisfiable only if this automaton accepts a string. Indeed the accepted string effectively determines the satisfying interpretation of atomic propositions b_i,p_j.

Since there is an algorithm to determine whether a finite automaton accepts the empty set, there is an algorithm to determine whether the original formula F is satisfiable. A result similar to this related logic was discovered by FISCHER and LADNER (1977).

These results can be considerably strengthened to monadic formulas with quantifiers by looking deeper at the computational models.

Theorem (CONSTABLE, 1977). *There is an algorithm to decide validity in the monadic logic of regular programs.*

To classify programming logics, it is useful to distinguish between the sublanguage of commands (i.e. the *programming language*) and the sublanguage of formulas which do not contain commands (the predicate language or *assertion* language). When the assertion language of the monadic programming logic is extended to include equality, it is easy to see that the decision problem is undecidable because arithmetic truth is definable. When the command language of the monadic programming logic is extended to include recursive procedures (e.g. definitions of the form $f := b \rightarrow \alpha, \beta$ where α or β may contain f), then the decision problem is again undecidable. In this case the undecidability can be proved by looking at pushdown automata. Essentially monadic recursive definitions correspond to deterministic pushdown automata. For these devices the containment problem is undecidable; this implies that the validity problem in the monadic recursive programming logic is undecidable (see CONSTABLE (1977).

Acknowledgements

I would like to thank several people for generously providing me with background information on the history of automata theory, in particular B. Howard, S. C. Kleene, J. Myhill, A. Nerode, M. Rabin, D. Scott, J. C. Shepherdson. I also want to thank my computer science colleagues John Hopcroft and Richard Ladner for their advice in preparing this talk.

References

BRAINERD, W. S., and L. LANDWEBER
[1974] *Theory of Computation* (John Wiley, New York).
BUCHI, J. R.
[1962] Turing machines and the Entscheidungs problem, *Math. Ann*, **148**, 201-213.
CONSTABLE, R. L.
[1977] On the theory of programming logics, in: *Proc. 9th Annual ACM Symposium on Theory of Computing*, Boulder, CO, May 1977, pp. 269-285.
CONSTABLE, R. L., and M. J. O'DONNELL
[1978] *A Programming Logic* (Winthrop Publishers, Cambridge).
DIJKSTRA, E. W.
[1976] *A Discipline of Programming* (Prentice-Hall, Englewood Cliffs, NJ).
EILENBERG, S.
[1974] *Automata, Languages and Machines*, Vol. A (Academic Press, New York).
ENGELER, E.
[1968] *Formal Languages: Automata and Structures*, (Markham, Chicago, IL).

FISHER, M. J., and R. LADNER

[1974] Propositional model logic of programs, in: *Proc. Ninth Annual ACM Symp. on Theory of Computing*, Boulder, CO, May 1977, pp. 286-294.

FLOYD, R. W.

[1967] Assigning meaning to programs, in: *Proceedings of Symposia in Applied Mathematics*, **19**, (Amer. Math. Soc., Providence, RI), pp.19-32.

GOLDSTINE, H. H., and J. VON NEUMANN,

[1963] Planning and coding of problems for an electronic computing instrument, in: *John von Neumann Collected Works*, *Part* 2, *Vol.* 1-3, edited by A. H. Taub (Pergamon, New York) (original published in 1947).

GREIBACH, S.

[1975] *Theory of Program Structures: Schemes, Semantics, Verification, Lecture Notes in Computer Science*, **36** (Springer–Verlag, New York).

HARTMANIS, J., and H. B. HUNT

[1974] The LBA problem and its importance in the theory of computing, in: *SIAM-AMS Proceedings*, Vol.7, (Am. Math. Soc., Providence, RI).

HARTMANIS, J., and R. E. STEARNS

[1961] On the computational complexity of algorithms, *Trans. Am. Math. Soc.*, **117**, 285-306.

HOARE, C. A. R.

[1969] An axiomatic basis for computer programming, *Comm. ACM,* **12**, 576-581.

HOPCROFT, J. E.

[1971] An $n \log n$ algorithm for minimizing states in a finite automaton, in: *Theory of Machines and Computations*, (Academic press, New York) pp. 189-196.

HOPCROFT, J. E., and D. ULLMAN

[1969] *Formal Languages and Their Relation to Automata* (Addison-Wesley, Reading, MA)

[1979] *Introduction to Automata Theory, Languages, and Computation* (Addison-Wesley, Reading, MA).

HUFFMAN, D. A.

[1954] The synthesis of sequential switching circuits, *J. Franklin Inst.*, **257** (3), 161-190 and (u), 275-303 (also see E. F. MOORE, *Sequential Machines: Selected Papers)*

HUNT, H. B.

[1974] On the time and tape complexity of languages, in: *Proc. 5th Annual ACM Symp. on Theory of Computing* (ACM, New York) pp. 64-74.

KLEENE, S. C.

[1951] Representation of events in nerve nets and finite automata, *Rand Research Memorandum*, RM-704.

[1956] Representation of events in nerve nets and finite automata, in: *Automata Studies* (Princeton University Press, Princeton, NJ) pp. 129-156.

MCCARTHY, J.

[1963] A basis for mathematical theory of computation, in: *Computer Programming and Formal System*, edited by P. Braffort and D. Hirschberg (North-Holland, Amsterdam) pp. 33-70.

MCCULLOCK, W. S., and W. PITTS

[1943] A logical calculus of ideas immanent in nervous activity, *Bull. Math. Biophys.*, **5**, pp. 115-133.

MEYER, A. R.

[1975] Weak monadic second-order theory of successor is not elementary-recursive, in: *Lecture Notes in Mathematics*, **453**, (Springer-Verlag, Berlin) pp. 132-154.

MEYER, A. R., and M. J. FISCHER

[1971] Economy of description by automata, grammars and formal systems, in: *Conf. Record of 12th Annual Symposium on Switching and Automata Theory*, (IEEE, New York) pp. 188-191.

MEYER, A. R., and L. STOCKMEYER

[1972] The equivalence problem for regular expressions with squaring requires exponential space, in: *Conf. Records, IEEE 13th Annual Symposium on Switching and Automata Theory* (IEEE, New York) pp. 125-129.

MOORE, E. F.

[1956] Gedanken-experiments in sequential machines, in: *Automata Studies*, (Princeton University Press, Princeton, NJ) pp. 129-156.

[1964] *Sequential Machines: Selected Papers* (Addison-Wesley, Reading, MA).

MYHILL, J.

[1957] Finite automata and the representation of events, Section 5 of Fundamental Concepts in the Theory of Systems, WADC Technical Report 57-624, ASTIA Document No. AD 155741, U. S. Air Force, Wright-Patterson Air Force Base, OH, May 1957.

NERODE, A.

[1958] Linear automaton transformation, *Proc. Am. Math. Soc.*, **9**, 541-544.

RABIN, M. O.

[1967] Mathematical theory of automata, in: *Proceedings of Symposia in Applied Mathematics*, **19**, (Am. Math. Soc., Providence, RI) pp. 153-175.

[1969] Decidability of second order theories and automata on infinite trees, *Trans. Am. Math. Soc.*, **141**, 1-35.

RABIN, M. O., and D. SCOTT

[1979] Finite Automata and their Decision Problems, *I. B. M. J. Res.*, **3** (2), 115-124 (also in: *Sequential Machines: Selected papers*, edited by E. F. Moore (Adison-Wesley, Reading, MA) pp. 63-91.)

TURING, A. M.

[1949] On checking a large routine, in: *Report of a Conference on High-Speed Automatic Calculating Machines*, University Mathematics Laboratory, Cambridge, pp. 67-69.

VON NEUMANN, J.

[1948] The general and logical theory of automata Hixon Symposium, September 1948, in: *Collected Works*, **5**, edited by A. H. Taub (Macmillan, New York) pp. 288-328.

This page intentionally left blank

J. Barwise, H. J. Keisler and K. Kunen, eds., *The Kleene Symposium*

Some Philosophical Aspects of Combinatory Logic

Haskell B. Curry
The Pennsylvania State University (retired)

Dedicated to Professor S. C. Kleene on the occasion of his 70th birthday

Abstract: This is a discussion of some philosophical criticisms of combinatory logic, preceded by a brief survey to give background.

0. Introduction

Combinatory Logic is a branch of mathematical logic which is concerned with the ultimate foundations. It is not an independent system of logic, competing with the theory of types, abstract set theory, mereology, or what not; nor does it attempt to form a consistent system adequate for this or that portion of classical mathematics. Rather it forms a common substratum for a variety of such theories. This paper will discuss some philosophical questions which have been raised concerning it; but first I shall survey its principal features in order to give a background for the discussion.[1]

1. Origins

It began, so far as I am concerned, with an attempt to analyze the process of substitution. In practically all logical systems before 1930—and many since—substitution is taken as a completely unanalyzed operation. As such it is a highly complex process—vastly more so than such processes as modus ponens.[2] In order to express it in terms of processes comparable in complexity to modus ponens, a new approach is necessary.

[1]Revised from a draft of a lecture given March 29, 1978 at the Florida State University at the invitation (originally) of the Philosophy Department. The material on recursive arithmetic has been added since. Use has also been made of outlines of similar, but more mathematical, lectures given before the Logic Seminar at the Pennsylvania State University on April 11, 1978; at the University of Porto, Portugal, on May 17, 1978; at the Technical University of Berlin, Germany, on May 24, 1978; also of suggestions made in correspondence with J. R. Hindley and J. P. Seldin.

[2]For some illustrations of this complexity see Curry and Feys (1958, pp. 2–3); one can also see it if one considers programming substitution on a computer.

The first steps in this direction were made by a Russian named Moses Schönfinkel. He gave a lecture in 1920 to the Mathematische Gesellschaft in Göttingen; this lecture was later written up and published by Heinrich Behmann as SCHÖNFINKEL (1924). In it Schönfinkel introduced a binary operation since called *application*. He wrote the application of an object f to an object a as (fa). This has the following interpretation. If f is a one-place function, (fa) represents the value which f assigns to the argument a. If f is an n-place function, with $n>1$, then (fa) represents the $n-1$-place function formed by giving the first argument the value a, the result being a function of the remaining arguments. If f is not a function, then (fa) can be regarded as a new object not otherwise defined. In further constructions by this operation, omission of parentheses was allowed according to the rule of association to the left; so that $fabc$ is the same as $(((fa)b)c)$. Thus what is ordinarily written $f(a_1, a_2, \ldots, a_n)$ becomes $fa_1a_2 \cdots a_n$. Some contemporary logicians call this way of looking at a function "currying", because I made extensive use of it; but Schönfinkel had the idea some 6 years before I did.

2. The system H

Now suppose we have a formal system H concerning certain objects which I shall call *obs* (some persons prefer to call them terms). The obs are generated from certain primitive ones, the *atoms*, by the operation of application. The (elementary[3]) statements of the system are formed by the binary predicate (statement function)$=$. This has the usual properties[4], as follows: for arbitrary obs X. Y, Z,

$(\rho)\ X=X$	(reflexiveness),
$(\sigma)\ X=Y \rightarrow Y=X$	(symmetry),
$(\tau)\ X=Y \,\&\, Y=Z \rightarrow X=Z$	(transitivity)
$(\mu)\ X=Y \rightarrow ZX=ZY$	(left monotony),
$(\nu)\ X=Y \rightarrow XZ=YZ$	(right monotony).

[3]The term "elementary" comes from my previous writings on formal systems (see CURRY (1963, Chapter 2) and works cited there); for a comparison with more orthodox notions see CURRY ET AL. (1972, Section 11A1). All that it is necessary to understand here is that further statements concerning the formal system can be formed from the elementary ones by the connectives of the U-language. Statements of derivability, consistency, etc. are examples.

[4]Of these $(\sigma), (\tau), (\mu)$ are primitive rules of the system; (ρ) is redundant in view of (1) below; and (ν) can be derived as shown in CURRY and FEYS (1958, Section 6B1).

Suppose the system H contains three atoms I[5], K, S, and that the system can be extended by adjoining additional atoms which, when no properties are postulated for them (other than being obs), are called *indeterminates*. For these I shall use letters such as "x", "y", "z". Obs formed by application from I, K, S alone are called *combinators*; those involving additional atoms are called simply *combinations* or combinatory obs. Concerning I, K, and S we postulate the axiom schemes[6]

$$\mathsf{I}x = x, \qquad \mathsf{K}xy = x, \qquad \mathsf{S}xyz = xz(yz), \tag{1}$$

where x, y, and z are arbitrary combinations. The combinators I, K, and S were introduced by Schönfinkel; but he used a different notation and terminology.

Now suppose Y is a combination of an indeterminate x and certain other atoms. Let us call $[x]Y$ a combination of combinators and these other atoms such that we have $Xx = Y$. Such an X can be defined in various ways; for example, if Y is ax, X can be $a, \mathsf{I}a, \mathsf{S}(\mathsf{KI})a, \mathsf{S}(\mathsf{K}a)\mathsf{I}$, and various others. We can, however, settle on a standard way of defining such an X. Consider the cases given by the lines in Table 1, in which U and V denote combinations which do not contain x.

Table 1. Algorithm for $X \equiv [x]Y$

	Y	X
(a)	V	$\mathsf{K}V$
(b)	x	I
(c)	Ux	U
(f)	$Y_1 Y_2$	$\mathsf{S}([x]Y_1)([x]Y_2)$

These cases overlap, and we get different versions of $[x]Y$ by taking them in different orders, understanding, as is usual with algorithms, that in case of conflict the clause which appears first in the ordering is to be followed. The simplest X's seem to come by taking all four of them in the order as listed. The $[x]Y$ so defined can then be regarded as the function which Y is of x. For functions of n variables we can use an induction on n as follows:

$$[x_1, x_2, \ldots, x_n, x_{n+1}]Y \equiv [x_1, x_2, \ldots, x_n][x_{n+1}]Y,$$

where "$\equiv$" is the symbol for identity by definition. Then if $[M/x]Y$ is the

[5]Actually I can be defined, as Schönfinkel showed, as SKK (or indeed as any S$\mathsf{K}X$); but there are advantages in taking it as primitive (see CURRY ET AL. (1972, p.23, note 2)). The equation $\mathsf{SK}X = \mathsf{I}$ is then a consequence of (2) below.

[6]Since arbitrary combinations can be substituted for any indeterminates throughout a derivation, it suffices to postulate (1) for indeterminates x, y, z.

result of substituting M for x in Y, we have

$$[M/x]Y=([x]Y)M.$$

This is derivable from a recursive definition of the left hand side.

3. Lambda-conversion

Let us now consider a related system, the Church calculus of lambda-conversion. The best account of this system is in CHURCH (1941), but it first appeared in print in his (1932) and (1933). However, I recall seeing in Göttingen in 1928, but not comprehending, an uncirculated manuscript of his containing λ's which was doubtless a progenitor. I shall describe its main features; but there are many fussy details which I shall not go into. It has no atoms except variables, of which there is an infinite sequence; the letters "x","y","z" will denote unspecified ones. It has two operations: application, as in Schönfinkel; and functional abstraction, forming from a variable x and an ob M the obλxM. The single predicate (statement functor) is convertability, which is a form of equality—I shall use "=" for it, and assume that it has the usual properties. These include, besides those stated above for H, with (to agree with the usual notation for λ-conversion) "M,"N","P" for "X","Y","Z", respectively, the rule

$$M=N\rightarrow\lambda xM=\lambda xN. \qquad (\xi)$$

It has the following axiom schemes:

$$\lambda xM=\lambda y[y/x]M \text{ if } y \text{ does not occur in } M, \qquad (\alpha)$$

$$(\lambda xM)N=[N/x]M. \qquad (\beta)$$

Here $[N/x]M$ is, as before, the result of substituting N for x in M; this is defined by recursion. In this system, unlike H, variables can be bound (note that $[x]Y$ is defined as an ob which does not contain x at all), and complex restrictions have to be introduced to avoid collisions of bound variables; these restrictions I am ignoring here. There are various forms of the theory. In Church's original theory λxM was an ob only if M actually contained x. This form has been called λI-conversion; if the restriction is dropped, Church called the theory λK-conversion. Again if V is an ob which does not contain x, one does not, in general, have

$$\lambda x(Vx)=V. \qquad (\eta)$$

If we want this, in order to have a strictly extensional system, one must add it to the assumptions. The theory with this assumption is called λη-conversion; the original theory is called λβ-conversion; either of these can be in either the I or K form.

4. Relations between the systems

Evidently the notions $[x]M$ and λxM have similar meanings. ROSSER's thesis (1935) initiated a study of their interrelations. One cannot simply identify them. In λ-conversion we have the property (ξ); in the system H, although we have (μ) and (ν), we do not have the property analogous to (ξ). One can get this analogous property by adjoining a few particular axioms, of which a simple example is

$$\mathsf{SK} = \mathsf{KI}, \tag{2}$$

changing the algorithm for $[x]Y$ so that U is subject to a special restriction, and splitting (f) into four clauses according to whether Y_1 and Y_2 are or are not of the form Vx.[7] The system with these changes is called H_β, whereas the original system is called H_0. There is also a system H_η in which we have the analogues of both (ξ) and (η). This system is even simpler. It does not require any changes in the algorithm for $[x]Y$, nor any distinction between U and V. Conversely the restrictions of λI-conversion require the omission of the combinator K; we must then postulate two new atomic combinators B and C with the axiom schemes

$$\mathsf{B}xyz = x(yz), \qquad \mathsf{C}xyz = xzy. \tag{3}$$

Appropriate changes must of course be made in the algorithm for $[x]Y$. If K is present, B and C can be defined, thus

$$\mathsf{B} \equiv \mathsf{S(KS)K}, \qquad \mathsf{C} \equiv \mathsf{S(S(KS)(S(KK)S))(KK)}$$

(these definitions can be obtained by the algorithm above mentioned). The upshot of all this is that for each of the stated variations of λ-conversion there is a variant of the system H which is equivalent to it, and vice-versa. There are algorithms for transforming an ob of any system of λ-conversion into an ob of the equivalent system H, and vice versa. In this transformation combinators transform into λ-obs without any free variables, and again vice versa. Thus a system H and its corresponding λ-system are different aspects of essentially the same system. Consequently λ-calculus is an aspect of combinatory logic, and what I have to say may apply to either aspect. When it is desired to emphasize that we are dealing with a system H I shall use the term "*synthetic combinatory logic*". The synthetic aspect gives the more profound foundational analysis; but the λ-aspect is more intuitive, and thus better suited for applications (e.g. to computer programming).

[7] A paper on this topic, tentatively called "Beta strong reduction" is in preparation.

5. The Church–Rosser theorem

Are these systems consistent? Note that, although the symbols are capable of taking meanings of a very general nature, there is nothing akin to a theory of types. It is, therefore, necessary to show that not every statable equation can be proved. This was done in a rather crude fashion in my thesis (1930). But in 1936 Church and Rosser proved a more elegant theorem. If we read (β) as a transformation from left to right only, and omit the rule (σ), we have a reduction, which I shall symbolize by $\geqslant$. Then CHURCH and ROSSER [1936] proved that in λ-conversion.

$$M=N\rightarrow(\exists P)M\geqslant P\&N\geqslant P. \tag{4}$$

Since there are different obs which are irreducible—in *normal form* is the technical term—and these obs are convertible only if they are identical, the theory is consistent (in the Post sense). The theorem has been extended to include other forms of λ-conversion—e.g., the addition of (η)—and, by means of a mapping of reduction onto systems H, to those systems also.

6. Combinatory arithmetic

One of Kleene's early accomplishments was the demonstration that combinatory logic contains a complete recursive arithmetic. The initial steps in this direction were made in CHURCH (1933). If we define the nth iterate f^n of an ob f by

$$f^0\equiv\lambda x(x),\ldots,f^{n+1}\equiv\lambda x(f(f^n x)),$$

then there exist combinators[8] Z_n such that, for any natural number n and any ob $f, \mathsf{Z}_n f=f^n$. Church introduced Z_1, and also a combinator, later shown to be SB, such that $\mathsf{SBZ}_n=\mathsf{Z}_{n+1}$; so that, in effect, he defined the Z_n for $n\geqslant 1$(Z_0 cannot be defined in λI-conversion, although the above equation is true in λK-conversion for $n=0$); thus the Z_n are appropriately called the *Church numerals*. KLEENE, in his thesis (1935), showed that a variety of numerical functions can be represented in the system. The climax of this development came in KLEENE (1936). A numerical function f of k arguments is called ***λ-definable*** just when there is a combinator $[\![f]\!]$ such that whenever

$$f(n_1,n_2,\ldots,n_k)=m, \tag{5}$$

[8]Note that in λ-conversion one can define a combinator as an ob with no free variables; these are then precisely the images, under the above described transformation, of combinators in the corresponding synthetic system.

then

$$[\![f]\!] Z_{n_1} Z_{n_2} \cdots Z_{n_k} = Z_m^{9}, \qquad (6)$$

and if the left side of (5) is undefined, that of (6) has no normal form. Then Kleene's result is that a necessary and sufficient condition that f be λ-*definable* is that it be partial recursive. Thus representability in combinatory logic is a criterion of constructiveness which is equivalent to other well-known ones.

This result has been generalized in various ways. Kleene's result was in λI-conversion; and he used, as already noted, a modified form of (6). The extension to λK-conversion (and to the above form of (6)) is easy. Also, in the synthetic theory, the relation (6) (in either form) is a weak reduction, i.e., it holds in H_0 with (σ) deleted. These generalizations are relatively trivial. A more interesting generalization comes from the existence of numerals other than the Church numerals—such numerals have been proposed by SCOTT (1963), BARENDREGT (1976/77), VAN DER POEL (1975), and others. This has suggested treating the numerals abstractly. i.e., taking 0 (zero), σ (the successor function) as additional atoms together with a new atom Z such that $Zn = Z_n$. For these "*abstract numerals*" the Kleene theory can be carried through; this is shown in CURRY ET AL. (1972, Section 13A), with a hiatus which is filled in CURRY (1975a). It has also been shown (see CURRY (1975b) and papers there cited) that a Markov algorithm over an alphabet whose letters are indeterminates can be represented by a combinator.

7. Illative combinatory logic

So far—except for the abstract arithmetic—we have been considering only what is called *pure combinatory logic*. In order to base logic on this we need to introduce new atoms standing for the usual logical notions, such as implication, quantification, etc. Combinatory logic with the addition of such primitives has been called illative. For this we need to introduce a new predicate (statement functor), a unary one, which I indicate by prefixing the Frege assertion sign "⊢". Thus if X is an ob,

$$\vdash X \qquad (7)$$

is the statement that X belongs to a category of obs called *assertions*. Concerning this new predicate we postulate the rule

RULE Eq. $\quad X = Y \,\&\, \vdash X \rightarrow \vdash Y.$

[9] Actually Kleene, since he has dealing with λI-conversion, and wished to include 0 among the natural numbers, used Z_{j+1} instead of Z_j for $j = n_1, \ldots, n_k, m$.

Let P, ,Π, Ξ, F be new obs, not necessarily all atoms, representing implication, universality, restricted generality (or formal implication), and functionality (i.e., $\mathsf{F}\alpha\beta$ represents being a function from a domain α into a category β), respectively; then these will have the rules

RULE P. $\vdash X \& \vdash \mathsf{P}XY \to \vdash Y$ (modus ponens),

RULE Π. $\vdash \Pi X \to \vdash XY$ for any ob Y,

RULE Ξ. $\vdash \Xi XY \& \vdash XU \to \vdash YU$ for any ob U,

RULE F. $\vdash \mathsf{F}XYZ \& \vdash XU \to \vdash Y(ZU)$.

These are examples of rules in illative combinatory logic. Naturally there may be some axioms and further properties. The term "*illative combinatory logic*" has now been extended to include any theory based on combinators in which there are atoms which are not combinators. We already have such a theory for arithmetic with abstract numerals; the new atoms are 0, σ, and Z. "Illative" in the previous sense is called "properly illative" in CURRY ET AL. (1972, Section 12A)[10].

It is not my purpose to go into illative combinatory logic here, but simply to mention its existence, and to give just enough information about it to give background for the philosophical criticisms which are the main purpose of this paper.

8. The assertion sign

The first criticism relates to the assertion sign. BARENDREGT (1977a), reviewing CURRY ET AL. (1972), has misunderstood this sign; and somewhat similar misunderstandings occur in CHURCH (1957) and ROSSER (1967). Thus Barendregt says: "If X is an ob, then $\vdash X$ is a formula, provable or not, in the object language. But in the expression '$X_1,\ldots,X_n \vdash X$', '$\vdash$' is used in the metalanguage, the meaning being that X is derivable from $X_1,\ldots,X_n$". This is a serious misunderstanding. Indeed, (7) is certainly not a formula of the object language, for the simple reason that there is no object language. Actually there is only one language in the main part of the book; this language is called the U-language. One could of course call that language the metalanguage, in which case

$$X_1, X_2, \ldots, X_n \vdash Y \tag{8}$$

[10]There exist some investigations concerning theories involving both kinds of extension. See for example CURRY ET AL. (1972, Section 13D) and HINDLEY ET AL. (1972, Chapter 11).

would quite properly be an expression of the metalanguage. But this is not done, and hence it is not clear what Barendregt means by "the metalanguage". Concerning the obs, all that was said was that they were objects of some sort. Thus they are essentially blanks into which one can fit any objects one pleases,[11] whether they are expressions of an object language or some other objects. If one chooses these objects in such a way that the postulates of the system become significant statements, one has what I call an *interpretation,*[12] which is valid if those statements are all true; if one leaves the predicates unspecified so that we have a more or less concrete structure isomorphic to the obs, I call it a *representation*. In CURRY (1952) I showed that one can have a representation of quite a general type of formal system in terms of certain (well-formed) expressions of an object language whose alphabet consists of two letters. A way of doing this for combinatory logic is given in CURRY ET AL. (1972, Section 11A2, p.9). Another such representation is one where the object-alphabet has only one letter; since such words are essentially numbers, this is a Gödel representation. No doubt one can find other object languages which are more advantageous. None of these, however, are part of the formulation of combinatory logic; they are options which one can exercise if one chooses. If one does exercise such an option, and calls the language being used the metalanguage, then it is obvious that (7) is not a formula of the object language, but is a sentence in the metalanguage just as (8) is.

Furthermore, one should note that (7) is the special case of (8) in which $n=0$ (it means that $\vdash X$ is derivable without adjoining any extra hypotheses). Furthermore, the symbol "$\vdash$" has meaning only in connection with some formal system. If S is a given system (of assertional type, of course) one might make this explicit by writing (7) as $\mathsf{S}\vdash X$. Let S be such a system, and let $X_1,\ldots,X_n$ be obs of S, then let S′ be the system formed by adjoining $\vdash X_1,\ldots,\vdash X_n$ to the axioms of S. Then (8) means the same as $\mathsf{S}'\vdash X$. That is (8) is, in a sense, a special case of (7).

The fundamental fallacy in BARENDREGT (1977a), and also in ROSSER (1967), is ignoring the distinction between a statement and a name. Thus neither of the expressions (7) or (8) is the name of anything; they are both statements affirming a state of affairs. Thus "$\vdash$" is a verb in the *U*-language, on a par with "=". To be sure it is unusual in many natural languages to use symbols as verbs; thus a German would never say "dass $x=y$," but "dass $x=y$ gilt". The letters "I", "K", "S" are also symbols of the *U*-language; in the representation of CURRY ET AL. (1972, p.9) they are

[11] Provided, of course, that distinctions between different obs are maintained.

[12] Strictly, an interpretation is any method of assigning meanings to the statements of the system. This may be done without assigning fixed objects to the atoms; e.g. interpreting $\vdash X$ in classical propositional calculus as "X is a tautology". See CURRY (1963, Section 2D5).

the names of the object-expressions "$+cc$", "$+c+c+cc$", "$+c+c+c+c+cc$", respectively, and the sentence (2) is the same as

$$\text{"}++c+c+c+c+cc+c+c+cc\text{"}=\text{"}++c+c+cc+cc\text{"}.$$

In the theory of formal systems it can be shown that one can reduce any system to one with a single unary predicate like "⊢". In combinatory logic there are two predicates, ⊢ and =. One can eliminate equality by introducing a new atom Q, and defining $X=Y$ as $\vdash \mathsf{Q}XY$. In that case there is another source of confusion, in that such terms as "axiom" have a double meaning. Thus one can think of a system as having the axioms $\vdash X_1,\ldots,\vdash X_n$ or as having the axioms $X_1,\ldots,X_n$. A proof, in general, is a series of statements; but with assertion as the only predicate, one can think of a sequence of the obs asserted in those statements as constituting a proof. Indeed it is pedantic to insist on always expressing the "⊢". But a certain amount of care is necessary, and this may be partly responsible for the confusions just mentioned.

9. Types

A second point is that pure combinatory logic is a type-free structure, and some persons object to this on the ground that it involves a generality which transcends the theory of types. Indeed a student of mine once suggested that it was improper to assume the eqs. (1) for absolutely unrestricted x, y, and z. Rather the x, y, and z must be restricted to range over certain types, and one must postulate separate combinators for each possible choice of these types. Criticisms of this kind have been made by Dana Scott. Thus in his paper (1969), he says that without the theory of types there is no way to make sense of the foundations of mathematics, and a type-free theory makes no sense whatever. Now, although he has repudiated this paper, saying that it no longer represents his current views, yet the criticism has appealed to other persons, and deserves to be answered.

One answer is that the consistency of pure combinatory logic is guaranteed by the Church—Rosser theorem. Type theory was invented by Russell to explain the paradoxes; since inconsistencies (in the Post sense) cannot arise, types are unnecessary. However, pure combinatory logic is not a closed system. It is intended to be a basis to which we can adjoin the entities of ordinary logic; and in illative combinatory logic we can indeed have contradictions. In fact, let P be implication, and suppose that Rule P, i.e. modus ponens, holds. Suppose further that we have, for arbitrary obs X and Z,

$$\vdash \mathsf{P}(\mathsf{P}X(\mathsf{P}XZ))(\mathsf{P}XZ). \tag{9}$$

In more ordinary notation this is

$$\vdash (X \supset (X \supset Z)) \supset (X \supset Z),$$

which is a thesis of propositional algebra, even the intuitionistic one. Let Y be the combinator

$$[x]UU, \quad \text{where } U \equiv [y]x(yy).$$

Let Z be an arbitrary ob, and let N be $[x](\mathsf{P}x(\mathsf{P}xZ))$. Let X be $\mathsf{Y}N$. Then we can argue as follows:

$$\vdash \mathsf{P}(\mathsf{P}X(\mathsf{P}XZ))(\mathsf{P}XZ) \quad \text{by}\,(9),$$

But $\mathsf{P}X(\mathsf{P}XZ) = NX = N(\mathsf{Y}N) = \mathsf{Y}N = X$. Hence

$\vdash \mathsf{P}X(\mathsf{P}XZ)$	by Rule Eq.
$\vdash X$	by Rule Eq.
$\vdash \mathsf{P}XZ$	by Rule P.
$\vdash Z$	by Rule P.

Thus the above theory is inconsistent since every ob is an assertion.

This form of the Russell paradox (with N taken as negation) shows that we cannot have (9) for all obs. We must formulate, somehow or other, the category of propositions, and restrict the variables in (9) to that category. Now, as I have said, a theory of types is essentially a device for saying that certain combinations are not propositions. Thus the paradox shows that we must formulate such a category, and in that sense we must have a theory of types of some kind. But it by no means follows that we must have the kind of type theory advanced by Russell. Thus the types might be cumulative (Quine). Furthermore, as Whitehead himself pointed out in a seminar at Harvard in the 1920's, the combinations excluded by the theory of types are by no means absolutely meaningless. There may indeed be some sort of reasoning which can be applied to them. General forms of type theory can be formulated in terms of functionality (F), or, even more generally, in terms of the "generalized functionality" (G) being studied by Seldin.

In his later papers, SCOTT has retreated somewhat, as already mentioned, from the position of his (1969). Thus in two papers in the Rome symposium (BÖHM, 1975), Scott takes a position which is more agreeable to that taken here. The second of these papers, his (1975a), has a title very similar to this one. It is, however, addressed to experts on combinatory logic and λ-calculus, and is concerned with pointing out directions for future growth; whereas here my purpose is to explain for nonspecialists what combinatory logic has already done. Thus there is not as much overlap as one might suppose. In the rest of this paper I shall discuss a few points which seem to me to deserve emphasis.

10. Conceptualization

Scott seems to argue that formalization without "conceptualization" is useless. What he means by this is not entirely clear; but it seems to mean an interpretation in terms of notions of ordinary mathematics. One reason for his change of opinion seems to be that, after some effort, he discovered such interpretations—in terms of lattices, sets of integers, etc. In so doing he added immensely to our knowledge about combinatory logic; but he did not, as he seems to imply, save it from damnation. While it is true that concocting formalisms entirely without regard to interpretation is probably fruitless, yet it is not necessary that there be "conceptualization" in terms of current mathematical intuitions. In fact, mathematical intuition is a result of evolution. Mathematicians depend on their intuitions a great deal; let us hope they always will. But the mathematical intuitions of today are not the same as those of a thousand years ago. Combinatory logic may not have had a conceptualization in what seems to be Scott's sense; but it did have an interpretation by which it was motivated. The formation of functions from other functions by substitution does form a structure, and this structure it analyzed and formalized. For progress we need the freedom to let our intuitions develop further; this includes the possibility of formalizing in new ways.

11. Priority of the function notion

Another point about combinatory logic is the priority of the notion of function. In current mathematics the tendency is to think of a function in the sense of Dirichlet, i.e. essentially as a set of ordered pairs, whereby the first elements range over a set, the domain of the function, and when the first element of a pair is given, the second is uniquely determined. Thus the notion of set is more fundamental than that of function, and the domain of a function is given before the function itself. In combinatory logic, on the other hand, the notion of function is fundamental; a set is a function whose application to an argument may sometimes be an assertion, or have some other property; its members are those arguments for which the application has that property. The function is primary; its domain, which is a set, is another function. Thus it is simpler to define a set in terms of a function than vice versa;[13] but the idea is repugnant to many mathematicians, and probably to Scott. This has been a great handicap and source of misunderstanding.

[13]For a similar idea cf. the set theory of VON NEUMANN (1928).

12. Finiteness of structure

Another point worth mentioning is its essentially finite character. If equality is defined in terms of Q, then pure combinatory logic can be formulated in terms of a finite number of atoms, a finite number of axioms, and a finite number of simple rules. If one has an illative atom, say Π, the reflexive property can be stated as an axiom, $\vdash \Pi(\mathsf{SQI})$; if one has the rules Eq., (μ), and (ν), then the axiom schemes (1) and the properties of equality follow from the following eight rules[14] (here $\leftrightarrows$ means inferability in either sense, so each line counts as two rules):

$$\vdash \mathsf{I}X \leftrightarrows X,$$
$$\vdash X(\mathsf{I}Y) \leftrightarrows XY,$$
$$\vdash X(\mathsf{K}YZ) \leftrightarrows XY,$$
$$\vdash X(\mathsf{S}YZU) \leftrightarrows \vdash X(YU(ZU)).$$

All other systems of logic and mathematics involve either a complex substitution process or axiom schemes involving infinitely many axioms, and, of course, infinitely many atoms. This is true even for propositional algebra. This finiteness of structure presumably holds for various forms of illative combinatory logic. It does not hold, however, even in pure combinatory logic, for theories of reduction; for Hindley has shown (CURRY ET AL., 1972, Section 11E7) that no finite set of axioms, not even a finite set of axiom schemes, will suffice for the form of reduction in synthetic combinatory logic which includes (ξ) and (η).

13. Pertinence to logic

Another criticism by Scott is that the use of the word "logic" in "combinatory logic" is premature. Although this is a question of the usage of terms, and therefore not strictly debatable, yet it may be worthwhile to explain in just what sense the use of the term is justified. In CURRY (1963, p.1), I defined mathematical logic as the study of mathematical structures having some connection with the "analysis and criticism of thought" (W.E. Johnson), and the analogues of such structures. Since combinatory logic analyzes a process used in practically all such structures, it is a part of mathematical logic so defined, just as propositional logic, relational logic, or even Aristotelian logic is. Scott's point, however, is that pure combinatory logic—which, at least the synthetic aspect of it, might perhaps better

[14]Rules of this kind were first suggested by ROSSER (1935).

be called the theory of combinators—excludes the propositional concepts which are central in logic as ordinarily understood. However, combinatory logic is broader than pure combinatory logic; it includes illative theories, some of which do contain those concepts. To be sure, illative combinatory logic is not completely settled. In view of the Gödel theorem, we cannot prove its consistency by methods which it itself formalizes. But it is still a well-defined field for investigation. Moreover significant work has been done in it. Scott himself proposes one system of it in his paper (1975b). It is plausible that some systems proposed by Bunder will prove acceptable; and there are nonconstructive consistency proofs of some systems proposed by Fitch. Indeed, given any of the usual systems of logic, if one formalizes the substitution process by pure combinatory logic, and the theory of types by functionality, one has a system of illative combinatory logic. Although this has not yet been done in most cases, it seems unreasonable to doubt that it is possible. In this field, in which our previously developed intuitions are not a safe guide, we make progress by trial and error.

14. Conclusion

One final point must be mentioned. The history of combinatory logic shows that progress can result from the interaction of different philosophies. One who, as I do, takes an empirical view of mathematics and logic, in the sense that our intuitions are capable of evolution, and who prefers constructive methods, would never discover the models which Scott proposed. On the other hand, it is doubtful if anyone with what seems to be Scott's philosophy, would have discovered combinatory logic. Both of these approaches have added to the depth of our understanding, and their interaction has produced more than either would have alone.

References

Barendregt, H. P.

[1976/77] A global representation of recursive functions in the λ-calculus, *Theoret. Comput. Sci.*, **3** (2), 225–242. [Also appeared (1963) in Locos no.5, an informal seminar publication of the University of Utrecht.]

[1977a] Review of Curry et al. (1972), *J. Symbolic Logic*, **42** (1), 109–110.

[1977b] The type free lambda calculus, in: *Handbook of Mathematical Logic* edited by J. Barwise (North-Holland, Amsterdam) pp. 1091–1132. (An exposition of some recent technical developments in the pure theory.)

BARWISE, J. (editor)

[1977] *Handbook of Mathematical Logic* (North-Holland, Amsterdam).

BÖHM, C. (editor)

[1975] *Lambda Calculus and Computer Science Theory, Proceedings of the Symposium held in Rome, March 25–29, 1975, Lecture Notes in Computer Science*, **37** (Springer, Berlin).

CHURCH, A.

[1932] A set of postulates for the foundation of logic, *Ann. of Math.* (2), **33** (2), 346–366. (See comment on his (1933).)

[1933] A set of postulates for the foundation of logic (second paper), *Ann. of Math.* (2), **34** (4), 839–864. (Continuation of (1932). Although the system in these two papers was later proved inconsistent, it is the original source of λ-conversion, and of the Church numerals. The papers were written in a notation obsolete since 1940.)

[1941] *The Calculi of Lambda Conversion* (Princeton University Press, 2nd edn, 1951, reprinted 1963 by University Microfilms, Ann Arbor, MI).

[1957] Review of CURRY (1939) (actually of 1954 reprint), *J. Symbolic Logic*, **22** (1), 85–86.

CHURCH, A. and J. B. ROSSER

[1936] Some properties of conversion, *Trans. Am. Math. Soc.*, **39**(3), 472–482.

CURRY, H. B.

[1930] Grundlagen der kombinatorischen Logik, *Am. J. Math*, **52**(3), 509–536, and (4), 789–834. (Source of the term "combinatory logic"; now largely superseded.)

[1939] Remarks on the definition and nature of mathematics, *J. Unified Science (Erkenntnis)*, **9**, 164–169. (Paper presented to the 5th International Congress of Unified Science, Harvard University, Sept. 5, 1939. An abstract of CURRY (1951). Except for preprints distributed at the congress all copies were destroyed in the bombing of Rotterdam. Reprinted without some corrections, 1954, in: *Dialectica* **8**(3), 228–233.)

[1951] *Outlines of a Formalist Philosophy of Mathematics* (North-Holland, Amsterdam).

[1952] *Leçons de logique algébrique* (Gauthier-Villars, Paris and E. Nauwelaerts, Louvain).

[1963] *Foundations of Mathematical Logic* (McGraw-Hill Book Co., New York; reprinted 1977 by Dover Publications, New York).

[1967] Logic, combinatory, in: *Encyclopedia of Philosophy*, **4**

(Macmillian Co. and The Free Press, New York) pp. 504–509.
[1968a] Combinatory logic, in: *Contemporary Philosophy, a Survey* edited by R. Klibansky (La Nuova Italia, Editrice, Firenze). Vol. 8, pp. 295–307. (A historical survey.)
[1968b] Recent advances in combinatory logic, *Bull. Soc. Math. Belg.* **20** (3), 233–298. (A survey for mathematicians somewhat more detailed than [1968a].)
[1975a] A study of generalized standardization in combinatory logic, in *ISILC Proof Theory Symposium* edited by J. Diller and G. H. Müller. *Lecture Notes in Mathematics,* **500** (Springer Verlag, Berlin) pp. 44–55.
[1975b] Representation of Markov algorithms by combinators, in: *The Logical Enterprise* edited by A. R. Anderson, R. B. Marcus and R. M. Martin (Yale University Press) pp. 109–119, 250–251.

CURRY, H. B. and R. FEYS
[1958] *Combinatory Logic, Vol. I* (for Vol. II see CURRY EL AL. (1972)) (North-Holland, Amsterdam).

CURRY, H. B., J. R. HINDLEY and J. P. SELDIN
[1972] Combinatory Logic, Vol. II (for Vol. I see CURRY and FEYS (1958)) (North-Holland, Amsterdam). (A research monograph, in places difficult reading. The introductory portions of both volumes are readable, particularly Sections A and B of Chapter 11, which bear on the subject of this paper.)

HINDLEY, J. R., B. LERCHER and J. P. SELDIN
[1972] *Introduction to Combinatory Logic* (Cambridge University Press, Cambridge.) (Recommended as general introduction which goes fairly deep; primarily mathematical.)

KLEENE, S. C.
[1935] A theory of positive integers in formal logic, *Am. J. Math.*, **57**(1), 153–173, and (2), 219–244.
[1936] λ-definability and recursiveness, *Duke Math. J.*, **2**(2), 340–353.
[1952] *Introduction to Metamathematics* (North-Holland, Amsterdam and P. Noordhof, Groningen)

ROSSER, J.B.
[1935] A mathematical logic without variables, *Ann. of Math.* (2), **36**, 127–150 and *Duke Math. J.*, **1**, 328–355.
[1955] Deux esquisses de logique (Gauthier-Villars, Paris and E. Nauwelaerts, Louvain).
[1967] Review of CURRY and FEYS (1958), *J. Symbolic Logic*, 32(2), 267f.

SCHÖNFINKEL, M.

[1924] Uber die Bausteine der mathematischen Logik, *Math. Annalen*, **92**, 305–316. (An account by Heinrich Behmann of a lecture delivered in 1920. The appendix by Behmann contains an error.)

SCOTT, D. S.

[1963] A system of functional abstraction, lectures delivered at the University of California, Berkeley, CA, 1962–63. (Photocopy of hand-written manuscript, issued by Stanford University, Sept. 1963, received from author in 1968.)

[1969] A type theoretical alternative to CUCH, ISWIM, OWHY, unpublished memorandum, privately circulated at Oxford in the early fall of 1969.

[1975a] Some philosophical issues concerning theories of combinators, in: *Lambda Calculus and Computer Science Theory* edited by C. Böhm, *Proceedings of the Symposium held in Rome, March 25–26, 1975, Lecture Notes in Computer Science*, **37** (Springer, Berlin) pp. 346–370.

[1975b] Combinators and classes, in: *Lambda Calculus and Computer Science Theory* edited by C. Böhm, *Proceedings of the Symposium held in Rome, March 25–29, 1975, Lecture Notes in Computer Science*, **37** (Springer, Berlin) pp. 1–26.

VAN DER POEL, W. L.

[1975] Combinatoren en lambdavormen, *Nederl. Akad. Wetensch. Verslag, Afd. Natuurk.*, 84(10), 172–179.

VON NEUMANN, J.

[1928] Die Axiomatisierung der Mengenlehre., *Math. Z.* 27, 669-752.

This page intentionally left blank

J. Barwise, H. J. Keisler and K. Kunen, eds., *The Kleene Symposium*
©North-Holland Publishing Company (1980)103–111

A Class of Intuitionistic Connectives

Dick H. J. de Jongh
Department of Mathematics, University of Amsterdam, Amsterdam. The Netherlands

Dedicated to Professor S. C. Kleene on the occasion of his 70th birthday

Abstract: It is shown that in intuitionistic logic unrestricted use of infinite disjunction will even with two propositional variables give rise to a proper class of equivalence-classes of formulae.

1. Introduction

One way to obtain new connectives for the intuitionistic propositional calculus is to use infinite disjunction $\bigvee$ (or conjunction $\bigwedge$) to form new formulae in a restricted number of variables and then showing the non-equivalence of some of the new formulae to any standard finite formula (cf. KREISEL (1979, Part III a)). In the case of one propositional variable this leads (at least in a classical metamathematical context) to exactly one new equivalence class of formulae (DE JONGH, 1968; GOAD, 1978). We will show here that with two propositional variables unrestricted use of infinite disjunction will lead to a proper class of mutally non-equivalent "formulae". The method for showing non-equivalence will be the method of Kripke-models used in a classical way. The formation of new formulae is approached classically too.

The result will be obtained by constructing a proper class of indices and assigning to each index a Kripke-model and a formula in such a way that no two formulae adhering to different indices will be equivalent (on the associated Kripke-models). There is also an algebraic interpretation of the result: there is no complete completely free pseudo-Boolean algebra on two generators, or to put it in a different way: there are complete pseudo-Boolean algebras generated by two elements larger than any given cardinality. This corresponds to the analogous result for Boolean algebras on $\aleph_0$ generators (HALES, 1964; GAIFMAN, 1964, improved by SOLOVAY, 1966 and KRIPKE, 1967).

2. Finite formulae and Kripke-models

As a starting point for the construction mentioned in the introduction we will take a countably infinite set of standard formulae (i.e. formulae built from the propositional variables p and q by repeated application of $\wedge, \vee, \neg$ and $\rightarrow$) none of which implies any of the others in the intuitionistic propositional calculus. Such a set was constructed in JANKOV (1968) and DE JONGH (1968). We will give in this section a construction which generalizes directly to the main result as given in the next section.

Since we will be concerned with formulae in two propositional variables (p,q) only, it will be convenient to use Kripke-models adapted to this purpose.

Definition. 2.1. A *Kripke-model* $\mathbf{K}$ is a function with as its domain a partially ordered set $\langle K, \leqslant_K \rangle$ with a least element and as its range the set $\{(1,1),(1,0),(0,1),(0,0)\}$, with the property that, if $k \leqslant_K k'$, then $\mathbf{K}(k) \leqslant \mathbf{K}(k')$, where $(0,0) \leqslant (0,1) \leqslant (1,1)$ and $(0,0) \leqslant (1,0) \leqslant (1,1)$.

We will usually write $\leqslant$ for $\leqslant_K$ and 00, 01, 10, 11 for $(0,0),(1,0),(0,1),(1,1)$.

Associated with K is a forcing-relation $\Vdash_{\mathbf{K}}$ (usually abbreviated as $\Vdash$) between nodes of K and formulae. It is defined by induction on the length of formulae as follows (writing $\nVdash$ for "it is not the case that $\Vdash$"):

Definition 2.2.

$$k \Vdash p \quad \Leftrightarrow K(k) = 11 \text{ or } K(k) = 10,$$
$$k \Vdash q \quad \Leftrightarrow K(k) = 11 \text{ or } K(k) = 01,$$
$$k \Vdash \phi \wedge \psi \Leftrightarrow k \Vdash \phi \text{ and } k \Vdash \psi,$$
$$k \Vdash \phi \vee \psi \Leftrightarrow k \Vdash \phi \text{ or } k \Vdash \psi,$$
$$k \Vdash \phi \rightarrow \psi \Leftrightarrow \forall k' (k \leqslant k' \Rightarrow (k' \nVdash \phi \text{ or } k' \Vdash \psi)),$$
$$k \Vdash \neg \phi \Leftrightarrow \forall k' (k \leqslant k' \Rightarrow k' \nVdash \phi).$$

Definition 2.3. A Kripke-model L is a *restriction of* a Kripke-model $\mathbf{K}$ ($\mathbf{L} \equiv \mathbf{K}$), if, for some $k \in K, L = \{k' \in K \mid k \leqslant k'\}$, $\leqslant_L = \leqslant_K \upharpoonright L$ and $\mathbf{L} = \mathbf{K} \upharpoonright L$; more specifically $\mathbf{L}$ is then called the *restriction of* $\mathbf{K}$ to $\{k' \in K \mid k \leqslant k'\}$.

We write $\mathbf{K} \Vdash \phi$, if for each $k \in K, k \Vdash \phi$ ($\Leftrightarrow$ for the least element k_0 of $K, k_0 \Vdash \phi$). It is now clear that, if $\mathbf{K} \Vdash \phi$ and $\mathbf{L} \equiv \mathbf{K}$, then $\mathbf{L} \Vdash \phi$. Also, that $\mathbf{K} \Vdash \phi \rightarrow \psi$ iff, for each $\mathbf{L} \equiv \mathbf{K}, \mathbf{L} \Vdash \phi \Rightarrow \mathbf{L} \Vdash \psi$; and that $\mathbf{K} \Vdash \neg \phi$ iff, for each $\mathbf{L} \equiv \mathbf{K}, \mathbf{L} \nVdash \phi$.

Definition 2.4. Two Kripke-models $\mathbf{K}$ and $\mathbf{L}$ are called *compatible*, if, for each $k \in K \cap L$, the restriction of $\mathbf{K}$ to $\{k' \in K | k \leqslant k'\}$ is the same as the restriction of $\mathbf{L}$ to $\{1 \in L | k \leqslant 1\}$.

It is clear that, if $\mathbf{K}$ and $\mathbf{L}$ are compatible and $k \in K \cap L$, then, for any $\phi, k \Vdash_{\mathbf{K}} \phi \Leftrightarrow k \Vdash_{\mathbf{L}} \phi$.

Definition 2.5. Let B be a set of mutally compatible Kripke-models. We associate with B a Kripke-model B' with domain $\cup_{\mathbf{K} \in B} K \cup \{b\}$ for some $b \notin \cup_{\mathbf{K} \in B} K$, the partial ordering being the partial ordering of the elements of B with b added on as a least element, and the values of B' being the values of the $\mathbf{K}$ in B with $B'(b) = 00$ added on. (In the sequel we can always take $b =$ index of B' so that B' will be uniquely determined.)

We now give a countable set I of indices, for each $i \in I$ a finite Kripke-model $\mathbf{K}_i$ and formulas ϕ_i and ϕ_i^* such that, for each $i, j \in I$:

(i) $\quad \mathbf{K}_j \Vdash \phi_i \Leftrightarrow \mathbf{K}_j \equiv \mathbf{K}_i$,

(ii) $\quad \mathbf{K}_j \Vdash \phi_i^* \Leftrightarrow \mathbf{K}_i \not\equiv \mathbf{K}_j$.

In the definition of the indices we will consider 0, 1, 2 to be atoms, not sets. If S is a set HC(S) will be the hereditary closure of S. I and $\mathbf{K}_i$ for $i \in I$ are defined inductively:

Definition 2.6. (a) $\quad 0, 1, 2 \in I$.

(b) If $i_1, \ldots, i_n \in I$ $(n \geqslant 2)$ and, for no $m, m' \leqslant n, i_m = i_{m'}$, or $i_m \in \mathrm{HC}(i_{m'})$, then $\{i_1, \ldots, i_n\} \in I$.

Definition 2.7 (a) $\quad K_m = \{m\}, \mathbf{K}_0(0) = 10, \mathbf{K}_1(1) = 01, \mathbf{K}_2(2) = 00$.

(b) $\quad \mathbf{K}_{\{i_1, \ldots, i_n\}} = \{\mathbf{K}_{i_1}, \ldots, \mathbf{K}_{i_n}\}'$.

A picture for 2.7 (b):

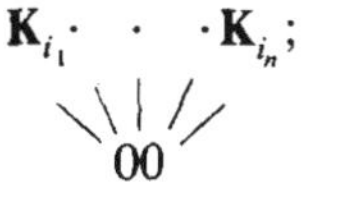

e.g.

$\mathbf{K}_{\{\{0,1\},2\}}$ is

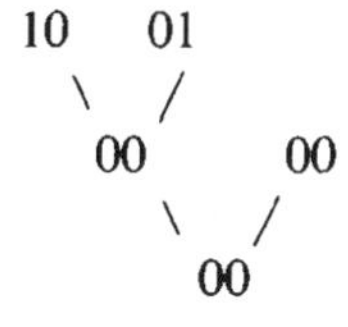

Note that by the construction it is clear that $\mathbf{K}_i \equiv \mathbf{K}_j$ iff $i \in \mathrm{HC}(j)$ or $i=j$ and $\mathbf{K}_i = \mathbf{K}_j$ iff $i=j$. This insures that, for no different $j', j'' \in j, \mathbf{K}_{j'} \equiv \mathbf{K}_{j''}$. It is also clear that the definition is correct in the sense that, if $\{i_1, \ldots, i_n\} \in I$, then $\mathbf{K}_{i_1}, \ldots, \mathbf{K}_{i_n}$ are mutually compatible. Finally, if $i \in I$ and $\mathbf{L} \equiv \mathbf{K}_i$, then $\mathbf{L} = \mathbf{K}_j$ for some $j \in I$.

Definition 2.8 (a) ϕ_0, ϕ_1, ϕ_2 are respectively $p \wedge \neg q, \neg p \wedge q, \neg p \wedge \neg q$, $\phi_0^{\bullet}, \phi_1^{\bullet}, \phi_2^{\bullet}$ respectively $\neg(p \wedge \neg q), \neg(\neg p \wedge q), \neg(\neg p \wedge \neg q)$.

(b) For $i \neq 0,1,2$,

$$\phi_i = \bigvee_{j \in i} \phi_j^{\bullet} \to \bigvee_{j \in i} \phi_j$$

and $\phi_i^{\bullet}$ is

$$\phi_i \to \bigvee_{j \in i} \phi_j$$

where $\bigvee_{j \in i}$ denotes the obvious iterated finite disjunction.

2.9. Proof of (i),(ii), by induction on i. We will leave it to the reader to check $i = 0,1,2$. So, let us assume that, for some $i \in I$, (i) and (ii) hold for each $i' \in i$. We will prove (i) and (ii) for i.

(i)$\Leftarrow$It is sufficient to prove $\mathbf{K}_i \Vdash \phi_i$. So, let $\mathbf{L} \equiv \mathbf{K}_i$,

$$\mathbf{L} \Vdash \bigvee_{i' \in i} \phi_{i'}^{*}.$$

Thus $\mathbf{L} \Vdash \phi_{i'}^{*}$, for some $i' \in i$. Therefore, by induction hypothesis, $\mathbf{K}_{i'} \not\equiv \mathbf{L}$. So, $\mathbf{L} \equiv \mathbf{K}_{i''}$ for some $i'' \in i, i'' \neq i'$. This, by induction hypothesis, implies $\mathbf{L} \Vdash \phi_{i''}$, whence

$$\mathbf{L} \Vdash \bigvee_{i' \in i} \phi_{i'}.$$

So, indeed $\mathbf{K}_i \Vdash \phi_i$.

$\Rightarrow$We will prove this by induction on j. Again we leave the cases $j = 0,1,2$, to the reader. So, we assume $j \neq 0,1,2, \mathbf{K}_j \Vdash \phi_i$. If $j' \in j$, then $\mathbf{K}_{j'} \Vdash \phi_i$. So, by the induction hypothesis on $j, \mathbf{K}_{j'} \equiv \mathbf{K}_i$ for each $j' \in j$. Since, for no two different $j', j'' \in j$, $\mathbf{K}_{j'} \equiv \mathbf{K}_{j''}$ and since $|j| \geqslant 2$, this implies that $\mathbf{K}_{j'} \neq \mathbf{K}_i$ for all $j' \in j$. So, $j' \in \mathrm{HC}(i)$ for each $j' \in j$. Hence, $\mathrm{HC}(j) \subseteq \mathrm{HC}(i)$. If $\mathbf{K}_j \equiv \mathbf{K}_{i'}$ for some $i' \in i$, we are done. So, we assume, for each $i' \in i$, $\mathbf{K}_j \not\equiv \mathbf{K}_{i'}$. It will now be sufficient to prove $\mathbf{K}_j = \mathbf{K}_i$, i.e. $j = i$. From the fact that $\mathbf{K}_j \Vdash \phi_i$ it follows that, if

$$\mathbf{K}_j \Vdash \bigvee_{i' \in i} \phi_{i'}^{*},$$

then

$$\mathbf{K}_j \Vdash \bigvee_{i' \in i} \phi_{i'}.$$

By induction hypothesis on $i, \mathbf{K}_j \nVdash \phi_{i'}$ for any $i' \in i$ and, therefore,

$$\mathbf{K}_j \nVdash \bigvee_{i' \in i} \phi_{i'}.$$

So,

$$\mathbf{K}_j \nVdash \bigvee_{i' \in i} \phi_{i'}^*,$$

whence $\mathbf{K}_j \nVdash \phi_{i'}^*$, for any $i' \in i$. By induction hypothesis on i, it then follows that $\mathbf{K}_{i'} \equiv \mathbf{K}_j$ for each $i' \in i$ which gives us by the same reasoning as above $\mathrm{HC}(i) \subseteq \mathrm{HC}(j)$, whence $\mathrm{HC}(i) = \mathrm{HC}(j)$ from which $i = j$ immediately follows.

(ii)$\Rightarrow$This is immediate from the obvious fact that $\mathbf{K}_i \nVdash \phi_i^*$.

$\Leftarrow$Again the cases $j = 0, 1, 2$ are left to the reader. So, assume $j \neq 0, 1, 2$, $\mathbf{K}_j \Vdash \phi_i^*$. Then, for some j', $\mathbf{K}_{j'} \equiv \mathbf{K}_j$ and $\mathbf{K}_{j'} \Vdash \phi_i$ and $\mathbf{K}_{j'} \nVdash \phi_{i'}$ for any $i' \in i$. By (i), $\mathbf{K}_{j'} \equiv \mathbf{K}_i$ and $\mathbf{K}_{j'} \not\equiv \mathbf{K}_{i'}$ for any $i' \in i$.

So $\mathbf{K}_i = \mathbf{K}_{j'} \equiv \mathbf{K}_j$.

To obtain a countable sequence of formulae none of which implies any of the others we will take a subsequence $\{\mathbf{L}_n\}_{n \in \omega}$ of $\{\mathbf{K}_i\}_{i \in I}$:

Definition 2.10 .

$$\mathbf{L}_0 = \mathbf{K}_{\{0,1,2\}}.$$

$$\mathbf{L}_1 = \mathbf{K}_{\{\{0,1\},\{0,2\},\{1,2\}\}},$$

$$\mathbf{L}_2 = \mathbf{K}_{\{\{\{0,1\},\{0,2\}\},\{\{0,1\},\{1,2\}\},\{\{0,2\},\{1,2\}\}\}}.$$

In general, if $\mathbf{L}_n = \mathbf{K}_i$, then

$$\mathbf{L}_{n+1} = \mathbf{K}_{\{j \subseteq i \mid |j| = 2\}}.$$

For any $n \in \omega$, if $L_n = K_i$, then $|i| = 3$ and, if $j \in \mathrm{HC}(i)$, then $|j| = 2$. From this it is clear that, for no two different $m, n \in \omega$, $\mathbf{L}_m \equiv \mathbf{L}_n$. In pictures:

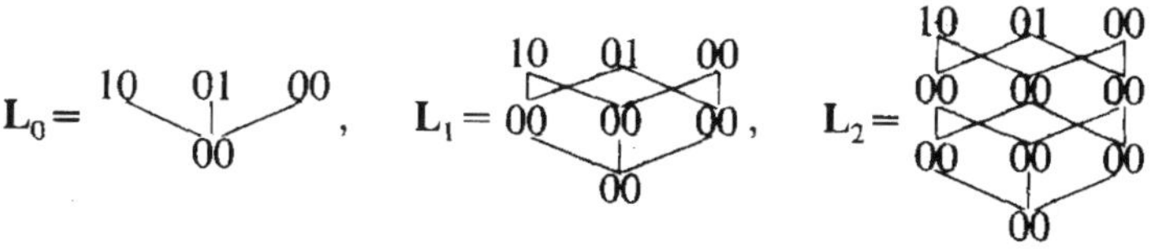

Let, for each $n \in \omega, \psi_n = \phi_i$ and $\psi_n^* = \phi_i^*$ if $\mathbf{L}_n = \mathbf{K}_i$. It is then clear that, for no $m, n \in \omega (m \neq n), \vdash \psi_m \to \psi_n$ in the intuitionistic propositional calculus.

3. Infinite formulae and Kripke-models

Henceforth formulae will be considered as set-theoretic objects with $\wedge, \vee, \rightarrow, \neg, \bigvee$ operations such as to assure unique readibility: for any set X of formulae $\bigvee X$ will be considered to be a formula. If X is indexed by J(i.e. $X=\{\phi_j \mid j\in J\}$), then we will write $\bigvee_{j\in J}\phi_j$ for X. The forcing relations on the Kripke-models will be extended by: $k \Vdash \bigvee X \Leftrightarrow \exists \phi \in X\ (k \Vdash \phi)$.

We will now construct, for each ordinal α, a set of indices I_α and, for each $i\in I_\alpha$, a Kripke-model K_i and formulae ϕ_i and ϕ_i^* such that

(i) $$\forall i,j \in \bigcup_{\alpha} I_\alpha \quad (\mathbf{K}_j \Vdash \phi_i \Leftrightarrow \mathbf{K}_j \equiv \mathbf{K}_i),$$

(ii) $$\forall i,j \in \bigcup_{\alpha} I_\alpha \quad (\mathbf{K}_j \Vdash \phi_i^* \Leftrightarrow \mathbf{K}_i \equiv \mathbf{K}_j),$$

(iii) $$|I_\alpha| = \beth_\alpha.$$

3.1 Definition of I_α, $\mathbf{K}_i$, ϕ_i and ϕ_i^* by induction on α:

0-step: $I_0=\{i\in\omega \mid i>2\}$. For $i\in I_0$, $\mathbf{K}_i=\mathbf{L}_{i-2}$, $\phi_i=\psi_{i-2}$, $\phi_i^*=\psi_{i-2}^*$. (0, 1 and 2 are not used as indices to insure that the sets of indices used in Sections 2 and 3 are disjoint. Again it will be convenient to consider natural numbers as being atoms rather than sets.)

Successor-step:

$$I_{\alpha+1}=\{j\subseteq I_\alpha \mid |j|\geqslant 2\}.$$

For $i\in I_{\alpha+1}$,

$$\mathbf{K}_i=(\{\mathbf{K}_j \mid j\in i\})',$$

$$\phi_i=\bigvee_{j\in i}\phi_j^* \rightarrow \bigvee_{j\in i}\phi_j,$$

$$\phi_i^*=\phi_i \rightarrow \bigvee_{j\in i}\phi_j.$$

Limit-step: Let λ be a limit ordinal: We first define, for each $j\in \bigcup_{\alpha<\lambda} I_\alpha$, $X_j^\lambda=\{j'\in \bigcup_{\alpha<\lambda} I_\alpha \mid j\notin \mathrm{HC}(j')\}$. Then $I_\lambda=\{X_j^\lambda \mid j\in \bigcup_{\alpha<\lambda} I_\alpha\}$ and, for $i\in I_\lambda$,

$$\mathbf{K}_i=(\{\mathbf{K}_j \mid j\in i\})',$$

$$\phi_i=\bigvee_{j\in i}\phi_j^* \rightarrow \bigvee_{j\in i}\phi_j,$$

$$\phi_i^*=\phi_i \rightarrow \bigvee_{j\in i}\phi_j.$$

It is clear that again $\mathbf{K}_j \equiv \mathbf{K}_j$ iff $i \in \mathrm{HC}(j)$ or $i=j$ and $\mathbf{K}_i = \mathbf{K}_j$ iff $i=j$. Together with (i) and (iii) this immediately implies the main result mentioned in the introduction.

3.2 Proof of (i)–(iii). By induction on α we will first prove (iii) together with (iv) and (v):

(iv) For each $\beta < \alpha$, $\quad I_\alpha \cap I_\beta = \varnothing$.

(v) $\mathrm{HC}(i) = \mathrm{HC}(j)$ implies $i=j$ for each $i, j \in \bigcup_{\alpha>0} I_\alpha$.

0-*step*: trivial.

Successor-step: (iii) trivial.

(iv) Take $j \in I_\beta, \beta \leqslant \alpha, i \in I_{\alpha+1}$. Assume contrary to (iv) that $i=j$. Clearly $\beta \neq 0$. So we can take $j' \in j, j' \in I_\gamma$ for some $\gamma < \beta$. Then also $j' \in i$, whence $j' \in I_\alpha$. But this is impossible, since it would imply $I_\alpha \cap I_\gamma \neq \varnothing$ contrary to the induction hypothesis. So $I_\beta \cap I_{\alpha+1} = \varnothing$.

(v) Assume $\mathrm{HC}(j) = \mathrm{HC}(i), j \in I_\beta, i \in I_{\alpha+1}, \beta \leqslant \alpha+1$. Take $i' \in i$. Then $i' \in I_\alpha$. Now $i' \in \mathrm{HC}(j)$ implies that $i' \in j'$ for some $j' \in \mathrm{HC}(j)$ or $i' \in j$.

In the first case $i' \in I_\gamma$ for some $\gamma < \alpha$ contrary to (iv). Therefore it has to be the case that $i' \in j$. By (iv) this immediately implies that $j \in I_{\alpha+1}$. Moreover, since we have shown it for each $i' \in i, i \subseteq j$. Since $j \in I_{\alpha+1}$, however, the reasoning also pertains in the other direction, whence $i=j$.

Limit step: (iv) It is sufficient to prove that each $X_j^\lambda \in I_\lambda$ contains elements from I_β for each $\beta < \lambda$. Let $j \in I_\alpha, \alpha < \lambda$. By induction hypothesis (iii), $|I_\alpha| = \beth_\alpha$. Therefore $|I_\alpha - \{j\}| = \beth_\alpha$. If we now define, for $\alpha < \beta < \lambda, I'_\beta = \{i \in I_\beta \mid j \notin \mathrm{HC}(i)\}$, then the construction of the I'_β from $I_\alpha - \{j\}$ runs exactly the same course as the construction of the I_β from I_α. So, $|I'_\beta| = \beth_\beta \geqslant 1$. Since clearly $I'_\beta \subseteq X_j^\lambda$ the result now follows.

(v) Let $i,j \in \bigcup_{\alpha<\lambda} I_\alpha$ and $\mathrm{HC}(X_i^\lambda) = \mathrm{HC}(X_j^\lambda)$. If $i' \in \mathrm{HC}(X_i^\lambda)$, then $i' \in i_1 \in \cdots \in i_n \in X_j^\lambda$ for some $i_1, \ldots, i_n$. So $i \notin \mathrm{HC}(i_n)$, whence $i \notin \mathrm{HC}(i')$ and $i' \in X_j^\lambda$. So, we can conclude $X_i^\lambda = X_j^\lambda$. Assume now $i \neq j, i \in I_\alpha, j \in I_\beta, \alpha \leqslant \beta \leqslant \lambda$. Take some $i' \in I_\alpha$ such that $i' \neq i, i' \neq j$. Then $i \in \mathrm{HC}(\{i,i'\})$, so $\{i,i'\} \notin X_i^\lambda$, but, from $\alpha \leqslant \beta$ it follows that $j \notin \mathrm{HC}(i')$ and therefore $j \notin \mathrm{HC}(\{i,i'\})$, whence $\{i,i'\} \in X_j^\lambda$. So $i \neq j$ is impossible as was to be proved.

(iii) It is sufficient to show that, if $i \neq j$, then $X_i^\lambda \neq X_j^\lambda$, but this was shown under (iv).

We are now able to prove (i) and (ii) in the same way as in section (ii). The only slight gaps are in the proof of (i)$\Rightarrow$. The first gap is where we concluded from $\mathbf{K}_{j'} \equiv \mathbf{K}_i$ for each $j' \in j$ to $\mathbf{K}_{j'} \neq \mathbf{K}_i$ for each $j' \in j$ (and similarly with i and j reversed). Assume to the contrary that $\mathbf{K}_{j'} \equiv \mathbf{K}_i$ for each $j' \in j$ and $\mathbf{K}_{j'} = \mathbf{K}_i$ for some $j' \in j$. It follows, that, for each $j'' \in j, \mathbf{K}_{j''} \equiv \mathbf{K}_{j'}$, i.e. $j'' \in \mathrm{HC}(j')$ or $j'' = j'$, and hence $j'' \in I_\beta$ for $\beta \leq \alpha$. This would imply that $j \in I_{\alpha+1}$ and $j = \{i\}$, but this is incompatible with the definition of $I_{\alpha+1}$. The second and last gap is in the final step of the proof and is to be filled by (v).

Remark 3.3 The constructive content of the result is by no means clear yet. The proof can be improved somewhat from the constructive point of view by taking $I_{\alpha+1}$ to be $\{j \subseteq i \mid |j| = 2\}$. Then we would get $|I_\alpha| = |\omega \cdot \alpha|$. So we would get where we wanted without using the power set axiom. However, essential uses of the excluded third seem to occur at several places in the proof. Yet to be investigated is how much of this can be interpreted in the theory of lawless sequences to show intuitionistically that one gets really new connectives. A final remark is that the fact that even classically seen Kripke-models make for more valid formulae than other methods (cf. NADEL, 1978) does not matter, since less valid formulae would give even more equivalence-classes.

References

DE JONGH, D. H. J.
  [1968] Investigations on the intuitionistic propositional calculus, Ph.D. Thesis, University of Wisconsin, Madison, WI, University Microfilms, Ann Arbor, MI.
GAIFMAN, H.
  [1964] Infinite Boolean polynomials, I, *Fund. Math.*, **54**, 230–250.
GOAD, C.
  [1978] Monadic infinitary propositional logic: a special operator, *Reports on Math. Logic*.
HALES, A. W.
  [1964] On the non-existence of free complete Boolean algebras, *Fund. Math.*, **54**, 45–66.
JANKOV, V. A.
  [1968] Constructing a sequence of strongly independent superintuitionistic propositional calculi, *Sov. Math. Dokl.*, **9**, 806–807.
KREISEL, G.
  [1979] Constructivistic approaches to logic, in: *Proceedings of the Conference on Modern Logic, Rome, Sept, 1977* (to appear).
KRIPKE, S. A.
  [1967] An extension of a theorem of Gaifman–Hales–Solovay, *Fund. Math.*, **61**, 29–32.

NADEL, M. A.
[1978] Infinitary intuitionistic logic from a classical point of view, *Ann. Math. Logic*, **14**, 159–192.

SOLOVAY, R. M.
[1966] New proof of a theorem of Gaifman and Hales, *Bull. Am. Math. Soc.* **72**, 282–284.

This page intentionally left blank

J. Barwise, H. J. Keisler and K. Kunen, eds., *The Kleene Symposium*
©North-Holland Publishing Company (1980) 113–122

A Strong Conservative Extension of Peano Arithmetic*

Harvey Friedman

Department of Mathematics, The Ohio State University, Columbus, OH 43220, U.S.A.

Dedicated to Professor S. C. Kleene on the occasion of his 70th birthday

Abstract: A system ALPO of set theory is presented and proved to form a conservative extension of Peano arithmetic. The system has sufficient strength to allow a very substantial amount of analysis to be directly formalized, including Lebesgue integration theory, without resorting to wholesale padding of objects with extra information.

In FRIEDMAN (1977), we presented a fragment **B** of Zermelo set theory with intuitionistic logic, and proved that any arithmetic sentence provable in **B** is provable in PA. (It is now known that **B** is a conservative extension of HA.) We indicated how constructive analysis can be directly formalized in **B** in a manner analogous to the way classical analysis is formalized in Zermelo set theory.

Another approach has been developed by S. FEFERMAN based on his paper of 1975. Subsequent to our work on **B**, Feferman gave a weakened form T_0 of his systems in FEFERMAN (1975), and he and Beeson established that T_0 is a conservative extension of HA. Unlike **B**, which is an extensional system, his systems rely on intensionality, and the formalization of the constructive analysis in these systems requires wholesale padding of objects with extra information. (For instance, uniformly continuous functions on [0, 1] cannot be handled as points without assuming that they are a pair consisting of an operation on [0, 1] together with a modulus of continuity. In particular, the space $C[0, 1]$ must be defined as such a space of pairs.)

At the end of FRIEDMAN (1977), we claimed that systems such as **B** together with the principle Bishop calls *the limited principle of omniscience* (LPO), suffices to directly formalize extensive classical analysis. LPO asserts that any sequence of 0's and 1's is either everywhere nonzero or somewhere zero. Here we present such a system, ALPO for "analysis based

*Research partially supported by NSF grants P038823 and MCS78-02558.

on the limited principle of omniscience", and prove that this is a conservative extension of PA.

A difficult part of classical analysis to formalize in conservative extensions of PA is the theory of Lebesgue integration. What axioms suffice to prove that for convergent sequences of Lebesgue integrable functions on [0, 1], the limit is Lebesgue integrable and the integral of the limit is the limit of the integrals?

There are three known approaches. One can formulate the appropriate choice principle (for choosing a sequence of appropriate open coverings) and use classical logic. All known choice principles lead to very strong systems under classical logic. One can interpret the notion "convergent sequence of Lebesgue integrable functions on [0, 1]" as a sequence of functions together with the extra information that demonstrates that they are Lebesgue integrable and convergent. I.e., that the appropriate open coverings are given as a sequence in the hypotheses, and use classical logic. However, this approach requires use of wholesale padding of objects with extra information. Indeed, this is the approach taken by TAKEUTI (1978) and FEFERMAN (1975, 1977, 1980).

The third approach is to formulate the appropriate choice principles and use intuitionistic logic with the nonintuitionistic LPO. This is the approach we take here, and leads to the formal system ALPO. This approach is based on the thesis that classical analysis is nonconstructive at type 0 (natural numbers), but constructive at higher types (i.e., functions and functionals).

It should be noted that some care has to be taken in ALPO in the handling of sets (of real numbers). One distinguishes between *pointsets of real numbers*, where analysis is to be conducted, and the more general *sets of real numbers*, in the sense of extensions of predicates. Thus a *pointset* is a set which has a characteristic function (on the whole space), or alternatively, a set in which membership is decidable. Another way of handling things is to never talk of pointsets but always of functions; i.e., never do measure theory, but instead do integration theory.

The system ALPO is based on the intuitionistic predicate calculus with identity, and the nonlogical symbols $\in$, N, M, S, 0 (see the introduction in FRIEDMAN (1977)). Since the axioms of extensionality, pairing, infinity, successor, union, and Δ_0-separation are included in ALPO, we introduce the auxiliary function symbols $\{x\}$, $\{x,y\}$ $\langle x,y\rangle$, Dom(x), Rng(x), U, $'$, and the constants $\emptyset$, ω, by explicit definition as in FRIEDMAN (1977).

We now introduce the axioms of ALPO. We take the ordinary equality axioms as understood.

A. *Ontological axioms.* $N(a) \vee M(a)$, $N(a) \rightarrow \sim M(a)$, $N(0)$, $S(x,y) \rightarrow (N(x) \& N(y))$, $x \in y \rightarrow M(y)$.

B. *Urelement extensionality.* $(M(a)\ \&\ M(b)\ \&\ (\forall x)(x \in a \leftrightarrow x \in b)) \rightarrow a = b$.

C. *Successor axioms.* $(S(x,y)\ \&\ S(x,z)) \rightarrow y = z$, $(S(y,x)\ \&\ S(z,x)) \rightarrow y = z$, $\sim S(x,0)$.

D. *Infinity.* $(\exists x)(\forall y)(y \in x \leftrightarrow N(y))$.

E. *Sequential induction.* $(\mathrm{Fcn}(a)\ \&\ \mathrm{Dom}(a) = \omega\ \&\ \mathrm{Rng}(a) \subset \omega\ \&\ \langle 0,0\rangle \in a\ \&\ (\forall x)(\langle x,0\rangle \in a \rightarrow \langle x',0\rangle \in a)) \rightarrow (\forall x)(x \in \omega \rightarrow \langle x,0\rangle \in a)$.

F. *Sequential recursion.* $(\mathrm{Fcn}(a)\ \&\ \mathrm{Dom}(a) = \omega\ \&\ \mathrm{Rng}(a) \subset \omega\ \&\ b \in \omega) \rightarrow (\exists x)(\mathrm{Fcn}(x)\ \&\ \mathrm{Dom}(x) = \omega\ \&\ \mathrm{Rng}(x) \subset \omega\ \&\ x(0) = b\ \&\ (\forall z \in \omega)(x(z') = a(x(z))))$.

G. *Pairing.* $(\exists x)(\forall y)(y \in x \leftrightarrow (y = a \vee y = b))$.

H. *Union.* $(\exists x)(\forall y)(y \in x \leftrightarrow (\exists z)(y \in z\ \&\ z \in a))$.

I. *Exponentiation.* $(\exists x)(\forall y)(y \in x \leftrightarrow (\mathrm{Fcn}(y)\ \&\ \mathrm{Dom}(y) = a\ \&\ \mathrm{Rng}(y) \subset b))$.

J. *Countable choice.* $(\forall x)(x \in \omega \rightarrow (\exists y)(\varphi)) \rightarrow (\exists z)(\mathrm{Fcn}(z)\ \&\ (\forall x)(x \in \omega \rightarrow (\exists y)(\langle x,y\rangle \in z\ \&\ \varphi)))$, where z is not free in φ.

K. Δ_0*-separation.* $(\exists x)(\forall y)(y \in x \leftrightarrow (y \in a\ \&\ \varphi))$, where φ is Δ_0 and x is not free in φ.

L. *Strong collection.* $(\forall x)(x \in a \rightarrow (\exists y)(\varphi)) \rightarrow (\exists z)((\forall x)(x \in a \rightarrow (\exists y)(y \in z\ \&\ \varphi))\ \&\ (\forall y)(y \in z \rightarrow (\exists x)(x \in a\ \&\ \varphi)))$, where z is not free in φ.

M. *Limited principle of omniscience.* $(\mathrm{Fcn}(a)\ \&\ \mathrm{Dom}(a) = \omega\ \&\ \mathrm{Rng}(a) \subset \omega) \rightarrow ((\exists x)(\langle x,0\rangle \in a) \vee (\forall x)(\langle x,0\rangle \notin a))$.

It is worth noting that ALPO suffices to prove induction for decidable predicates of natural numbers, and LPO in the form

$$(\forall x)(N(x) \rightarrow (\varphi(x) \vee \sim\varphi(x))) \rightarrow ((\exists x)(N(x)\ \&\ \varphi(x)) \vee (\forall x)(N(x) \rightarrow \sim\varphi(x))).$$

It also suffices to develop full primitive recursion on ω.

As in FRIEDMAN (1977), the first step is to interpret the system ALPO in an appropriate system of intuitionistic arithmetic with enumeration.

As in FRIEDMAN (1977), the language of second order arithmetic is based on predicate calculus with variables n; (a,b,c,p,q,k,m,n) over natural numbers, variables x; (u,v,w,y,z) over sets (of natural numbers), the numerical constant 0, the numerical function symbols $'$, $+$, $\cdot$, the identity relation $=$ between numbers, and the $\in$ relation between numbers and sets. As in FRIEDMAN (1977), we can introduce the notation $(z)_m=\{n: 2^n3^m\in z\}$.

The axioms of Y_1 are as follows.

a. *Successor axioms.* $\sim n'=0$, $n'=m'\to n=m$.

b. *Recursion equations.* $n+0=n$, $n+(m')=(n+m)'$, $n\cdot 0=0$, $n\cdot(m')=(n\cdot m)+n$.

c. *Decidable induction.* $((\forall n)(\varphi(n)\vee\sim\varphi(n))\ \&\ \varphi(0)\ \&\ (\forall n)(\varphi(n)\to\varphi(n')))\to(\forall n)(\varphi(n)$.

d. *Arithmetic separation.* $(\exists x)(\forall n)(n\in x\leftrightarrow\varphi)$, where φ has no bound set variables, and x is not free in φ.

e. *Enumeration.* $(\exists w)(\forall x)(((\forall n\in y)(\forall m)(\forall r)(2^n3^r\in x\leftrightarrow(z)_m=(z)_r)\ \&\ ((\forall k\in x)(\exists n\in y)(\exists r)(k=2^n3^r))\to(\exists e)(x=(w)_e))$.

f. *Countable choice.* $(\forall n)(\exists x)(\varphi(n,x))\to(\exists y)(\forall n)(\exists x)(\varphi(n,x)\ \&\ x=(y)_n)$, where y is not free in φ.

g. *Strong collection.* $(\forall n\in x)(\exists y)(\varphi(n,y))\to(\exists z)(\exists w)((\forall n\in x)(\exists m\in w)(\varphi(n,(z)_m)\ \&\ (\forall m\in w)(\exists n\in x)(\varphi(n,(z)_m))$, where z,w do not occur in $\varphi(n,y)$.

h. *Limited principle of omniscience.* $(\forall n)(\varphi(n)\vee\sim\varphi(n))\to((\exists n)(\varphi(n))\vee(\forall n)(\sim\varphi(n)))$.

Lemma 1. *Every arithmetic sentence provable in ALPO is provable in* Y_1.

Proof. By interpreting ALPO within Y_1 as in FRIEDMAN (1977, Section 1).

As in FRIEDMAN (1977), the next step is to interpret Y_1 into a system Y_2 of intuitionistic second order arithmetic without the classically false axiom of enumeration. Only this time we cannot use the recursive realizability interpretation of FRIEDMAN (1977) because of the presence of LPO in Y_1. Instead, we use what amounts to arithmetic sets of natural numbers as the realizing objects.

The language of Y_2 is the same as the language of Y_1 except that we introduce a predicate symbol A (for arithmetic) on sets of natural numbers.

The axioms of Y_2 are as follows:

1. *Successor axioms.*

2. *Recursion equations.*

3. *A-induction.* $(\forall x)((A(x)\ \&\ 0\in x\ \&\ (\forall n)(n\in x\rightarrow n'\in x))\rightarrow(\forall n)(n\in x)$.

4. *Atomic decidability.* $(\forall n)(\forall x)(n\in x\vee\sim n\in x)$.

5. *Relativized separation.* Separation for formulas whose set quantifiers range over sets obeying A.

6. *A-arithmetic separation.* $(A(x_1)\ \&\ \cdots\ \&A(x_k))\rightarrow(\exists y)(A(y)\ \&(\forall n)(n \in y\leftrightarrow\varphi(n,x_1,\dots,x_k)))$, where $x_1,\dots,x_k$ are the only set variables appearing in φ, and φ has no bound set variables.

7. *Extensionality.* $(A(x)\ \&\ (\forall n)(n\in x\leftrightarrow n\in y))\rightarrow A(y)$.

8. *A-arithmeticity* $(\forall x)(A(x)\rightarrow(\exists y)(\exists n)(A(y)\ \&\ (y)_0=\emptyset\ \&\ (\forall j< n)((y)_{j+1}$ is the Turing jump of $(y)_j)\ \&\ x$ is recursive in $(y)_n))$.

9. *$A-\Sigma_1^1$ axiom of choice.* $(A(x_1)\ \&\ \cdots\ \&\ A(x_k)\ \&\ (\forall n)(\exists y)(A(y)\ \&\ \varphi(n,y,x_1,\dots,x_k)))\rightarrow(\exists z)(A(z)\ \&\ (\forall n)(\exists y)(y=(z)_n\ \&\ \varphi(n,y,x_1,\dots,x_k)))$, where $x_1,\dots,x_k$ are the only set variables appearing in φ, and φ has no bound set variables.

10. *Countable choice.* Here, as above, A is allowed in φ.

Within Y_2, an $A-\Sigma$ relation is a relation among sets satisfying A, which is given by a Σ formula whose set parameters must obey A, and whose existential set quantifiers must range over sets obeying A.

It is easy to see that, in Y_2, every $A-\Sigma$ relation can be uniformized by an $A-\Sigma$ relation. By uniformizing a universal $A-\Sigma$ relation, we obtain a metarecursion theory of the kind we need within Y_2. Let $\langle x\rangle(y)=z$ be the result of this uniformization, which will serve as our universal partial binary $A-\Sigma$ function.

For each formula φ of second order arithmetic we define another formula $x\,r\,\varphi$ of second order arithmetic as follows. (For technical reasons, assume that x does not appear in φ.)

A. $xrs=t$ is $s=t$ & $A(x)$.

B. $xrs\in y$ is $(\exists e)(x=\{e\}$ & $2^e3^s\in y)$.

C. $xr(\varphi$ & $\psi)$ is $((x)_0r\varphi)$ & $((x)_1r\psi)$.

D. $xr(\varphi\vee\psi)$ is $((x)_0=\{0\}$ & $(x)_1r\varphi)\vee((x)_0=\{1\}$ & $(x)_1r\psi)$.

E. $xr(\varphi\rightarrow\psi)$ is $(\forall y)(yr\varphi\rightarrow\langle x\rangle(y)r\psi)$ & $A(x)$.

F. $xr(\sim\varphi)$ is $(\forall y)(A(y)\rightarrow\sim\langle x\rangle(y)r\varphi)$ & $A(x)$.

G. $xr(\forall n)(\varphi)$ is $(\forall n)(\langle x\rangle(\{n\})r\varphi)$.

H. $xr(\exists n)(\varphi)$ is $(\exists n)((x)_0=\{n\}$ & $(x)_1r\varphi)$ & $A(x)$.

I. $xr(\forall y)(\varphi)$ is $(\forall y)(xr\varphi)$.

J. $xr(\exists y)(\varphi)$ is $(\exists y)(xr\varphi)$.

The interpretation of Y_1 into Y_2 sends φ into $(\exists x)(xr\bar{\varphi})$, where $\bar{\varphi}$ is the universal closure of φ.

The interpretations of axioms a and b are trivially provable in Y_2.

The interpretation of decidable induction relies firstly on the Σ_1^1 axiom of choice in order to make use of a realizer for $(\forall n)(\varphi(n)\vee\sim\varphi(n))$. Using atomic decidability and A- arithmetic separation, one can produce a set z with $A(z)$ such that for all n, if $\varphi(n)$ is realizable, then $(z)_n r\varphi(n)$. Also, evidently $\{n:\varphi(n)$ is realizable$\}$ obeys A. Therefore, from A-induction only, we see that if $\varphi(0)$ & $(\forall n)(\varphi(n)\rightarrow\varphi(n'))$ is realizable, then each $\varphi(n)$ is realizable. Therefore, z essentially provides a realization of $(\forall n)(\varphi(n))$.

The interpretation of arithmetic separation is straightforwardly proved using relativized separation.

The interpretation of enumeration proceeds exactly in FRIEDMAN (1977, Section 2], except that Lemma 2.3 is modified as follows. By A-arithmeticity and relativized separation, there is an enumeration of all sets obeying A; i.e., a set u such that the $(u)_e$ comprise exactly the sets obeying A. Lemma 2.3 is modified to read: There is an α such that for each e with $(u)_e r(\exists x)(x:y\rightarrow\bar{z})$, we have $(u)_e rx:y\rightarrow\bar{z}$, where $x=(\alpha)_e$.

The interpretation of countable choice is easily proved using countable choice.

The interpretation of strong collection proceeds as in FRIEDMAN (1977, 2), using choice.

The interpretation of the limited principle of omniscience is easily proved using atomic decidability and A-induction.

Since as usual, every arithmetic sentence is provably equivalent to its own realizability, we have the following:

Lemma 2. *Every arithmetic sentence provable in Y_1 is provable in Y_2.*

As in FRIEDMAN (1977), the next step is to interpret Y_2 in a system Y_3 of *classical* second order arithmetic (with the predicate symbol A and the set constant symbol c).

The system Y_3 will include the axioms and rules of classical predicate calculus appropriate for its language. The proper axioms of Y_3 are as follows. Below, "n is well-founded" means that every nonempty subset of $[0,n]$ has a least element.

i. *Successor axioms.*

ii. *Recursion equations.*

iii. *A-induction.*

iv. *Relativized separation.*

v. *A-arithmetic separation.*

vi. *Extensionality.*

vii. *A-arithmeticity.*

viii. *$A-\Sigma_1^1$ axiom of choice.*

ix. Σ_1^1 *collection.* $(\forall n)(\exists x)(\varphi(n,x)) \rightarrow (\exists y)(\forall n)(\exists x)(\exists m)(x=(y)_m$ & $\varphi(n,x))$, where y is not free in φ and φ has no bound set quantifiers.

x. *Special definability.* Every set can be defined by an arithmetic formula whose only set parameter is c and whose quantifier depth is well-founded.

xi. *c-selection scheme.*The set $b=\{2^n 3^m\colon \ (\exists y)(A(y)$ & $(y)_0=\emptyset$ & $(\forall y< n)((y)_{j+1}$ is the Turing jump of $(y)_j)$ & $m\in(y)_n)\}$ exists. For each specific arithmetic formula $\varphi(b,m)$ with only the free variable m, whose Gödel number is n, we have $(2^{\bar{n}}3^m\in c$ & $2^{\bar{n}}3^r\in c)\rightarrow m=r$, and $\varphi(b,m)\rightarrow(\exists r)(2^{\bar{n}}3^r \in c)$.

We remark that axioms x and xi are formalized in terms of the Tarski truth definition for formulas of bounded quantifier depth.

Within Y_3, a Σ relation is a relation among sets which is given by a Σ formula, possibly with parameters, and possibly with A.

Unfortunately, because of the lack of induction for sets in Y_3, we cannot show in Y_3 that every Σ relation can be uniformized by a Σ relation. However, let $R(x,y,z)$ be a universal Σ relation (which can be constructed in Y_3). In Y_3, we can at least construct a Σ relation $S(x,y,z)$ such that

$$(\exists! z)(R(x,y,z)) \to (\exists z)(R(x,y,z) \,\&\, S(x,y,z)),$$

and

$$(S(x,y,z) \,\&\, S(x,y,w)) \to z = w.$$

Let $[x](y) = z$ if and only if $S(x,y,z)$.

For each formula φ of second order arithmetic we define another formula $x r^+ \varphi$ of second order arithmetic as follows. (For technical reasons, assume that x does not appear in φ.)

i. $x r^+ s = t$ is $s = t$.

ii. $x r^+ s \in y$ is $s \in y$.

iii. $x r^+ A(y)$ is $A(y)$.

iv. $x r^+ (\varphi \,\&\, \psi)$ is $((x)_0 r^+ \varphi) \,\&\, ((x)_1 r^+ \psi)$.

v. $x r^+ (\varphi \vee \psi)$ is $((x)_0 = \{0\} \,\&\, (x)_1 r^+ \varphi) \vee ((x)_0 = \{1\} \,\&\, (x)_1 r^+ \psi)$.

vi. $x r^+ (\varphi \to \psi)$ is $(\forall y)(y r^+ \varphi) \to [x](y) r^+ \psi$.

vii. $x r^+ (\sim \varphi)$ is $(\forall y)(\sim [x](y) r^+ \varphi)$.

viii. $x r^+ (\forall n)(\varphi)$ is $(\forall n)([x](\{n\}) r^+ \varphi)$.

ix. $x r^+ (\exists n)(\varphi)$ is $(\exists n)((x)_0 = \{n\} \,\&\, (x)_1 r^+ \varphi)$.

x. $x r^+ (\forall y)(\varphi)$ is $(\forall y)([x](y) r^+ \varphi)$.

xi. $x r^+ (\exists y)(\varphi)$ is $(\exists y)((x)_0 = y \,\&\, (x)_1 r^+ \varphi)$.

The interpretation of Y_2 into Y_3 sends φ into $(\exists x)(x r^+ \bar{\varphi})$, where $\bar{\varphi}$ is the universal closure of φ.

It is straightforwardly seen that the interpretation of every theorem of Y_2 is a theorem of Y_3. Axiom xi is used to interpret relativized separation under a lack of induction.

Lemma 3. *Every arithmetic sentence provable in Y_2 is provable in Y_3.*

Let Y_4 be the following system of classical second order arithmetic (with no extra predicate symbol).

A. *Successor axioms.*

B. *Recursion equations.*

C. *Set induction.* $(0 \in x \mathbin{\&} (\forall n)(n \in x \rightarrow n' \in x)) \rightarrow (\forall n)(n \in x)$.

D. *Arithmetic separation.*

E. Σ_1^1 *axiom of choice.* $(\forall n)(\exists x)(\varphi(n,x)) \rightarrow (\exists y)(\forall n)(\exists x)(x=(y)_m \mathbin{\&} \varphi(n,x))$, where y is not free in φ and φ has no bound set quantifiers.

F. *Arithmeticity.* $(\forall x)(\exists y)(\exists n)((y)_0=\emptyset \mathbin{\&} (\forall_j<n)((y)_{j+1}$ is the Turing jump of $(y)_j) \mathbin{\&} x$ is recursive in $(y)_n))$.

The system Y_4 is closely related to the system K_3 of FRIEDMAN (1977). The same model theoretic argument used there to show that every arithmetic sentence provable in K_3 is provable in PA shows that every arithmetic sentence provable in Y_4 is provable in PA.

Lemma 4. *Every arithmetic sentence provable in Y_4 is provable in PA.*

It remains to show that every arithmetic sentence provable in Y_3 is provable in Y_4.

Arguing in PA, let φ be an arithmetic sentence which is consistent with Y_4. Let M be a model of $Y_4+\varphi$ whose complete diagram is Δ_2^0. Define c in the obvious way so that axiom xi will hold (with $A(y)$ vacuously true), using externally least (external to M) witnesses m. Then (M,c) satisfies all of $Y_3+\varphi$ except axioms ix, x (with $A(y)$ vacuously true), and the truth predicate for all subformulas of these satisfied formulas is Δ_2^0. Hence let (U,c) be in the language of Y_3, satisfy all of $Y_3+\varphi$ except axioms ix, x, be recursively saturated, and have Δ_2^0 complete diagram. Finally let (V,c) have the same arithmetic part and A-part as (U,c), but whose sets of integers are those sets of integers first order definable over (U,c). Then (V,c), but not its complete diagram, is Δ_2^0. Now axioms ix and x, which are finitely axiomatizable, can be verified in (V,c), using recursive saturation. Hence $Y_3+\varphi$ is consistent.

Lemma 5. *Every arithmetic sentence provable in Y_3 is provable in Y_4.*

Theorem. *Every arithmetic sentence provable in ALPO is provable in PA.*

In fact, we have given a proof of the Theorem within PA. Further considerations show that it is provable in primitive recursive arithmetic.

If induction for sets is added to ALPO, then we obtain a theory whose arithmetic theorems are the same as PA augmented with transfinite induction on all ordinals below the ϵ_0-th critical number (or $R(<\epsilon_0)$ of FRIEDMAN (1977)). If induction for all formulas (or induction for all formulas together with relativized dependent choice for all formulas) is added to ALPO, then we obtain a theory whose arithmetic theorems are the same as PA augmented with transfinite induction on all ordinals below the $\kappa(\epsilon_0)$-th critical number, where $\kappa(\epsilon_0)$ is the ϵ-th critical number (or $R(<\kappa(\epsilon))$).

References

FEFERMAN, S.

[1970] Ordinals and functionals in proof theory, *Proc. Int. Congr. Math., Nice 1970*, **1**, 229–233.

[1975] A language and axioms for explicit mathematics, in: *Lecture Notes in Mathematics* **450** (Springer, Berlin) pp. 87–139.

[1977] Theories of finite type related to mathematical practice, in: *Handbook of Mathematical Logic* (North-Holland, Amsterdam) pp. 913–971.

[1980] Forthcoming book of the Ω-group, in preparation.

FRIEDMAN, H.

[1977] Set theoretic foundations for constructive analysis, *Ann. of Math.*, **105**, 1–28.

TAKEUTI, G.

[1978] A conservative extension of Peano arithmetic, Part II of "Two applications of logic to mathematics," *Publ. Math. Soc. Japan*, **13**.

J. Barwise, H. J. Keisler and K. Kunen, eds., *The Kleene Symposium*
©North-Holland Publishing Company (1980) 123–148

Church's Thesis and Principles for Mechanisms

Robin Gandy
Mathematical Institute, St. Giles, Oxford OX1 3LB, Great Britain

For Stephen Kleene

Beware! Beware!
His flashing eyes, his floating hair!
Weave a circle round him thrice,
And close your eyes in holy dread,
For he on honeydew hath fed,
And drunk the milk of paradise.

Abstract: After a brief review of Church's thesis and Gödel's objection to it, it is argued that Turing's analysis of computation by a human being does not apply directly to mechanical devices. A set-theoretic form of description for discrete deterministic machines is elaborated and four principles (or constraints) are enunciated, which, it is argued, any such machine must satisfy. The most important of these, called "the principle of local causality" rejects the possibility of instantaneous action at a distance. Although the principles are justified by an appeal to the geometry of space-time, the formulation is quite abstract, and can be applied to all kinds of automata and to algebraic systems. It is proved that if a device satisfies the principles then its successive states form a computable sequence. Counter-examples are constructed which show that if the principles be weakened in almost any way, then there will be devices which satisfy the weakened principles and which can calculate *any* number-theoretic function.

1. Introduction

Throughout this paper we shall use "calculable" to refer to some intuitively given notion and "computable" to mean "computable by a Turing machine"; of course many equivalent definitions of "computable" are now available.

Church's Thesis. *What is effectively calculable is computable.*

TURING, by this analysis of the process of calculation in his paper (1936) on computable numbers, gave cogent arguments in support of this thesis. Both Church and Turing had in mind calculation by an abstract human being using some mechanical aids (such as paper and pencil). The word "abstract" indicates that the argument makes no appeal to the existence of practical limits on time and space. The word "effective" in the thesis serves to emphasize that the process of calculation is deterministic— not dependent on guesswork—and that it must terminate after a finite time.

Gödel has objected, against Turing's arguments, that the human mind may, by its grasp of abstract objects, be able to transcend mechanism. (This objection is stated briefly in footnote** on p. 72 of "The Undecidable" and at greater length on pp. 325–326 of WANG's "From Mathematics to Philosophy"; also in a typescript "Footnote*** to be added at the word 'mathematics' on p. 73, line 3 of: The Undecidable…"). Examples where we appear to use our insight and imaginative grasp to arrive at decisions in advance of any process of mechanical verification are well-known: we recognize that certain formal systems are consistent by imagining models for them; when we have gained a little familiarity with a system of ordinal notions we perceive that it is pre-well-ordered. The question is whether these examples are inspired guesswork or lucky accidents, or whether, as Gödel believed, they are the result of the workings of a non-mechanical intelligence. Gödel's objection can only be properly justified by a theory of intelligence. As he admits, our present understanding of the human mind is far from being penetrating enough for the construction of such a theory. For this purpose the knowledge provided by introspection, the history of ideas, experimental psychology, neurophysiology and artificial intelligence seems meagre indeed. One can only keep an open mind.

What we can say is that Turing outlined a proof of the following:

Theorem T. *What can be calculated by an abstract human being working in a routine way is computable.*

We shall return to Turing's proof and Gödel's objection at the end of this paper. Our chief purpose is to analyze mechanical processes and so to provide arguments for the following:

Thesis M. *What can be calculated by a machine is computable.*[1]

Although some of Turing's arguments can be applied indifferently to men or machines, there are crucial steps in Turing's analysis where he appeals to the fact that the calculation is being carried out by a human being. One such appeal is used to justify the assumption that the calculation proceeds as a sequence of elementary steps. A human being can only write one symbol at a time. But, if we abstract from practical limitations,

[1]I have been interested in the problem of how to justify this thesis for a long time. My earlier attempts were unsatisfactory. I owe to a conversation with Harvey Friedman the renewal of interest which led to the writing of this paper.

we can conceive of a machine which prints an arbitrary number of symbols simultaneously. (I owe to conversations with J. C. Shepherdson a realization that proofs of Thesis M must take parallel working into account. See SHEPHERDSON (1975) for a discussion of some of the problems this raises.) Turing's arguments do not suffice, nor do I think he would have claimed that they suffice, to justify Thesis M. The reader may (like several of those with whom I have talked) feel that, once Turing's ideas have been grasped, Thesis M is so unproblematic as to make arguments for it uninteresting or even unnecessary. This feeling has I think two sources. Firstly, actual machines which calculate fall under a narrow range of stereotypes to each of which Turing's arguments may rather easily be adapted. And the design of the most successful calculating machines—digital computers—was, at least in the early stages of their development, significantly influenced by Turing's ideas. But a slight effort of imagination will suggest devices which differ radically from the practical stereotypes. Conway's construction of a universal machine from his game of life is a good example.[2] There can be no guarantee that a further effort of imagination may not result in a device to which Turing's analysis is inapplicable.

The second source for a lack of interest in Thesis M is the belief that it can *only* function as a definition: if some imagined device can calculate what is not computable it is no machine. But I shall propose criteria for "being a machine" which, on the face of it, differ significantly from the criterion "works in a computable manner". At the very least, then, I hope to explain to the reader who believes that Thesis M is unproblematic the grounds for his belief.

For vividness I have so far used the fairly nebulous term "machine". Before going into details I must be *rather* more precise. Roughly speaking I am using the term with its nineteenth century meaning; the reader may like to imagine some glorious contraption of gleaming brass and polished mahogany, or he may choose to inspect the parts of Babbage's "Analytical Engine" which are preserved in the Science Museum at South Kensington.

(1) In the first place I exclude from consideration devices which are *essentially* analogue machines. In his paper "A notion of mechanistic theory" (1974) KREISEL discusses the ways in which physical theories (including even Newtonian theory) might give rise to non-computable functions. A more extreme possibility than those considered by Kreisel is the following: could some physical theory lead to a linear operator which has an infinite discrete spectrum and which is such that the multiplicity of

[2] An account of "the game of life" will be found in GARDNER (1970 and 1971). J. H. Conway, who invented the game, described how, with an appropriate initial configuration, it could be made to mimic the action of any Turing machine at the Logic Summer School held in Cambridge (U.K.) in 1971.

an eigenvalue is not a computable function of its place in the spectrum? So I shall distinguish between "mechanical devices" and "physical devices" and consider only the former.[3] The only physical presuppositions made about mechanical devices (Cf. Principle IV below) are that there is a lower bound on the linear dimensions of every atomic part of the device and that there is an upper bound (the velocity of light) on the speed of propagation of changes.

(2) Secondly we suppose that the progress of calculation by a mechanical device may be described in discrete terms, so that the devices considered are, in a loose sense, digital computers.

(3) Lastly we suppose that the device is deterministic; that is, the subsequent behaviour of the device is uniquely determined once a complete description of its initial state is given.

After these clarifications we can summarize our argument for a more definite version of Thesis M in the following way.

Thesis P. *A discrete deterministic mechanical device satisfies principles I–IV below.*

Theorem. *What can be calculated by a device satisfying principles I–IV is computable.*

The principles were arrived at by considering, schematically, examples that might at least in principle be realized by mechanical or electrical means. But (not too surprisingly in view of the generality aimed at) there are many abstract structures for which they also hold. Hence the principles and the theorem may be of interest even to those who are not concerned with Thesis M. Some examples are given in the concluding discussion.

Note on terminology. In the above statement we used "discrete deterministic mechanical device" to emphasize the somewhat restricted significance we are giving to the term "machine". Now that the point has been made we shall, for brevity, revert to the word "machine"; for the sake of variety, and for their flavor, we shall also sometimes use the words "device" and "mechanism" (for an object, not for a tenet).

2. The form of description

Since we are considering discretely acting machines, we may without loss of generality suppose that the action of a machine is described by

[3] Pour-El (1974) investigates the computing power of a particular class of analog machines. But, in principle, any physical phenomenon might be used for analog computation.

describing the sequence $S_0, S_1, \ldots,$ of its states; the input or initial arguments are encoded in S_0. It is of little importance whether we designate certain states as "halt" states encoding the output, or whether we consider, as Turing did, an infinite sequence which enumerates a (possibly empty) set whose members are encoded by certain of the S_i.

Our use of the term "discrete" presupposes that each state of the machine can be adequately described in finite terms. In order that we can apply any insights which we may have about mechanisms we want this description to reflect the actual, concrete, structure of the device in a given state. On the other hand, we want the form of description to be sufficiently abstract to apply uniformly to mechanical, electrical or merely notional devices. We have chosen to use hereditarily finite sets; other forms of description might be equally acceptable. We suppose that labels are chosen for the various parts of the machine—e.g., for the teeth of cog wheels, for a transistor and its electrodes, for the beads and wires of an abacus. Labels may also be used for positions in space (e.g., for squares of the tape of a Turing machine) and for physical attributes (e.g., the color of a bead, the state of a transistor, the symbol on a square). Starting from a potentially infinite set L of labels we form sets, sets of sets and so on; but we do not use the empty set in this construction, since it is an abstract object while our descriptions by sets may be of completely concrete structures. (We do, however, allow the empty structure.)

2.1 Definition.

$$\mathbf{P}_F(X) =_{\mathrm{Df}} \{ Y : Y \subseteq X \wedge Y \neq \varnothing \wedge Y \text{ is finite} \}.$$

$$\mathrm{HF}_0 =_{\mathrm{Df}} \varnothing,$$

$$\mathrm{HF}_{n+1} =_{\mathrm{Df}} \mathbf{P}_F(L \cup \mathrm{HF}_n).$$

$$\mathrm{HF} =_{\mathrm{Df}} \bigcup \{ \mathrm{HF}_n : n \in \omega \} \cup \{ \varnothing \}.$$

The variables a, b, c, d (perhaps decorated with subscripts etc.) will always range over L, while s, t, u, v, w, x, y, z range over HF or designated subsets of HF. Note that $\varnothing \notin \mathrm{HF}_n$, $a \notin \mathrm{HF}$, and $\mathrm{HF}_n \subseteq \mathrm{HF}_{n+1}$.

2.2 It is convenient at this point to introduce various definitions and notations which will be used later. Since $\in$ is a well-founded relation on HF, we can use $\in$-recursion as a method of definition.

(1) $\quad \operatorname{Sup} x =_{\mathrm{Df}} \{ a : a \in x \vee \exists y \in x . a \in \operatorname{Sup} y \}.$

$\operatorname{Sup} x$, the *support* of x is the set of labels which occur in the construction of x.

(2) $\quad |x| =_{\mathrm{Df}} \operatorname{Card}(\operatorname{Sup} x).$

$|x|$, the *size* of x is the number of labels occuring in x. It is to be distinguished from $\bar{\bar{x}}(=\text{Card}(x))$

Let $A \subseteq L$; we define the *restriction* $x \upharpoonright A$ of x to A by

(3) $$x \upharpoonright A =_{\text{Df}} (x \cap A) \cup \{ y \upharpoonright A : y \in x \wedge \text{Sup}\, y \cap A \neq \varnothing \}.$$

It is the structure whose construction follows that of x omitting all labels not in A. The condition $\text{Sup}\, y \cap A \neq \varnothing$ is necessary to ensure that $x \upharpoonright A \in$ HF. If $\text{Sup}\, x \cap A = \varnothing$, then $X \upharpoonright A = \varnothing$.

The *transitive closure* of $x, \text{TC}(x)$ is defined by

(4) $$\text{TC}(x) =_{\text{Df}} \bigcup \{ \text{TC}(y) : y \in x \} \cup (x \cap L).$$

2.3 As our terminology suggests, labels, like co-ordinates, are necessary for the description of concrete devices but do not by themselves have direct physical reference. A particular state of a machine corresponds not to a particular $x \in \text{HF}$, but rather to the $\in$-isomorphism-type (which we shall call the *stereotype*) of x. All the information about that state which is relevant to the operation of the machine must be encoded in any structure x which is used to describe it. Relations and functions over HF which have concrete reference must be invariant under isomorphisms.

However, it is natural to suppose that if a label refers to a particular element of a mechanism in the state described by x, then it will refer to the same element in the next state; most things preserve their identity as time passes. In general the next state will also contain some new elements; for example, when a Turing machine moves left from the leftmost square so far used a new square must be created. A transition function F which determines the description Fx of the next state must specify new labels for the new elements, but no physical significance attaches to that specification. We now incorporate these ideas in a series of definitions.

(1) The variable π ranges over permutations of L. The effect of such a permutation on a structure x is defined in the obvious way:

$$a^{\pi} = \pi(a); \qquad x^{\pi} = \{ y^{\pi} : y \in x \} \cup \{ a^{\pi} : a \in x \}.$$

(2) Two structures x, y are *isomorphic over a set A of labels*, written "$x \simeq_A y$" just in case there is a permutation π which is the identity on A (i.e., $\forall a \in A . \pi(a) = a$) and which carries x into y (i.e., $x^{\pi} = y$). They are *isomorphic* ("$x \simeq y$") if they are isomorphic over the empty set. We shall write "$x \simeq_z y$" for "$x \simeq_{\text{Sup}\, z} y$". Note that if $x \simeq_A y$, then $x \upharpoonright A = y \upharpoonright A$.

(3) A property P of structures and the corresponding class P' ($= \{ x : P(x) \}$) are *structural* iff they are closed under isomorphism; i.e., if

$$P(x) \wedge x \simeq y . \rightarrow P(y).$$

(4) $X \subseteq \mathrm{HF}$ is a *stereotype* iff

$$\exists x.X = \{y : y \simeq x\}.$$

(5) A function $F: \mathrm{HF} \to \mathrm{HF}$ is *structural* iff for all π

$$(Fx)^{\pi} \simeq_{x^{\pi}} Fx^{\pi}.$$

This is stronger than the condition $x \simeq y \to Fx \simeq Fy$, but not as strong as requiring invariance (i.e., $(Fx)^{\pi} = Fx^{\pi}$). It expresses precisely the requirement that F describes the transition between physical states with some persistent elements.

2.4. *We can now state:*

Principle I ***The form of description***

Any machine M can be described by giving a structural set $S_M \subseteq \mathrm{HF}$ of state-descriptions together with a structural function $F: S_M \to S_M$. If $x_0 \in S_M$ describes an initial state, then $Fx_0, F(Fx_0), \ldots,$ describe the subsequent states of M.

2.5. Examples. (1) The state of a Turing machine can be described by a structure of the form

$$\langle x, y, z, a_i, c_j \rangle$$

where x is the graph of the relation "a' is the label for the square standing immediately to the right of the square with label a", y is the graph of the relation "b is the label for the symbol printed on the square with label a", z codes the programme for the machine, a_i is the label for the scanned square and c_j is the label for the current instruction. Notice that the persistence of symbols (even those which do not occur on the tape at a given instant) is guaranteed by the occurrence of their labels in the (fixed) structure z.

(2) In describing an abacus one cannot treat the beads on a wire as an unstructured set, since, if one did, the label for a bead to be removed could not be structurally determined and the transition function would not be structural.

(3) An example which played an important part in our development of the theory is "The game of life" (Gardner, 1970, 1971) or, more generally, the crystalline automata of Von Neumann (1966). Such an automaton consists of a (finite portion of a) rectangular lattice of cells, each of which may be in one of a fixed number of states (or, equivalently, may bear one of a fixed list of symbols). The next state of each cell is determined, by a fixed table, from its own state and that of its immediate neighbors. Initially only finitely many cells are in non-quiescent states, so that the state can

always be adequately specified by giving the states of the finitely many cells which have so far been brought into play. This example is important both because it involves parallel action (the symbols on all cells in play may change simultaneously) and because it raises problems about identifications between new labels (see 4.8).

Of course, one can also consider three-dimensional crystalline automata. In principle the state of any concrete device can be adequately described by specifying, to a certain degree of approximation, the relevant physical parameters (chemical composition, pressure, current flow and so on) of sufficiently small regions of space. (By using the words "concrete", "mechanical" and the like, we intend that these regions will always be very much larger than the size of an atom). This would not be a good way of describing most devices, but we shall appeal to the possibility of so doing when we come to justify Principle IV.

(4) For a given machine one can always arrange that each label in a fixed finite list is structurally distinguished from all others; for example in $x' = \langle x, a, b, c \rangle$ the labels a, b, c are distinguished. We shall use such *distinguished labels* in our examples without explicitly giving the (additional) structure which distinguishes them.

3. Conditions on S

The remaining three principles place certain restrictions on the set S of state-descriptions and on the transition function F. We shall show that if any of the principles be significantly weakened in (almost) any way then every function becomes calculable. More precisely, let α be an arbitrary predicate of natural numbers; we shall exhibit a machine (S, F) which satisfies the weakened condition and all the other conditions and which calculates α in some obvious sense. We put this picturesquely by saying that the machine displays free will. (Actually it is the class of machines satisfying the weakened conditions which displays free will.)

Each of the principles involves certain finiteness or boundedness conditions. We say that a quantity is *bounded* if there is a number k which may depend on the machine considered, but which does not depend on the state for which the quantity is being evaluated, such that in all states the value of the quantity is less than k.

Principle II ***The Principle of Limitation of Hierarchy***

The set-theoretic rank of the states is bounded. I.e.

$$\exists k. S \subseteq \mathrm{HF}_k.$$

It may well be natural or convenient to describe a machine or a method of storing information in hierarchical terms. The multiplier in a computer, or a series of parallel processors may be subordinate to the control unit; some kinds of data are treated as lists of lists. It is natural when describing such a hierarchical structure by a member of HF to think of a high level part as having its subordinates as members. The principle asserts that for a given machine the maximum height of its hierarchical structure must be bounded.

In describing real or imagined devices one tends to keep the overall rank low. I think that this corresponds to a real limitation of the human intellect. Anyone who has worked with the theory of types knows that above about type 4 one can only work formally, not conceptually; whereas there is no such difficulty in handling functions of large numbers of arguments. One gets over the conceptual block by thinking of a function of high type as operating on *names* for its arguments; in just the same way, a bibliography of bibliographies lists their titles in preference to transcribing their contents. But because of the block it is hard to think of devices which would make good use of objects of high rank.

Of course it is always possible to give a first-order description of HF: one treats the sets as objects rather than as structures. But it does not follow, as Counterexample 3.1 shows, that Principle II is vacuous. The first-order description requires the introduction of extra labels, and this, in conjunction with the other principles reduces the computational power.

3.1. Counterexample. Let $S=\{\iota^n(a): a\in L,\ 1\leqslant n\}$ (where $\iota(x)=\{x\}$), and let F be defined by

$$F(\iota^n(a))=\begin{cases}\iota a & \text{if } \alpha(n),\\ \iota^2 a & \text{otherwise.}\end{cases}$$

It is trivial to verify that (S,F) satisfies the remaining principles; plainly this machine displays free will.

3.2. The next principle says, roughly, that any device can be assembled from parts of bounded size, and that these parts can be so labelled that there is a unique way of putting them together. Model construction-kits aim, not always successfully, to satisfy this principle. However, unlike construction-kits, we shall consider parts which overlap. Thus the tape of a Turing machine can be uniquely reassembled from the collection of *all* pairs of consecutive squares with their symbols; two such pairs are glued together with overlap if they both contain the same label for some square.

The proper formulation of this idea in terms of HF requires care. The simplest way (which I at one time thought was the correct way) would be to take as the parts of a structure x the restrictions $x\restriction A$ of x to subsets of the support of x of bounded size. However, these parts are insufficiently structured to insure unique assembly.

3.3. Counterexample. Let

$$x=\{\langle a,b_1\rangle,\langle a,b_2\rangle,\ldots,\langle a,b_n\rangle\}.$$

Let $\langle a\rangle=\langle a,b_i\rangle\restriction\{a\}$; for any reasonable definition of ordered pair this will be independent of b_i. Let $y=x\cup\{\langle a\rangle\}$. It is easily seen that if $A\subseteq \operatorname{Sup}x$ and $\overline{\overline{A}}<n$, then $x\restriction A=y\restriction A$. Thus x cannot be uniquely assembled from parts of the form $x\restriction A$ with A of bounded size.

What has gone wrong here is that, since

$$\langle a,b_i\rangle\restriction\{a,b_j\}=\{\langle a\rangle\}\quad \text{for } i\neq j,$$

$x\restriction\{a,b_j\}$ contains not only the intended $\langle a,b_j\rangle$ but also the "floating" part $\langle a\rangle$. To put matters right we need to know the restrictions of x to *structured* parts (such as $\langle a,b_j\rangle$) not merely its restrictions to lists of labels.

3.4. Definition. Let $P\subseteq L\cup \mathrm{HF}$.

(i) The set Part(x,P) of parts of x from the list P is defined by

$$\operatorname{Part}(x,P)=\begin{cases}\{x\} & \text{if } x\in P,\\ \bigcup\{\operatorname{Part}(y,P):y\in x\}\cup(x\cap P\cap L) & \text{otherwise.}\end{cases}$$

(ii) The *restriction* $x\restriction P$ *of* x *to the list of parts* P is defined by

$$x\restriction P=\begin{cases}x & \text{if } x\in P,\\ \{y\restriction P:y\in x\wedge \operatorname{Part}(y,P)\neq\varnothing\}\cup(x\cap P\cap L) & \text{otherwise.}\end{cases}$$

(iii) The list P *covers* x iff $x\restriction P=x$; if in addition $P\subseteq \mathrm{TC}(\{x\})$, then P is a set of parts *for* x.

3.5. Remarks. (1) If $P\subseteq L$, then Part$(x,P)=\operatorname{Sup}x\cap P$.

(2) P covers x iff every $\in$-chain $a\in x_n\in\cdots\in x_1\in x$ contains a member of P.

(3) If $P\subseteq Q$, then $x\restriction P=(x\restriction Q)\restriction P$. But this equation is not a consequence of $P\subseteq \mathrm{TC}(Q)$.

(4) $x\restriction P\cup Q$ is not, in general, determined by $x\restriction P$ and $x\restriction Q$ (see 3.7).

(5) The sets $\{x\},x$, Sup$x$ are all sets of parts for x.

(6) The word "part" is used ambiguously to denote the stereotype, a particular set z belonging to this stereotype, and the located part $x\restriction\{z\}$. This is illustrated by: "I need a new sparking plug since the ones in the

garage are all duds and something is wrong with the plug in the first cylinder".

3.6. Let P be a set of parts for x of bounded size; it is not to be expected that x will be uniquely determined by all the $x \upharpoonright \{z\} (z \in P)$. But many structures can be uniquely reassembled from lists of bounded size of parts of bounded size (see Theorem 3.8). We shall refer to such lists (located or unlocated) as *sub-assemblies*.

Definition. Let $Q \subseteq \mathbf{P}_F(\mathrm{TC}(x))$. The structure x can be *uniquely reassembled* from the set Q of sub-assemblies iff x is the unique object y satisfying

(i) $y \in \mathrm{HF}$;
(ii) $\bigcup Q$ covers y;
(iii) $\forall s \in Q.\ x \upharpoonright s = y \upharpoonright s$.

Principle III ***The Principle of Unique Reassembly***

There is a bound q and for each $x \in S$ a set $Q \subseteq \mathbf{P}_F(\mathrm{TC}(x))$ from which x can be uniquely reassembled such that $|s| < q$ for each $s \in Q$.

Remarks. (1) If S satisfies II, then q determines a bound for the number of members of each $s \in Q$ as well as for its size. If II is not satisfied, then the cardinality of s might be unbounded.

(2) Observe that if x can be uniquely reassembled from Q, then $\bigcup Q$ covers x.

3.7. A good idea of what S must be like if it is to satisfy III can be gained from the following

Example. Let $S_n = \{x : x \subseteq \mathbf{P}_F(L) \wedge \bar{\bar{x}} = n\}$. Then $n < q$ is a necessary and sufficient condition for S_n to satisfy III with the bound q.

Proof. Since the members of $x \in S_n$ may be arbitrarily large, the only sub-assemblies we need consider are subsets of L of size q. Suppose $q \leqslant n$. Let $x = \{\{a, b_i\} : 1 \leqslant i \leqslant n\}$, where $a, b_1, \cdots, b_n$ are distinct labels. Then for any $s \subseteq L$ with $\bar{\bar{s}} \leqslant q$,

$$x \upharpoonright s = (x \cup \{\{a\}\}) \upharpoonright s,$$

and unique reassembly fails. To show the sufficiency of the condition, suppose that $x \in S_n$, $x \neq y$, $v \in y - x$ and $x = \{u_1, \ldots, u_n\}$.[1] For $1 \leqslant i \leqslant n$ pick $a_i \in u_i / v$ (the symmetric difference of u_i and v) and let $b \in v$. Set

[1]See 'Notes added in proof' on p. 147.

$s=\{a_1,\ldots,a_n,b\}$. Then $v\restriction s\neq\emptyset$ and for each $1\leqslant i\leqslant n$, $u_i\restriction s\neq v\restriction s$. Thus

$$x\restriction s\neq y\restriction s$$

and the result follows by contraposition.

The proof of sufficiency may readily be generalized to other situations. One can replace L by some other list of parts, and one can consider structures which have sets like S_n in their transitive closures. In particular one readily proves:

3.8. Theorem. *Let σ be a finite similarity type of a first-order structure. Let S be the set of sets which code, in some natural way, first order-structures whose similarity type is σ and whose domain is a finite subset of L. Then S satisfies* III.

Thus almost any kind of automaton can be described in such a way that III is satisfied.

3.9. Counterexample. Let $A_n=\{a_1,\ldots,a_n\}$ be a set of n distinct labels, and e, 0 be distinguished labels. Let

$$y^{n,r}=\left\{v:v\subseteq A_n\wedge\bar{\bar{v}}=n-r\right\}\quad\text{for } 0\leqslant r<n;$$

$$z_\alpha^{n,r}=\begin{cases} y^{n,r} & \text{if } r=0,\text{ or if } r\neq 0 \text{ and } \alpha(n),\\ y^{n,r}\cup y^{n,r-1} & \text{if } 0<r<n \text{ and } \neg\alpha(n);\end{cases}$$

$$x_\alpha^{n,r}=\begin{cases}\{\langle a_1,a_2\rangle,\langle a_2,a_3\rangle,\ldots,\langle a_{n-r},e\rangle\}\cup z_\alpha^{n,r} & \text{if } 0\leqslant r<n,\\ \{e\} & \text{if } r\geqslant n \text{ and } \alpha(n),\\ \{0\} & \text{if } r\geqslant n \text{ and } \neg\alpha(n).\end{cases}$$

Let $S_\alpha=\bigcup\{\{y:y\simeq x_\alpha^{n,p}\}:\ n,p\in\omega\}$, and let F_α be defined by $F_\alpha x_\alpha^{n,p}=x_\alpha^{n,p+1}$. It is obvious that this is a machine displaying free will and which satisfies I and II. We shall verify (see Example 4.4) that it satisfies IV. III fails because, for example, if $s\subseteq A_n$ and $\bar{s}<n$, then

$$y^{n,1}\restriction s=(y^{n,1}\cup y^{n,0})\restriction s.$$

Further, if we weaken III by substituting "$y\in S$" for "$y\in\mathrm{HF}$" in the definition of unique reassembly, then S_α satisfies the weakened principle. For if $y\in S_\alpha$ and $\operatorname{Sup}y=\operatorname{Sup}x_\alpha^{n,p}$ and for each $1\leqslant i\leqslant n$

$$y\restriction\{e,a_i\}=x_\alpha^{n,p}\restriction\{e,a_i\},$$

then $y=x_\alpha^{n,p}$. One might say that the weakened principle is satisfied because S_α is freely determined. It is therefore worth remarking that the principles do not require that S be computable. For example

$$S=\{y:y\subseteq L\wedge\alpha(\bar{\bar{y}})\}$$

satisfies I-III for any α.

4. Local causation

We now come to the most important of our principles. In Turing's analysis the requirement that the action depend only on a bounded portion of the record was based on a human limitation. We replace this by a physical limitation which we call the *principle of local causation.* Its justification lies in the finite velocity of propagation of effects and signals: contemporary physics rejects the possibility of instantaneous action at a distance.

Principle IV (Preliminary version). *The next state, Fx, of a machine can be reassembled from its restrictions to overlapping "regions" s and these restrictions are locally caused. That is, for each region s of Fx there is a "causal neighborhood" $t\subseteq \mathrm{TC}(x)$ of bounded size such that $Fx\restriction s$ depends only on $x\restriction t$.*

4.1. Complications in formulating this principle precisely arise from three sources.

(1) Since we wish our analysis to apply to abstract as well as to concrete devices, we do not wish to distinguish spatial structure from other structure.

(2) We have to be able to determine the causal neighborhoods of x without knowing in advance what are the regions s of Fx.

(3) If Fx has new labels these cannot be determined by $x\restriction t$ (see 2.3).

We cannot require that for *every* $s\subseteq \mathrm{TC}(Fx)$ of bounded size there be a causal neighborhood. For example, if $s=\{a\}$ where a is a particular cell-state of a crystalline automaton then whether or not $Fx\restriction\{a\}$ is empty depends globally, not locally, on x. Because of this and because of (2) above it turns out to be easiest first to decide what are causal neighborhoods.

4.2. In his analysis Turing writes "The new observed squares must be immediately recognizable by the computer". For us the natural analogue of this requirement is that the causal neighborhoods be structurally determined. Thus there must be a list T of stereotypes of bounded size; t is a causal neighborhood of x if $t \subseteq \mathrm{TC}(x)$ and $t \in \bigcup T$. It is, however, very convenient to introduce a slight complication. We permit some of the stereotypes in T to be parts of others, and then we take as causal neighborhoods only those $t \subseteq \mathrm{TC}(x)$ which are included in no larger stereotype. For example for a Turing machine we need not specifically indicate which is the leftmost square of the tape; if a causal neighborhood contains $\langle a_i, a_{i+1} \rangle$, then a_i cannot be the leftmost square and we do not get another causal neighborhood by willfully omitting $\langle a_i, a_{i+1} \rangle$. For further examples see Example 4.4 below.

4.2.1. Definition. (i) We say that t' *subsumes* t, and write $t \preccurlyeq t'$, iff $t \subseteq \mathrm{TC}(t')$.

(ii) The set $\mathrm{CN}_1(x, T)$ of causal neighborhoods of x determined by the set T of stereotypes is defined by

$$\mathrm{CN}_1(x,T) =_{\mathrm{Df}} \{ t : t \subseteq \mathrm{TC}(x) \wedge t \in \bigcup T \\ \wedge \forall t' \subseteq \mathrm{TC}(x).\, t' \in \bigcup T \wedge t \preccurlyeq t' \rightarrow t \preccurlyeq t' \}.$$

Since T is fixed for a given device we shall usually omit references to it.

4.3. We first consider the case $\mathrm{Sup}\, Fx \subseteq \mathrm{Sup}\, x$, that is, no new labels are introduced. The effect of the located causal neighborhood $x \restriction t$ will be given by $G(x \restriction t)$ for some structural function G; this effect, which we call a *determined region*, is $Fx \restriction s$ for some $s \subseteq \mathrm{TC}(Fx)$. Because we are supposing $\mathrm{Sup}\, Fx \subseteq \mathrm{Sup}\, x$ we must require

(1) $\quad \mathrm{Sup}\, G(x \restriction t) \subseteq \mathrm{Sup}\, t \quad$ for all $t \in \mathrm{CN}_1(x)$;

that is a causal neighborhood must include the region it affects. Typically $Fx \restriction s$ will describe the state of some bounded region V space, and $x \restriction t$ will describe the state at the previous instant of the region consisting of all points (or cells) within a distance ct of V, where c is the velocity of light and t is the time between instants. At this point it is convenient to introduce a bit of notation:

$$u \subseteq^* y \leftrightarrow_{\mathrm{Df}} \exists s \subseteq \mathrm{TC}(y).\, u = y \restriction s.$$

(If one considers y as a tree of its $\in$-chains, then $u \subseteq^* y$ implies that u is a subtree with the same vertex as y). Now let

$$D = \{ v : v \subseteq^* Fx \wedge \exists t \in \mathrm{CN}_1(x).\, v = G(x \restriction t) \};$$

this is the set of determined regions. It might happen that not every causal neighborhood determined a region in D. But if that were allowed, we could construct a machine displaying free will by deciding arbitrarily which of two causal neighborhoods was to take effect (see Counterexample 4.9.1).

Hence we require

(2) $\quad \forall t \in \mathrm{CN}_1(x).\ \exists v \in D.\ v = G(x \upharpoonright t).$

This can be expressed picturesquely as "every cause has an effect".

Finally we require that Fx can be reassembled (perhaps uniquely) from D: every region of Fx must have a cause. D consists of *located* subassemblies, and s is not uniquely determined by $Fx \upharpoonright s$. We might have required that G determines s as well as $Fx \upharpoonright s$, but that is not necessary. Instead we modify Definitions 3.4(iii) and 3.6.

4.3.1. Definition. (i) A set C of located subassemblies *covers* x if

$$\exists Q. \bigcup Q \text{ covers } x \wedge C = \{x \upharpoonright s : s \in Q\}.$$

(ii) Let $C \subseteq \{v : v \subseteq^* x\}$; then x can be *uniquely reassembled* from C iff

$$\forall y.\ [C \text{ covers } y \wedge \forall v \in C.\ v \subseteq^* y] \to y = x.$$

Then our last requirement for the case $\mathrm{Sup}\, Fx \subseteq \mathrm{Sup}\, x$ is

(3) $\quad D$ covers Fx, and if III is satisfied, then Fx can be uniquely reassembled from D.

4.4. Example. We show that the transition function of Counterexample 3.9 satisfies 4.3(1)–(3). First we give the stereotypes for the causal neighborhoods

$$t_1^i = \{a_i'\}, \qquad t_2^i = \{\langle a_i, a_{i+1}\rangle\}, \qquad t_3^i = \{\langle a_i, a_{i+1}\rangle, \langle a_{i+1}, e\rangle\},$$

$$t_4^i = \{\langle a_i, e\rangle\}, \qquad t_5^{i,j} = \{\langle a_i, e\rangle, \{a_i'\}, \{a_i', a_j'\}\}, \qquad t_6 = \{0\},$$

$$t_7 = \{e\},$$

where $a_i' = \{0, a_i\}$ and in the definition of $y^{n,r}$ A_n is to be replaced by $\{a_i' : 1 \leqslant i \leqslant n\}$. But $t_2^i \leqslant t_3^i, t_4^i \leqslant t_5^{i,j}, t_6^i \leqslant t_1^i, t_7 \leqslant t_3^i, t_7 \leqslant t_4^i, t_7 \leqslant t_5^{i,j}, t_1^i \leqslant t_5^{i,j}, t_1^j \leqslant t_5^{i,j}, t_4^{i+1} \leqslant t_3^i, t_6 \leqslant t_5^{i,j}$. From these we compute:

$$\mathrm{CN}_1(x_\alpha^{n,r}) = \{t_1^i : 1 \leqslant i \leqslant n\} \cup \{t_2^i : 1 \leqslant i < n-r-1\} \cup \{t_3^{n-r-1}\} \quad \text{if } r < n-1;$$

$$\mathrm{CN}_1(x_\alpha^{n,n-1}) = \begin{cases} \{t_4^1\} & \text{if } \alpha(n), \\ \{t_5^{1,j} : 1 \leqslant j \leqslant n\} & \text{otherwise;} \end{cases}$$

$$\mathrm{CN}_1(x_\alpha^{n,r}) = \begin{cases} \{t_6^1\} & \text{if } \alpha(n) \text{ and } r \geqslant n, \\ \{t_7^1\} & \text{if } \neg\alpha(n) \text{ and } r \geqslant n. \end{cases}$$

Then define $G(x \upharpoonright t_3^i) = \{\langle a_i, e\rangle\}$, $G(x \upharpoonright t_4^i) = \{0\}$, $G(x \upharpoonright t_5^{i,j}) = \{e\}$ and $G(x \upharpoonright t_k^i) = x \upharpoonright t_k^i$ in all other cases. It is straightforward to verify that conditions

(1)–(3) of 4.3 are satisfied. Note that in the case of t_5 a number of different causal neighborhoods all have the same effect.

4.5. We now turn to the case $\operatorname{Sup} Fx \nsubseteq \operatorname{Sup} x$. The ground was prepared in 2.3. We cannot require $Fx \restriction s = G(x \restriction t)$ but only $Fx \restriction s \simeq_x G(x \restriction t)$; and even this is too strong since the new labels ($\notin \operatorname{Sup} t$) in $G(x \restriction t)$ might by accident belong to $\operatorname{Sup} x$. And we want the new labels in $G(x \restriction t)$ to correspond to new labels in Fx; that is we require $\operatorname{Sup} s \cap \operatorname{Sup} x \subseteq \operatorname{Sup} t$. This condition includes 4.3(1) as a special case. We are thus led to the following definition of the determined regions of Fx (cf. the definition of D in 4.3).

4.5.1. Definition. The set $\mathrm{DR}_1(Fx,x)$ of determined regions of Fx is given by

$$\mathrm{DR}_1(Fx,x) =_{\mathrm{Df}} \{v : v \subseteq^* Fx \wedge \exists t \in \mathrm{CN}_1(x).\ v \simeq_t G_1(x \restriction t) \wedge \operatorname{Sup} v \cap \operatorname{Sup} x \subseteq \operatorname{Sup} t\}.$$

Corresponding to 4.3(2) we give the following formulation of "every cause has an effect".

4.5.2. $\forall t \in \mathrm{CN}_1(x).\ \exists v \in \mathrm{DR}_1(Fx,x).\ v \simeq_t G_1(x \restriction t)$. Notice that, in view of the last clause of the definition of DR_1, 4.5.2 implies the apparently stronger version obtained by replacing "$\simeq_t$" by "$\simeq_x$" provided the function G_1 satisfies $\operatorname{Sup} G_1(x \restriction t) \cap \operatorname{Sup} x \subseteq \operatorname{Sup} t$. In the future we shall suppose, without explicit mention, that G_1 satisfies this proviso.

4.6. Example. *A simple meitotic machine.* For any x let $\pi[x]$ be a permutation of L satisfying $\operatorname{Sup} x \cap \operatorname{Sup} x^{\pi[x]} = \varnothing$. Let Dx, the duplicate of x be defined by

$$Dx = x \cup x^{\pi[x]}.$$

Fix some y_1; set

$$S = \{x : \exists n.\ x \simeq D^n(\{y_1\})\}, \qquad T = \{t : t \simeq \{y_1\}\},$$

$$Gu = Du.$$

Then if we take $F = D$, F satisfies 4.5.2. For if $x \in S$, then $x = \{y_1, y_2, \dots, y_{2^n}\}$ where each y_i is a distinct isomorph of y, and $\mathrm{CN}_1(x,T) = \{\{y_i\} : y_i \in x\}$, and $x \restriction \{y_i\} = \{y_i\}$ and $G(x \restriction \{y_i\}) = \{y_i, y'\}$ where y' is a distinct isomorph of y_i.

A comparison of this extremely simple self-reproducing device with von Neumann's complex 29-state crystalline self-reproducing automaton shows both the power and the limitation of our abstract approach. Unlike, say, the action of a Turing machine, the copying process is done in one stroke.

But the mechanism gives no guidance for the construction of a concrete self-reproductive machine situated in space.

4.7. The example also illustrates an inadequacy of our formulation thus far. If we took, for example, F to be D^2, then the same S, T, G as above could still be used; 4.5.2 would still be satisfied, and $\mathrm{DR}_1(Fx,x)$ would still provide an assembly for x. The definition of DR_1 does not put any bounds on the number of distinct regions which arise from a given causal neighborhood. This lack of determination allows free will to be displayed (see Counterexample 4.9.2). To prevent this we require, roughly speaking, that every cause have a unique effect. More precisely:

4.7.1.

$$\forall v \in \mathrm{DR}_1(Fx,x).\ \forall s \subseteq \mathrm{TC}(Fx).\ Fx \restriction s \simeq_x v \rightarrow Fx \restriction s = v.$$

This requirement could be met, for example, by a meitotic machine if instead of building the daughters of x from new labels, we built them from, say, pairs $\langle a,x \rangle$ so that each daughter cell showed, so to speak, its ancestry within itself. For concrete devices the requirement is met naturally. New labels refer to new regions of space or to new states and these have a distinctive geometrical or structural connection to parts of the causal neighborhood which gave rise to them. All that is required is that this connection be encoded in the description Fx.

4.8. We now come to the last, and most complex, condition. In general new determined regions may overlap. For example, a crystalline automaton may introduce unboundedly many new cells at a given instant. The geometric relations (in particular the neighborhood relation) between these new cells must be causally determined; since the determined regions will be of bounded size, there must be overlaps between them. It is obvious (see Counterexample 4.9.3) that such overlaps must be determined, or else free will will be displayed. Further this determination must be local. It would, I suppose, be possible to specify exactly which new labels of the regions corresponding, say, to $G_1(x \restriction t_1), G_1(x \restriction t_2)$ should be identified. But it seems easier to determine, to within isomorphism over x, $Fx \restriction s$ for a region s which includes $s_1, \ldots, s_r$ where $Fx \restriction s_1, \ldots, Fx \restriction s_r$ are overlapping determined regions. $Fx \restriction s$ is to be determined by an appropriate causal neighborhood. Thus there must be a list T_2 of stereotypes from which the set $\mathrm{CN}_2(x) = \mathrm{CN}(x, T_2)$ is got by Definition 4.2.1, and a new structural function G_2 such that for suitable r the following condition is satisfied.

4.8.1$_r$.

$$\forall V \subseteq \mathrm{DR}_1(Fx,x).\ \left[\, \overline{\overline{V}} \leqslant r \wedge \bigcap \{\mathrm{Sup}\, v : v \in V\} \nsubseteq \mathrm{Sup}\, x \right]$$
$$\rightarrow \exists v \in \mathrm{DR}_2(Fx,x).\ \forall v_1 \in V.\ v_1 \subseteq^* \bar{\bar{v}}.$$

Remarks. (1) Here DR_2 is got from $CN_2(x)$ and G_2 as in 4.5.1., i.e.

$$DR_2(Fx,x) =_{Df} \{v : v \subseteq^* Fx \wedge \exists t \in CN_2(x).\ v \simeq_t G_2(x \upharpoonright t)$$
$$\wedge \operatorname{Sup} v \cap \operatorname{Sup} t \subseteq \operatorname{Sup} t\}.$$

(2) Counterexample 4.9.4 shows that it will not suffice to take $r=2$. In Lemma 5.1 we shall show that an appropriate value for r is one greater than the maximum number of new labels in any determined region.

(3) By substituting $\{v_1\}$ for V in the condition we see that T_2 and G_2 must in effect include T_1 and G_1. However, since we need to determine the overlaps between members of DR_1, but not between those of DR_2, and since moreover we do not require the uniqueness condition 4.7.1 for DR_2, we still need to single out T_1 and G_1.

(4) A weaker form of $4.8.1_r$ could be got by requiring V to range only over subsets of a subset of DR_1 from which Fx could be (uniquely) reassembled. In the absence of III this is too weak. I do not know if the weaker form would suffice to avoid free will in the case of unique reassembly.

Exactly as in 4.5 we need to insist that every cause has an effect; we simply take 4.5.2 with subscript 1 replaced by 2.

The problems of identifying labels in new regions correspond to real problems in real life. Embryologists investigate what causes sheets of tissue to join up, and how migrating groups of cells recognize that they have reached their destination; the problem is to discover local causes (e.g., a gradient of concentration of some substance) for these phenomena. For crystalline automata 4.8.1 is not automatically satisfied. Consider, for example, two adjacent perpendicular blocks of cells already in play forming, as it were, two sides of a rectangular courtyard. Suppose the extremities of the blocks start to "grow" new cells so as to make the other two sides of the courtyard. When these sides meet identifications will have to be made, but these cannot be locally determined. Free will could be displayed; if one chose not to make the identification, the automaton would be growing on a Riemann surface rather than on the plane. One way of ensuring local determination would be to supplement the automaton with a computer which kept track of the coordinates of all cells in play; at each step the computer would output instructions (part of the bounded $t \in CN_2$) about identifications to be made. This technique was used by Eupalinus in the sixth century B.C. so as to ensure that the excavations of a tunnel from both sides of Mount Castro on Samos should meet in the middle (see VAN DER WAERDEN (1954)).

Another way around the difficulty is to arrange that the cells in play of a crystalline automaton always form a convex set. Suppose two cells Q, R in play both require the bringing in to play of a new cell at P. The distances

PQ and PR, and hence also QR are bounded. And since the cells already in play form a convex set, the description x will include a neighborhood of bounded size which determines the spatial relations between Q and R, and hence can be used to prescribe the necessary identification at P. Applying this idea in three dimensions we see that 4.8.1 can be justified for concrete machines (see 2.5(3)).

4.9. We sum up all the requirements we have made:

Principle IV ***The Principle of Local Causality***

There are sets T_1, T_2 *of stereotypes of bounded size and structural functions* G_1, G_2 *such that the conditions below are satisfied.* $\mathrm{CN}_k = \mathrm{CN}(x, T_k)$ *and* $\mathrm{DR}_k = \mathrm{DR}_k(Fx, x)$ *are given, for* $k=1,2$, *by definitions* 4.2.1 *and* 4.5.1.

(1) $\forall t \in \mathrm{CN}_k.\ \exists v \in \mathrm{DR}_k.\ v \simeq_t G_k(x \upharpoonright t) \quad \text{for } k=1,2.$

(2) $\forall v \in \mathrm{DR}_1.\ \forall v' \subseteq^* Fx.\ v' \simeq_x v \rightarrow v' = v.$

For each r

$(3)_r$ $\forall V \subseteq \mathrm{DR}_1.\ \left(\overline{\overline{V}} \leqslant r \wedge \cap \{\mathrm{Sup}\, v : v \in V\} \nsubseteq \mathrm{Sup}\, x\right)$
$\rightarrow \exists v \in \mathrm{DR}_2.\ \forall v_1 \in V.\ v_1 \subseteq^* v.$

(4) DR_1 covers Fx, and if principle III is satisfied, then Fx can be uniquely reassembled from DR_1.

Counterexamples. We hope that by now the interested reader will be able to flesh out our rather schematic treatment. In particular we shall not specify S nor T_1 and T_2, but rather give x, Fx, $\mathrm{CN}_1(x)$ etc. explicitly; it should be obvious that they correspond to structural properties and functions. We use n^* to stand for some structure which encodes the number n, and $0, 1, 2$ for distinguished labels.

4.9.1. (Necessity of IV(1) for $k=1$).
Let $x = \{n^*, 0, 1\}$,

$$Fx = \begin{cases} \{0\} & \text{if } \alpha(n), \\ \{1\} & \text{otherwise.} \end{cases}$$

Take $\mathrm{CN}_1(x) = \{\{0\}, \{1\}\}$ and G_1 the identity function. Evidently II, III and IV(2), (4) are satisfied.

4.9.2. (Necessity of IV(2)).

Let $x_0=\{n^*,0,1\}$. If $\alpha(n)$, let

$$Fx_0=\{0,1,\{a\}\}\ (=y,\text{ say})\quad\text{and}\quad F(Fx_0)=\{0\};$$

while, if $\neg\alpha(n)$,

$$Fx_0=\{0,1,\{a\},\{b\}\}\ (=z,\text{ say})\quad\text{and}\quad F(Fx_0)=\{1\}.$$

Let $t_1=\{0,1\}, t_2=y, t_3=z$; so $t_1\preccurlyeq t_2\preccurlyeq t_3$. Let $G(\{0,1\})=\{0,1,c\}, G(y)=\{0\}, G(z)=\{1\}$. Then II, III, IV(1), (4) are satisfied. The conditions on CN_2 and G_2 are not required as there is no overlapping. The second application of F, and t_2, t_3 were used only to reduce the two non-isomorphic values of Fx_0 to some standard output. In the remaining counterexamples we omit this step.

4.9.3. (Necessity of IV(1) for $k=2$).

Let $x_0=\{n^*,0,1\}$; let

$$Fx_0=\begin{cases}\{\{0,a\},\{1,a\}\}\ (=y\text{ say}) & \text{if } \alpha(n),\\ \{\{0,a\},\{1,b\}\}\ (=z\text{ say}) & \text{if } \neg\alpha(n).\end{cases}$$

Let $CN_1(x_0)=\{\{0\},\{1\}\}$, $G_1(\{0\})=\{\{0,c\}\}$, $G_1(\{1\})=\{\{1,d\}\}$.
Let $CN_2(x_0)=\{\{0\}\}$ and $G_2(\{0\})=\{\{0,c\},\{1,c\}\}$.
The overlap causal neighborhood takes effect only when $\alpha(n)$ is true. II, III, IV(1) (for $k=1$), (2), (3) and (4) are satisfied.

4.9.4. (Necessity of IV(3)$_r$ with $r>2$).

Let $x_0=\{n^*,0,1,2\}$, $p_0=\{a_1,a_2\}$, $p_1=\{a_0,a_2\}$,

$$p_2=\begin{cases}\{a_0,a_1\} & \text{if } \alpha(n),\\ \{a_2,a_3\} & \text{otherwise.}\end{cases}$$

Let

$$Fx_0=\{0,1,2,p_0,\iota p_1,\iota^2 p_2\}.$$

(Fx_0 describes a triangle with vertices a_0,a_1,a_2 if $\alpha(n)$ and a star with center a_2 otherwise).

Take $CN_1(x_0)=\{\{0\},\{1\},\{2\}\}$ and $CN_2(x_0)=\{\{0,1\},\{1,2\},\{2,0\}\}$. Let

$$G_1(\{k\})=\{k,\iota^k\{b,c\}\}\quad(k<3),$$

and

$$G_2(\{j,k\})=\{j,k,\iota^j\{b,c\},\iota^k\{c,d\}\}\quad(j\neq k, j,k<3).$$

One can compute that

$$\mathrm{DR}_1 = \{\{k, \iota^k p_k\} : k < 3\},$$

$$\mathrm{DR}_2 = \{\{j, k, \iota^j p_j, \iota^k p_k\} : j, k < 3, j \neq k\}.$$

All the principles are satisfied if in IV(3) r is given the value 2.

5. The Main Theorem

First we prove a key lemma which is a consequence of IV alone.

5.1. The Key Lemma. *Let S, T_1, T_2, G_1, G_2 be given and let G_1 satisfy*

$$\forall u.\ \mathrm{Card}(\mathrm{Sup}\, G_1 u - \mathrm{Sup}\, u) \leqslant r.$$

Suppose that F, F' satisfy IV (1), (2), (3)$_{r+1}$, (4); *then*

$$\mathrm{DR}_1(Fx, x) \simeq_x \mathrm{DR}_1(F'x, x).$$

We first prove several subsidiary lemmas. Throughout the proofs, the placing of a prime on any expression indicates the result of replacing Fx by $F'x$ and priming all other introduced constants. In particular $\mathrm{DR}_1' = \mathrm{DR}_1(F'x, x)$.

5.2. Lemma. *Let $\mathrm{DR}_1 = \{v_i : i < m\}$; then $\mathrm{DR}_1' = \{v_i' : i < m\}$ where v_i' is the unique $v' \subseteq^* F'x$ which is isomorphic over x to v_i.*

Proof. By the definition of DR_1, given v_i there is some $t_i \in \mathrm{CN}_1$ such that $v_i \simeq_x G_1(x \upharpoonright t_i)$. By IV(1) ($k = 1$) and (2) there is a unique $v' \subseteq^* F'x$ such that $v' \in \mathrm{DR}_1'$ and $v' \simeq_x v_i$. This argument is symmetrical between DR_1 and DR_1' so the result follows.

5.3. Lemma. *Let r and m be as in Lemmas* 5.1 *and* 5.2. *If $K \subseteq \{0, 1, \ldots, m-1\}$, $\overline{\overline{K}} \leqslant r+1$ and*

$$\bigcap \{\mathrm{Sup}\, v_i : i \in K\} \nsubseteq \mathrm{Sup}\, x,$$

then there is a permutation π, which is the identity on x, such that $v_i' = v_i^\pi$ for $i \in K$.

Proof. By IV(3) there is a $v \in \mathrm{DR}_2$ with $v_i \subseteq^* v$ ($i \in K$). By IV(1) (with $k = 2$) there is a $v' \simeq_x v$. Let $v' = v^\pi$. Then $v_i^\pi \subseteq^* v'$, since $\subseteq^*$ is a structural relation. But then, by Lemma 5.2, $v_i^\pi = v_i'$.

We define $\mu(a)=\{i\colon a\in \operatorname{Sup} v_i\}$ and call it the *signature* of a. We write A for $\operatorname{Sup} Fx - \operatorname{Sup} x$.

5.4. Lemma. *For any* $a_0\in A$, $b_0\in A'$,

(i) $\operatorname{Card}\{a\in A\colon \mu(a_0)\subseteq\mu(a)\}\leqslant \operatorname{Card}\{b\in A'\colon \mu(a_0)\subseteq\mu'(b)\}$;

(ii) $\operatorname{Card}\{b\in A'\colon \mu'(b_0)\subseteq\mu'(b)\}\leqslant \operatorname{Card}\{a\in A\colon \mu'(b_0)\subseteq\mu(a)\}$.

Proof. The slight tiresomeness of this lemma and its proof arises from the fact that DR_2 does not tell us which v_i's do *not* overlap. By symmetry it is sufficient to prove (i).

Since, by IV(4), DR_1 covers $Fx, \mu(a_0)\neq\varnothing$; for simplicity, suppose $a_0\in v_0$ so that $0\in\mu(a_0)$. For each $b\in \operatorname{Sup} v_0'\cap A'$ such that $\mu(a_0)\not\subseteq\mu'(b)$ let i_b be the least number such that $i_b\in\mu(a_0)-\mu'(b)$, and let $I=\{i_b\colon b\in \operatorname{Sup} v_0'\cap A'\}$. Now the condition on r implies that $\operatorname{Card}(\operatorname{Sup} v_0'\cap A')\leqslant r$ and so $\bar{\bar{I}}\leqslant r$. Hence by Lemma 5.3, (taking $K=I\cup\{0\}$ and observing that $a_0\in\operatorname{Sup} v_i$ for $i\in K$) there is a π, the identity on $\operatorname{Sup} x$, with $v_i^\pi=v_i'$ for $i\in K$.

We claim that if $a\in \operatorname{Sup} v_0\cap A$ and $\mu(a_0)\subseteq\mu(a)$, then $\mu(a_0)\subseteq\mu'(a^\pi)$. For if not, since $a^\pi\in\operatorname{Sup} v_0^\pi=\operatorname{Sup} v_0'$, there is an $i\in K$ such that $i\notin\mu'(a^\pi)$ and $i\in\mu(a_0)$. Thus $a^\pi\notin\operatorname{Sup} v_i'=\operatorname{Sup} v_i^\pi$, and so $a\notin\operatorname{Sup} v_i$ which contradicts $i\in\mu(a_0)\subseteq\mu(a)$. Thus

$$\{a\in A\colon \mu(a_0)\subseteq\mu(a)\}^\pi\subseteq\{b\in A'\colon \mu(a_0)\subseteq\mu'(b)\}.$$

5.5. Lemma. *For all* $K\subseteq\{0,1,\ldots,m-1\}$

$$\operatorname{Card}\{a\in A\colon \mu(a)=K\}=\operatorname{Card}\{b\in A'\colon \mu'(b)=K\}.$$

Proof. Let ν_K, ν_K' stand for the LHS and RHS of the equation. Then

$$\nu_K=\operatorname{Card}\{a\in A\colon K\subseteq\mu(a)\}-\sum\{\nu_{K'}\colon K\subset K'\}.$$

The result follows readily from Lemma 5.4 by downward induction on $\bar{\bar{K}}$.

Proof of 5.1. μ partitions A into disjoint sets $\{a\in A\colon \mu(A)=K\}$ for those K with $\nu_K\neq 0$. By Lemma 5.5 there is an exactly corresponding partition of A'. Hence we can define a permutation π which is the identity on x, such that

$$\mu'(a^\pi)=\mu(a)\quad \text{for all } a\in A.$$

But then $a\in\operatorname{Sup} v_i\leftrightarrow a^\pi\in\operatorname{Sup} v_i'$, for all $i<m$. So $v_i'=v_i^\pi$ and hence $DR_1^\pi= DR_1'$.

5.6. Corollary. *If S satisfies* III, *then the conclusion of Lemma* 5.1 *can be strengthened to*: $Fx\simeq_x F'x$.

5.7. The Main Theorem. *If S and F satisfy* I–IV, *then, to within isomorphism, F is computable.*

Proof. There are only a finite number of stereotypes of a given rank and given size. Hence the lists T_1 and T_2 may be taken as finite, and there are also only finitely many stereotypes for $\{x \upharpoonright t\colon x \in S,\ t \in T_1 \cup T_2\}$. Thus, to within isomorphism, the domains of G_1 and G_2 are finite and these functions can be given by enumeration. This also gives a bound r to Card ($\mathrm{Sup}\,Gu - \mathrm{Sup}\,u$). By III and Corollary 5.6 the stereotype of Fx (over x) is uniquely determined by the conditions of IV (with $(3)_{r+1}$). Hence it can be computed by a search procedure; for all the conditions of IV involve quantification only over finite lists.

6. Discussion

(1) It is perhaps worth emphasizing how unrestrictive the principles are. Unlike most automata and algorithms which have been proposed, our treatment does not depend on singling out any set of "elementary" operations. The concept of an algorithm introduced by KOLMOGOROV and USPENSKY (1953) shares this feature; but at each step only a bounded portion of the whole state is changed. The "elementary" operations of most procedures can be carried out in a single step by a device satisfying I–IV. For example it is not too hard to construct a machine which will carry out all the outermost reductions in a formula of the λ-calculus in a single step. One exception (pointed out to me by William Boone) is a Markov algorithm. The process of deciding whether a particular substitution is applicable to a given word is essentially global.

(2) I am sorry that Principle IV does not apply to machines obeying Newtonian mechanics. In these there may be rigid rods of arbitrary lengths and messengers travelling with arbitrary large velocities, so that the distance they can travel in a single step is unbounded. I tried to construct a Newtonian device which should calculate some non-computable function without displaying free will, but was quite unsuccessful. Perhaps some elliptic equation—e.g. Laplace's equation—would permit the construction of such a device.

Problem. Find an alternative to IV which would be satisfied by Newtonian machines, but which would not allow free will to be displayed.

(3) I think it fair to say that the main theorem provides a better proof of Turing's Theorem T (see the Introduction) than any given so far. Turing's own analysis makes clear that calculation by a human being will satisfy I;

II and III can always be satisfied by using a suitable form of description for the record and the program of the calculation; and Turing's arguments *almost* forces one to accept IV (local causation) without any further investigation of particulars of the record and its description.

(4) Since abandoning any of II–IV allows free will, Gödel's objection might be met by abandoning any of them. And GÖDEL himself (1958) showed how the use of functionals of unbounded type could be used to transcend finitistic reasoning (though not computability). But I think it plain that a theory of non-mechanical human intelligence would in fact need only to conflict with IV. The non-mechanical intelligence would, so to speak, see the state x as a *Gestalt*, and by abstract thought make global determinations which could not be got at by local methods.

(5) Despite the liberality advertised in (1) above there is a limit to what a machine can do in a single step. The number of stereotypes in HF_n of size not greater than q can be computed from n and q by a function in Grzegorczek's class $\mathcal{E}_3$ (in which the number of iterations of exponentiation allowed is bounded). Hence, as in the proof of the main theorem, one can compute in $\mathcal{E}_3$ a bound for the number of possible stereotypes of values of Fx from the size of x. In particular, if every number can be uniquely coded by some member of S, then any numerical function which can be calculated in a single step must be bounded by a function in $\mathcal{E}_3$.

Problem. Find a weakening of II which does not permit free will to be displayed.

(6) Because our analysis is not tied to any particular choice of elementary operations it serves to emphasize the familiar view that recursive procedures can be characterized without reference to any kind of recursion or inductive process other than iteration. What is essential is the total exclusion of the infinite. POST (1936) entitled his paper "Finite combinatory processes" and TURING wrote (in (1936)) "The computable numbers may be briefly described as the real numbers whose expressions as a decimal are calculable by finite means". Such characterizations may exorcise intimations of the supernatural from theorems (such as those of BOONE and HIGMAN (1974) and HIGMAN (1961)) which define recursive or recursively enumerable structures in a purely algebraic way. In passing we may note an easily proved, trivial, generalization of Higman's theorem.

Theorem. *Let L, L' be two disjoint sets of labels. Let $X \subseteq \mathrm{HF}_n(L)$ be Σ_1 over* $\mathrm{HF}(L)$. *Then we can find a machine (S,F) with $S \subseteq \mathrm{HF}(L \cup L')$ satisfying* II-IV *and an $x_0 \in S$ such that*

$$X = \{x \colon \exists m\ x = F^m x_0 \wedge \mathrm{Sup}\, x \subseteq L\}.$$

(7) The heavy use made of restrictions, and the complications involved in fitting them together (for unique reassembly and the proof of Lemma 5.1) suggest that a treatment using concepts analogous to those of sheaf theory or topos theory might be worth developing. However, it seems to me that the concepts from category theory which would be necessary would be too abstract to allow one to use them (as we have used the more concrete notions of set theory) as a justification for the main thesis of this paper.

Notes added in proof

To 3.7: In the proof of sufficiency we must also consider the case $y \subset x$. Suppose $v = u_1$ and $v \notin y$. Let $s = \{a_2, \ldots, a_n, b\}$ where $a_i \in v/u_i$ for $2 \leqslant i \leqslant n$. Then $x \restriction s \neq y \restriction s$.

To 4.2.1: Observe that if $x \subseteq \mathrm{TC}(t) - t$ and $t' = t \cup x$, then $t \preccurlyeq t'$ and $t' \preccurlyeq t$.

To 4.3.1: It is not clear whether the condition given for unique reassembly from located sub-assemblies is equivalent to that given in 3.6 for unique reassembly from (unlocated) sub-assemblies. Our counter-examples do in fact satisfy the principle in either sense. But I would now favour adopting the suggestion made in the last paragraph of 4.3.

To 4.8: The statement that if unboundedly many new labels are added, then there *must* be overlaps is false. It is sufficient for our purpose that there *may* be overlaps.

References

BOONE, W. W. and G. HIGMAN
[1974] An algebraic characterization of groups with soluble word problem, *J. Austral. Math. Soc.*, **18**, 41–53.
DAVIS, M. (Editor)
[1965] *The Undecidable* (New York)
GARDNER, M.
[1971] Mathematical games, *Sci. Am.*, **223**(4), 120–123; **224**(2), 112–116.
GÖDEL, K.
[1958] Über eine bisher noch nicht benützte Erweiterung des finiten Standpunkts, *Dialectica*, **12**, 280–287.
HIGMAN, G.
[1961] Subgroups of finitely presented groups, *Proc. Roy. Soc. Ser. A*, **262**, 455–475.
KOLMOGOROV, A. N. and USPENSKI, V. A.
[1953] On the definition of an algorithm, *Uspehi Math. Nauk.*, **8**, 125–176; *Am. Math. Soc. Translations*, **29** (1963), 217–245.

KREISEL, G.
[1974] A notion of mechanistic theory, *Synthese*, **29**, 11–26.
POST, E. L.
[1936] Finite combinatory processes. Formulation I, *J. Symbolic Logic*, **1**, 103–105; reprinted in Davis (1965).
POUR-EL, M. B.
[1974] Abstract computability and its relation to the general purposes analog computer, *Trans. Am. Math. Soc.*, **199**, 1–28.
SHEPHERDSON, J. C.
[1975] Computation over abstract structures: serial and parallel procedures and Friedman's effective definitional schemes, in: *Logic Colloquium '73*, edited by H. E. Rose and J. C. Shepherdson (Amsterdam), pp. 445–513.
TURING, A. M.
[1936–7] On computable numbers, with an application to the Entscheidungs problem, *Proc. London Math. Soc. Ser. 2*, **42**, 230–265; reprinted in Davis (1965).
VAN DER WAERDEN, B. L.
[1954] *Science Awakening* (Groningen).
VON NEUMANN, J.
[1966] *The Theory of Self-Reproducing Automata*, edited by A. W. Burks (Urbana and London).
WANG, H.
[1974] *From Mathematics to Philosophy* (London).

J. Barwise, H. J. Keisler and K. Kunen, eds., *The Kleene Symposium*
 149–166

Recent Advances in the Theory of Higher Level Projective Sets*

Alexander S. Kechris
California Institute of Technology, Pasadena, CA 91125, U.S.A.,
and University of Paris VII, Paris, France

Dedicated to Professor S. C. Kleene on the occasion of his 70th birthday

Abstract: An outline is given of certain aspects of the current theory of higher level projective sets, especially concentrating on the structure theory of Π^1_3 sets, from determinacy hypotheses. Some of the key open problems in this area are also discussed.

1. Introduction

Many decades of work in descriptive set theory, from the beginning of this century until the early sixties, resulted in an extensive theory of the projective sets of the first two levels of the projective hierarchy and its effective analog, the analytical hierarchy of Kleene. As it is by now well-known, this work was done in two almost consecutive and originally independent stages. The first one, which today we call *classical descriptive set theory*, lasted until the late 1930's. The second one, which we call today *effective descriptive set theory*, originated, independently of the classical work, in Kleene's pioneering researches in recursion and definability theory during an almost 20 year period starting in the mid-1930's, but it was not before the work of Mostowski and mainly Addison in the early 1950's, that it was recognized as providing an effective refinement and strengthening, and as a consequence an extension, of the classical theory.

Despite these considerable achievements in comprehending the nature of sets of the first two levels of the projective hierarchy, those of level higher than two remained totally inaccessible. Why this was the case was explained by the work in the metamathematics of set theory, originated by Gödel and Cohen. As it turned out, most of the important questions that

*Preparation of this paper was partially supported by NSF Grant MCS76-17254 A01. The author is an A. P. Sloan Foundation Fellow.

were left unanswered concerning higher level projective sets until the early 1960's were unsolvable within the limits of classical ZFC set theory.

Thus, starting from the important discovery of Solovay in the mid-1960's, that the existence of measurable cardinals solves the Lebesgue measurability problem for Σ_2^1 sets, much of the recent work in descriptive set theory uses hypotheses of set theory which transcend the limits of classical ZFC set theory, *large cardinal axioms* and, mainly during the last 10 years, the hypothesis or *Projective Determinacy* (PD) (or more generally various forms of definable determinacy).

This hypothesis was shown by the early 1970's to have dramatic consequences for the structure theory of projective sets, one of the most fundamental ones being the *periodicity of order two* in the basic structure of the levels of the projective (and analytical) hierarchy, established by Martin and Moschovakis.

In fact, about 10 years ago, with the work of Addison, Martin, Moschovakis, Solovay and others, the foundations of a program were laid with ultimate goal to obtain a complete structure theory of all projective sets, using determinacy hypotheses. A lot of progress has been made towards the realization of this program and we can say today that it has been brought to a high degree of conclusion for the *first four* levels of the projective (and analytical) hierarchy. Most of this paper will be devoted to an outline of the theory of Π_3^1 sets as it has now evolved, from which the theory of Σ_4^1 sets can be easily deduced. At the end, we shall try also to give an idea of the main problems lying ahead, as one attempts to progress to the fifth level of the hierarchy and beyond, in order to fully complete the aforementioned program.

2. Review of the theory of Π_1^1 sets

The best way to understand the theory of Π_3^1 sets is by looking at the analogies, similarities and differences, which it exhibits when compared with the well-known theory of Π_1^1 sets. So we shall first review some of the basic principles underlying this last theory, choosing our particular line of presentation so that the comparison can be made as transparent as possible. All the results in this section are standard and their proofs, as well as extensive historical references can be found in MOSCHOVAKIS (1979). For the concepts and facts from the theory of positive elementary induction, which are also needed below, the reader is referred to MOSCHOVAKIS (1974). Also as a general convention in this paper we shall understand that we work in ZF + DC and we will state explicitly all further set theoretical hypotheses that we are using each time.

We can view the theory of Π^1_1 sets as starting with the following

Theorem 2.1 (Basic Representation Theorem for Π^1_1). *A set $A \subseteq \omega^\omega$ is Π^1_1 iff there is a recursive tree T on $\omega \times \omega$ such that*

$$\alpha \in A \Leftrightarrow T(\alpha) \text{ is well-founded.}$$

Our terminology and notation here are standard: A *tree* on $\omega \times \lambda$, where λ is an ordinal, is a set T of pairs (u, v) of finite sequences $u \in \omega^{<\omega}, v \in \lambda^{<\omega}$ with $\text{length}(u) = \text{length}(v)$, closed under subsequences, i.e. if $(u, v) \in T$ and $n < \text{length}(u)$, then $(u \upharpoonright n, v \upharpoonright n) \in T$. For such a tree T and for every real $\alpha \in \omega^\omega$, we let

$$T(\alpha) = \{v : (\alpha \upharpoonright \text{length}(v), v) \in T\},$$

so that $T(\alpha)$ is a tree on λ.

Another important representation of Π^1_1 sets is in terms of inductive definability.

Theorem 2.1′ *A set $A \subseteq \omega^\omega$ is Π^1_1 iff it is inductive in the structure $\hat{\omega} = \langle \omega, < \rangle$.*

Let us explain here the concept of inductive second order relation in a structure. Let $\mathfrak{M} = \langle M, R_1 \cdots R_n \rangle$ be a structure, which for our purposes we can always assume that it contains ω i.e. $\omega \subseteq M$. We will be actually looking only at such $\mathfrak{M}$ with $M = \lambda$, an infinite ordinal. A second order relation $\mathcal{P}(\bar{x}, \bar{S})$, where $\bar{x} = (x_1 \cdots x_m)$ is a tuple of elements of M and $\bar{S} = (S_1 \cdots S_k)$ a tuple of relations, of various arguments, on M, is called *inductive* in $\mathfrak{M}$ if there is a formula $\varphi(\bar{u}, P^+, \bar{x}, \bar{S})$, positive (in P) elementary over $\mathfrak{M}$ with parameters from M, where $\bar{u}$ varies over some M^n and P is an n-ary relation variable, such that for some fixed $\bar{a}$ from M

$$\mathcal{P}(\bar{x}, \bar{S}) \Leftrightarrow \varphi^\infty(\bar{a}; \bar{x}, \bar{S}).$$

Here $\varphi^\infty(\bar{u}; \bar{x}, \bar{S})$ is the least fixed point of the operator determined by φ, viewing $\bar{x}, \bar{S}$ as parameters carried along in the following induction:

$$\varphi^\xi(\bar{u}; \bar{x}, \bar{S}) \Leftrightarrow \varphi\big(\bar{u}, \{\bar{u}' : \exists \eta < \xi \varphi^\eta(\bar{u}'; \bar{x}, \bar{S})\}, \bar{x}, \bar{S}\big)$$

$$\varphi^\infty(\bar{u}; \bar{x}, \bar{S}) \Leftrightarrow \exists \xi \varphi^\xi(\bar{u}; \bar{x}, \bar{S}).$$

If we do not allow φ to contain parameters from M and we take only $\bar{a}$ from ω, then $\mathcal{P}$ is called *absolutely* or *lightface inductive* in $\mathfrak{M}$. In the case $\mathfrak{M} = \hat{\omega}$, these two concepts coincide but this is not the case in practically every other structure, and the distinction between inductive with or without parameters relations will play an important role in understanding some of the basic differences between the theory of Π^1_1 and Π^1_3 sets.

And one minor remark: Although a set of reals $A \subseteq \omega^\omega$ is not literally a second order relation on $\hat{\omega} = \langle \omega, < \rangle$, it can be obviously thought of as one, by identifying each $\alpha \in \omega^\omega$ with its graph.

Starting from these representation theorems, we can proceed now to develop further the theory of Π^1_1 sets. Their first important structural property is summarized in the following result.

Theorem 2.2 (The Prewell-ordering Theorem for Π^1_1). *The class of* Π^1_1 *sets has the prewell-ordering property.*

Recall here that a class Γ has the *prewell-ordering property* if for each $A \subseteq \omega^\omega$ in Γ there is a norm $\sigma : A \to$Ordinals such that the following two relations are in Γ:

$$\alpha \leqslant^*_\sigma \beta \Leftrightarrow \alpha \in A \wedge [\beta \notin A \vee \sigma(\alpha) \leqslant \sigma(\beta)],$$

$$\alpha <^*_\sigma \beta \Leftrightarrow \alpha \in A \wedge [\beta \notin A \vee \sigma(\alpha) < \sigma(\beta)].$$

Such a norm is called a Γ-*norm*.

The proof of Theorem 2.2. can be deduced from either Theorem 2.1 or Theorem 2.1′. If $A \in \Pi^1_1$ and for some recursive T, $\alpha \in A \Leftrightarrow T(\alpha)$ is well-founded, then we can put for $\alpha \in A$:

$$\sigma(\alpha) = \text{rank}(T(\alpha)),$$

and we can then compute directly that $\leqslant^*_\sigma$, $<^*_\sigma$ are Π^1_1. Or, using Theorem 2.1′, let $\alpha \in A \Leftrightarrow \varphi^\infty(\bar{n}; \alpha)$, where $\varphi(\bar{u}, P^+; \alpha)$ is positive elementary on $\hat{\omega}$. Put then for $\alpha \in A$:

$$\tau(\alpha) = \text{least } \xi \text{ such that } \varphi^\xi(\bar{n}; \alpha).$$

By the general theory of inductive definability now, $\leqslant^*_\tau$, $<^*_\tau$ are inductive on $\hat{\omega}$, therefore Π^1_1.

From the Prewell-ordering Theorem one can obtain abstractly reduction for Π^1_1, separation for Σ^1_1, boundedness, coding for Δ^1_1 sets, closure of Π^1_1 under the quantifier "$\exists \alpha \in \Delta^1_1(\beta)$" etc. so that a large portion of the basic properties of Π^1_1 sets is encapsulated in Theorem 2.2.

In dealing, however, with uniformization and basis problems a considerably stronger structural property is required.

Theorem 2.3 (The Scale Theorem for Π^1_1). *The class of* Π^1_1 *sets has the scale property.*

Again a class Γ has the *scale property* if for each $A \subseteq \omega^\omega$ in Γ there is a sequence $\{\sigma_n\}$ of norms, $\sigma_n : \text{A} \to$Ordinals, such that:

(i) If for all i, $\alpha_i \in A$ and $\alpha_i \to \alpha$, while for each $n, \sigma_n(\alpha_i) = \text{constant} \equiv \lambda_n$ for all large enough i, then $\alpha \in A$ and $\sigma_n(\alpha) \leq \lambda_n, \forall n$.

(ii) The relations $P(n,\alpha,\beta) \Leftrightarrow \alpha \leqslant_{\sigma_n}^* \beta$, $Q(n,\alpha,\beta) \Leftrightarrow \alpha <_{\sigma_n}^* \beta$ are in Γ.

One can derive a proof of Theorem 2.3 from Theorem 2.1 by representing a given Π_1^1 set $A \subseteq \omega^\omega$ in the form, $\alpha \in A \Leftrightarrow T(\alpha)$ is well-founded, and then essentially letting $\sigma_n(\alpha) = \text{rank of } s_n$ in $T(\alpha)$, where $\{s_n\}$ is a recursive enumeration of $\omega^{<\omega}$. Actually, one needs to modify these $\{\sigma_n\}$ a bit to make sure that (ii) above is satisfied, but this is irrelevant here.

From Theorem 2.3 follow the Novikov–Kondo–Addison Uniformization Theorem, the Shoenfield-type tree representation of Π_1^1 sets as ω_1-Souslin sets and other important results about Π_1^1 sets.

As far as the basis problem for the dual class of Σ_1^1 sets is concerned, we have the solution in the following

Theorem 2.4 (Kleene Basis Theorem for Σ_1^1). *Let $a \subseteq \omega$ be a complete Π_1^1 set of integers. Then the set of reals Δ_1^1 in a is a basis for Σ_1^1 (i.e. if $A \subseteq \omega^\omega$ is Σ_1^1 and nonempty, it contains a real Δ_1^1 in a).*

For the proof, represent $\emptyset \neq A \subseteq \omega^\omega$ in Σ_1^1 as $\alpha \in A \Leftrightarrow T(\alpha)$ is not well-founded $\Leftrightarrow \exists \beta \forall n (\alpha \restriction n, \beta \restriction n) \in T$, and let (α^*, β^*) be the leftmost branch of T. Clearly $\alpha^* \in A$ and $\alpha \in \Delta_1^1(a)$, as the definition of (α^*, β^*) involves only questions of well-foundedness of recursive trees, which can be answered recursively from a (so that actually α^* is recursive in a).

As opposed to the preceding three results, which express purely analytical structural properties of the class of Π_1^1 sets, the next result provides an important link between Π_1^1 sets on the one hand and admissible set theory on the other and is the basis of an important "duality theory", that reveals basic set theoretic (as opposed to analytic) aspects of the theory of Π_1^1 sets.

Theorem 2.5 (The Companion Theorem for Π_1^1). *A set $A \subseteq \omega^\omega$ is Π_1^1 iff there is a Σ_1 formula $\varphi(x)$ of the language of set theory such that*

$$\alpha \in A \Leftrightarrow L_{\omega_1}\alpha[\alpha] \models \varphi(\alpha).$$

Here ω_1^α = supremum of the recursive in α ordinals = supremum of the lengths of $\Delta_1^1(\alpha)$ well-orderings of ω = the first α-admissible ordinal $> \omega$.

A proof of Theorem 2.5 can be based on Theorem 2.1 or on Theorem 2.1′ and the general companion theory of inductive relations.

Beyond these principles underlying the theory of Π_1^1 sets, there is also a number of basic objects and concepts which play a fundamental role in this theory. These are the concepts of

2.6 (Reducibility, jump operation and associated ordinal assignment). The basic reducibility in the theory of Π_1^1 sets is given by the relation

$$\alpha \leqslant_{\Delta_1^1} \beta \Leftrightarrow \alpha \in \Delta_1^1(\beta).$$

This gives rise to the concept of Δ_1^1-*degree* (or *hyperdegree*), where the Δ_1^1-degree of a real α is defined to be

$$[\alpha]_{\Delta_1^1} = \{\beta : \alpha \leqslant_{\Delta_1^1} \beta \wedge \beta \leqslant_{\Delta_1^1} \alpha\}.$$

There is also an appropriate notion of jump operation corresponding to this notion of degree. If d is a Δ_1^1-degree, then its *jump* d' is given by

$$d' = [W_1^\alpha]_{\Delta_1^1}, \quad \text{if } d = [\alpha]_{\Delta_1^1},$$

where W_1^α is a complete $\Pi_1^1(\alpha)$ set of integers. Finally, to each Δ_1^1-degree d, we associate the ordinal

$$\lambda_1^d = \lambda_1^\alpha = \omega_1^\alpha, \quad \text{if } d = [\alpha]_{\Delta_1^1}.$$

Note also here that λ_1^α = supremum of the lengths of $\Delta_1^1(\alpha)$ prewell-orderings of ω^ω.

These concepts are nicely connected in the following relationship, known as *Spector's Criterion*:

$$d \leqslant e \Rightarrow [d' \leqslant e \Leftrightarrow \lambda_1^d < \lambda_1^e].$$

(Here d, e are Δ_1^1-degrees, say of α, β respectively and $d \leqslant e \Leftrightarrow \alpha \leqslant_{\Delta_1^1} \beta$).

In conclusion, let us also recall how the notion of reducibility $\leqslant_{\Delta_1^1}$ comes in a couple of other disguises, which will turn out to be significant in the next section. We shall only mention the equivalent versions for the absolute notion "$\alpha \in \Delta_1^1$" as those of the reducibility "$\alpha \in \Delta_1^1(\beta)$" can be obtained by direct relativization.

First consider the *Gandy–Kreisel–Tait Theorem*. It states that

$$\alpha \in \Delta_1^1 \Leftrightarrow \alpha \textit{ belongs to every } \omega\textit{-model of analysis} + \mathrm{DC}.$$

Also from 2.1′ we have the following, which can be considered as a version of the Souslin–Kleene Theorem:

$$\alpha \in \Delta_1^1 \Leftrightarrow \alpha \in \mathrm{HYP}(\hat{\omega}) (\Leftrightarrow \alpha \in \mathbf{HYP}(\hat{\omega})).$$

Here for any structure $\mathfrak{M} = \langle M, R_1 \cdots R_n \rangle$, $\mathrm{HYP}(\mathfrak{M}) = \{R \subseteq M^k$: both R and its negation are absolutely inductive on $\mathfrak{M}\}$ and $\mathbf{HYP}(\mathfrak{M})$ is the corresponding concept which allows parameters from M i.e. replaces above "absolutely inductive" by "inductive".

3. The theory of Π^1_3 sets

The first breakthrough in the study of higher than level two projective sets came around 1967, when Martin and Solovay proved their basis theorem for Σ^1_3 sets, assuming that a measurable cardinal exists or actually only that $\forall\alpha$ ($\alpha^{\#}$ exists). From their proof the following semi-representation theorem for Π^1_3 emerged.

Theorem 3.1 (MARTIN and SOLOVAY (1969)). *Assume* $\forall\alpha$ ($\alpha^{\#}$ *exists*). *If* $A\subseteq\omega^\omega$ *is* Π^1_3, *then there is a tree* T *on* $\omega\times u_\omega$ *such that*

$$\alpha\in A\Leftrightarrow T(\alpha) \text{ is well-founded.}$$

Let us explain first what u_ω is. By I^α we denote the class of *Silver indiscernibles* for $L[\alpha]$. Let then

$$\mathcal{U}=\bigcap_{\alpha\in\omega^\omega} I^\alpha,$$

be the class of *uniform indiscernibles*, and let

$$\mathcal{U}=\{u_1,u_2,u_3,\ldots,u_\xi,\ldots,\}_{\xi\in\mathrm{ORD}},$$

be its increasing enumeration. Thus u_ω is the ωth uniform indiscernible. Clearly $u_1=\omega_1$ and $u_2\leqslant\omega_2$. By work of Solovay (see e.g. MARTIN (198?)) cofinality(u_n)=cofinality(u_2), for $n\geqslant 2$, so using AC, $u_\omega<\omega_3$.

There is, however, a better way to understand u_ω. Let $L[\mathcal{R}]$ be the smallest inner model of ZF containing $\mathcal{R}=\omega^\omega$ (so that also $L[\mathcal{R}]\models$DC, using AC in the real world) and let us assume that all games in $L[\mathcal{R}]$ are determined or equivalently that $L[\mathcal{R}]\models$AD. As the questions about the structure of projective sets we are interested in are obviously absolute between the real world and $L[\mathcal{R}]$, we can clearly consider ourselves as working entirely *inside* $L[\mathcal{R}]$. In this case the ordinals u_n for $n\leqslant\omega$ become more familiar, as MARTIN (198?) (for $n=\omega$) and Kunen, Solovay (for $n<\omega$) have shown that

$$\mathrm{ZF}+\mathrm{DC}+\mathrm{AD}\Rightarrow\forall n\leqslant\omega(u_n=\omega_n);$$

see KECHRIS (1979) for an exposition. Thus working in $L[\mathcal{R}]$ (under the stated hypothesis), u_ω is just ω_ω, so that in the real world

$$u_\omega=(\omega_\omega)^{L[\mathcal{R}]}$$

i.e. u_ω can be thought of as "definable" ω_ω.

Although Theorem 3.1 can be used to obtain important results about Σ^1_3 sets (the basis theorem being one of them), it is not sufficient for developing a theory of Π^1_3 sets. For example, it does not imply that Π^1_3 has the prewell-ordering property. Although every Π^1_3 set A can be written in the form $\alpha \in A \Leftrightarrow T(\alpha)$ is well-founded, so that there is a natural norm on A, given by $\sigma(\alpha) = \text{rank}(T(\alpha))$, it is not clear that this defines a Π^1_3-norm. Indeed, as $\forall\alpha$ ($\alpha^{\#}$ exists) holds in $L[\mu]$, where μ is a normal measure on a measurable cardinal, κ and in this model Σ^1_3 has the prewell-ordering property, it follows that Π^1_3 does not have the prewell-ordering property in $L[\mu]$, so that one cannot hope to show using only that $\forall\alpha$ ($\alpha^{\#}$ exists) that the norm provided by Theorem 3.1 establishes that Π^1_3 has the prewell-ordering property. So it is clear that a stronger hypothesis than the existence of sharps (or even measurable cardinals) is needed for the development of a theory of Π^1_3 sets.

Then determinacy burst into the scene and the theory of higher level projective sets developed rapidly in the following years but (perhaps surprisingly) along lines very different than those suggested by the Martin–Solovay result.

First Martin and Moschovakis, independently, proved around 1968 the following theorem.

Theorem 3.2 (MARTIN (1968), Moschovakis (see ADDISON and MOSCHOVAKIS (1968)). *Assume Determinacy* ($\mathbf{\Delta}^1_{2n}$). *Then the class of* Π^1_{2n+1} *sets has the prewell-ordering property.*

The interesting point is that the proofs of this result, which used game theoretic ideas, did not utilize any direct representation of Π^1_{2n+1} sets (so that specializing to $n=0$ they provide new proofs for the prewell-ordering property for Π^1_1).

A few years later, in 1971, Moschovakis proved the analogous result for the scale property.

Theorem 3.3 (MOSCHOVAKIS (1971)). *Assume Determinacy* ($\mathbf{\Delta}^1_{2n}$). *Then the class of* Π^1_{2n+1} *sets has the scale property.*

Again the preceding remarks apply to this theorem as well.

Also the same year Moschovakis established another important fact, first conjectured by Martin, namely

Theorem 3.4 (MOSCHOVAKIS (1979)). *Assume Determinacy* ($\mathbf{\Delta}^1_{2n}$). *Then every* Σ^1_{2n} *game won by player I has a* Δ^1_{2n+1} *winning strategy.*

On the basis of these results a fairly extensive theory was developed, which extended to all higher levels of the analytical hierarchy, *with a periodicity of order 2*, most of the "structural analytical properties" (e.g. those avoiding representations, companion theory etc.) of Π^1_1 (and Σ^1_2 sets). This was already done by 1972 and at that time the prevailing feeling was that "everything" in this analytical context known for Π^1_1, ought to generalize with periodicity of order 2 to higher levels, assuming PD.

Among the few important results of the theory of Π^1_1 sets that had not been yet extended at that time higher up, was the Kleene Basis Theorem 2.4. Periodicity suggested that the following should be true assuming PD:

Let $a \subseteq \omega$ be a complete Π^1_{2n+1} set of integers. Then the set of reals Δ^1_{2n+1} in a is a basis for Σ^1_{2n+1}.

This, however, was proven to be false at the end of 1972 by Martin and Solovay, who showed

Theorem 3.5 (MARTIN and SOLOVAY (198?)). *Assume Determinacy* (Δ^1_{2n}). *If* $n > 1$, *then the set of reals* Δ^1_{2n+1} *in the complete* Π^1_{2n+1} *set of integers is not a basis for* Σ^1_{2n+1}.

Thus the first important fault in the hitherto perfect periodicity picture of the analytical hierarchy was discovered. And this soon led to the discovery of the so called *Q-theory*, developed independently by Kechris and Martin–Solovay, which succeeded in identifying what went wrong with the straightforward generalization of the Kleene Basis Theorem, and developed the notions from which the correct versions of the extensions of this and other results to the higher levels emerged. In the discussion below, we will in general assume PD, when discussing concepts and results at levels higher than two.

In trying to give a brief overview of the key ideas associated with Q-theory, let us first recall that for each odd level $2n+1$ we have the straightforward analogs of the basic reducibility, jump operation and ordinal assignment of the first level, discussed in Remark 2.6. The standard reducibility at level $2n+1 (n \geqslant 0)$ is of course

$$\alpha \leqslant_{\Delta^1_{2n+1}} \beta \Leftrightarrow \alpha \in \Delta^1_{2n+1}(\beta),$$

from which the notion of Δ^1_{2n+1}-degree is obtained in the usual way. If

$$d = [\alpha]_{\Delta^1_{2n+1}}$$

is the Δ^1_{2n+1}-degree of α, its jump d' is defined by

$$d' = [W^\alpha_{2n+1}]_{\Delta^1_{2n+1}},$$

where W^α_{2n+1} is a complete $\Pi^1_{2n+1}(\alpha)$ set of integers. Finally, the corresponding ordinal assignment to the Δ^1_{2n+1}-degree d of α is given by

$$\lambda^d_{2n+1} = \lambda^\alpha_{2n+1} = \textit{supremum of the lengths of the } \Delta^1_{2n+1}(\alpha) \textit{ prewell-orderings of } \omega^\omega.$$

As usual the Spector Criterion also goes through.

In Q-theory now, one develops a new notion of reducibility, which for level 1 coincides with $\leqslant_{\Delta_1^1}$ but for higher levels $2n+1$, where $n \geqslant 1$ it diverges from it. For this we define first, for each $n \geqslant 0$ and each real β, a countable set of reals $Q_{2n+1}(\beta)$ and then put

$$\alpha \leqslant_{Q_{2n+1}} \beta \Leftrightarrow \alpha \in Q_{2n+1}(\beta).$$

The set $Q_{2n+1}(\beta)$ can be given various equivalent (not always trivially) definitions, among which we shall present the following, in order to facilitate comparison with the case $n=0$. (We also give only the absolute case as the relativization is straightforward.) We put

$$\alpha \in Q_{2n+1} \Leftrightarrow \alpha \textit{ belongs to every } \omega\textit{-model of analysis} + \mathrm{DC}$$
$$+ \textit{ Determinacy } \left(\boldsymbol{\Delta}_{2n}^1\right), \textit{ which is}$$
$$\times\ \Sigma_{2n}^1\textit{-absolute} \quad (\text{i.e. } a\ \beta_{2n}\textit{-model}).$$

Thus from the Gandy–Kreisel–Tait Theorem, we see immediately that Δ_1^1-reducibility and Q_1-reducibility coincide i.e. $\leqslant_{\Delta_1^1} = \leqslant_{Q_1}$. But for $n \geqslant 1$, these two notions diverge badly and $\Delta_{2n+1}^1(\beta)$ is just a "tiny" subset of $Q_{2n+1}(\beta)$ (for example $W_{2n+1}^{\beta} \in Q_{2n+1}(\beta)$), so that $\leqslant_{\Delta_{2n+1}^1} \subsetneqq \leqslant_{Q_{2n+1}}$, if $n \geqslant 1$, and each Q_{2n+1}-degree consists of a lot of Δ_{2n+1}^1-degrees bunched together.

The next step is to define an appropriate jump operation. Let d be the Q_{2n+1}-degree of α. Then its jump is defined to be

$$d' = \left[y_{2n+1}^{\alpha}\right]_{Q_{2n+1}}.$$

Here y_{2n+1}^{α} is "the first nontrivial $\Pi_{2n+1}^1(\alpha)$-singleton". Let us explain now the meaning of this term. We shall present again only the absolute case i.e. define "the first nontrivial Π_{2n+1}^1-singleton" y_{2n+1}.

For that consider the set of all Π_{2n+1}^1-singletons $\mathbb{S}_{2n+1}$. It is not very hard to see that if $\alpha \in \mathbb{S}_{2n+1}$, then $[\alpha]_{\Delta_{2n+1}^1} \subseteq \mathbb{S}_{2n+1}$, so that one can view $\mathbb{S}_{2n+1}$ as a set of Δ_{2n+1}^1-degrees. Next one can show that this set is actually well-ordered under its natural partial ordering relation, so it can be enumerated in a transfinite sequence in increasing order $d_0, d_1, d_2, \dots, d_{\xi}, \dots$. Clearly d_0 is the Δ_{2n+1}^1-degree of the Δ_{2n+1}^1 reals i.e. the "trivial" Π_{2n+1}^1-singletons. By definition, y_{2n+1} is any Π_{2n+1}^1-singleton in d_1—this explains obviously its name.

As ought to be expected from our general pronouncements in the beginning of this discussion, for $n=0$ this notion of jump coincides with the old notion of jump for Δ_1^1-degrees. This is equivalent to saying that if $d_0, d_1, \dots, d_{\xi}, \dots$ is the increasing enumeration of the Δ_1^1-degrees of Π_1^1-singletons, then $W_1 =$ the complete Π_1^1 set of integers, is in d_1. This is indeed the case as one can verify by standard methods. On the other hand for $n>0$ the complete Π^1_{2n+1} set of integers W_{2n+1} is *not* a Π^1_{2n+1}-singleton and as we mentioned before it belongs to Q_{2n+1}, so it has Δ^1_{2n+1}-degree much smaller than that of y_{2n+1}, and the notions of jump for Δ^1_{2n+1}-degrees and Q_{2n+1}-degrees diverge wildly for $n>0$.

We can now return to the Kleene Basis Theorem, equipped with these fresh notions of Q-theory. Read in the context of reducibilities and jump operators, it asserts that the reals of Δ^1_1-degree $\leqslant$ the Δ^1_1-jump of $\mathbf{0}(=$least Δ^1_1-degree) form a basis for Σ^1_1. Theorem 3.5 tells us that this fails when $2n+1>1$ and Δ^1_1-reducibilities are replaced by Δ^1_{2n+1}-reducibilities. What turns out to be the correct generalization, is to replace throughout Δ^1_{2n+1}-reducibilities by Q_{2n+1}-reducibilities. Indeed then we have the following extension of the Kleene Basis Theorem to all odd levels of the analytical hierarchy.

Theorem 3.6 (Martin and Solovay (198?)). *Assume Determinacy* ($\mathbf{\Delta}^1_{2n}$). *Then the set of reals which are Q_{2n+1} in the first nontrivial Π^1_{2n+1}-singleton y_{2n+1} is a basis for Σ^1_{2n+1}.*

(In fact "Q_{2n+1} in" can be even replaced by "recursive in" in Theorem 3.6)

Finally we discuss briefly the relevant ordinal assignment for the Q-degrees. If d is the Q_{2n+1}-degree of α, we let

$$\kappa^d_{2n+1}=\kappa^\alpha_{2n+1}=\sup\{\lambda^{\langle\alpha,\beta\rangle}_{2n+1}: y^\alpha_{2n+1}\nleq_{Q_{2n+1}}\langle\alpha,\beta\rangle\}.$$

Again it can be shown that for $n=0$, $\kappa^d_1=\lambda^d_1$ but for $n>0$, $\kappa^\alpha_{2n+1}\gg\lambda^\alpha_{2n+1}$.

The theory outlined from the beginning of this section, undoubtedly provides a very satisfactory and comprehensive extension of the "purely analytical" structural theory of Π^1_1 sets, not only to level 3 but to all levels $2n+1$ as well, from PD. It does not give us, however, any concrete representation theorems of higher level Π^1_{2n+1} sets (such as those of Theorem 2.1 and Theorem 2.1′ for Π^1_1), which would seem to be needed for a better understanding of the nature of analytic sets at higher levels, the solution of several problems which the "pure analytical" theory appears inadequate to deal with and also, quite importantly, for the development of an appropriate companion theory at higher levels. Also there remains the problem of clarifying the concepts and explaining the workings of Q-theory, which undoubtedly are of a somewhat unusual nature, at least from the point of view of the standard theory of Π^1_1 sets.

As it turned out, these problems could be given a satisfactory answer for the third level, on the basis of the following result, proved in 1977, which, following the extensive developments via the alternative route described after Theorem 3.1 and until now, brings us back to the original context of this Martin–Solovay result.

Recall first the definition of the uniform indiscernibles $u_1, u_2, \ldots$, granting that $\forall\alpha$ ($\alpha^{\#}$ exists). Consider then the shift map

$$s_k(u_i) = \begin{cases} u_i, & \text{if } 1 \leqslant i < k, \\ u_{i+1}, & \text{if } i \geqslant k. \end{cases}$$

This can be extended to an embedding $j_k : u_\omega \to u_\omega$ by letting

$$j_k\left(t^{L[\alpha]}(u_1 \cdots u_m)\right) = t^{L[\alpha]}(s_k(u_1) \cdots s_k(u_m)),$$

since by a result of Solovay every ordinal $< u_\omega$ can be written as $t^{L[\alpha]}(u_1 \cdots u_m)$ for some term t and some real α. Let R be the relation on u_ω coding all these embeddings i.e.

$$R(k, \xi, \eta) \Leftrightarrow k \in \omega \wedge j_k(\xi) = \eta.$$

Then put

$$\hat{u}_\omega = \langle u_\omega, <, R \rangle.$$

We now have the following representation theorems for Π^1_3 sets, which are clearly the analogs of Theorem 2.1 and Theorem 2.1′.

Theorem 3.7 (Kechris and Martin (1978)). *Assume Determinacy* ($\mathbf{\Delta}^1_2$). *Then*

(i) *a set* $A \subseteq \omega^\omega$ *is* Π^1_3 *iff there is a tree* T *on* $\omega \times u_\omega$, *recursive in* $\hat{u}_\omega$, *such that*

$$\alpha \in A \Leftrightarrow T(\alpha) \textit{ is well-founded};$$

(ii) *a set* $A \subseteq \omega^\omega$ *is* Π^1_3 *iff it is absolutely inductive in* $\hat{u}_\omega$.

(Here "recursive in $\hat{u}_\omega$" can be interpreted in any reasonable sense of generalized recursion theory.)

Now the necessary conditions of the above equivalences are of course nothing else but essentially Theorem 3.1, the converses being the new facts here. These need also the stronger hypothesis of Determinacy ($\mathbf{\Delta}^1_2$), as the representation Theorem 3.7 easily implies directly that Π^1_3 has the prewell-ordering property, exactly as in the case of Π^1_1 sets using Theorem 2.1′. It also shows from a new point of view, how naturally this determinacy hypothesis enters in picking up the development of the theory of Π^1_3 sets, where the sharp hypothesis or measurable cardinals have left it off.

It is now a routine matter, for everyone familiar with inductive definability to see that Theorem 3.7 yields immediately a corresponding companion theorem.

Let for each structure $\mathfrak{M}$ $\kappa^{\mathfrak{M}}$ be the closure ordinal of positive elementary inductions on $\mathfrak{M}$ or equivalently (for most reasonable structures) the supremum of the hyperelementary in $\mathfrak{M}$ prewell-orderings of the universe of $\mathfrak{M}$ or the ordinal of the smallest admissible set above $\mathfrak{M}$. Thus

$$\kappa^{\hat{\omega}} = \textit{Church–Kleene } \omega_1$$

and

$$\kappa^{\langle \hat{\omega}, \alpha \rangle} = \omega_1^{\alpha}.$$

Similarly

$$\kappa^{\langle \hat{u}_\omega, \alpha \rangle} = \kappa^{\langle u_\omega, <, R, \alpha \rangle}$$

is the smallest $\langle \hat{u}_\omega, \alpha \rangle$-admissible ordinal$>u_\omega$. Working in $L[\mathfrak{R}]$ and assuming AD+DC in this model, we have that

$$u_\omega = \omega_\omega < \kappa^{\langle \hat{u}_\omega, \alpha \rangle} < \omega_{\omega+1}$$

and

$$\sup_\alpha \kappa^{\langle \hat{u}_\omega, \alpha \rangle} = \omega_{\omega+1}.$$

We have now

Corollary 3.8 (KECHRIS and MARTIN (1978)). *Assume Determinacy* ($\boldsymbol{\Delta}_2^1$). *Then a set* $A \subseteq \omega^\omega$ *is* Π_3^1 *iff there is a* Σ_1 *formula* $\varphi(x)$ *such that*

$$\alpha \in A \Leftrightarrow L_{\kappa^{\langle \hat{u}_\omega, \alpha \rangle}}[\hat{u}_\omega, \alpha] \models \varphi(\alpha).$$

Let us see now how Q-theory at the third level finds a nice justification and explanation on the basis of the representation Theorem 3.7 and its consequences.

First it is immediate from Theorem 3.7 that granting Determinacy ($\boldsymbol{\Delta}_2^1$), for any real α,

$$\alpha \in \Delta_3^1 \Leftrightarrow \alpha \in \mathrm{HYP}(\hat{u}_\omega).$$

On the other hand it can be proved from the same hypothesis that

$$\alpha \in Q_3 \Leftrightarrow \alpha \in \mathbf{HYP}(\hat{u}_\omega).$$

Thus the distinction between Δ_3^1 and Q_3 and the split between these two notions, is explained by the absence or presence of ordinal parameters $<u_\omega$ in inductive definability over $\hat{u}_\omega$ i.e. by the divergence of the concepts of "inductive" and "absolutely inductive" over $\hat{u}_\omega$. This of course does not happen at level 1, where we have the structure $\hat{\omega}$, thus the coincidence of Δ_1^1 and Q_1.

Now we come to the jump operation of Q_3-degrees and try to understand the meaning of y_3^α. Taking again the absolute case of y_3, we have seen that it is not the same as W_3, the complete Π_3^1 set of integers. Of course W_3 is absolutely inductive over $\hat{u}_\omega$, but it is not hard to see that it is in $\mathbf{HYP}(\hat{u}_\omega)$. (We have also already mentioned before that it can be seen independently of these ideas that $W_3 \in Q_3$). What y_3 turns out to be equivalent to, however, is this: Let $P \subseteq u_\omega$ be a complete absolutely inductive in $\hat{u}_\omega$ subset of u_ω and let us denote by $\equiv_{\mathrm{HYP}}$ the concept of HYP($\hat{u}_\omega$)-equivalence for subsets of u_ω. Then it turns out that

$$y_3 \equiv_{\mathrm{HYP}} P,$$

using Determinacy ($\mathbf{\Delta}^1_2$) again. Thus y_3 "is" the complete absolutely inductive in $\hat{u}_\omega$ *subset of* u_ω, *not* ω! This again explains the meaning of the jump operation for Q_3-degrees in terms of concepts analogous to those of the standard theory of Π^1_1 sets. It also provides a new way of proving Theorem 3.6 (for $n=1$), which is very similar to that of the Kleene Basis Theorem we sketched in Section 2. (The original one was based on totally different ideas even for the case $n=0$.) Indeed by Theorem 3.7, given $\emptyset \neq A \subseteq \omega^\omega$ in Σ^1_3 we can represent it as $\alpha \in A \Leftrightarrow \exists f \forall n (\alpha \upharpoonright n, f \upharpoonright n) \in T$, where T is a tree on $\omega \times u_\omega$, recursive in $\hat{u}_\omega$. Then we let (α^*, f^*) be the leftmost branch of $T(f^* : \omega \to u_\omega)$. Now $\alpha^* \in A$ and notice that (α^*, f^*) is in HYP($\langle \hat{u}_\omega, P \rangle$), as its definition involves asking well-foundedness questions of **HYP**($\hat{u}_\omega$) trees, which P can easily answer. Thus α^* is in HYP($\langle \hat{u}_\omega, P \rangle$) so by the above in HYP($\langle \hat{u}_\omega, y_3 \rangle$) i.e. it is $\leqslant_{Q_3} y_3$ as asserted in Theorem 3.6.

Finally, we can illustrate the meaning of the ordinal assignment $\alpha \mapsto \kappa^\alpha_3$. As it turns out

$$\kappa^\alpha_3 = \kappa^{\langle \hat{u}_\omega, \alpha \rangle},$$

while on the other hand

$$\begin{aligned} \lambda^\alpha_3 &= \textit{supremum of the lengths of} \\ &\quad \mathrm{HYP}(\langle \hat{u}_\omega, \alpha \rangle)\ \textit{prewell-orderings} \\ &\quad \textit{of } u_\omega \\ &< \textit{supremum of the lengths of} \\ &\quad \mathbf{HYP}(\langle \hat{u}_\omega, \alpha \rangle)\ \textit{prewell-orderings} \\ &\quad \textit{of } u_\omega \\ &= \kappa^{\langle \hat{u}_\omega, \alpha \rangle} \\ &= \kappa^\alpha_3. \end{aligned}$$

So in conclusion, one has now, by combining the various ideas and results mentioned in this section, a fairly complete picture of how Π^1_3 sets look like. Many facts about Π^1_3 are analogous to those about Π^1_1, but the transition from ω to u_ω expressed vaguely in the analogy

$$\frac{\Pi^1_1}{\omega} \sim \frac{\Pi^1_3}{u_\omega},$$

introduces new phenomena. One of them is Q-theory, which is accounted at this level by the fact that Π^1_3 is just the reduct to ω^ω of the absolutely inductive second order relations on $\hat{u}_\omega$, a fact that brings unavoidably in the theory of Π^1_3 the study of ordinals and subsets of u_ω, which remain undetected when one just looks at the purely analytical theory of Π^1_3 sets.

And it is appropriate before we embark into discussing the questions left open for the future, to summarize what has been said in Sections 2 and 3 in the following Table 1. (In the rightmost column the relevant determinacy hypotheses are being assumed.)

Table 1
Comparison of Π^1_1 and Π^1_3.

	Π^1_1	Π^1_3	
Representation Theorems	$A \subseteq \omega^\omega$ is Π^1_1 iff (i) $\alpha \in A \Leftrightarrow T(\alpha)$ is w.f., T on $\omega \times \omega$, recursive. (ii) A is inductive in $\hat{\omega}$	$A \subseteq \omega^\omega$ is Π^1_3 iff (i) $\alpha \in A \Leftrightarrow T(\alpha)$ is well-founded, T on $\omega \times u_\omega$, recursive in $\hat{u}_\omega$. (ii) A is absolutely inductive in $\hat{u}_\omega$.	
Prewell-ordering Theorem	The Prewell-ordering Property holds for Π^1_1	The Prewell-ordering Property holds for Π^1_3	
Scale Theorem	The Scale Property holds for Π^1_1	The Scale Property holds for Π^1_3	
Kleene Basis Theorem	The set of reals $\leqslant_{\Delta^1_1} W_1$ forms a basis for Σ^1_1	The set of reals $\leqslant_{\Delta^1_3} W_3$ forms a basis for Σ^1_3, is false!	The set of reals $\leqslant_{Q_3} y_3$ forms a basis for Σ^1_3
Companion Theorem	$A \subseteq \omega^\omega$ is Π^1_1 iff $\alpha \in A \Leftrightarrow L_{\omega_1^\alpha}[\alpha] \models \phi(\alpha)$, where ϕ is Σ_1	$A \subseteq \omega^\omega$ is Π^1_3 iff $\alpha \in A \Leftrightarrow L_{\kappa_3^\alpha}[\hat{u}_\omega, \alpha] \models \phi(\alpha)$, where ϕ is Σ_1	
Reducibility, jump operation and associated ordinal assignment	(i) $\alpha \leqslant_{\Delta^1_1} \beta \Leftrightarrow \alpha \in \Delta^1_1(\beta)$. (ii) The jump of $[\alpha]_{\Delta^1_1}$ is $[W_1^\alpha]_{\Delta^1_1}$, where W_1^α is the complete $\Pi^1_1(\alpha)$ subset of ω. (iii) $d \mapsto \lambda_1^d$, where if $d = [\alpha]_{\Delta^1_1}, \lambda_1^d = \lambda_1^\alpha = \sup$ of the lengths of $\Delta^1_1(\alpha)$ prewell-orderings of $\omega^\omega (= \omega_1^\alpha)$.	(i) $\alpha \leqslant_{\Delta^1_3} \beta \Leftrightarrow \alpha \in \Delta^1_3(\beta)$. (ii) The jump of $[\alpha]_{\Delta^1_3}$ is $[W_3^\alpha]_{\Delta^1_3}$, where W_3^α is the complete $\Pi^1_3(\alpha)$ subset of ω. (iii) $\mathrm{d} \mapsto \lambda_3^{\mathrm{d}}$, where if $d = [\alpha]_{\Delta^1_3}, \lambda_3^{\mathrm{d}} = \lambda_3^\alpha = \sup$ of the lengths of $\Delta^1_3(\alpha)$ prewell-orderings of ω^ω.	(i) $\alpha \leqslant_{Q_3} \beta \Leftrightarrow \alpha \in Q_3(\beta)$. (ii) The jump of $[\alpha]_{Q_3}$ is $[y_3^\alpha]_{Q_3}$, where y_3^α is the first non-trivial $\Pi^1_3(\alpha)$-singleton. (iii) $d \mapsto \kappa_3^d$, where if $d = [\alpha]_{Q_3}$, $\kappa_3^d = \kappa_3^\alpha = \sup \{\lambda_3^{\langle \alpha, \beta \rangle} : y_3^\alpha \nleqslant_{\Delta^1_3} \langle \alpha, \beta \rangle\}$ $(= \kappa^{\langle \hat{u}_\omega, \alpha \rangle})$.
	$\Delta^1_1 = \mathrm{HYP}(\omega) \cap \omega^\omega$	$\Delta^1_3 = \mathrm{HYP}(\hat{u}_\omega) \cap \omega^\omega$	$Q_3 = \mathbf{HYP}(\hat{u}_\omega) \cap \omega^\omega$

4. To level 5 and beyond

So we have seen that from determinacy hypotheses, one has a fairly good understanding of the Π^1_3 and consequently the Σ^1_4 sets. We address ourselves now to the problems lying ahead, when one climbs up to higher levels.

From the experience gained from the work described in Section 3, it seems that the realization of the full program of establishing a similarly complete structure theory for all analytical sets, essentially comes down to a precise computation of the projective ordinals $\boldsymbol{\delta}^1_n$. Let us recall their definition first:

$$\boldsymbol{\delta}^1_n = \sup\{\xi:\ \xi \textit{ is the length of a } \boldsymbol{\Delta}^1_n \textit{ prewell-ordering of the reals}\}.$$

Now to explain what we mean by a precise computation of $\boldsymbol{\delta}^1_n$, we shall embed ourselves once again in the model $L[\mathfrak{R}]$ and we will assume that AD + DC holds in this model. Obviously the meaning of $\boldsymbol{\delta}^1_n$ remains unchanged in this transition. Now MOSCHOVAKIS (1970) has shown that in these circumstances i.e. assuming AD + DC, each $\boldsymbol{\delta}^1_n$ is a cardinal, and then work of Moschovakis, Martin, Kunen and others revealed the exciting possibility that one might actually be able to compute precisely the position of the $\boldsymbol{\delta}^1_n$'s in the series of alephs. This has been already done for $n \leqslant 4$. It is classical (in ZF + DC only) that

$$\boldsymbol{\delta}^1_1 = \omega_1$$

and MARTIN (198?) showed that

$$\boldsymbol{\delta}^1_2 = \omega_2, \qquad \boldsymbol{\delta}^1_3 = \omega_{\omega+1}, \qquad \boldsymbol{\delta}^1_4 = \omega_{\omega+2}$$

(the last computation also obtained independently by Kunen). It is also known that the following general rules must be always obeyed:

$$\boldsymbol{\delta}^1_{2n+2} = (\boldsymbol{\delta}^1_{2n+1})^+ \quad (\text{Kunen, Martin } (198?))$$

$$\boldsymbol{\delta}^1_{2n+1} = (\rho_{2n+1})^+, \quad \textit{where } \rho_{2n+1} \textit{ is a cardinal of cofinality } \omega \ (\text{Kechris } (1974)).$$

Clearly $\rho_1 = \omega$, $\rho_3 = \omega_\omega = u_\omega$ and the computation of all $\boldsymbol{\delta}^1_n$ is reduced to that of the $\boldsymbol{\delta}^1_{2n+1}$'s or equivalently ρ_{2n+1}'s.

Now from the results in Sections 2 and 3 it is natural to make the following

Conjecture 4.1 *A set $A \subseteq \omega^\omega$ is Π^1_{2n+1} iff it is absolutely inductive in $\hat{\rho}_{2n+1} = \langle \rho_{2n+1}, <, R_{2n+1}\rangle$, for some appropriate R_{2n+1}. Similarly for a tree representation.*

Proof of such a conjecture would lift the known structure theory of the first four levels all the way up. From the experience in proving this conjecture for $n=1$, it seems that it will require a precise computation of ρ_{2n+1}, partly because the methods and techniques developed to achieve that, would seem to be essential for completing such a project. In this sense the precise computation of $\boldsymbol{\delta}_n^1$ for $n \geqslant 5$, features as a key question, apart from its own intrinsic importance.

Now $\boldsymbol{\delta}_5^1$ is the first unknown of these cardinals. Kunen has developed in 1971 very important methods towards computing it, which essentially suggested suprisingly the following value:

$$\boldsymbol{\delta}_5^1 = \omega_{\omega^3+1}.$$

If the picture suggested by the work of Kunen is correct, the value of $\boldsymbol{\delta}_{2n+1}^1$ would depend in an essential way on the number of different regular cardinals, i.e. cofinalities, below $\boldsymbol{\delta}_{2n-1}^1$. For $n=1$, there is only one cofinality $<\boldsymbol{\delta}_1^1 = \omega_1$, i.e. ω, so $\boldsymbol{\delta}_3^1$ has the smallest possible value allowed by the general rules we mentioned before i.e. $\boldsymbol{\delta}_3^1 = \omega_{\omega+1}$. For $n=2$, $\boldsymbol{\delta}_3^1 = \omega_{\omega+1}$ has 3 cofinalities below it, namely ω, ω_1, ω_2 and this may explain the guess for the high value ω_{ω^3+1} for $\boldsymbol{\delta}_5^1$, instead of the minimal allowed one i.e. $\omega_{\omega\cdot 2+1}$.

In any case this is at the moment only speculation, and the first priority at this stage is to actually find the precise value of $\boldsymbol{\delta}_5^1$. The task does not seem easy, but there is no doubt that the effort spent in accomplishing it will be surely amply rewarded (see KECHRIS and MOSCHOVAKIS (1978)).

Added in proof

Martin has now shown that $\boldsymbol{\delta}_5^1 \geqslant \omega_{\omega^3+1}$ (from AD).

References

ADDISON, J. W. and Y. N. MOSCHOVAKIS

[1968] Some consequences of the axiom of definable determinateness, *Proc. Nat. Acad. Sci. U.S.A.*, **59**, 708–712

KECHRIS, A. S.

[1974] On projective ordinals, *J. Symbolic Logic*, **39**, 269–282.

[1978] AD and projective ordinals, in: *Cabal Seminar 76-77, Proc. of Caltech-UCLA Logic Seminar 1976–1977*, edited by A. S. Kechris and Y. N. Moschovakis, *Lecture Notes in Mathematics* (Springer-Verlag, Berlin), Vol. 689.

KECHRIS, A. S. and D. A. MARTIN

[1978] On the theory of Π_3^1 sets of reals, *Bull. Am. Math. Soc.*, **84**, 149–151.

KECHRIS, A. S. and Y. N. MOSCHOVAKIS, editors

[1978] *Cabal Seminar 76–77, Proc. of Caltech-UCLA Logic Seminar 1976–77, Lecture Notes in Mathematics* (Springer-Verlag, Berlin), Vol. 689.

MARTIN, D. A.

[1968] The axiom of determinateness and reduction principles in the analytical hierarchy, *Bull. Am. Math. Soc.*, **74**, 687–689.

[198?] Projective sets and cardinal numbers: Some questions related to the continuum problem, *J. Symbolic Logic*, to appear.

MARTIN, D. A. and R. M. SOLOVAY

[1969] A basis theorem for Σ_3^1 sets of reals, *Ann. Math.*, **89**, 138–160.

[198?] Basis theorems for Π_{2k}^1 sets of reals, to appear.

MOSCHOVAKIS, Y. N.

[1970] Determinacy and prewell-orderings of the continuum, in: *Mathematical Logic, and Foundations of Set Theory*, edited by Y. Bar-Hillel (North-Holland, Amsterdam), pp. 24–62.

[1971] Uniformization in a playful universe, *Bull. Am. Math. Soc.*, **77**, 731–736.

[1974] *Elementary Induction on Abstract Structures* (North-Holland, Amsterdam).

[1979] *Descriptive Set Theory* (North-Holland, Amsterdam), to appear.

J. Barwise, H. J. Keisler and K. Kunen, eds., *The Kleene Symposium*

Kleene's Realizability and "Divides" Notions for Formalized Intuitionistic Mathematics

Joan Rand Moschovakis
Occidental College, Los Angeles, CA 90041, U.S.A.

Dedicated to Professor S.C. Kleene on the occasion of his 70th birthday

Abstract: S.C. Kleene's two major continuing research interests have been recursive function theory and the foundations of intuitionistic mathematics. They come together in his various notions of recursive realizability, which in turn give rise to his "divides" notions. We review the original definitions and results, and indicate a few further developments.

0. Introduction

Stephen Cole Kleene's two major continuing research interests have been recursive function theory and the foundations of intuitionistic mathematics. Both stem from his basic concern with the constructive aspects of mathematics. We begin by exploring Kleene's various notions of recursive realizability, which draw together the two major branches of his research and span the years from 1940 to 1969. In (1973), on which we have relied heavily in the preparation of Sections 1 and 2 of this paper, Kleene has given a personal view of the development of these concepts.

Realizability, like most of Kleene's notions, has been adapted for various purposes by other researchers. In Section 3 we indicate some of these modifications of realizability, some of which (unlike the original notions) are highly non-constructive. A more complete treatment is in Chapter III of TROELSTRA'S (1973).

1. Kleene's number-realizability and "divides" for intuitionistic first-order arithmetic HA

To motivate the notions, we can first consider the intuitionistic interpretation of the logical connectives and quantifiers. For an intuitionist, only the simplest statements (such as prime sentences of number theory) are considered to be complete in themselves, and hence immediately recognizable as true or false. More complicated statements are regarded as incomplete communications of information. For instance, a disjunction $A \vee B$ is an incomplete communication which can be completed only by specifying one of A, B and giving information which completes it. An implication $A \to B$ can be completed by giving a method which, from any bit of information completing A, produces information which completes B (with an equal degree of certainty). An existential statement $(Ex)A(x)$ can be completed by specifying a particular x, and providing information which completes $A(x)$. And a statement $(x)A(x)$ can be completed by giving a method which, for each x, provides information completing $A(x)$.

It follows that a statement of the form $(x)(Ey)A(x,y)$ can be completed by giving a method which, for each x, specifies a particular y and provides information completing the statement $A(x,y)$. This method can obviously be thought of as a function with argument x, and with value the pair $\langle y$, information completing $A(x,y)\rangle$. The idea which occurred to Kleene early in 1940 was that, if $(x)(Ey)A(x,y)$ is an intuitionistically correct number-theoretic statement, then the function completing it should be effectively calculable, and hence (by Church's Thesis) recursive.

So far we have been considering *informal* statements. Although Brouwer himself explicitly rejected formalization of mathematical reasoning, in (1930) HEYTING had published an axiomatization of intuitionistic mathematics; and in (1936) KLEENE had developed an axiomatization of the number-theoretic part **HA** of Heyting's system. So it seemed natural to Kleene to conjecture that for an intuitionistic formal system at least as strong as **HA**:

(*) *If* $\forall x \exists y\, A(x,y)$ *is closed and* $\vdash \forall x \exists y\, a(x,y)$, *then there is a general recursive function* φ *such that, for each* x, $A(\mathbf{x},\mathbf{y})$ *is true for* $y = \varphi(x)$.

Since φ is recursive, it can be given by a Gödel number e.

In attempting to prove this conjecture Kleene was led to seek a property of arbitrary formulas of **HA** such that, on the one hand, its possession by a closed formula $\forall x \exists y\, A(x,y)$ would guarantee the conclusion of (*); and, on the other hand, its possession by all closed theorems of **HA** could be established by an inductive argument on the structure of proofs in **HA**. He was successful in 1941 with the definition of "e realizes E", by induction on the logical form of E, using the recursive functions $(e)_i$(= the exponent of the $(i+1)$st prime in the prime factorization of e). Here (especially in Clause 4) it is essential that the realizing objects only be required to be Gödel numbers of *partial* recursive functions.

Definition 1.1 (KLEENE (1945)). A natural number e *realizes* a closed formula E of **HA**, as follows.

(1) e *realizes* a closed prime formula P, if P is true.

(2) e *realizes* A & B, if $(e)_0$ *realizes* A and $(e)_1$ *realizes* B.

(3) e *realizes* $A \vee B$, if $(e)_0 = 0$ and $(e)_1$ *realizes* A, or $(e)_0 \neq 0$ and $(e)_1$ *realizes* B.

(4) e *realizes* $A \supset B$, if (e is the Gödel number of a partial recursive function $\{e\}$ of one variable such that) whenever a *realizes* A, then $\{e\}(a)$ is defined and *realizes* B.

(5) e *realizes* $\neg A$, if e *realizes* $A \supset 1 = 0$ (i.e. if no number *realizes* A).

(6) e *realizes* $\forall x A(x)$, if for every x, $\{e\}(x)$ is defined and *realizes* $A(\mathbf{x})$.

(7) e *realizes* $\exists x A(x)$, if $(e)_1$ *realizes* $A(\mathbf{x})$ for $x = (e)_0$.

A closed formula E is said to be *realizable* if there is a natural number e which realizes E. A formula $E(y_1, \ldots, y_m)$ containing free at most the variable shown is *realizable* if there is a general recursive function φ such that, for each $y_1, \ldots, y_m$: $\varphi(y_1, \ldots, y_m)$ realizes $E(\mathbf{y}_1, \ldots, \mathbf{y}_m)$.

Because the definition was published in 1945, this notion is sometimes called "1945-realizability", or (more obviously) "number-realizability". David Nelson, who was then a doctoral student, verified the following conjecture of Kleene:

Theorem 1.2 (Nelson (1947)). *For each sequence* Γ *of realizable formulas and each formula* E *of* **HA**: *if* $\Gamma \vdash E$, *then* E *is realizable.*

Corollary 1.3. (∗) *holds for* **HA** *with* "$A(\mathbf{x}, \mathbf{y})$ *is realizable*" *instead of* "$A(\mathbf{x}, \mathbf{y})$ *is true*".

Observe that $1 = 0$ is not realizable. Hence:

Corollary 1.4. *Any extension of* **HA** *obtained by adjoining realizable axioms is consistent.*

One interesting example of a classically true but unrealizable formula (Kleene (1945), Nelson (1947)) is $\forall x(\exists z T_1(x,x,z) \vee \neg \exists z T_1(x,x,z))$, where $T_1(x,y,z)$ numeralwise expresses the predicate "x is a Gödel number of the partial recursive function $\{x\}$ and z is the Gödel number of a computation of $\{x\}(y)$." Hence $\neg \forall x(\exists z T_1(x,x,z) \vee \neg \exists z T_1(x,x,z))$ is classically false but realizable and hence, by Corollary 1.4, consistent with **HA**. Thus also $\neg\neg \forall x(A(x) \vee \neg A(x))$ is unprovable in intuitionistic predicate calculus **Pd**.

The 1945-realizability notion is expressible within **HA**. To each formula A of **HA** another formula $\exists e(e\mathbf{R}A)$ (briefly **R**A) of **HA** is associated such that A is realizable if and only if **R**A is true. Nelson (1947) showed that every instance of $A \backsim \mathbf{R}A$ is realizable, every instance of $\mathbf{R}A \backsim \mathbf{R}(\mathbf{R}A)$ is provable in **HA**, and whenever $\vdash A$, then $\vdash \mathbf{R}A$; however, neither $A \supset \mathbf{R}A$ nor $\mathbf{R}A \supset A$ is a theorem schema for **HA**. By the realizability the system **HA** + $(A \backsim \mathbf{R}A)$(in which realizability and truth are identified) is a consistent extension of **HA**. At the time of KLEENE (1945) it seemed not beyond the bounds of possibility that this was *the* correct intuitionistic system of arithmetic; however, the later discovery of alternative realizability notions has tended against this conclusion (cf. Section 2).

In (1932) Gödel had proved that for formulas A, B of intuitionistic propositional calculus **Pp**: if $\vdash$ A$\vee$B, then $\vdash$A or $\vdash$B. GENTZEN (1934–5) extended this result to closed formulas A, B of **Pd**. By modifying slightly the definition of number-realizability, Kleene was able to obtain for **HA** the corresponding result.

Definition 1.5 (KLEENE (1945 for Γ empty; 1952)). The natural number e *realizes*-($\Gamma\vdash$) E, where Γ is a list of formulas and E a closed formula of **HA**, as follows. We give only the clauses which differ from those of Definition 1.1 (other than by having "realizes-($\Gamma\vdash$)" in place of "realizes").

(3′) e *realizes*-($\Gamma\vdash$) A$\vee$B, if $(e)_0=0$ and $(e)_1$ *realizes*-($\Gamma\vdash$) A and $\Gamma\vdash$A, or $(e)_0\neq 0$ and $(e)_1$ *realizes*-($\Gamma\vdash$)B and $\Gamma\vdash$B.

(4′) e *realizes*-($\Gamma\vdash$) A$\supset$B, if whenever a *realizes*-($\Gamma\vdash$) A and $\Gamma\vdash$A, then $\{e\}(a)$ *realizes*-($\Gamma\vdash$) B.

(7′) e *realizes*-($\Gamma\vdash$)$\exists$x A(x), if $(e)_1$ *realizes*-($\Gamma\vdash$) A(**x**) and $\Gamma\vdash$A(**x**) for $x=(e)_0$.

A closed formula E is *realizable*-($\Gamma\vdash$) if there is a number e which realizes-($\Gamma\vdash$) E. An open formula E($y_1,\ldots,y_m$) is *realizable*-($\Gamma\vdash$) if there is a general recursive function φ such that, for each $y_1,\ldots,y_m$: $\varphi(y_1,\ldots,y_m)$ realizes-($\Gamma\vdash$) E($\mathbf{y}_1,\ldots,\mathbf{y}_m$).

Theorem 1.6 (KLEENE, 1945, 1952). *If* $\Gamma\vdash$E *in* **HA**, *and the formulas* Γ *are realizable*-($\Gamma\vdash$), *then E is realizable*-($\Gamma\vdash$).

Corollary 1.7. *For closed formulas* A, B, $\exists x$ A(x) *of* **HA**:
(a) if $\vdash$A$\vee$B, then $\vdash$A or $\vdash$B;
(b) if $\vdash\exists$x A(x), then $\vdash$A(**x**) for some numeral **x**.

Corollary 1.8. (*) *holds with the conclusion* "$\Gamma\vdash$A(**x**,**y**) *and hence* A(**x**,**y**) *is true*," *for any extension of* **HA** *by axioms* Γ *which are all realizable*-($\Gamma\vdash$) *and true*.

After reading HARROP (1956), Kleene realized that by suppressing the realizing objects in Definition 1.5 he could get an easier proof of Corollary 1.7 and a new characterization of **Pd**. We give Kleene's original definition of "$\Gamma|$E" for a (possibly empty) list Γ of assumption formulas. If Γ is empty, the "$\Gamma\vdash$P" in Clause 1 can be replaced by "P is true" (KLEENE, 1973), making the definition of "$|$E" closer to that of "E is realizable."

Definition 1.9 (KLEENE, 1962, 1963). $\Gamma|\mathrm{E}$ as follows, for Γ a list of closed formulas and E a closed formula of **HA**. Here "$\Gamma|\!\vdash \mathrm{E}$" abbreviates "$\Gamma|\mathrm{E}$ and $\Gamma\vdash \mathrm{E}$".

(1) If P is prime, $\Gamma|\mathrm{P}$ if $\Gamma\vdash\mathrm{P}$.
(2) $\Gamma|\mathrm{A}\,\&\,\mathrm{B}$, if $\Gamma|\mathrm{A}$ and $\Gamma|\mathrm{B}$.
(3) $\Gamma|\mathrm{A}\vee\mathrm{B}$, if $\Gamma|\!\vdash\mathrm{A}$ or $\Gamma|\!\vdash\mathrm{B}$.
(4) $\Gamma|\mathrm{A}\supset\mathrm{B}$, if, if $\Gamma|\!\vdash\mathrm{A}$, then $\Gamma|\mathrm{B}$.
(5) $\Gamma|\neg\mathrm{A}$, if not ($\Gamma|\!\vdash\mathrm{A}$).
(6) $\Gamma|\forall x\,\mathrm{A}(x)$, if, for each numeral $\mathbf{x}$, $\Gamma|\ \mathrm{A}(\mathbf{x})$.
(7) $\Gamma|\exists x\,\mathrm{A}(x)$, if, for some numeral $\mathbf{x}$, $\Gamma|\!\vdash\mathrm{A}(\mathbf{x})$.

(Kleene reads "$\Gamma|\mathrm{E}$" as "Γ divides E", not as "Γ slash E".) If $\mathrm{E}(y_1,\ldots,y_m)$ contains free at most $y_1,\ldots,y_m$, then $\Gamma\|\mathrm{E}(y_1,\ldots,y_m)$ if, for each m-tuple of numerals $\mathbf{y}_1,\ldots,\mathbf{y}_m$, $\Gamma|\mathrm{E}(\mathbf{y}_1,\ldots,\mathbf{y}_m)$.

Theorem 1.10 (KLEENE, 1962, 1963). *If* Γ *is a list of closed formulas and* E *is a formula of* **HA** *such that* $\Gamma\vdash\mathrm{E}$, *then* $\Gamma\|\mathrm{E}$.

Corollary 1.7 is an immediate consequence. Also:

Corollary 1.11. *For closed formulas* A, B, C *of* **Pp**: *if* $\mathrm{C}|\mathrm{C}$ *and* $\vdash\mathrm{C}\supset\mathrm{A}\vee\mathrm{B}$, *then* $\vdash\mathrm{C}\supset\mathrm{A}$ *or* $\vdash\mathrm{C}\supset\mathrm{B}$.

Corollary 1.12. *For closed formulas* C, $\exists x\,\mathrm{A}(x)$ *of* **Pd**: *if* $\mathrm{C}|\mathrm{C}$ *and* $\vdash\mathrm{C}\supset\exists x\,\mathrm{A}(x)$ *then, for some numeral* $\mathbf{x}$, $\vdash\mathrm{C}\supset\mathrm{A}(\mathbf{x})$.

Kleene conjectured, and DE JONGH (1970) proved, that **Pp** is maximal with respect to the property of Corollary 1.11.

2. Kleene's function-realizability for intuitionistic analysis FIM

When Kleene began to develop an axiomatization for intuitionistic analysis about 1950, he naturally thought of extending his realizability interpretation to the new system. There were two obvious obstacles. First, there was no recursive way of adding enough function constants to the formal system to allow the clauses for $\forall\alpha\,\mathrm{A}(\alpha)$ and $\exists\alpha\,\mathrm{A}(\alpha)$ to simply imitate those for $\forall x\,\mathrm{A}(x)$ and $\exists x\,\mathrm{A}(x)$. This problem was easily solved by allowing free function variables in E, and interpreting them informally by particular functions. For symmetry, the number variables could also be interpreted informally, rather than formally by substituted numerals. Thus realizability of a formula E containing free at most the (number and function) variables $\mathbf{\Psi}$ would be relative to a particular list Ψ of natural numbers and number-theoretic functions.

A more serious problem arose in connection with the interpretation of implication. In 1945-realizability, in order for e to realize $A \supset B$, it was only necessary that $\{e\}(a)$ should realize B whenever (A itself was realizable and) a realized A. Now, however, the information completing $A(\boldsymbol{\beta})$ relative to a particular nonrecursive interpretation β of $\boldsymbol{\beta}$ may well depend essentially on β, and hence not be recursive; thus even when $A(\boldsymbol{\beta})$ is not recursively realizable, a realizing object (relative to β) for $A(\boldsymbol{\beta}) \supset B(\boldsymbol{\beta})$ must extract from *any* information a completing $A(\boldsymbol{\beta})$ (relative to β) information completing $B(\boldsymbol{\beta})$ (relative to β) at least as constructively as a completes $A(\boldsymbol{\beta})$. To solve this difficulty Kleene brought in relative recursiveness. In 1951 he first succeeded in working out a realizability definition for his formal system **FIM** of intuitionistic analysis; this was published in (1957). The exposition was greatly simplified, in (1959), by the use of indices ε of continuous functionals, rather than relative Gödel numbers e, as realizing objects. It is this form of the definition which appears in KLEENE and VESLEY (1965).

Definition 2.1 (KLEENE, 1959). For ε, $\alpha \in 2^\omega$ and $t \in \omega$, define $\{\varepsilon\}[\alpha](t) \simeq \varepsilon(\langle t\rangle * \bar{\alpha}(\mu y \varepsilon(\langle t\rangle * \bar{\alpha}(y)) > 0)) \dot{-} 1$. (Here $\langle t\rangle$ is the code 2^{t+1} for a sequence consisting of just t, $*$ is a primitive recursive concatenation operator, and $\bar{\alpha}(y) = \langle \alpha(0), \alpha(1), \ldots, \alpha(y \dot{-} 1)\rangle$ is a primitive recursive code for the sequence of the first y values of α). Then ε is an index of a continuous partial functional, which is total if and only if $(\alpha)(t)(Ey)\varepsilon(\langle t\rangle * \bar{\alpha}(y)) > 0$.

Definition 2.2 (KLEENE, 1957; KLEENE and VESLEY, 1965). If E is a formula of **FIM** containing free at most the variables $\boldsymbol{\Psi}$, and Ψ interpret $\boldsymbol{\Psi}$, then the number-theoretic function ε *realizes*-Ψ E as follows.

(1) ε *realizes*-Ψ a prime formula P, if P is true when $\boldsymbol{\Psi}$ are interpreted by Ψ.

(2) ε *realizes*-Ψ A & B, if $(\varepsilon)_0(=\lambda t(\varepsilon(t))_0)$ *realizes*-Ψ A and $(\varepsilon)_1(=\lambda t(\varepsilon(t))_1)$ *realizes*-Ψ B.

(3) ε *realizes*-Ψ A$\vee$B, if $(\varepsilon(0))_0 = 0$ and $(\varepsilon)_1$ *realizes*-Ψ A, or $(\varepsilon(0))_0 \neq 0$ and $(\varepsilon)_1$ *realizes*-Ψ B.

(4) ε *realizes*-Ψ $A \supset B$, if, for each α, if α *realizes*-Ψ A, then $\{\varepsilon\}[\alpha]$ is defined and *realizes*-Ψ B.

(5) ε *realizes*-Ψ $\neg$A, if ε *realizes*-Ψ $A \supset 1 = 0$.

(6) ε *realizes*-Ψ $\forall x\, A(x)$, if, for each x, $\{\varepsilon\}[\lambda t x]$ is defined and *realizes*-Ψ, x A(x).

(7) ε *realizes*-Ψ $\exists x\, A(x)$, if $(\varepsilon)_1$ *realizes*-Ψ, $(\varepsilon(0))_0$ A(x).

(8) ε *realizes*-Ψ $\forall \boldsymbol{\alpha}\, A(\boldsymbol{\alpha})$, if, for each α, $\{\varepsilon\}[\alpha]$ is defined and *realizes*-Ψ, α $A(\boldsymbol{\alpha})$.

(9) ε *realizes*-Ψ $\exists \boldsymbol{\alpha}\, A(\boldsymbol{\alpha})$, if $\{(\varepsilon)_0\}(=\{(\varepsilon)_0\}[\lambda t0])$ is defined and $(\varepsilon)_1$ *realizes*-Ψ, $\{(\varepsilon)_0\}$ $A(\boldsymbol{\alpha})$.

A closed formula E is *realizable* if there is a general recursive function ε which realizes E. A formula E containing free at most the variables Ψ is *realizable* if there is a general recursive function φ such that, for each Ψ, $\varphi[\Psi]$ realizes-Ψ E.

With this definition Kleene was able to obtain for **FIM** the analog of Nelson's Theorem 1.2:

Theorem 2.3. (Kleene, in KLEENE and VESLEY (1965)). *For each sequence* Γ *of realizable formulas and each formula* E *of* **FIM**: *if* $\Gamma \vdash$ E, *then* E *is realizable.*

The method of the proof uses only reasoning which is correct both classically and intuitionistically, so since $1=0$ is unrealizable:

Corollary 2.4. FIM *is consistent.*

This result is nontrivial because **FIM** contains a postulate, "Brouwer's Principle", which contradicts classical analysis. Intuitively, this principle says that every function from number-theoretic functions to number-theoretic functions is continuous in the natural topology on Baire space.

Corollary 2.5. *Any extension of* **FIM** *obtained by adjoining realizable axioms is consistent.*

For instance, Markov's Principle

$$M_1\colon \forall z \forall x [\neg \forall y \neg T_1(z,x,y) \supset \exists y T_1(z,x,y)]$$

is realizable by a classical argument, so that (classically) **FIM** + M_1 is consistent.

Variants of this basic function-realizability notion were also introduced by KLEENE in (1965) to obtain the independence e.g. of the "fan theorem" (an intuitionistic version of König's lemma) from the other postulates of **FIM**, and of Markov's Principle from **FIM**. He defined "ε C/realizes-Ψ E", for C a class of functions closed under general recursiveness, like "ε realizes-Ψ E" except that ε, α, and the functions in Ψ in the definition are restricted to vary over C. Then a closed formula E is C/realizable/T, for $T \subset C$, if ε C/realizes E for some ε recurive in T. Taking C to be the general recursive functions, Kleene gave an instance of the fan theorem which is not C/realizable, while all the other axioms of **FIM** are, and the rules of inference preserve C/realizability.

To show the independence of M_1 Kleene defined "special realizability" or"$_s$realizability" by assigning orders to formulas (based on their logical form) and corresponding orders to the realizing functions. This changes e.g. the clause for $A \supset B$ (and hence the clause for $\neg A$); now ε ${}_{S}$realizes-Ψ $A \supset B$ if: $\{\varepsilon\}[\alpha]$ is defined and has the same order as B whenever α has the same order as A, and if α ${}_{S}$realizes-Ψ A, then also $\{\varepsilon\}[\alpha]$ ${}_{S}$realizes-Ψ B. Theorem 2.3 still holds for ${}_{S}$realizability, but M_1 is not ${}_{S}$realizable.

Thus not every number-realizable formula of **HA** is ${}_{S}$realizable as a formula of **FIM**. Even ordinary function-realizability disagrees with number-realizability concerning the interpretation of number-theoretic formulas; for instance, the number-realizable formula $\neg \forall x(\exists z T_1(x,x,z) \vee \neg \exists x T_1(x,x,z))$ (cf. Section 1) is not function-realizable, since $\forall x(\exists z T_1(x,x,z) \vee \neg \exists z T_1(x,x,z))$ can be realized by a suitably chosen nonrecursive function.

The scarcity of function constants in the language made it impossible to obtain (∗) and the analog of Corollary 1.7 directly for **FIM** by an informal notion corresponding to "*e* realizes-(Γ⊢)E" or "Γ|E." However, in (1967) KLEENE correctly conjectured that formalizing function-realizability within **FIM** would give both these results; the details are given in (1969). To each formula E of **FIM** Kleene associated two formulas $\varepsilon \, \mathrm{r} \, E$ (expressing simply that E is realized by ε) and $\varepsilon \, \mathrm{q} \, E$ (defined below). Here $!\{\varepsilon\}[\alpha]$ abbreviates $\forall t \exists y \, \varepsilon(\langle t \rangle * \bar{\alpha}(y)) > 0$.

Definition 2.6. (KLEENE, 1967, 1969). If ε is a function variable and E is a formula of **FIM**, then $\varepsilon \, \mathrm{q} \, E$ is again a formula, as follows.

(1) $\varepsilon \, \mathrm{q} \, P$ is P, for P a prime formula.
(2) $\varepsilon \, \mathrm{q}(A \& B)$ is $(\varepsilon)_0 \, \mathrm{q} \, A \& (\varepsilon)_1 \, \mathrm{q} \, B$.
(3) $\varepsilon \, \mathrm{q}(A \vee B)$ is $((\varepsilon(0))_0 = 0 \supset (\varepsilon)_1 \, \mathrm{q} \, A \& A) \& ((\varepsilon(0))_0 \neq 0 \supset (\varepsilon)_1 \, \mathrm{q} \, B \& B)$.
(4) $\varepsilon \, \mathrm{q}(A \supset B)$ is $\forall \alpha(\alpha \, \mathrm{q} \, A \& A \supset !\{\varepsilon\}[\alpha] \& [\{\varepsilon\}[\alpha] \, \mathrm{q} \, B])$.
(5) $\varepsilon \, \mathrm{q} \, \neg A$ is $\forall \alpha \neg(\alpha \, \mathrm{q} \, A \& A)$.
(6) $\varepsilon \, \mathrm{q} \, \forall x A(x)$ is $\forall x(!\{\varepsilon\}[\lambda t x] \& [\{\varepsilon\}[\lambda t x] \, \mathrm{q} \, A(x)])$.
(7) $\varepsilon \, \mathrm{q} \, \exists x A(x)$ is $(\varepsilon)_1 \mathrm{q} A((\varepsilon(0))_0) \& A((\varepsilon(0))_0)$.
(8) $\varepsilon \, \mathrm{q} \, \forall \alpha A(\alpha)$ is $\forall \alpha(!\{\varepsilon\}[\alpha] \& [\{\varepsilon\}[\alpha] \, \mathrm{q} \, A(\alpha)])$.
(9) $\varepsilon \, \mathrm{q} \, \exists \alpha A(\alpha)$ is $!\{(\varepsilon)_0\} \& [(\varepsilon)_1 \, \mathrm{q} \, A(\{(\varepsilon)_0\}) \& A(\{(\varepsilon)_0\})$.

In Theorem 2.7 and Corollaries 2.8 and 2.9, U(y) expresses a primitive recursive function which extracts the value of $\{e\}(x)$ from the computation with Gödel number y.

Theorem 2.7. (KLEENE, 1969). *In* **FIM**: *if* ⊢E, *then there is a numeral* **f** *such that* $\vdash \exists \varepsilon[\forall x \exists y T_1(\mathbf{f}, x, y) \& U(y) = \varepsilon(x) \& \varepsilon \, \mathrm{q} \, \forall E]$.

Corollary 2.8. *For closed formulas* A, B, $\exists x A(x)$, $\exists \alpha A(\alpha)$ *of* **FIM**:

(a) *if* $\vdash A \vee B$, *then* $\vdash A$ *or* $\vdash B$;
(b) *if* $\vdash \exists x A(x)$, *then* $\vdash A(\mathrm{x})$ *for some numeral* **x**;

(c) *if* $\vdash \exists\alpha A(\alpha)$, *then for a suitable numeral* **e**, $\vdash \exists\alpha[\forall x \exists y[T_1(\mathbf{e}, x, y) \& U(y) = \alpha(x)] \& A(\alpha)]$.

Corollary 2.9. (∗)*holds for* **FIM**, *and in fact* φ *has a Gödel number* e *such that* $\vdash \forall x \exists y[T_1(\mathbf{e}, x, y) \& A(x, U(y))]$.

A similar theorem holds for "εr E" (obtained from "εq E" by omitting the "&A", "&B", "&$A((\varepsilon(0))_0)$", "&$A(\{(\varepsilon)_0\})$" in Clauses 3, 4, 5, 7, 9 of the definition):

Theorem 2.10. (KLEENE, 1969). *If* $\vdash$E *in* **FIM**, *then in the (classically as well as intuitionistically correct) subsystem* **B** *of* **FIM** *obtained by dropping Brouwer's Principle*: $\vdash \exists\varepsilon[\forall x \exists y[T_1(\mathbf{f}, x, y) \& U(y) = \varepsilon(x)] \& \varepsilon r \forall E]$ *for some numeral* **f**.

This verifies that the realizability interpretation provides a consistency proof of **FIM** relative to **B** (cf. Corollary 2.4).

3. Realizability: further developments

Since 1969, realizability has evolved in many directions. It has been applied to high-order intuitionistic arithmetics $\mathbf{HA}^n$ and $\mathbf{HA}^\omega$ (cf. TROELSTRA (1971), GOODMAN (1978), to quasiintuitionistic set theories (cf. FRIEDMAN (1973)), and to an extension of **FIM** which is sufficient to formalize the Gödel *Dialectica* interpretation (VESLEY, 1970). Troelstra has given numerous proof-theoretic characterizations of formalized realizability, and LAÜCHLI (1970) defined an abstract notion of realizability for which **Pd** is complete.

Rather than attempt a comprehensive survey, we prefer to focus now on two particular recent developments: Friedman's "realizability" for **HAS** (Heyting's arithmetic with species variables and the Comprehension Axiom) and Krol's "δ-realizability". Neither notion involves recursiveness, and only by stretching the imagination can one consider Krol's neighborhoods δ to be realizing objects.

In order to extend Kleene's "divides" notion to partial systems of intuitionistic analysis, the present author (1967) was forced to consider uncountable conservative extensions of the system in question. FRIEDMAN (1973) showed how to extend the notion also to **HAS** by using conservative extensions. We give TROELSTRA'S (1973) form of Friedman's definition, to facilitate comparison with Definition 1.9.

First define an extension **H** of **HAS** as follows. Let $\mathbf{H}_0 = \mathbf{HAS}$. Then $\mathbf{H}_{n+1}$ is obtained from $\mathbf{H}_n$ by adding, for each formula F of $\mathbf{H}_n$ containing free at most the number variables $x_1,\ldots,x_n$ (and no free species variables) and each collection V of n-tuples of closed terms of $\mathbf{H}_n$, a new species constant $C_{F,V}$ with the axiom $C_{F,V}(x_1,\ldots,x_n)\sim F(x_1,\ldots,x_n)$. Let $\mathbf{H}=\bigcup_{n\in\omega}\mathbf{H}_n$. Clearly **H** is an uncountable conservative extension of **HAS**.

Definition 3.1. (Friedman (1973), as restated by Troelstra (1973)). If E is a closed formula of **H**, then | E as follows. Here "$\vdash$ E" means that E is provable in **H**. Clauses 2–7 are exactly as in Definition 1.9. There are three new clauses.

(1) If t, s are closed (numerical) terms of **H**, then | t=s if $\vdash$t=s. If $t_1,\ldots,t_n$ are closed terms, then | $C_{F,V}(t_1,\ldots,t_n)$ if $\langle t_1,\ldots,t_n\rangle\in V$.

(8) $|\forall X\,A(X)$, if $|A(C_{F,V})$ for all $C_{F,V}$.

(9) $|\exists X\,A(X)$, if $|\vdash A(C_{F,V})$ for some $C_{F,V}$.

Theorem 3.2 (Friedman, 1973). *If* E *is a closed formula of* **HAS** *such that (in* **H** *or* **HAS**) $\vdash$ E, *then* | E.

Corollary 3.3. *For closed formulas* $A\vee B$, $\exists x\,A(x)$, $\exists X\,B(X)$ *of* **HAS**:

(a) *if* $\vdash A\vee B$, *then* $\vdash A$ *or* $\vdash B$;

(b) *if* $\vdash\exists x\,A(x)$, *then* $\vdash A(\mathbf{n})$ *for some numeral* **n**;

(c) *if* $\vdash\exists X\,B(X)$, *then, for some predicate* C,

$$\vdash\exists X\left[\forall x_1\cdots\forall x_n\,(X(x_1,\ldots,x_n)\backsim C(x_1,\ldots,x_n))\,\&\,B(X)\right].$$

Friedman's trick can be adapted to intuitionistic analysis and combined with methods of the present author (1979) to give another proof of Krol's result 3.5 below. Observe that Theorem 3.2 is only claimed for formulas of **HAS**; in particular, it is not true that $|C_{F,V}(t_1,\ldots,t_n)$ for all V, even if $\vdash F(t_1,\ldots,t_n)$ so $\vdash C_{F,V}(t_1,\ldots,t_n)$ for all V.

Krol (1977) introduces a localized "divides" notion, which he calls "δ-realizability", for a system **K** of intuitionistic analysis with Kripke's Schema and a weakened form of Brouwer's Principle ("Myhill's system"). The definition uses an uncountable system $\mathbf{K}^*$ which is in a sense a conservative extension of **K**. Formulas of $\mathbf{K}^*$ are of the form $\boldsymbol{\delta}\subseteq\|A\|$, where δ represents a clopen subset δ of the space $\mathfrak{B}$ of functions $f:\omega\to\{0,1\}$ with the topology inherited from Baire space; $\mathbf{K}_0^*$ is intermediate between **K** and $\mathbf{K}^*$. The definition of "$|(\delta,A)$" is by induction on the logical form of A (for example, $|(\delta,A\vee B)$ if there exist clopen $\delta_1,\delta_2\subseteq\mathfrak{B}$ such that $\delta_1\cup\delta_2=\delta$, $|(\delta_1,A)$ and $\mathbf{K}_0^*\vdash\boldsymbol{\delta}_1\subseteq\|A\|$, and $|(\delta_2,B)$ and $\mathbf{K}_0^*\vdash\boldsymbol{\delta}_2\subseteq\|B\|$).

Theorem 3.4. (Krol, 1977). *If* E *is a closed formula of* **K** *such that* $\vdash$E, *then* $|(\delta,E)$ *For all clopen* $\delta\subseteq\mathfrak{B}$.

For formulas E of K, Krol proves that $\vdash$E if and only if **K***$\vdash \boldsymbol{\delta} \subseteq \|E\|$ for some nonempty clopen $\delta \subseteq \mathfrak{B}$. Hence:

Corollary 3.5. *For closed formulas* A, B, $\exists$x A(x) *of* **K**:
(a) *if* $\vdash$A$\bigvee$B, *then* $\vdash$A *or* $\vdash$B;
(b) *if* $\vdash\exists$x A(x), *then* $\vdash$A(**x**) *for some numeral* **x**.

Since $|(\delta, 0=1)$ implies $\delta = \emptyset$, he also proves

Corollary 3.6. K *is consistent.*

Realizability, in its many variations, has been and no doubt will continue to be a valuable tool for establishing consistency and independence results, and disjuction and existence properties, for intuitionistic formal systems. Its simplest forms also provide effective intuitive interpretations, or "outer models," of these systems. Because the realizability interpretations themselves can be understood either classically or constructively, they help to bridge the gap between intuitionistic and classical mathematics.

Acknowledgements

We thank A.S. Troelstra for bringing Krol's work to our attention, and Y.N. Moschovakis for helpful remarks on the exposition.

References

De Jongh, D. H. J.
[1970] A characterization of the intuitionistic propositional calculus, in: *Intuitionism and Proof Theory* (North-Holland, Amsterdam), pp. 211-217.
Friedman, H.
[1973] Some applications of Kleene's methods for intuitionistic systems, in: *Cambridge Summer School in Mathematical Logic* (Springer-Verlag Berlin), pp. 113–170.
Gentzen, G.
[1934–5] Untersuchungen über das logische Schliessen, *Math. Z.*, **39**, 176–210, 405–431.
Gödel, K.
[1932] Zum intuitionistischen Aussagenkalkul, reprinted in *Ergebnisse eines math. Kolloquiums*, Heft 4, p. 40.

GOODMAN, N.

[1978] Relativized realizability in intuitionistic arithmetic of all finite types, *J. Symbolic Logic*, **43**, 23–44.

HARROP, R.

[1956] On disjunctions and existential statements in intuitionistic systems of logic, *Math. Ann.*, **132**, 347–361.

HEYTING, A.

[1930] Die formalen Regeln der intuitionistischen Mathematik, *Sitz. Preuss. Akad. Wiss., Physik. Math. Klasse*, **1930**, 158–169.

KLEENE, S.C.

[1945] On the interpretation of intuitionistic number theory, *J. Symbolic Logic*, **10**, 109–124.

[1952] Introduction to Metamathematics (North-Holland, Amsterdam); (Noordhoff, Groningen); (Van Nostrand, New York and Toronto), 550 pp.

[1957] Realizability, reprinted in *Constructivity in Mathematics* (North-Holland, Amsterdam) 1959, pp. 285–289.

[1959] Countable functionals, in: *Constructivity in Mathematics* (North-Holland, Amsterdam), pp. 81–100.

[1962] Disjunction and existence under implication in elementary intuitionistic formalisms, *J. Symbolic Logic*, **27**, 11–18.

[1963] An addendum, *J. Symbolic Logic*, **28**, 154–156.

[1967] Constructive functions in "The foundations of intuitionistic mathematics", in: *Logic, Methodology and Philosophy of Science III* (North-Holland, Amsterdam), pp. 137–144.

[1969] Formalized recursive functionals and formalized realizability, *Memoirs Am. Math. Soc.*, **89**, 106 *pp*.

[1973] Realizability: a retrospective survey, in: *Cambridge Summer School in Mathematical Logic* (Springer-Verlag, Berlin), pp. 95–111.

KLEENE, S. C., and R. E. VESLEY

[1965] *The Foundations of Intuitionistic Mathematics, Especially in Relation to Recursive Functions* (North-Holland, Amsterdam), 206 pp.

KROL, M.D.

[1977] Disjunctive and existenial properties of intuitionistic analysis with Kripke's scheme, *Soviet Math. Dokl.*, **18**, (3), 755-758.

LAUCHLI, H.

[1970] An abstract notion of realizability for which intuitionistic predicate calculus is complete, in: *Intuitionism and Proof Theory* (North-Holland, Amsterdam), pp. 227–234.

MOSCHOVAKIS, J.R.

[1967] Disjunction and existence in formalized intuitionistic analysis, in:

Sets, Models and Recursion Theory (North-Holland, Amsterdam), pp. 309–331.

[1979] Disjunction and existence properties for intuitionistic analysis with Kripke's schema (abstract), *J. Symbolic Logic*, to appear.

NELSON, D.

[1947] Recursive functions and intuitionistic number theory, *Trans. Am. Math. Soc.*, **61**, 307–368.

TROELSTRA, A.

[1971] Notions of realizability for intuitionistic arithmetic and intuitionistic arithmetic in all finite types, in: *Proceedings of the Second Scandinavian Logic Symposium* (North-Holland, Amsterdam), pp. 369–405.

[1973] *Metamathematical Investigation of Intuitionistic Arithmetic and Analysis* (Springer-Verlag), Berlin, 485 pp.

VESLEY, R.E.

[1970] A palatable substitute for Kripke's schema, in: *Intuitionism and Proof Theory* (North-Holland, Amsterdam), pp. 197–207.

J. Barwise, H. J. Keisler and K. Kunen, eds., *The Kleene Symposium*
©North-Holland Publishing Company (1980) 181–200

Second order logic and first order theories of reducibility orderings*

Anil Nerode and Richard A. Shore
Department of Mathematics, Cornell University, Ithaca, N.Y. 14853, U.S.A.

Dedicated to Professor S. C. Kleene on the occasion of his 70th birthday

Abstract: We show that the first order theories of many reducibility orderings are recursively isomorphic to second order logic on countable sets (and so to true second order arithmetic). The reduction procedure uses some initial segment results and Spector's theorem on countable ideals in the degrees to code quantification over symmetric irreflexive binary relations. This is known to be enough to obtain full second order logic. Applications to other theories are mentioned as are several to problems of definability in, and automorphisms of, the Turing degrees.

0. Introduction

The first order theory of relational systems has been a mainstay of mathematical logic since FREGE (1879). For many of the common relational systems arising in classical mathematics, the degrees of unsolvability of their complete theories are easy to calculate. They are usually decidable, or full first order arithmetic, or at least analytic in the sense of KLEENE (1955). It was, therefore, somewhat surprising when SIMPSON (1977) proved that the first order theory of the Turing degrees with just the ordering relation has the same degree as the full second order theory of true arithmetic. There were, however, some precedents for such results. In particular within recursion theory itself ELLENTUCK (1970), (1973) showed that the first order theory of the isols based on addition, multiplication (and weak exponentiation) was equivalent to second order arithmetic. In model theory a number of results have been derived by showing that the theory of the infinitely generic or existentially universal models of some theory T (e.g., division rings or groups) is equivalent to second order arithmetic. (See, for example, HIRSCHFELD and WHEELER (1975), Ch. 16, CHERLIN (1976), Ch. IV, Section 3.) All of these results were proved by finding special devices to first code in a countable set equipped with

*The preparation of this paper was partially supported by NSF Grant MCS 77-04013. We would also like to thank C. Jockusch and M. Lerman for helpful conversations and correspondence.

operations coding + and × on the natural numbers and by then using some other device to code monadic second order quantification over this set. We will present a quite different and apparently simpler plan of attack on the general problem of coding second order arithmetic. The main idea is to avoid arithmetic entirely and go directly back to general relational systems. That is we will code full second order logic by coding in quantification over arbitrary relations. The major simplification over coding arithmetic that this allows is based on the fact that it actually suffices to code all symmetric irreflexive binary (s.i.b.) relations. In the setting of first order arithmetic itself this idea first appears in CHURCH and QUINE (1952) who showed how to reduce first order arithmetic to one such relation. Later LAVROV (1963) (see also ERSHOV et al. (1965, Theorem 3.3.3)) and RABIN and SCOTT (n.d.) showed in a general setting how all relations could be coded by s.i.b. ones. They then exploited this argument (RABIN, 1965) to give simple proofs of undecidability results for first order theories by coding such relations in their models. For an exposition of many such results see ERSOV (1965). RABIN (1965, p. 58, and 1977, pp. 614-615), observes that analogous results hold for higher order languages as well but as he did not see any interesting applications, he omitted detailed consideration of second order problems. We will fill this gap by providing such a treatment with suitable applications. The point will be that, as all relations can be coded by s.i.b. ones, quantification over s.i.b. relations gives full second order logic.

Although we will touch on other applications of these methods we will deal in detail only with the case of reducibility orderings. The key ingredient of our coding is the theorem of SPECTOR (1956) that every countable ideal I of degrees has an exact (i.e., minimal) pair of degrees $(\mathbf{a}, \mathbf{b})$ over it (i.e., $I = \{\mathbf{x} | \mathbf{x} \leqslant \mathbf{a} \& \mathbf{x} \leqslant \mathbf{b}\}$). We want to exploit it by devising a coding that converts quantification over arbitrary countable s.i.b. relations to quantification over ideals in the degrees. Spector's theorem then allows us to convert this into first order quantification over pairs of elements thereby giving an interpretation of second order logic on countable structures in the first order theory of the degrees.

In addition to avoiding the particularities of arithmetic a second type of simplication is achieved. Unlike SIMPSON (1977) we need no results using the jump operator nor even any new structural lemmas about the degrees. Indeed other than Spector's theorem on ideals, we only need some old initial segment results. Lachlan's theorem (LACHLAN, 1968) that every countable distributive lattice can be embedded as an initial segment of the degrees is more than sufficient.

It is fitting for this symposium that the structural facts we need about the degrees can be traced quite directly to the seminar in logic conducted by Professor Kleene in Madison in 1953. The proof of Spector's theorem on ideals actually is essentially contained in Section 4 of KLEENE and POST (1954) on which the seminar was based. It first appears in SPECTOR (1956) which was a direct outgrowth of that seminar. Of course, this paper also contains the direct ancestor of all initial segment results: There is a minimal degree. This theorem answered a major open question in the Kleene and Post paper.

As is frequently the case, the rewards of simplication are greater generality and wider applicability. Since our constructions rely on only these two properties of the Turing degrees we can almost immediately deduce the same result for many other reducibilities. Thus the first order theories of one–one, many–one, truth-table, weak-truth-table and arithmetic degrees as well as Turing degrees are all recursively isomorphic to true second order arithmetic (or equivalently second order logic on countable sets).

Analogous results also hold for reasonable substructures of these orderings as well. The proofs, however, are a bit more technical and will be included in NERODE and SHORE (1980). We will, however, discuss these results and the key ingredients of their proofs along with many other applications in the fourth section of this paper. The applications will give much information about automorphisms of the Turing degrees and general definability questions for many reducibility orderings. For the first area, for example, we can show that every automorphism of the Turing degrees is the identity on a cone. As to definability in the Turing degrees we can show that every class closed under the jump which is definable in true second order arithmetic is definable in $\langle \mathcal{D}, \leqslant, \mathbf{0}'' \rangle$ as is the ω-jump and every such relation on degrees above $\mathbf{0}^{(\omega)}$. Again these results will be discussed in Section 4 but the detailed computations and proofs will appear elsewhere.

In this paper we take the liberty of proceding at a leisurely pace. As we view this paper as propaganda for a particular approach to coding second order logic or arithmetic as well as an expository work, we will go into more detail than might otherwise be necessary in the first two sections. Readers familiar with undecidability results in general should just skim these sections. The first section will consist of a review of material on codings using s.i.b. relations with enough pictures to convince the reader that quantification over such relations suffices for all of second order logic. In Section 2 we will show how to describe in a first order way some distributive lattices such that quantification over lattice ideals codes quantification over all s.i.b. relations on the atoms of the lattice. Thus full second order logic can be interpreted in the theory of distributive lattices with monadic quantification over ideals. It is then an easy matter to translate this into the first order theory of the Turing degrees using

Spector's theorem on countable ideals and Lachlan's embedding theorem. This will be done in Section 3 along with similar treatments for the other reducibility orderings. The last section will, as we have mentioned, discuss applications and generalizations of these methods.

1. s.i.b. relations and second order logic

We will begin by essentially reproducing pictures and explanations from an unpublished paper of Rabin and Scott (n.d) (some of which appears in Rabin (1964)) that show how to code arbitrary relations by s.i.b. ones. We will then try to make explicit the use of these codings for interpreting full second order logic that is implicit in Rabin (1965). (It was intentionally omitted from that account and that of Rabin (1977, §3) (which we also recommend) as it was irrelevent to the concerns of those papers.)

1.1. Binary relations. Suppose we are given a structure $\mathfrak{A}=\langle A,R\rangle$ with R an arbitrary binary relation. We wish to construct $\mathfrak{A}'=\langle A',S\rangle$ so that $A'\supseteq A$ and to code R by the symmetric irreflexive binary relation S on a subset of A' definable from S. Relativization will then supply an effective transformation F_1 taking a formula φ involving R to one φ^{F_1} involving S such that $\mathfrak{A}\models\varphi\Leftrightarrow\mathfrak{A}'\models\varphi^{F_1}$.

For each $x\in A$ we will add two new elements $t_1(x)$ and $t_2(x)$ tagging it as being in the field of R. For each pair (x,y) such that $R(x,y)$ we also add three new elements $u(x,y)$, $v(x,y)$ and $w(x,y)$ to form the set A'. The s.i.b. relation S is then defined as indicated by the connecting lines in Fig. 1.1.

To be quite explicit we say that $S(x,t_i(x))$ holds for each $x\in A$. Also, for each x,y such that $R(x,y)$, all of $S(x,u(x,y))$, $S(x,v(x,y))$, $S(w(x,y),v(x,y))$, $S(w(x,y),y)$ and their converses hold. We now define our transformation F_1 on φ by relativizing the quantifiers and free variables (as if they were universally quantified) to the elements of A defined by the tagging:

$$T_S(x)\equiv(\exists t_1,t_2)\big[t_1\neq t_2\,\&\,S(x,t_1)\,\&\,S(x,t_2)\\ \&\,\forall y(S(y,t_1)\vee S(y,t_2)\rightarrow y=x)\big].$$

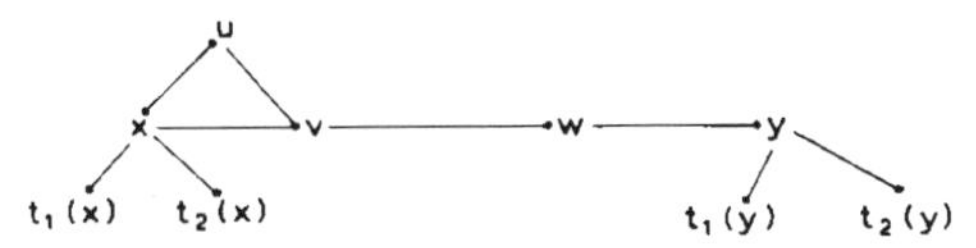

Figure 1.1

Figure 1.2

We then replace $R(x,y)$ in the matrix by $(R(x,y))^{F_1}$:

$$(R(x,y))^{F_1} \equiv \exists u,v,w\left[S(x,u)\,\&\,S(x,v)\,\&\,S(u,v)\right.$$
$$\left.\&\,S(v,w)\,\&\,S(w,y)\,\&\,v\neq w\right].$$

It is clear that $\mathfrak{A}'\models(R(x,y))^{F_1}\Leftrightarrow\mathfrak{A}\models R(x,y)$ for $x,y\in A$ and so our transformation F_1 is as required. (Note that we use two tags rather than one to distinguish the elements of A not in the field of R as well. The point w is included to allow for either $R(x,x)$ or $\neg R(x,x)$ for each $x\in A$.)

The next step in our translation project is to show how to code an n-ary relation P as a binary one.

The pictures for $n=3$ and $n=4$ are given in Fig. 1.2.
Here the arrows $u\rightarrow y_1$ indicate that $R(u,y_1)$ holds. Given then a structure $\mathfrak{A}=\langle A,P\rangle$ with P n-ary we construct A' by adding on to A $1+\Sigma_{i=1}^{n}i$ new elements for each n-tuple $\vec{x}$ such that $P(\vec{x})$ holds and defining R as indicated by the arrows above. Again, for those with a love of indices, for each $\vec{x}$ such that $P(\vec{x})$ we add new elements $u(\vec{x})$ and $y_j^i(\vec{x})$ for each $1\leqslant j\leqslant i$, $1\leqslant i\leqslant n$. We then define R to hold only of the pairs $(u(\vec{x}),y_j^i,(\vec{x}))$, $(y_j^i(\vec{x}),y_{j+1}^i(\vec{x}))$ and $(y_i^i(\vec{x}),x_i)$ for $i\leqslant n,j<i$. This then defines $\mathfrak{A}'=\langle A',R\rangle$.

We now transform a sentence φ about $\mathfrak{A}$ into one φ^{F_2} about $\mathfrak{A}'$ by relativizing all quantifiers and free variables to the subset A of A' which is defined by $T_R(x)\equiv\forall y\neg R(x,y)$ and then replacing $P(x_1,\ldots,x_n)$ by $(P(x_1,\ldots,x_n))^{F_2}$:

$$(P(x_1,\ldots,x_n))^{F_2}\equiv\exists u\,\exists y_1^1\,\exists y_1,y_2^2,\exists y_1^3\cdots$$
$$\times\left[\underset{i<n}{\&}R(u,y_1^i)\,\&\underset{j<i<n}{\&}R(y_j^i,y_{j+1}^i)\,\&\underset{i<n}{\&}R(y_i^i,x_i)\right].$$

Again it should be clear that for $x_i\in A$ $\mathfrak{A}\models P(x_1,\ldots x_n)$ iff $\mathfrak{A}'\models(P(x_1,\ldots,x_n)^{F_2})$ and so $\mathfrak{A}\models\varphi\Leftrightarrow\mathfrak{A}'\models\varphi^{F_2}$. Indeed, although we do not need it here, an only slightly more complicated procedure (attaching an i-cycle to u to indicate the ith predicate) gives a transformation coding an arbitrary string $P_1, P_2, \ldots P_i \ldots$ of n_i-ary predicates as a single binary one. This procedure appears in RABIN and SCOTT (n.d.) .

Let us turn now to second order logic by which we mean the usual first order logical language augmented by variables $P_{n,i}$ ranging over *all* n-ary predicates. We call the language of second order logic $\mathfrak{L}^2$.

Note first that our transformations F_2 and F_1 could equally well be applied to structures with many predicates translating $P_1, P_2, \ldots, P_m$ into $R_1, \ldots, R_m$ and then into $S_1, \ldots, S_m$ respectively except that we must arrange for the domains to coincide. We can therefore slightly modify a combined transformation to handle our second order language. Without loss of generality we may start with a formula φ in a prenex form in which all the second order variables are at the beginning of the formula. We wish to specify one subset as the intended domain for all our coded predicates and another as our stock of "new" elements for filling in the required configurations. To simplify matters we will only consider infinite structures. We build all of these ideas into one sentence:

$$\psi \equiv \forall y \exists ! x \left[x \neq y \,\&\, R(x,x) \,\&\, R(y,x) \right] \&\, \forall x \exists ! y \left[y \neq x \,\&\, R(y,x) \right]$$

$$\&\, \forall y \exists ! x \left[x \neq y \,\&\, \neg R(x,x) \,\&\, R(y,x) \right].$$

We will think of $R' = \{x | R(x,x)\}$ as our domain and $R'' = \{x | \neg R(x,x)\}$ as our stock of coding elements. We transform a formula φ of $\mathfrak{L}^2$ into one φ^{F_4} with only binary relation symbol by setting $\varphi^{F_4} \equiv \exists R(\psi \,\&\, \varphi^{F_3})$. φ^{F_3} is gotten by first replacing all second order quantifiers $\exists P_{n,i} (\forall P_{n,i})$ by $\exists R_{n,i} (\forall R_{n,i})$ which range over binary relations only. Then all first order quantifiers and free variables are relativized to R' and all atomic formulas $P_{n,i}(\vec{x})$ are replaced by $(P_{n,i}(\vec{x}))^{F_2}$ with the new quantifiers restricted to R''. If one believes in the first order translation given by F_2 it is not hard to see that φ is satisfiable (valid) in some infinite structure A iff φ^{F_4} is also. (Note that a second order formula is satisfiable in one infinite structure if it is satisfiable in any structure of the same cardinality. Thus we may keep the set A fixed in the second order case.)

Finally we reduce a formula φ of $\mathfrak{L}^2$ to one φ^F with only s.i.b. relations. φ^F is just $\exists S(\psi^{F_1} \& \varphi^{F_5})$. φ^{F_5} is gotten from φ^{F_3} (which we think of as being fully written out, i.e. $(\exists x)^{R'} \cdots$ is $(\exists x)(R(x,x) \& \cdots))$ as one might expect. We replace the second order quantifiers $\exists R_{n,i} (\forall R_{n,i})$ by $\exists S_{n,i} (\forall S_{n,i})$. First order quantifiers are relativized to the elements appropriately tagged by S. Atomic formulas are replaced by their images under F_1. Again the correctness of the translation in the first order case easily convinces one that for any infinite A and any φ of $\mathfrak{L}^2$, $A \models \varphi$ iff $A \models \varphi^F$.

2. s.i.b. relations and distributive lattices

In this section we will show how to pick out in a first order way certain distributive lattices for which quantification over ideals can be used to replace quantification over all s.i.b. relations on the minimal elements (atoms) of the lattice. Combining this with the results of Section 1 will give us a transformation G such that for any sentence φ of $\mathfrak{L}^2$ there is an infinite $A \models \varphi$ if and only if there is a distributive lattice $\mathfrak{L}$ (always assumed to have a least and a greatest element) with $\mathfrak{L} \models \varphi^G$. ($\varphi^G$ will be in the language LI, i.e. lattice theory augmented by second order monadic quantification over ideals $I_{n,i}$.) This is really just an example of a more general phenomenon of restricted versions of monadic quantification giving full second order logic. Another example that we will discuss briefly below is the theory of commutative rings with quantification over ideals. These results for rings and lattices contrast quite sharply with those of Rabin for Boolean algebras. RABIN (1969) shows that the theory of countable Boolean algebras with quantification over ideals is in fact decidable.

Let us consider some (distributive) lattice $\mathfrak{L}$ with atoms $\{a_i\}$. The key property we need is independence:

Definition 2.1. A subset X of $\mathfrak{L}$ is *independent* if for any $x \in X$ and any finite $F \subseteq X, x \leqslant \bigvee F$ iff $x \in F$.

The point is that if some set X such as the atoms of $\mathfrak{L}$ is independent then quantification over ideals of $\mathfrak{L}$ gives monadic quantification over X. One just translates a subset of X as the ideal it generates.

To move up to quantification over s.i.b. relations over A, the set of atoms, we want our lattice to have some sort of code $c(a_1, a_2)$ for each unordered pair of distinct atoms $\{a_1, a_2\}$. If these codes form an independent set, then quantification over ideals gives monadic quantification over the set of codes which in turn is obviously equivalent to quantification over s.i.b. relations on A: $S(a_i, a_j) \equiv c(a_i, a_j) \in I_S$ where I_S is the ideal generated by $\{c(a_i, a_j) | S(a_i, a_j)\}$. Of course we will want the codes to be effectively definable to give our final translation. To this end we will specify that

(2.1) for each pair of distinct atoms a_1 and a_2 there will be a unique element of the lattice which is strictly above $a_1 \bigvee a_2$ but not above any elements other than a_1, a_2, $a_1 \bigvee a_2$, and zero.

The next step is to guarantee that A and C, the set of codes, are independent subsets of L. We can do so directly in LI (see 4.3) or we can even do it in a first order way by requiring that L contain "pseudo-complements" $\bar{a}, \bar{c}$ (not necessarily unique) for each $a \in A$ and $c \in C$. By this we mean the following:

$$(\forall a_1 \in A)(\exists \bar{a}_1)(\forall a_2 \in A)(a_2 \leqslant \bar{a}_1 \leftrightarrow a_1 \neq a_2) \tag{2.2}$$

and

$$(\forall c_1 \in C)(\exists \bar{c}_1)(\forall c_2 \in C)(c_2 \leqslant \bar{c}_1 \leftrightarrow c_1 \neq c_2). \tag{2.3}$$

As both A and C are definable these are sentences of lattice theory.

$$\big(C(x) \equiv (\exists a_1, a_2 \in A)\big[\, a_1 \neq a_2 \,\&\, a_1 \vee a_2 < x \,\&\, (\forall y > 0) \\ \big[\, y < x \rightarrow y = a_1 \text{ or } y = a_2 \text{ or } y = a_1 \vee a_2 \big]\big).$$

As there is a unique such x for each $\{a_1, a_2\}$ by (2.1) this also shows that the function $c(a_1, a_2)$ is also definable as well. Thus if $x \in A(C)$ and F is a finite subset of $A(C)$ with $x \notin F$, then $x \nleqslant \bigvee F$ as required for independence.

We now begin to describe our translation of second order logic into LI. The first step is to define a transformation G on (prenex) formulas φ of second order logic. One forms φ^F and then replaces every quantifier $\exists S_{m,i}(\forall S_{n,i})$ by $\exists I_{n,i}(\forall i_{n,i})$ and each occurrence of $S_{n,i}(x,y)$ by $c(x,y) \in I_{n,i}$. The first order quantifiers in φ^F are of course all relativized to A as well. Now for φ^G to faithfully represent φ it must be interpreted in a lattice satisfying (2.1)–(2.3). Note that any φ^F is satisfiable only in infinite domains and so if $\mathfrak{L} \models (2.1)\text{–}(2.3) \,\&\, \varphi^G$, I must have infinitely many atoms. Finally we define φ^H to be $(2.1) \,\&\, (2.2) \,\&\, (2.3) \,\&\, \varphi^G$.

It should be clear that a sentence φ is satisfiable over some [every] infinite (cardinality $\kappa \geqslant \omega$) domain iff φ^H is satisfiable in some [every] distributive lattice satisfying (2.1)–(2.3) (with κ many atoms). Or at least this should be clear assuming such distributive lattices exist. To answer even such concerns we give a concrete (effective) field of sets representation of the standard countable lattice we have in mind: Let N denote the natural numbers and let C be any (recursive) one–one onto map from $[E]^2$, the pairs of even natural numbers, to $\mathcal{O}$, the odd natural numbers. Our lattice $L_0 \subseteq 2^N$ is the one generated by $\cup$ and $\cap$ from the elements $\{x\}$, $\{x, y, C(x,y)\}$, $E - \{x\}$ and $N - \{C(x,y)\}$.

We have thus proven the following:

Theorem 2.2. *The theory of distributive lattices in the language LI is recursively isomorphic to that of full second order logic (on infinite structures).*

The restriction to infinite structures can of course be eliminated by slightly complicating the translation to restrict quantification to a definable subset of the atoms but this is not too important. We also immediately have similar results for many other restricted monadic theories in which distributive lattices can be interpreted. A rather trivial example is that of the full monadic theory of a binary relation. Actually this particular result follows quite easily from the first order codings alone and was surely known to many of the workers in this area. We mention it explicitly for the sake of our propagandistic intentions and a specific example: THOMASON (1975) showed that the monadic theory of one binary relation can be interpreted in the usual tense and model logics. He remarks that this is a substantial portion of second order logic. These observations show that it is in fact all of second order logic. We should also note that S. KRIPKE (private communication) proved in 1976 that various fragments of a number of modal systems ($S4, M, B, K, K4$, etc.) with propositional quantifiers are recursively isomorphic to full second order logic using some related coding schemes.

3. Reducibility orderings

Our next goal is to give a faithful translation of LI into the first order theories of certain types of partial orderings.

Let $\mathcal{P}$ be a partial ordering. By a translation J of LI into the first order theory of $\mathcal{P}$ we mean an effective map from formulas Ψ of LI to ones Ψ^J of the language of partial orderings. We call J faithful (with respect to countable structures) if for all Ψ, Ψ is satisfiable in some (countable) distributive lattice iff Ψ^J is satisfiable in $\mathcal{P}$. We will specify conditions on $\mathcal{P}$ which guarantee the existence of such a faithful translation but first some definitions for upper semi-lattices.

Definition 3.1. An *ideal* I in $\mathcal{P}$ is a non-empty subset of $\mathcal{P}$ such that $x \leqslant y \in I$ implies that $x \in I$ and $x,y \in I$ implies that $x \vee y \in I$. Every pair (x,y) from $\mathcal{P}$ determines an ideal $I_{x,y} = \{z \in \mathcal{P} \mid z \leqslant x \,\&\, z \leqslant y\}$ the pair (x,y) is called an *exact* pair for I if $I = I_{x,y}$. The *segment* $[u,v]$ of $\mathcal{P}$ is the set $\{z \in \mathcal{P} \mid u \leqslant z \,\&\, z \leqslant v\}$.

Now for our conditions on $\mathcal{P}$.

(3.1) $\mathcal{P}$ is an upper semi-lattice with least element 0.

(3.2) Every countable ideal in $\mathcal{P}$ has an exact pair in $\mathcal{P}$.

(3.3) Every countable distributive lattice $\mathcal{L}$ is isomorphic to some segment of $\mathcal{P}$.

Although we could get by with weaker conditions, (3.1)–(3.3) will simplify the exposition. For the translation from LI, condition (3.3) tells us that every countable distributive lattice $\mathfrak{L}$ can be realized as a segment of $\mathcal{P}$. Condition (3.2) then says that we can replace (second-order) quantification over ideals of $\mathfrak{L}$ by (first order) quantification over pairs from $\mathcal{P}$. More formally, given Ψ, a formula of LI, we transform it into $\Psi^J \equiv \exists u, v([u,v]$ is a distributive lattice and $\Psi')$ where Ψ' is obtained as follows: First replace $\bigvee$ and $\bigwedge$ by their usual definitions in terms of $\leqslant$. Then relativize all first order quantifiers to $[u,v]$. Finally replace all second order quantifiers $(\exists I)$ by $(\exists x)(\exists y)$ and atomic formulas $a \in I$ by $a \leqslant x \,\&\, a \leqslant y$. It is clear from conditions (3.1)–(3.3) that a sentence x is true in some countable distributive lattice iff Ψ^J is true in $\mathcal{P}$.

Our final transformation of second order logic to the first order theory of some partial orderings $\mathcal{P}$ is thus given by $\varphi \rightarrow \varphi^K = (\varphi^H)^J$. Stringing together our stated equivalences now shows that a sentence φ of second logic is true in some countable set iff φ^K is true in $\mathcal{P}$. It will be useful for some of the applications discussed in Section 4 to observe that we only need the embeddability of one lattice satisfying conditions (2.1)–(2.3) as a segment of $\mathcal{P}$.

We have thus proven the following:

Theorem 3.2. *If $\mathcal{P}$ satisfies conditions* (3.1), (3.2) *and* (3.3) (*at least for some distributive lattice satisfying* (2.1)–(2.3)), *then satisfiability* (*validity*) *in countable structures for second order logic is one—one reducible to the first order theory of $\mathcal{P}$.*

Remark 3.3 It is well-known that the theory of true full second order arithmetic ($\mathrm{Th}^2(\langle N, =, +, \times, \rangle)$) is reducible to second order logic on countable structures. One simply uses the usual second order definitions of an ω-ordering and the standard inductive definitions of $+$ and $\times$ from the ordering. Indeed if one codes n-tuples as numbers either is equivalent to the monadic second order theory of true arithmetic (i.e. $\mathrm{Th}(\langle N, 2^N, \in, \leqslant, +, \times \rangle)$) usually referred to as (true) second order arithmetic or (true) analysis. So another rephrasing of Theorem 3.2 is that the truth set of analysis is one–one reducible to $\mathrm{Th}(\langle P, \leqslant \rangle)$ for any partial ordering of the specified type.

Now we will apply known theorems on the structure of various reducibility orderings (with some minor variations) to characterize the exact degree of their theories as that of second order arithmetic.

Theorem 3.4 *The second order theory of true arithmetic is recursively isomorphic to the first order theories of the degrees of the following reducibilities: one–one, many–one, truth-table, weak-truth-table, Turing and arithmetic.*

Proof. In one direction it is obvious that all of these orderings are effectively definable in second order arithmetic and so their theories are one–one reducible to that of true analysis.

For the other direction we need only show, except for the one–one degrees, that the hypothesis of Theorem 3.2 are satisfied. (The one–one degrees present special problems which we will deal with later.) The other orderings are trivially upper-semi-lattices and so it suffices to verify conditions (3.2) and (3.3). The required embedding theorems are given by LACHLAN (1970) for many–one degrees, LACHLAN (1968) for Turing degrees and HARDING (1974) for the arithmetic degrees.

The embedding results for truth-table and weak-truth-table degrees (introduced in Friedberg and ROGERS (1959)) are also essentially given by most proofs of the embedding theorem for Turing degrees with perhaps some minor modifications. We assume familiarity with some exposition of this theorem (as in EPSTEIN (1979) or LERMAN (1981)) and sketch a proof of the following:

Lemma 3.5. *Every countable distributive lattice can be embedded as an initial segment of the tt and wtt degrees.*

Proof. (sketch): The key point is to use trees T_e such that for each n if $\varphi_e^{T(\sigma)}(n)$ is defined for any σ, then $\varphi_e^{T(\tau)}(n)$ is defined for all τ of length n. It is then easy to see that if A is the representative of the top degree constructed and φ_e^A is a characteristic function, then $\varphi_e^A \equiv_{tt} A(X)$ for some recursive $X \subseteq N$ representing an element x of the lattice ($n \in A(X)$ iff the nth element of X is in A.) The usual proof shows that $\varphi_e^A \equiv_T A(X)$ but the reduction procedures can be made total quite simply. In the recovery of $A(X)$ from φ_e^A if we hit a branching on the appropriate tree T_e which differs on X, then it must split for some y. If one computation agrees with φ_e^A follow it. If neither, then output zero for all further calculations. To compute $\varphi_e^A(n)$ from $A(X)$ go to a level of T_e of length n and compute $\varphi_e^{T(\tau)}$ for any τ agreeing with $A(X)$ up to this point if there is one to get the required answer. If all such $T(\tau)$ are incompatible with $A(X)$ output zero from now on. These procedures are clearly total and applied to φ_e^A give the same results as the usual ones. For more details we refer the matter to EPSTEIN (1979) or LERMAN (1981).

Thus Turing reducibility and truth-table reducibility coincide (and so also with weak truth table reducibility which is caught between them) on the sets T-reducible to A. The initial segments of these degrees below A of course are also then the same and we have our embedding theorem for *tt* and *wtt* degrees.

Turning now to condition (3.2) we note that for Turing degrees this is just the theorem of SPECTOR (1956). Essentially the same proof, however, gives the result for many–one, truth table and weak truth table degrees. (We include the proof so as to be able to refer to it for the case of one–one degrees.)

Let φ_e be a list of the appropriate reduction procedures and let $I=\{a_i\}$ with representative sets A_i. We construct X and Y in stages. At stage $s=\langle s_1,s_2\rangle$ we have X_s and Y_s defined on the first $s-1$ columns of X and Y and on at most finitely many other places. We ask if some finite extensions X' of X_s or Y' of Y_s guarantee that $\varphi_{s_1}^{X'}(x)\neq\varphi_{s_2}^{Y'}(x)$ for some x. If so choose such extension. If not let $X'=X_s$, $Y'=Y_s$. We now get X_{s+1} and Y_{s+1} by extending X' and Y' so that their sth column (the elements such that $\langle s,y\rangle$ belongs to them) is equal to A_s except perhaps at the finitely many points of the column which are already fixed. The point of course is that if $\varphi_{s_1}^X=\varphi_{s_2}^Y$, then at stage $s=\langle s_1,s_2\rangle$ any extension of X_s or Y_s would give the same set and so if we fill out X_s to Z by letting $Z(n)=0$ when X_s is not defined at n, then we have $\varphi_{s_1}^Z$ but Z is even many–one reducible to $\{A_i|i<s\}$ so that the degree of $\varphi_{s_1}^X$ is in I.

We now have to verify (3.2) for the arithmetic degrees. We assume familiarity with perfect forcing as in SACKS (1971):

Let $\{A_i\}$ be the sets with arithmetic degree in the given ideal I. Let $G_1\times G_2$ be generic with respect to the product of the partial ordering of pointed perfect closed sets arithmetic in some A_i with itself. As the conditions are pointed, A_i is arithmetric in G_1 and G_2 for each i. Standard arguments now show that if T is arithmetic in both G_1 and G_2, then T is arithmetic in some A_i: If so some condition

$$(P_1,P_2)\Vdash\forall x\left(\varphi_{e_1}^{G_1^{(n)}}(x)=\varphi_{e_2}^{G_2^{(n)}}(x)\right)$$

for some P_1,P_2 and n. By the definition of forcing

$$\varphi_{e_1}^{T_1^{(n)}}(x)=\varphi_{e_2}^{T_2^{(n)}}(x)$$

for any $T_1\in P_1, T_2\in P_2$. Thus T is arithmetic in any such T_1 or T_2 but we can choose T_1,T_2 arithmetic in P_1,P_2 and so in some A. Thus the degree of T is in I.

This completes our proof except for the one–one degrees which are a bit more complicated. The first problem is that by YOUNG (1964) they do not form an upper semi-lattice. We could have proven our main theorem for directed sets but this is not the real source of difficulties. The main problem arises in the verification of (3.2). Following the outline above we build sets X and Y for a given sequence of sets $\{A_i\}$ using 1–1 reduction procedures. We then get that if Z is one–one reducible to X and Y, then it is reducible to the disjoint union of B_i for $i\leqslant s$ some s where B_i differs from A_i only finitely. Thus we must assume that the sets in I are closed under disjoint union (in the notation of LACHLAN (1969) which we follow for this discussion $A\oplus B=\{2x|x\in A\}\cup\{2x+1|x\in B\}$) and finite modifications.

Now if $\mathbf{0} \leqslant_1 \mathbf{a}$ ($\mathbf{0}$ is the degree of the infinite coinfinite recursive sets), then $\mathbf{a}$ is closed under finite modifications. Moreover LACHLAN (1969) points out that each countable distributive lattice can be embedded as an initial segment of the one–one degrees above $\mathbf{0}$. As $\mathbf{0}$ is definable in the ordering of one–one degrees we can work as Lachlan does entirely above it without any loss of generality. (To see that $\mathbf{0}$ is definable argue as follows:

(1.) $\{\emptyset\}$ and $\{N\}$ are of minimal one–one degree.

(2.) $\{\emptyset\} \leqslant_1 \mathbf{a}$ iff for every $A \in \mathbf{a}$ $\bar{A}$ has an infinite recursive subset, and $\{N\} \leqslant_1 \mathbf{a}$ iff every $A \in \mathbf{a}$ has an infinite recursive subset. Thus $\{\emptyset\} \leqslant_1 \mathbf{a} \& \{N\} \leqslant_1 \mathbf{a}$ iff $\mathbf{0} \leqslant_1 \mathbf{a}$.

(3.) No other degree is minimal: If $\emptyset \nleqslant_1 B \& N \nleqslant_1 B$, then $\{x | 2x \in B\} <_1 B$.

Thus $\mathbf{0}$ is the least degree above the two minimal ones.)

To complete the argument for the faithfulness of our translation from second order logic to the first order theory of the partial ordering of the one–one degrees we need only show that for lattices L such as described in Section 2 embedded as initial segments above $\mathbf{0}$ and below some $\mathbf{a}$, any ideal (of the lattice) generated by a subset of the codes is closed under disjoint union. [We then can replace the quantification over ideals by quantification over pairs $\mathbf{x},\mathbf{y}$ such that $I_{\mathbf{x},\mathbf{y}}$ is an ideal of the embedded lattice (a first order restriction on $\mathbf{x}$ and $\mathbf{y}$).] To verify this condition it suffices to show that, for any codes $\mathbf{c}_1, \mathbf{c}_2, \ldots, \mathbf{c}_n$ in $\mathfrak{L}, \mathbf{c}_1 \oplus \mathbf{c}_2 \oplus \cdots \oplus \mathbf{c}_n$ is in fact the element $\mathbf{c}_1 \vee \mathbf{c}_2 \vee \cdots \vee \mathbf{c}_n$ where the join is taken in the sense of $[\mathbf{0}, \mathbf{a}]$. We rely here on the analysis of LACHLAN (1969) with which we assume familiarity. We argue by induction on n. As each $\mathbf{c}_i$ has only finitely many predecessors (above $\mathbf{0}$) so does $\mathbf{c}_i = \bigoplus \{\mathbf{c}_i | i \leqslant n\}$. Now each $\mathbf{c}_i$ is easily seen to be indecomposable while if $\mathbf{c} \leqslant_1 \bigvee \{\mathbf{c}_i | i \neq j\}$, then $\mathbf{c}_j \leqslant_1 \bigoplus \{\mathbf{c}_i | i \neq j\} = \bigvee \{\mathbf{c}_i | i \neq j\}$ by induction for a contradiction. Thus $\mathbf{c}_1 \oplus \mathbf{c}_2 \oplus \cdots \oplus \mathbf{c}_n$ is the canonial decomposition of $\mathbf{c}$. By Lemma 2 of LACHLAN (1969) $\mathbf{c} \leqslant \bigvee \mathbf{c}_i$ iff $\forall j \leqslant n (\mathbf{c}/\mathbf{c}_j \leqslant \bigvee \mathbf{c}_i/\mathbf{c}_j)$. Now

$$\mathbf{c}/\mathbf{c}_j = \sum_{i \leqslant n} \mathbf{c}_i/\mathbf{c}_j$$

as $\mathbf{c}_j$ is indecomposable. Of course $\mathbf{c}_i/\mathbf{c}_j = 0$ for $i \neq j$ and so $\mathbf{c}/\mathbf{c}_j = 1$ or ∞ if $\mathbf{c}_j$ is a cylinder. But clearly $\bigvee \mathbf{c}_i/\mathbf{c}_j \geqslant 1$ and is ∞ if $\mathbf{c}_j$ is a cylinder. Thus $\mathbf{c} \leqslant \bigvee \mathbf{c}_i$. As the initial segment below $\bigvee \mathbf{c}_i$ is a lattice and $\mathbf{c}_i \leqslant \mathbf{c}$ for each i we have the desired conclusion that $\mathbf{c} = \bigoplus \mathbf{c}_i = \bigvee \mathbf{c}_i$.

Theorem 3.2 can also be used to give a lower bound on the complexity of degree orderings which are not necessarily definable in second order arithmetic.

Corollary 3.6. *True-second order arithmetic is one–one reducible to the first order theory of α-degrees for every countable admissible α.*

Proof. By MACINTYRE (1974) (3.3) holds for each countable admissible. The standard proof of Spector's theorem using α-reduction procedures for the φ_e and α-finite extensions as well then gives (3.2) for α-degrees.

It is tempting to try to apply these arguments to hyperdegrees and constructibility degrees as well. With mild set theoretic assumptions (e.g. $\aleph_1^{L[x]}$ is countable for every real x) (3.1) holds for the constructibility degrees by ADAMOWICZ (1977). For hyperdegrees, finite distributive lattices have been embedded by THOMASON (1970), but it seems likely that the countable ones can also be embedded. The real problem is that (3.2) fails for these degrees. We believe that it is probably possible to find an embedding in the constructibility degrees of a lattice of the required sort with enough genericity to do something like the proof for the arithmetic case. Although this would suffice for some of our later applications it is not enough to get the main theorem. The problem is that we do not see how to guarantee the required genericity by any first order statement about the lattice that we embed.

4. Applications

We will give a couple examples of applications of the methods of Sections 1 and 2 to other subjects and a sketch of some further results that can be gotten for the Turing degrees.

Our first example is the theory of commutative rings in the language *RI* which augments the usual first order language of ring theory with quantification over ideals. In analogy to the codings of Section 2 we show how to code s.i.b. relations by ideals to give a reduction of second order logic to this theory. The role of the top element of our lattice is now played by an element a of a ring and the atoms are replaced by the set $D=\{x|x^2=a\}$ which will be the domain for our s.i.b. relations. Corresponding to condition (2.1)–(2.3) we have the following:

(4.1) For all distinct elements x_1, x_2, x_3, x_4 of D, $x_1x_2 \neq x_3x_4$.

(4.2) $(\forall x_1, x_2, x_3)^D(x_1x_2x_3=0)$.

(4.3) $(\forall x_1 \neq x_2)^D(\exists I)(\forall x_3 \neq x_4)^D(x_3x_4 \in I \,\&\, x_1x_2 \notin I)$.

Quantification over s.i.b. relations on D is replaced by quantification over ideals in the ring by the coding $S(x_i, x_j) \Leftrightarrow x_ix_j \in I$. Condition (4.3) guarantees that for each relation S there is an ideal I with $S(x_i, x_j) \Leftrightarrow x_ix_j$

$\in I$. Thus we can quantify over all s.i.b. relations on D in RI and so encode all of second order logic. (Again to see that rings satisfying (4.1)–(4.3) exist just consider the formal linear combinations over $\mathbb{Z}$ of $1, a, x_i, x_j x_k$ with the indicated multiplication rules i.e. the "semi-group" ring over $\mathbb{Z}$.)

We next consider the theory of the isols. ELLENTUCK (1970) first noticed that one could express certain second order properties of Λ (the Q_1-theory) in the first order language of $\langle\Lambda, +, \times, a^x\rangle$. He then (1973) eliminated exponentiation and also showed by a coding of arithmetic that the first order theory of $\langle\Lambda, +, \times\rangle$ is recursively isomorphic to second order arithmetic. NERODE and MANASTER (1971) later announced that this can also be proved for the theory of $(\Lambda, \leqslant)$ by using coding techniques from NERODE and MANASTER (1970). In fact it also follows immediately from the results of that paper and those of Section 1 much the way those of Section 3 do. Again the key ideas are ideals and covers for them although the terminology of NERODE and MANASTER (1970) is not so algebraic. We supply a glossary.

Definition 4.1. $I \subseteq \Lambda$ is a *normal ideal* iff

(1) $1, 0 \in I$,

(2) $x \leqslant y \in I \Rightarrow x \in I$, and

(3) $x, y \in I \Rightarrow x + y \in I$.

Definition 4.2. X is an *indecomposable cover* for I iff

(1) $(\forall y \in I)(x \geqslant y)$, and

(2) $x = y + z \Rightarrow y \in I$ or $z \in I$.

NERODE and MANASTER (1970, §1) shows that every normal ideal has $2^{\aleph_0}$ pairwise incomparable indecomposable covers (which are minimal elements modulo I). They then show how to describe an isol T (analogous to the top of our distributive lattice initial segment of $\mathfrak{D}$) with countably many minimal elements over the definable ideal ω below it as well as other covers coding pairs of these minimal elements. These "minimal" elements or indecomposable covers serve as the domain on which they code binary relations by covers of other ideals generated by sets of code elements. (Actually as they code arbitrary binary relations rather than s.i.b. ones they are forced up one level of complexity more than is necessary.) As these covers are elements of Λ and all the required predicates on Λ are definable from the ordering they actually have coded quantification over binary relations on a countable domain in the first order theory of $(\Lambda, \leqslant)$. It is therefore recursively isomorphic to second order arithmetic.

To close this paper we would like to sketch some applications to the Turing degrees that will be treated in full in NERODE and SHORE (1980). Using the first order codings of Section 1 we can extend the description of our lattice in Section 2 to require that certain elements of the lattice code (via the principal ideals they generate) an ω-like ordering on some subset of the atoms together with operators satisfying the inductive definitions of $+$ and $\times$. By quantifying over all of $\mathfrak{D}$ and so over all subsets of the ordering we can also require that it be a true ω-ordering. If $\mathbf{d}$ is the top element of such a lattice we say that $\mathbf{d}$ or the set D of degree $\mathbf{d}$ codes a model of true arithmetic.

A straightforward calculation shows that there is a function $f \leqslant D^{(5)}$ with $\mathbf{d}_n = \deg\{f(n)\}^D$ the degrees interpreted as n in this model of arithmetic. The key observation needed for our applications is the following:

Lemma 4.3. (a) *If the pair* $\mathbf{a}_1, \mathbf{a}_2$ *codes the set* A *in this model, i.e.* $n \in A \langle = \rangle \mathbf{d}_n \leqslant_T \mathbf{a}_1, \mathbf{a}_2$, *then* $A \leqslant f \vee (\mathbf{a}_1 \vee \mathbf{a}_2 \vee \mathbf{d})^3$

(b) *If* $f, \mathbf{d}' \leqslant A$, *then there are* $\mathbf{a}_1, \mathbf{a}_2 \leqslant A$ *coding* A *in the model below* $\mathbf{d}$. *Indeed it is known that there is a* $\mathbf{d}$ *coding a model of arithmetic with* $\mathbf{d}'' = \mathbf{0}''$ *and* $f \leqslant \mathbf{0}''$. *For such a model there are* $\mathbf{a}_1, \mathbf{a}_2 \leqslant A$ *coding the set* A *in it for any* $A \geqslant \mathbf{0}''$.

We give some sample applications of this observation. First a definability result. Let $\mathfrak{A}(\mathbf{x}) \equiv \mathbf{x}$ is arithmetic. By Lemma 4.3 (a) if any set A is coded in a model below $\mathbf{0}''$ by a pair recursive in $\mathbf{x} \vee \mathbf{0}''$, then A is arithmetic in $\mathbf{x}$. By Lemma 4.3 (b), if $\mathbf{x}$ is not arithmetic, then there is a model below $\mathbf{0}''$ and a pair recursive in $\mathbf{x} \vee \mathbf{0}''$ coding a non-arithmetic set on this model. Thus we have a first order definition of $\mathfrak{A}$ from the parameter $\mathbf{0}''$. Similar arguments prove the following:

Theorem 4.4. *If* $\mathfrak{C} \subseteq \mathfrak{D}$ *is closed downward and under jump and join, then* $\mathfrak{C}$ *is first order definable in* $\langle \mathfrak{D}, \leqslant, \mathbf{0}'' \rangle$ *iff* $\mathfrak{C}$ *is definable in second order arithmetic.*

A bit more work would enable us to give relativized versions of the above and replace the parameter $\mathbf{0}''$ by the predicate $\mathfrak{A}$. Somewhat different methods can be employed to prove the following strengthening of the main result (Theorem 3.12) of SIMPSON (1977):

Theorem 4.5. (a) *The* ω*-jump is definable in* $\langle \mathfrak{D}, \leqslant, \mathbf{0}'' \rangle$.

(b) *If* $\mathfrak{C} \subseteq \{\mathbf{x} | \mathbf{x} \geqslant \mathbf{0}^{(\omega)}\}$, *then* $\mathfrak{C}$ *is definable in second order arithmetic iff* $\mathfrak{C}$ *is definable in* $\langle \mathfrak{D}, \leqslant, \mathbf{0}'' \rangle$.

We now consider theories of the Turing degrees with the jump operator. Again by analyzing which sets can be coded in models of arithmetic below given degrees as in Lemma 4.3 we can show that for most degrees $\mathbf{a}, \mathcal{D}_a = \langle \{\mathbf{x} | \mathbf{x} \geqslant \mathbf{a}\}, \leqslant, ' \rangle$ is not elementarily equivalent to $\mathcal{D}' = \langle \mathcal{D}, \leqslant, ' \rangle$. A more complicated version of Spector's theorem gives quite a good estimate of what can be coded and lets us prove the following improvement of another result from SIMPSON (1977) and EPSTEIN (1978).

Theorem 4.6 *If* $\mathcal{D}_a' \equiv \mathcal{D}'$, *then* $\mathbf{a}^{(4)} \leqslant \mathbf{0}^{(5)}$.

We can also use these computations to show that $\langle \mathcal{D}, \leqslant, (n) \rangle \not\equiv \langle \mathcal{D}, \leqslant, (m) \rangle$ if $n \neq m$. ((n) is the nth jump operator.) The idea is that if $n > m$, then enough iterations gives us a large enough gap between $\mathbf{0}^{(kn)}$ and $\mathbf{0}^{(km)}$ so that certain definable sets can be coded by pairs in models of arithmetic below $\mathbf{0}^{(kn)}$ but not below $\mathbf{0}^{(km)}$. This answers a question of SELMAN (1972).

For our last application we consider automorphisms of $\mathcal{D}$. Now it is not known if there are any non-trivial automorphisms of $\mathcal{D}'$ or even of $\mathcal{D}$. For $\mathcal{D}'$ it is known that every automorphism is the identity on the cone above $\mathbf{0}^{(3)}$ (RICHTER (1977) after JOCKUSCH and SOLOVAY (1977) and YATES (1971)). By combining Lemma 4.3 with the completeness theorem of FRIEDBERG (1957) we can prove the following:

Theorem 4.7. *Every automorphism* φ *of* $\mathcal{D}$ *is the identity on a cone.*

Proof. If $\mathbf{x} \geqslant \varphi^{-1}(\mathbf{0}'') = \mathbf{c}$, then Lemma 4.3 tells us that $\varphi(x) \leqslant x^5$. Let $\mathbf{x} \geqslant \mathbf{c}^{(5)}$. So by Friedberg $\mathbf{x} = \mathbf{y}^{(5)} = \mathbf{y} \vee \mathbf{c}^{(5)}$ for some $\mathbf{y} \geqslant \mathbf{c}$. Thus

$$\varphi(\mathbf{x}) = \varphi(\mathbf{y}^{(5)}) = \varphi(\mathbf{y}) \vee \varphi(\mathbf{c}^{(5)}) \leqslant \mathbf{y}^{(5)} \vee \varphi(\mathbf{c}^{(5)}).$$

Thus if $\mathbf{x} \geqslant \varphi(\mathbf{c}^5)$ as well, $\varphi(\mathbf{x}) \leqslant \mathbf{x}$. Applying the same argument to φ^{-1} we see that $\varphi^{-1}(\mathbf{x}) \leqslant (\mathbf{x}$ (and so $\mathbf{x} \leqslant \varphi(\mathbf{x}))$ on a cone. Intersecting the cones gives one on which $\mathbf{x} = \varphi(\mathbf{x})$.

Other methods allow much finer calculations of the base of this cone but use much more sophisticated results such as LERMAN (1978) and JOCKUSCH and POSNER (1978). By employing such facts we can give a calculation providing the following strengthening of the theorem of RICHTER (1977):

Theorem 4.8. *Every automorphism of* $\mathcal{D}$ *fixing* $\mathbf{0}'$ *is the identity on the cone above* $\mathbf{0}^{(3)}$.

As our departing salvo we remark that although on its face picking out a copy of true ω depends on quantification over all subsets and so over all of $\mathfrak{D}$ this can in certain cases be avoided by calculating just how much we need to assert to guarantee well-foundedness. This allows us in NERODE and SHORE (1980) to prove analogs for all our results for substructures of the degrees closed downward and under jump and join.

References

ADAMOWICZ, Z.
[1977] Constructible semi-lattices of degrees of constructibility, in: *Set Theory and Hierarchy Theory V*, Lecture Notes in Mathematics, 619, edited by Lachlan, Srebny and Zarach (Springer-Verlag, Berlin), pp. 1–44.

CHERLIN, G.
[1976] *Model Theoretic Algebra, Selected Topics*, Lecture Notes in Mathematics, 521 (Springer-Verlag, Berlin).

CHURCH, A. and W. V. QUINE
[1952] Some theorems on definability and decidability, *J. Symbolic Logic*, **17**, 179–187.

ELLENTUCK, E.
[1970] A coding theorem for isols, *J. Symbolic Logic*, **35**, 378–382.
[1973] Degrees of isolic theories, *Notre Dame J. Formal Logic*, **14**, 331–340.

EPSTEIN, R.
[1978] Analysis and the degrees of unsolvability $\leqslant \mathbf{0}'$, *Notices Am. Math. Soc.*, **25**, A-441.
[1979] *Degrees of Unsolvability: Structure and Theory*, Lecture Notes in Mathmetics, 759 (Springer-Verlag, Berlin).

ERSHOV, YU., I. A. LAVROV, A. D. TAIMANOV and A. M. TAITSLIN
[1965] Elementary theories, *Russian Math Surveys*, **20**, 35–105.

FREGE, G.
[1879] Begriffschrift, eine der arithmetischen nachgebildete Formelsprache des reinen Denkens, Halle. Translated in: *From Frege to Gödel*, edited by J. van Heijenoort (Harvard Univ. Press, Cambridge, MA, 1967), pp. 1–82.

FRIEDBERG, R. M.
[1957] A criterion for completeness of degrees of unsolvability. *J. Symbolic Logic*, **22**, 159–160.

FRIEDBERG, R. M. and H. ROGERS
[1959] Reducibility and completeness for sets of integers, *Z. Math. Logik Grundlagen Math.*, **5**, 117–125.

HARDING, C. J.

[1974] Forcing in recursion theory, Thesis, University College of Swansea.

HIRSCHFELD, J. and W. H. WHEELER

[1975] *Forcing, Arithmetic, Division Rings*, Lecture Notes in Mathematics, 454 (Springer-Verlag, Berlin).

JOCKUSCH, C. G. and D. POSNER

[1978] Double jumps of minimal degrees, *J. Symbolic Logic*, **43**, 715–724.

JOCKUSCH, C. G. and R. M. SOLOVAY

[1977] Fixed points of jump preserving automorphisms of degrees. *Israel J. Math.*, **26**, 91–94.

KLEENE, S. C.

[1955] Hierarchies of number theoretic predicates, *Bull. Am. Math. Soc.*, **61**, 195–213.

KLEENE, S. C. and E. L. POST

[1954] The upper semi-lattice of degrees of recursive unsolvability, *Ann. Math.*, **59**, 379–407.

LACHLAN, A. H.

[1968] Distributive initial segments of the degrees of unsolvability, *Z. Math. Logik Grundlagen Math.*, **14**, 457–472.

[1969] Initial segments of one–one degrees, *Pacific J. Math.*, **29**, 351–366.

[1970] Initial segments of many–one degrees, *Can. J. Math.*, **22**, 75–85.

LAVROV, I. A.

[1963] Effective inseparability of the sets of identically true formulae and finitely refutable formulae for certain elementary theories, *Algebra i Logika*, **2**, 5–18.

LERMAN, M.

[1978] Initial segments of the degrees below $\mathbf{0}'$, *Notices Am. Math. Soc.*, **25**, A-506.

[1981] *The Degrees of Unsolvability* (Springer-Verlag, Berlin), to appear.

MACINTYRE, J. M.

[1973] Non-initial segments of α-degrees, *J. Symbolic Logic*, **38**, 368–388.

NERODE, A. and A. B. MANASTER

[1970] A universal embedding property of the RET's. *J. Symbolic Logic*, **35**, 51–59.

[1971] The degree of the theory of addition of isols, *Notices Am. Math. Soc.*, **18**, 83.

NERODE, A. and R. A. SHORE

[1980] Reducibility orderings: theories, definability and automorphisms, Ann. Math. Logic, **18**, 61–89.

RABIN, M.

[1965] A simple method for undecidability proofs and some applications, in *Logic, Methodology, and Philosophy of Science, Proc. 1964 Int. Congress*. edited by Y. Bar-Hillel (North-Holland, Amsterdam), pp. 58–68.

[1969] Decidability of second order theories and automata on infinite trees, *Trans. Am. Math. Soc.*, **141**, 1–35.

[1977] Decidable theories, in: *Handbook of Mathematical Logic*, edited by J. Barwise (North-Holland, Amsterdam), pp. 595–629.

RABIN, M. and D. SCOTT

[n.d.] The undecidability of some simple theories, mimeographed notes.

RICHTER, L. J. C.

[1977] Degrees of unsolvability of models, Ph. D. Thesis, University of Illinois, Urbana-Champaign.

SACKS, G. E.

[1971] Forcing with perfect closed sets, in: *Axiomatic Set Theory, Proc. Symp. Pure Math. vol XVII pt 1*, edited by D. Scott (Am. Math. Soc., Providence, RI).

SELMAN, A. L.

[1972] Applications of forcing to the degree theory of the arithmetic hierarchy, *Proc. London Math. Soc.*, **25**, 586–602.

SIMPSON, S. G.

[1977] First order theory of the degrees of recursive unsolvability, *Ann. Math*, **105**, 121–139.

SPECTOR, C.

[1956] On degrees of recursive unsolvability, *Ann. Math.*, **64**, 581–592.

THOMASON, S. K.

[1970] On initial segments of hyperdegrees, *J. Symbolic Logic*, **31**, 159–168.

[1975] Reduction of a second order logic to modal logic, *Z. Math. Logik Grundlagen Math.*, **21**, 107–114.

YATES, C. E. M.

[1971] Initial segments and implications for the structure of degrees, in: *Conference in Mathematical Logic—London 1970*, edited by W. Hodges, Lecture Notes in Mathematics, 255 (Springer-Verlag, Berlin), pp. 305–335.

YOUNG, P. R.

[1964] On reducibility by recursive functions, *Proc. Am. Math. Soc.*, **15**, 889–892.

J. Barwise, H. J. Keisler and K. Kunen, eds., *The Kleene Symposium*

Post's Problem, Absoluteness and Recursion in Finite Types

*Gerald E. Sacks**

Massachusetts Institute of Technology and Harvard University, Science Center, 1 Oxford Street, Cambridge, MA 02138, U.S.A.

Dedicated to Professor S. C. Kleene on the occasion of his 70th birthday

Abstract: The unresolved character of the power set operation stymies the solution of elementary problems arising in Kleene's theory of recursion in objects of finite type. E.g. Post's problem for ^{3}E has a positive solution if $V=L$ (Normann, 1975), and a negative if AD holds. Let $\mathbb{S}$ be the class of all sets $R \subseteq 2^\omega$ such that R is recursive in ^{3}E,b for some real b. A forcing construction shows $\mathbb{S}$ is not recursively enumerable in ^{3}E when there is a recursively regular well-ordering of 2^ω recursive in ^{3}E. It follows that the concepts of Σ^*, and weak Σ^*, definability differ. With the aid of the notions of indexicality and ordinal recursiveness, an absolute version of Post's problem for ^{3}E is devised, and studied when certain recursively enumerable projecta are equal.

The problem raised by Post (1944) concerning the Turing degrees of recursively enumerable sets of numbers has persisted in the study of generalizations of ordinary recursion theory. A seemingly temperate version of his problem in the context of Kleene's theory of recursion in objects of finite type (Kleene, 1959) is as follows.

Post's Problem for ^{n}E: Does there exist a type n object A such that A is recursively enumerable in ^{n}E; A is not complete; and A is not recursive in ^{n}E,b for any object b of type less than n?

Kleene's theory of recursion in normal objects of finite type is reviewed several pages ahead. Until then make do with the following. The objects of type 0 are the natural numbers. An *object of type n* ($n>0$) is a collection of objects of type less than n. ^{n}E is the equality functional for objects of type less than n. Thus

$$ {}^nE(x,y)=\begin{cases} 0 & \text{if } x=y,\\ 1 & \text{if } x\neq y,\end{cases} $$

*This paper is based on an address given at the Kleene Symposium in May 1978. Its preparation was supported in part by National Science Foundation Grant MCS 7807344. The author is indebted to S. Homer, D. Normann and T. Slaman.

where x and y are arbitrary objects of type less than n. Kleene's notion of reducibility for objects of finite type is written $\leqslant$, and read "recursive in". Thus $A \not\leqslant {}^nE, B$ means A is not recursive in ${}^nE, B$. Let A be type n and recursively enumerable in ${}^nE$; A is *complete* if $B \leqslant A, {}^nE$ for every type n B recursively enumerable in ${}^nE$. An important result of Kleene, and a key witness to the correctness of his theory, says

$$X \leqslant {}^2E \leftrightarrow X \text{ is hyperarithmetic,}$$

where X is any type 1 object, e.g. real number.

An affirmative answer to Post's problem faces two impedimenta. Both have to do with the computations developed in the course of recursively enumerating a collection of objects of finite type. The meaning of "computation" will have to await the promised review of Kleene's theory, but for now there is no harm in thinking of a computation as a well-founded tree that serves as a proof that some object belongs to some recursively enumerable collection. The two obstacles are:

(I) a violation of parity between elements of recursively enumerable collections and computations;

(II) lack of a suitable well-ordering of computations.

The first obstacle goes back to KREISEL's early work (1961) on the foundations of metarecursion theory. A violation of parity is at the least a difference in type or complexity. There is a parity (and it is quite useful) between elements and computations in ordinary recursion theory. In that case a computation has the form of a finite tree, hence can be effectively encoded by a number, and consequently there is little difference between a complete, recursively enumerable set of numbers and the set of all computations.

The next proposition is a classic example of obstacle (I) at the type 2 level.

Proposition 1. *Let A be a set of natural numbers. If A is recursively enumerable, but not recursive, in ${}^2E$, then A is complete* (SPECTOR, 1955).

Proof. Since A is not recursive in ${}^2E$, the natural enumeration of A is cofinal with the natural enumeration of C, the complete, recursively enumerable in ${}^2E$ subset of ω. Hence $m \in C$ iff

$$(\mathrm{E}n)[n \in A \;\&\; m \text{ is enumerated in } C \text{ before } n \text{ is enumerated in } A],$$

and the latter is recursive in $A, {}^2E$.

Kreisel was the first to point out the violation of parity inherent in Proposition 1. A number is enumerated in A iff an appropriate computation exists. In this case a computation has the form of a countable tree encoded by a real recursive in ^{2}E. Kreisel's remedy is almost immediate. Let REC(^{2}E) be the class of all reals recursive in ^{2}E. Enumerate subclasses of REC(^{2}E) rather than subsets of ω. The associated computations will still be encoded by elements of REC(^{2}E), and so obstacle (I) will have vanished. The remedy has a further prescription somewhat less immediate. The desire for parity requires some revision of the notion of completeness. Let $A \subseteq \mathrm{REC}({}^2\mathrm{E})$ be recursively enumerable in ^{2}E. A is said to be *complete* if B is recursive in $A, {}^2\mathrm{E}$ *on* REC(^{2}E) for every recursively enumerable (in ^{2}E) $B \subseteq \mathrm{REC}({}^2\mathrm{E})$. To be more precise, there is a Kleene partial recursive functional $\{e\}$ such that for all $R \in \mathrm{REC}({}^2\mathrm{E})$:

$$R \in B \leftrightarrow \{e\}({}^2\mathrm{E}, A, R) = 0;$$

$$R \notin B \leftrightarrow \{e\}({}^2\mathrm{E}, A, R) = 1.$$

Thus $\{e\}$ correctly answers all membership questions about B from A, so long as those questions are restricted to REC(^{2}E). To require $\{e\}$ to do more is to violate parity, because computations involving arguments outside REC(^{2}E) have little hope of being encoded by elements of REC(^{2}E). Note that $\{e\}$ is not required to be total.

The next theorem attests to the efficacy of Kreisel's remedy. It is a positive solution to Post's problem at the type 2 level analogous to the Friedberg (1957)–Muchnik (1956) solution at the type 1 level.

Theorem 2. *There exist A and B, each a recursively enumerable (in* ^{2}E*) subclass of* REC(^{2}E), *such that $A \nleq B, {}^2\mathrm{E}$ and $B \nleq A, {}^2\mathrm{E}$ on* REC(^{2}E).

Proof. Implicit in Sacks (1966). Let A_1 and B_1 be hyperregular, ω_1^{CK}-recursively enumerable sets such that neither is ω_1^{CK}-recursive (i.e. metarecursive) in the other. Let $\{H_\delta \mid \delta < \omega_1^{\mathrm{CK}}\}$ be the natural enumeration of the hyperarithmetic subsets of ω obtained by iterating the Turing jump. Define

$$A = \{H_\delta \mid \delta \in A_1\} \quad \text{and} \quad B = \{H_\delta \mid \delta \in B_1\}.$$

Let $\{e\}$ be a Kleene partial recursive functional, and suppose $B \leqslant A, {}^2\mathrm{E}$ on REC(^{2}E) via $\{e\}$. The hyperregularity of A implies that the computation of $\{e\}({}^2\mathrm{E}, A, H)$, where H is a hyperarithmetic real, is encodable by a hyperarithmetic real. The collection of all such computations is an ω_1^{CK}-recursively enumerable set. It follows that B_1 is ω_1^{CK}-recursive in A_1.

The next proposition states that impediment (I) occurs not only at the type 2 level, as noted by Spector, but also at all higher levels.

Proposition 3. *Assume $n \geqslant 2$. Let A be a type $(n-1)$ object. If A is recursively enumerable in ^{n}E, but not recursive in ${}^n\mathrm{E}, a$ for any object a of type less than $n-1$, then A is complete.*

Proof. Same as that of Proposition 1.

The Kreisel remedy is readily applied at all type levels beyond the first. Fix $n \geqslant 2$. An *individual* is defined to be any object of type less than $n-1$. By a *set* is meant any collection R of individuals such that $R \leqslant {}^n\mathrm{E}, a$ for some individual a. To remedy the violation of parity in Proposition 3, enumerate classes of sets rather than classes of individuals. The associated computations are encodable by sets, and so parity has been restored.

Impediment (II) is not manifest at the type 2 level. Computations arising from recursion in ${}^2\mathrm{E}$ have numerical codes, and consequently lend themselves to a suitable well-ordering. At the type 3 level the codes are reals, and there need not be any effective well-ordering of the reals in sight. The situation is especially grim with respect to Post's problem if AD, the axiom of determinateness, is assumed, as is seen in Proposition 4.

From now on it will be best to focus on recursion in ${}^3\mathrm{E}$. All that follows, save Proposition 4, applies equally well to ${}^n\mathrm{E}$ when $n > 3$, but is more easily grasped when $n = 3$.

In accord with the definition of set given above, a *set* R is a collection of reals such that $R \leqslant {}^3\mathrm{E}, a$ for some real a. A set R is said to be *indexical*[1] if there exists a nonempty set I of reals such that $I \leqslant {}^3\mathrm{E}, R$ and

$$a \in I \to R \leqslant {}^3\mathrm{E}, a$$

for all a. The notion of indexicality is less technical than a first glance suggests. Its intuitive meaning will be discussed after the review of Kleene recursion. For now it must suffice to say that the class of indexical sets is recursively enumerable in ${}^3\mathrm{E}$, but the class of all sets need not be.[2]

Proposition 4. *Assume AD. Let A be a recursively enumerable (in* ${}^3\mathrm{E}$*) class of sets. Then there exists a real b such that either* (i) *or* (ii) *holds.*

(i) $A \leqslant {}^3\mathrm{E}, b$ *on the class of all indexical sets.*

(ii) $C \leqslant {}^3\mathrm{E}, A, b$ *where C is the complete, recursively enumerable (in* ${}^3\mathrm{E}$*) sets of reals.*

Proof. Modeled on Simpson (1974, p.439). Let cA be the class of all indexical sets not in A. If $R \in cA$, then there is an e such that

$$I = \{e\}({}^3\mathrm{E}, R), \quad \varnothing \neq I \subseteq \omega \times 2^\omega,$$

and

$$\langle e_0, a\rangle \in I \to R = \{e_0\}({}^3\mathrm{E}, a) \quad \text{for all } \langle e_0, a\rangle \in I.$$

[1]Cf. the definition of indexical given in Webster's Third International Dictionary.

[2]This appears to be the reason that Normann (1975) introduces ${}^1M$.

Gandy's selection theorem for recursively enumerable (in ${}^3\mathrm{E}, R$) classes of numbers makes it possible to pass effectively from an indexical set R to a unique e (denoted by e_R) that witnesses the indexicality of R. Define

$$I_R = \{e_R\}({}^3\mathrm{E}, R) \quad \text{and} \quad i[cA] = \bigcup \{I_R | R \in cA\}.$$

According to Wadge, AD yields a continuous function f satisfying either (1) or (2).

(1) $\quad x \in i[cA] \leftrightarrow f(x) \in C \quad$ (for all real x).

(2) $\quad x \notin C \leftrightarrow f(x) \in i[cA].$

Let b be a real that encodes f, that is b encodes rational approximations of f on all rational arguments. Assume (1). Then

$$R \in cA \leftrightarrow f[I_R] \subseteq C$$

for every indexical set R, and so cA is recursively enumerable in ${}^3\mathrm{E}, b$. Hence (i) holds, since A is recursively enumerable in ${}^3\mathrm{E}$. Now assume (2). Then

$$x \notin C \leftrightarrow f(x) = \langle e_0, a \rangle \;\&\; R = \{e_0\}({}^3\mathrm{E}, a)$$
$$\& \; R \in cA,$$

and (ii) holds.

Two remedies have been found for impediment (II), one by HARRINGTON (1973), and the other by NORMANN (1975). The first approach is to restrict attention to sets and computations that of necessity lend themselves to a suitable well-ordering. In the case of ${}^3\mathrm{E}$ this means narrowing down to the class of sets recursive in ${}^3\mathrm{E}$, namely REC(${}^3\mathrm{E}$). For ${}^n\mathrm{E}$ ($n > 3$) the restriction is to the subrecursive sets, an uncountable, but bounded subfamily of the indexical sets (cf. SACKS (1977) for details). The second approach is to assume a suitable well-ordering of 2^ω recursive in ${}^3\mathrm{E}$.

Harrington's Remedy: There exist recursively enumerable (in ${}^3\mathrm{E}$) classes $A, B \subseteq \mathrm{REC}({}^3\mathrm{E})$ such that $A \nleq B, {}^3\mathrm{E}$ and $B \nleq A, {}^3\mathrm{E}$ on REC(${}^3\mathrm{E}$).

Normann's Remedy: Assume that the cardinality c of the continuum is regular and that some well-ordering of the real of height c is recursive in ${}^3\mathrm{E}$. Then there exist recursively enumerable (in ${}^3\mathrm{E}$) classes A and B of sets such that for every set R,

$$A \nleq B, {}^3\mathrm{E}, R \quad \text{and} \quad B \nleq A, {}^3\mathrm{E}, R$$

on the class of all sets.

The virtue of Harrington's remedy is its explicitness. No assumptions are made about ^{3}E. There is no reference, at least no direct reference, to an indeterminate continuum. All sets to be enumerated, as well as the associated computations, have numerical codes, and Post's problem is solved by an adaptation of the α-recursion theoretic methods that succeeded at the type 2 level. But a virtue pursued to the limit, particularly the virtue of explicitness, becomes a fault. Harrington's result says very little about recursively enumerable (in ^{3}E) classes of sets. It avoids the problem of the continuum, a problem that Normann confronts more directly. His assumption concerning the regularity of the continuum and its computability from ^{3}E is true in L and elsewhere. He in fact manages with a weaker regularity assumption made explicit just before Lemma 5 below. (This paper and its sequel deal with the singular case.) An unexpected consequence of Normann's regularity assumption is the non-recursive enumerability in ^{3}E of the class of all sets (Theorem 9), one of the reasons for introducing indexical sets.

The absolute approach to Post's problem for ^{3}E formulated below owes more to Normann's remedy than it does to Harrington's. It consists of staying as close as possible to the ordinals arising as lengths of computations generated by recursion in ^{3}E. These ordinals define sets that will take the place of the sets mentioned in Normann's remedy. Some of his sets are lost, sacrificed on the altar of absoluteness, but the loss will be seen to be less than one might think.

The time of the promised review of Kleene recursion in ^{3}E is at hand. Recall how the hyperarithmetic sets of numbers are obtained by iteration of the Turing jump through the recursive ordinals. The class of all sets is similarly obtained by iteration of the ^{3}E jump through the ordinals recursive in ^{3}E and an arbitrary real. Do not forget that "set" has been defined above to mean a collection of reals recursive in ^{3}E$,a$ for some real a. The sets and ordinals in question are defined by a simultaneous recursion. The hierarchy developed is equivalent to an initial segment of $L(2^\omega)$. Thus the sets recursive in ^{3}E$,a$ for some real a are obtained by starting with the continuum, iterating first order definability, and stopping at the first ordinal that cannot be justified from below in a certain predicative fashion.

Let $X \subseteq 2^\omega$. The ^{3}E jump of X, denoted by $j(X)$, is an encoding of certain definable collections of reals. Let $\{e\}^X_\pi$ be the eth function from 2^ω into ω definable over $\langle 2^\omega, \varepsilon, X\rangle$ by means of a first order formula whose only parameters are nonnegative integers, and whose atomic parts are of the form $y \in z$ or $y \in \chi$, where y and z are real variables and χ is a constant naming X.

$j(X)$ is the collection of all $\langle e,a,m\rangle$ such that $\{e\}^X_\pi(a)=m$.

$\mathcal{O}$ is the collection of notations for ordinals recursive in ^{3}E$,a$ for some real a. Notations are reals $a,b,c,\ldots$; if $b\in\mathcal{O}$, then $|b|$ is the ordinal represented by b. H_σ is the set obtained by σ iterations of the ${}^3\text{E}$ jump j. The simultaneous recursive definition of $\mathcal{O}$ and H_σ has three clauses.

(1) $\langle 1,a\rangle \in \mathcal{O}$ and $|\langle 1,a\rangle| = 0$ for all $a \in 2^\omega$. And $H_0 = \varnothing$.

(2) Suppose $\langle e,a\rangle \in \mathcal{O}$ and $|\langle e,a\rangle| = \sigma$. Then $\langle 2^e,a\rangle \in \mathcal{O}$, $|\langle 2^e,a\rangle| = \sigma+1$, and $H_{\sigma+1} = j(H_\sigma)$.

(3) Suppose $\langle m,a\rangle \in \mathcal{O}$ and $|\langle m,a\rangle| = \sigma$. Define

$$W_{e,a}^{H_\sigma} = \{ b | \{e\}^{H_\sigma}(a*b) = 0 \},$$

where $a*b$ is a real code for $\langle a,b\rangle$. Suppose $W_{e,a}^{H_\sigma} \subseteq \mathcal{O}$. Then $\langle 3^m\cdot 5^e,a\rangle \in \mathcal{O}$ and $|\langle 3^m\cdot 5^e,a\rangle| = \lambda$, where λ is the least limit ordinal greater than σ and greater than $|b|$ for all $b \in W_{e,a}^{H_\sigma}$. And H_λ is the set of $\langle b,c\rangle$ such that $b \in \mathcal{O}$, $|b| < \lambda$ and $c \in H_{|b|}$.

Let κ_1 be the supremum of $|b|$ for all $b \in \mathcal{O}$. The hierarchy $\{H_\sigma | \sigma < \kappa_1\}$ is equivalent to a hierarchy of all the computations that arise in the computing of partial functions from ${}^3\text{E}$ and real arguments.

$\{e\}({}^3\text{E},a)$ is defined and equal to m if $e = \langle e_0,e_1\rangle$, $\langle e_0,a\rangle \in \mathcal{O}$, $|\langle e_0,a\rangle| = \sigma$ and $\{e_1\}_\pi^{H_\sigma}(a) = m$.

Thus to compute $\{e\}({}^3\text{E},a)$ is to iterate the ${}^3\text{E}$ jump until $\langle e_0,a\rangle$ is seen to be a notation for an ordinal σ and then to pluck the value from H_σ in a first order fashion dictated by e_1.

A function f is recursive in ${}^3\text{E},a$ (in symbols $f \leqslant {}^3\text{E},a$) if there is an e such that $f(b) = \{e\}({}^3\text{E},a*b)$ for all real b.

A collection R of reals in recursive in ${}^3\text{E},a$ (in symbols $R \leqslant {}^3\text{E},a$) if the characteristic function of R is recursive in ${}^3\text{E},a$. (Such an R has been termed a "set" above.)

An ordinal γ is *constructive* in ${}^3\text{E},a$ if γ is $|\langle e,a\rangle|$ for some $\langle e,a\rangle \in \mathcal{O}$. γ is *recursive* in ${}^3\text{E},a$ if γ is the height of a prewell-ordering of 2^ω recursive in ${}^3\text{E},a$. If γ is constructive in ${}^3\text{E},a$, then γ is recursive in ${}^3\text{E},a$. If γ is recursive in ${}^3\text{E},a$, then γ is less than or equal to some ordinal constructive in ${}^3\text{E},a$.

Clause 3 of the definition of $\mathcal{O}$ and H implies a *bounding principle*: if $f: 2^\omega \to \mathcal{O}$ is recursive in ${}^3\text{E},a$, then there exists an ordinal γ constructive in ${}^3\text{E},a$ such that $|f(b)| < \gamma$ for all b. It follows from the bounding principle that R is recursive in ${}^3\text{E},a$ iff there is a σ constructive in ${}^3\text{E},a$ and an e such that

$$R = W_{e,a}^{H_\sigma}.$$

An *index* for a *set* R is $\langle d,e,a\rangle$, where $|\langle d,a\rangle| = \sigma$ and the above equation relating R and W is satisfied by σ, e and a. From now on assume that indices for sets are encoded by reals. Note that if b is an index for R, then $R \leqslant {}^3\text{E},b$.

Let S be an arbitrary collection of reals. Recursion relative to ${}^3\text{E},S$ is defined in the same hierarchic fashion as recursion in ${}^3\text{E}$. The only difference is beginning with S rather than $\varnothing$. Thus the simultaneous recursive definition of $\mathcal{O}^S$ and H_σ^S has three clauses identical with those given for $\mathcal{O}$ and H_σ save that $H_0^S = S$. κ_1^S is of course the supremum of $|b|$ for all $b \in \mathcal{O}^S$, and may exceed κ_1. $\{e\}({}^3\mathrm{E}, S, a)$ is defined and equal to m if $e = \langle e_0, e_1 \rangle$, $\langle e_0, a \rangle \in \mathcal{O}^S$, $|\langle e_0, a \rangle| = \sigma$ and $\{e_1\}_\pi^{H_\sigma^S}(a) = m$.

$\mathcal{O}$ and $\{H_\sigma | \sigma < \kappa_1\}$ can be imbedded effectively in $\mathcal{O}^S$ and $\{H_\sigma^S | \sigma < \kappa_1^S\}$. In other words recursion in ${}^3\mathrm{E}$ is effectively included in recursion in ${}^3\mathrm{E}, S$. One can pass effectively from e to e^* so that

$$\{e\}({}^3\mathrm{E}, a) \simeq \{e^*\}({}^3\mathrm{E}, S, a)$$

for all S and a. ($\simeq$ is Kleene's symbol for equality of graphs of partial functions.) The imbedding can be reversed if S is a set (i.e. recursive in ${}^3\mathrm{E}, a$ for some a). If S is a set, then $\kappa_1^S = \kappa_1$, and the H_σ^S hierarchy can be imbedded in the H_σ hierarchy by means of an effective transfinite recursion that begins at the location of S in the H_σ hierarchy. In short the computations for recursion in ${}^3\mathrm{E}, S$ are essentially the same as those for recursion in ${}^3\mathrm{E}$ if S is a set.

Let A be a class of collections of reals. A is *recursively enumerable* in ${}^3\mathrm{E}$ if there is an e such that

$$S \in A \leftrightarrow \{e\}({}^3\mathrm{E}, S) \quad \text{is defined}$$

(for every collection S of reals). As in Normann's remedy given above, interest tends to center on recursively enumerable classes of sets. Let A be a recursively enumerable (in ${}^3\mathrm{E}$) class of sets for the sake of sharpening the intuitions behind Kreisel's remedy for impediment (I) to the solution of Post's problem. A set R is put into A if and only if there exists a computation Z of $\{e\}({}^3\mathrm{E}, R)$. The existence of Z is equivalent to the existence of some H_σ^R with some first order property specified by e. Since R is a set, there is an a such that $R \leqslant {}^3\mathrm{E}, a$. Then Z, if it exists, will be recursive in ${}^3\mathrm{E}, a$, and so will be a set located in the H_σ hierarchy not far from R. Thus recursively enumerable (in ${}^3\mathrm{E}$) classes of sets are enumerated via computations which are themselves sets.

To carry the parity between elements of recursively enumerable classes and computations one step further, let A be a class of sets recursively enumerable in ${}^3\mathrm{E}, R_0$, where R_0 is a set; that is there is an e such that

$$R \in A \leftrightarrow \{e\}({}^3\mathrm{E}, R_0, R) \quad \text{is defined}$$

(for every set R). As in the previous paragraph, R lands in A if and only if there exists a computation Z to that effect. As above, if such a Z exists, it will be recursive in ${}^3\mathrm{E}, R_0, R$, and so will be a set. Now the requirements of parity have been fully met: the elements of A, the parameter R_0 that defines the procedure for enumerating A, and the associated computations all live on the same level of complexity; they are all sets, that is sets located in the H_σ hierarchy for recursion in ${}^3\mathrm{E}$.

But there is an unexpected fly in the ointment. The class of all sets is not recursively enumerable in ^{3}E when ^{3}E satisfies Normann's regularity assumption.

Let $\prec$ be a well-ordering of 2^ω. $\prec$ is said to be *recursively regular* if there is no function $f \leqslant {}^3\mathrm{E}, a$ for some real a such that the domain of f is a proper initial segment of $\prec$ and the range of f is unbounded in $\prec$.

NRA: *Normann's Regularity Assumption.* There exists a recursively regular well-ordering $\prec$ of 2^ω such that $\prec$ is recursive in ^{3}E.[3]

Lemma 5. *Suppose Normann's regularity assumption is satisfied by $\prec$. Let $A \subseteq 2^\omega$ be recursively enumerable in ${}^3\mathrm{E}, a$ for some real a, and let B be a proper initial segment of $\prec$. Then $A \cap B$ is recursive in ${}^3\mathrm{E}, b$ for some real b.*

Proof. An immediate consequence of the main theorem of NORMANN (1977).

The proof below that the class of all sets is not recursively enumerable in ^{3}E (when Normann's regularity assumption holds) is largely a proof of the existence of a non-indexical set. This notion of set, defined just before Proposition 4, is in need of some clarification. Let R be a set. Thus $R \lessdot {}^3\mathrm{E}, a$ for some real a, and R is first order definable over $\langle 2^\omega, \varepsilon, H_\sigma \rangle$, where $\sigma = |\langle e, a \rangle|$ for some e. Let σ_R be the least such σ. In short σ_R is the first level of the $\{H_\sigma | \sigma < \kappa_1\}$ hierarchy where the set R occurs.

Proposition 6. *Let R be a set. R is indexical iff σ_R is recursive in ${}^3\mathrm{E}, R$.*

Proof. First suppose R is indexical. Thus there is a nonempty $I \subseteq 2^\omega$ such that $I \leqslant {}^3\mathrm{E}, R$ and

$$(a)[a \in I \rightarrow R \leqslant {}^3\mathrm{E}, a].$$

It follows from Gandy selection and bounding that there exists an ordinal γ recursive in ${}^3\mathrm{E}, R$ such that

$$(a)[a \in I \rightarrow |a| < \gamma].$$

Consequently $\sigma_R < \gamma$, and so σ_R is recursive in ${}^3\mathrm{E}, R$.

Now suppose σ_R is recursive in ${}^3\mathrm{E}, R$. To see that R is indexical, let I be the set of all real a such that $|(a)_0| = \sigma_R$ and $|(a)_1|$ is the parameter needed in the first order definition of R over $\langle 2^\omega, H_{\sigma_R}, \varepsilon \rangle$.

The above proposition is intended to express the intuitive content of the concept of indexicality. Each set R has a natural location in the hierarchy

[3] Analogous to the assumption in α-recursion theory that α^* is a regular α-cardinal.

of all sets obtained by iterating the ^{3}E jump through all ordinals recursive in ^{3}E, a for some real a. To say R is indexical is to say R is located at some ordinal recursive in ^{3}E, R.

Proposition 7. (i) *Every H_σ is indexical.*
(ii) *The class of indexical sets is recursively enumerable in* ^{3}E.

Proof. (i) Let I be $\{a \mid |a| = \sigma\}$. Then $I \leqslant {}^3\mathrm{E}, H_\sigma$, and $H_\sigma \leqslant {}^3\mathrm{E}, a$ for all $a \in I$.
(ii) By Proposition 6, R is indexical iff there is an e such that

$$\langle e,0\rangle \in \mathcal{O}^R \ \&\ R \text{ is first order definable over } H_{|\langle e,0\rangle|_R}.$$

Observe that every recursively enumerable (in ^{3}E) class A of indexical sets has a natural enumeration tied to that of $\mathcal{O}$. The members of $\mathcal{O}$ are enumerated in κ_1 steps. At step σ, all $a \in \mathcal{O}$ such that $|a| = \sigma$ are enumerated, and also all $R \in A$ such that $\sigma_R = \sigma$. Some such enumeration of A seems to be an essential part of priority arguments in the study of recursively enumerable classes of sets. This is why Normann (1975, p.8) restricts his attention from M to ${}^1M$. If A were allowed to have some non-indexical member S, then the reason for putting S into A might not become evident until S was located at stage σ_S of the H_σ hierarchy. By then it would be too late to put S in A, because σ_S exceeds all ordinals recursive in ^{3}E, S, and the latter are the only ordinals allowed to serve as the height of a computation that puts S in a recursively enumerable class.

Proposition 8. *Assume there is a well-ordering W of 2^ω recursive in* ^{3}E. *Then* (i), (ii) *and* (iii) *are equivalent.*
(i) *R is an indexical set.*
(ii) *There exists a real a such that $R \leqslant {}^3\mathrm{E}, a$ and $a \leqslant {}^3\mathrm{E}, R$.*
(iii) *There exists a $b \in \mathcal{O}$ such that $\sigma_R = |b|$ and $b \leqslant {}^3\mathrm{E}, R$.*

Proof. Note that if S is a nonempty set, then the least member of S (least with respect to W) is recursive in ^{3}E, S. Assume (i) to show (ii). Thus there is a nonempty $I \leqslant {}^3\mathrm{E}, R$ such that $R \leqslant {}^3\mathrm{E}, b$ for every $b \in I$. Let a be the W-least member of I. Then a satisfies (ii).

Assume a satisfies (ii). Then σ_R is recursive in ^{3}E, a. The set of all b that satisfy $\sigma_R = |b|$ is recursive in ^{3}E, a. The least such b is recursive in ^{3}E, R.
(i) follows from (iii) by Proposition 6.

Theorem 9. *Assume* ^{3}E *satisfies Normann's regularity assumption. Then the class of all sets, namely*

$$\{R \mid R \subseteq 2^\omega \ \&\ (\mathrm{E}a)(a \in 2^\omega \ \&\ R \leqslant {}^3\mathrm{E}, a)\},$$

is not recursively enumerable in ^{3}E, c *for any* $c \in 2^\omega$.

Corollary 10. *Assume* ^{3}E *satisfies* NRA. *Then there exists a non-indexical set.*

Proof. Theorem 9 and Proposition 7(ii).

Corollary 11. *Assume the cardinality of the continuum is regular. Then there exists a normal type 3 object F such that*

$$\{R \mid R \subseteq 2^\omega \;\&\; (\mathsf{E}a)(a \in 2^\omega \;\&\; R \leqslant F, a)\}$$

is not recursively enumerable in F, b *for any* $b \in 2^\omega$.

Proof. Let W be a well-ordering of the continuum of minimal height. Define $F = \langle {}^3\mathrm{E}, W\rangle$ and imitate the proof of Theorem 9.

Normann (1975, p. 9) asks if the concepts of Σ^* and weak Σ^* definability coincide. Let A be a class of sets. Call A *weakly recursively enumerable* in ^{3}E if there is an e such that for every set R,

$$R \in A \leftrightarrow (a)\big[R \leqslant {}^3\mathrm{E}, a \to \{e\}({}^3\mathrm{E}, R, a) \quad \text{is defined}\big].$$

Clearly every class recursively enumerable in ^{3}E is also weakly recursively enumerable in ^{3}E. For ^{3}E, (weak) Σ^* definability is the same as (weak) recursive enumerability.

Corollary 12. *Assume* ^{3}E *satisfies Normann's regularity assumption. Then*

$$\{R \mid R \subseteq 2^\omega \;\&\; (\mathsf{E}a)(a \in 2^\omega \;\&\; R \leqslant {}^3\mathrm{E}, a)\}$$

is weakly Σ^* *definable, but not* Σ^* *definable, in* ^{3}E.

The Ramified Language $\mathfrak{L}(\kappa_1, \mathfrak{T})$. The proof of Theorem 9 is a forcing argument that ranges over the H_σ hierarchy of sets associated with recursion in ^{3}E. The result of adjoining a generic $T \subseteq 2^\omega$ to $\{H_\sigma \mid \sigma < \kappa_1\}$ will be studied. The resulting structure, $\{H_\sigma^T \mid \sigma < \kappa_1\}$, is of greatest interest when $\kappa_1^T = \kappa_1$. The language $\mathfrak{L}(\kappa_1, \mathfrak{T})$ contains just what is needed to analyze $\{H_\sigma^T \mid \sigma < \kappa_1\}$. It includes: number variables $(x_0, y_0, \ldots)$; real variables $(x, y, \ldots)$; ordinal variables $(\sigma, \tau, \ldots)$; ordinal constants $(\boldsymbol{\alpha}, \alpha < \kappa_1)$; real constants $(\mathbf{a}, a \in 2^\omega)$; quantifiers on numbers, reals and ordinals; and propositional connectives. It also includes predicates

$$t_2 \in H_{t_1}^{\mathfrak{T}}, \qquad |t_2|_{\mathfrak{T}} = t_1 \quad \text{and} \quad t_2 \in W\big(e, t_3, H_{t_1}^{\mathfrak{T}}\big),$$

where t_1 is an ordinal term, and t_2 and t_3 are real terms.

Each formula $\mathfrak{F}$ has one or more Gödel numbers of the form $\langle e, a_1, \ldots, a_m, b_1, \ldots, b_n\rangle$: e specifies the logical form of $\mathfrak{F}$; a_i specifies a real constant $\mathbf{a}_i$ occurring in $\mathfrak{F}$; and b_j is a notation from $\mathcal{O}$ that specifies an ordinal constant $|\mathbf{b}_j|$ occurring in $\mathcal{F}$. There is no need to choose a unique Gödel number for $\mathcal{F}$, since each is recursive in every other via ^{3}E. Thus the ^{3}E degree of the Gödel number of $\mathcal{F}$ is unique. Let $\ulcorner\mathcal{F}\urcorner$ denote a Gödel number of $\mathcal{F}$. The collection of all Gödel numbers is recursively enumerable in ^{3}E.

$\mathcal{F}$ is said to be a *bounded formula* if every ordinal quantifier in $\mathcal{F}$ is bounded (by some ordinal less than κ_1).

Every formula $\mathcal{F}$ has a *rank* based on the range of its ordinal quantifiers, its ordinal constants and its logical complexity. It is a routine matter to define a rank function whose range is well-founded with the following properties:

(r1) rank $\mathcal{F}(\boldsymbol{\alpha}) <$ rank$(\mathsf{E}\sigma)\mathcal{F}(\sigma)$ for all $\alpha < \kappa_1$.

(r2) rank $\mathcal{F}(\boldsymbol{\beta}) <$ rank$(\mathsf{E}\sigma)_{\sigma<\tau}\mathcal{F}(\sigma)$ for all $\beta < \tau$.

(r3) rank $\mathcal{F}(\mathbf{b}) <$ rank$(x)\mathcal{F}(x)$ for all $b \in 2^\omega$.

(r4) rank $\mathcal{F} <$ rank$(\mathcal{F}\ \&\ \mathcal{G})$.

(r5) rank$(\sim\mathcal{F}) <$ rank $\mathcal{F}$.

(r6) $|\langle\mathbf{m},\mathbf{a}\rangle|_{\mathcal{T}} = \boldsymbol{\alpha}$ has lower rank than $|\langle\mathbf{n},\mathbf{b}\rangle|_{\mathcal{T}} = \boldsymbol{\beta}$ if $\alpha < \beta$.

In addition the graph of the rank function can be taken to be recursively enumerable in ^{3}E, since the rank of $\mathcal{F}$ can be effectively obtained from a Gödel number of $\mathcal{F}$.

The Forcing Relation. Assume NRA holds. Thus $\prec$ is a well-ordering of 2^ω recursive in ^{3}E and recursively regular. Forcing conditions are denoted by $p, q, r, \ldots$, and specify a small part of the characteristic function of T. A typical forcing condition is a function $f: \{c \mid c \prec a\} \to 2$ such that f is recursive in ^{3}E$, b$ for some real b. A condition p is encoded by a pair $\langle e, b\rangle$ such that $p = \{e\}({}^3\mathrm{E}, b)$. p is extended by q (in symbols $p \geqslant q$) if graph $p \subseteq$ graph q.

The forcing relation $p \Vdash \mathcal{F}$ is defined by recursion on the rank of $\mathcal{F}$. Each clause of the definition either parallels some clause of the definition of $\{H_\sigma^T \mid \sigma < \kappa_1\}$ or is a standard feature of forcing technology. Some of the clauses are as follows:

(f1) $p \Vdash \mathbf{a} \in \mathcal{T}$ iff $p(a) = 1$. ($\mathcal{T}$ is $H_0^{\mathcal{T}}$.)

(f2) $p \Vdash \sim\mathcal{F}$ iff $(q)_{p\geqslant q} \sim [q \Vdash \mathcal{F}]$.

(f3) $p \Vdash (\mathsf{E}\sigma)\mathcal{F}(\sigma)$ iff $p \Vdash \mathcal{F}(\boldsymbol{\alpha})$ for some $\alpha < \kappa_1$.

(f4) $p \Vdash (\mathsf{E}x)\mathcal{F}(x)$ iff $p \Vdash \mathcal{F}(\mathbf{a})$ for some $a \in 2^\omega$.

(f5) $p \Vdash |\langle \mathbf{2}^e, \mathbf{a}\rangle|_{\mathcal{T}} = \boldsymbol{\alpha} + \mathbf{1}$ iff $p \Vdash |\langle \mathbf{e}, \mathbf{a}\rangle|_{\mathcal{T}} = \boldsymbol{\alpha}$.

(f6) Let $\lambda < \kappa_1$ be a limit ordinal.

$$p \Vdash |\langle 3^m \cdot 5^e, \mathbf{a}\rangle|_{\mathcal{T}} = \boldsymbol{\lambda} \text{ iff (f6.1)–(f6.3) hold for some } \alpha < \lambda.$$

(f6.1) $p \Vdash |\langle \mathbf{m}, \mathbf{a}\rangle|_{\mathcal{T}} = \boldsymbol{\alpha}$. Let W be $W(\mathbf{e}, \mathbf{a}, H_{\boldsymbol{\alpha}}^{\mathcal{T}})$.

(f6.2) $p \Vdash (x)(\mathsf{E}\sigma)_{\sigma<\lambda}[x \in W \to |x|_{\mathcal{T}} = \sigma]$.

(f6.3) $p \Vdash (\sigma)_{\sigma<\lambda}(\mathrm{E}x)(\mathrm{E}n)_{n<\omega}[x \in W \ \& \ |x|_{\mathfrak{T}} + \mathbf{n} = \sigma]$.

(f7) The above clauses indicate, it is hoped, how to force $\mathbf{b} \in W(\mathbf{e}, \mathbf{a}, H_{\alpha}^{\mathfrak{T}})$. Recall that $W(\mathbf{e}, \mathbf{a}, H_{\alpha}^{\mathfrak{T}})$ denotes a first order definable subset of $\langle \mathbf{2}^{\omega}, \varepsilon, H_{\alpha}^{\mathfrak{T}} \rangle$ whose definition is built up from the atomic predicate $x \in H_{\alpha}^{\mathfrak{T}}$, real variable quantifiers, and logical connectives. Thus clauses such as (f2) and (f4) can be used to reduce the forcing of reals into $W(\mathbf{e}, \mathbf{a}, H_{\alpha}^{\mathfrak{T}})$ to the forcing of reals into $H_{\alpha}^{\mathfrak{T}}$, and the latter situation reduces to events occurring below level $\boldsymbol{\alpha}$.

Proposition 13. *There exists a recursive* (*in* ^{3}E) *set of codes for the forcing conditions.*

Proof. Call a collection R of reals a *small* set if it is contained in $\{c \mid c \prec a\}$ for some real a, and is recursive in ${}^3\mathrm{E}, b$ for some real b. A forcing condition is in essence a small set. According to Lemma 1 of NORMANN (1977), every small set occurs in the H_{σ} hierarchy for ^{3}E prior to level $|\prec|$, the ordertype of $\prec$, an ordinal recursive in ^{3}E. Thus there is a recursive (in ^{3}E) set of codes for the conditions defined at level $|\prec|$.

From now on assume that the forcing conditions are codes as in Proposition 13. It will be convenient to allow $p, q, r, \ldots$ to denote both forcing conditions and their codes.

The ordinal rank of a bounded formula $\mathfrak{F}$ is the greatest ordinal constant that occurs in $\mathfrak{F}$, either as a strict upper bound on some ordinal variable or as a parameter in some predicate.

Sublemma 14. *If* $a \in \mathcal{O}$, *then the relation*

$$p \Vdash \mathfrak{F}, \text{ restricted to } \mathfrak{F}\text{'s of ordinal rank } \leqslant |a|,$$

is recursive in ${}^3\mathrm{E}, a$ (*uniformly in* a).

Proof. By effective transfinite induction on the rank of $\mathfrak{F}$. Suppose by way of illustration that $\mathfrak{F}$ is $(\sigma)_{\sigma<\alpha}\mathfrak{G}(\sigma)$. Then $p \Vdash \mathfrak{F}$ means

$$(\beta)_{\beta<\alpha}(q)_{p>q}(\mathrm{E}r)_{q>r}[r \Vdash \mathfrak{G}(\boldsymbol{\beta})].$$

The ordinal rank of $\mathfrak{G}(\boldsymbol{\beta})$ is at most that of $\mathfrak{F}$. The rank of $\mathfrak{G}(\boldsymbol{\beta})$ is less than that of $\mathfrak{F}$. Proposition 13 justifies the induction step up to $\mathfrak{F}$ from sentences of lower rank.

The next lemma will establish bounds on the ordinals recursive in ${}^3\mathrm{E}, T, a$ when T is generic. It will be seen that

$$\text{if } p \Vdash [\,|\mathbf{a}|_{\mathfrak{T}} \leqslant \kappa_1\,], \quad \text{then } p \Vdash |\mathbf{a}|_{\mathfrak{T}} < \boldsymbol{\beta}$$

for some β recursive in ${}^3\mathrm{E},p,a$. The proof of bounding seems to require the notion of weak forcing. Say $p \Vdash^* \mathfrak{F}$ (read p weakly forces $\mathfrak{F}$) if there is no q such that $p \geq q$ and $q \Vdash \sim\mathfrak{F}$. Note that Sublemma 14 remains true when $\Vdash$ is replaced by $\Vdash^*$, thanks to Proposition 13.

Lemma 15. *Suppose* $p \Vdash^* (\mathsf{E}\sigma)[|\mathbf{a}|_{\mathfrak{T}} = \sigma]$. *Then there exists a q and a γ such that*

$$p \geq q, \quad q \Vdash |\mathbf{a}|_{\mathfrak{T}} = \gamma, \quad q \leq {}^3\mathrm{E},p,a$$

and γ is recursive in ${}^3\mathrm{E},p,a$.

Proof. q and γ, regarded as partial recursive functions of ${}^3\mathrm{E}$, p and a, are defined by effective transfinite induction on the rank of $\langle p,a\rangle$. A rank can be assigned to $\langle p,a\rangle$, when

(1) $$p \Vdash^* (\mathsf{E}\sigma)\big[\,|\mathbf{a}|_{\mathfrak{T}} = \sigma\big],$$

by appealing to the inductive definition of $\Vdash^*$. (1) becomes

(2) $$(q)_{p>q}(\mathsf{E}r)_{q>r}(\mathsf{E}\gamma)\big[\,r \Vdash |\mathbf{a}|_{\mathfrak{T}} = \gamma\big].$$

The pair $\langle r,\gamma\rangle$, if $r \Vdash |\mathbf{a}|_{\mathfrak{T}} = \gamma$, of course has rank γ. The rank of $\langle p,a\rangle$, if (1) holds, is the supremum of the ranks of the pairs $\langle r,\gamma\rangle$ needed to verify the truth of (2). Logical complexity also plays a part in the assignment of rank to $\langle p,a\rangle$. For example, suppose a is $\langle 2^e,b\rangle$ and (1) holds. Then $\langle p,a\rangle$ has higher rank than $\langle p,\langle e,b\rangle\rangle$ because it has greater logical complexity. The logical difference has to be considered, since $\langle p,\langle 2^e,b\rangle\rangle$ and $\langle p,\langle e,b\rangle\rangle$ may have the same ordinal rank. Thus a pair $\langle p,a\rangle$, if (1) holds, will have an ordinal rank less than or equal to κ_1, and a logical rank developed in accord with clauses (f5)–(f7) of the definition of $\Vdash$.[4]

To see how q and γ are defined, let a be $\langle 3^m, 5^e, c\rangle$ and assume (1) holds. Then

$$p \Vdash^* (\mathsf{E}\sigma)\,]\,|\langle \mathbf{m},\mathbf{c}\rangle|_{\mathfrak{T}} = \sigma\big].$$

By induction there exist q_0 and γ_0, each computed from ${}^3\mathrm{E},p,a$, such that

$$p \geq q_0 \quad \text{and} \quad q_0 \Vdash |\langle \mathbf{m},\mathbf{c}\rangle|_{\mathfrak{T}} = \gamma_0.$$

Let W be $W(\mathbf{e},\mathbf{c},H_{\gamma_0})$. Then

$$q_0 \Vdash^* (x)(\mathsf{E}\sigma)\big[\,x \in W \to |x|_{\mathfrak{T}} = \sigma\big],$$

and so for each $b \in 2^\omega$,

$$(q)_{q_0>q}(\mathsf{E}r)_{q>r}\big[\,r \Vdash (\mathsf{E}\sigma)(\mathbf{b} \in W \to |\mathbf{b}|_{\mathfrak{T}} = \sigma)\,\big].$$

Fix b and $q \leq q_0$ in order to define $r(q,b)$ and $\gamma(q,b)$. If there is an r such

[4] T. Slaman has noted that the wellfoundedness of the rank relation for $\langle p,a\rangle$ requires a proof based on the countable closure property of the forcing conditons. Cf. Sacks (1980).

that $q \geqslant r$ and $r \Vdash \mathbf{b} \notin W$, then $r(q,b)$ is the least such r (least with respect to $\prec$), and $\gamma(q,b)=0$. If there is no such r, then

$$q \Vdash^* (\mathrm{E}\sigma)\big[\,|\mathbf{b}|_{\mathfrak{T}} = \sigma\big].$$

By induction there exist $r(q,b)$ and $\gamma(q,b)$, each computed from ${}^3\mathrm{E}, q, b$ such that

$$q \geqslant r(q,b) \quad \text{and} \quad r(q,b) \Vdash |\mathbf{b}|_{\mathfrak{T}} = \gamma(\mathbf{q},\mathbf{b}).$$

By Sublemma 14, the functions $r(q,b)$ and $\gamma(q,b)$ are recursive in ${}^3\mathrm{E}, p, a, q, b$. It follows that $\gamma(q,b)$ is bounded, for all $q \leqslant q_0$ and all b, by some γ_1 recursive in ${}^3\mathrm{E}, p, a$. Hence

(3) $\qquad q_0 \Vdash^* |\mathbf{a}|_{\mathfrak{T}} \leqslant \gamma_1.$

The desired q and γ can now be obtained from (3) by an application of Sublemma 14 similar to the one above.

Lemma 16. *Suppose* $p \Vdash (x)(\mathrm{E}\sigma)[x \in W(\mathbf{e},\mathbf{a},H_\gamma) \rightarrow |x|_{\mathfrak{T}} = \sigma]$. *Then there exists an* $\alpha < \kappa_1$ *such that* $p \Vdash (x)(\mathrm{E}\sigma)_{\sigma<\alpha}[\cdots]$.

Proof. Let W be $W(\mathbf{e},\mathbf{a},H_\gamma)$. Then for each b,

$$(q)_{p>q}(\mathrm{E}r)_{q>r}\big[\,r \Vdash (\mathrm{E}\sigma)(\mathbf{b} \in W \rightarrow |\mathbf{b}|_{\mathfrak{T}} = \sigma)\big].$$

Now proceed as in the last part of the proof of Lemma 15 to obtain the bound on σ.

Let $T \subseteq 2^\omega$ and $\mathfrak{F}$ be a sentence of $\mathfrak{L}(\kappa_1, \mathfrak{T})$. T is said to be *generic over* $\mathfrak{F}$ if there is a p such that T satisfies p, and either $p \Vdash \mathfrak{F}$ or $p \Vdash \sim\mathfrak{F}$. T is said to be *generic* if T is generic over $\mathfrak{F}$. A familiar argument that proceeds by induction on the rank of sentences shows: if T is generic, then $\mathfrak{F}$ is a true statement about T if and only if there is a p satisfied by T such that $p \Vdash \mathfrak{F}$. It follows from Lemma 16 that $\kappa_1^T = \kappa_1$ when T is generic. Unfortunately the existence of a generic T is far from certain. If $|\prec|$ (the ordinal height of $\prec$, the well-ordering of 2^ω assumed to be recursive in ${}^3\mathrm{E}$) is a regular cardinal, then a straightforward recursion on $|\prec|$ yields a generic T. It seems unlikely that the regularity assumption of Theorem 9, namely that $|\prec|$ is recursively regular, is strong enough to yield a generic T, but it is enough to develop a T generic over all bounded sentences. The next two lemmas are needed to control κ_1^T when T is boundedly generic.

Lemma 17. *Suppose* $p \Vdash (\sigma)[|\mathbf{b}|_{\mathfrak{T}} \neq \sigma]$. *Then there exist a* q, *a real* c *and a bounded sentence* $\mathfrak{F}$ *with the following properties.*

(i) $p > q$ *and* $q \Vdash \mathfrak{F}$.

(ii) $q \Vdash (\sigma)[|\mathbf{c}|_{\mathfrak{T}} \neq \sigma]$.

(iii) *For all* $T \subseteq 2^\omega$: *if* $\mathfrak{F}$ *is a true statement about* T *and* $b \in \mathcal{O}^T$, *then* $c \in \mathcal{O}^T$ *and* $|b|_T > |c|_T$.

Proof. The definition of q, c and $\mathfrak{F}$ has three cases, only the last of which is of interest.

Case I: b is not of the form $\langle 2^e, a\rangle$ or $\langle 3^m \cdot 5^e, a\rangle$. Let $q=p$, $c=b$ and $\mathfrak{F}$ be $1=1$.

Case II: b is of the form $\langle 2^e, a\rangle$. Let $q=p$, $c=\langle e,a\rangle$ and $\mathfrak{F}$ be $1=1$.

Case III: b is $\langle 3^m \cdot 5^e, a\rangle$.

Subcase III$_1$: $p\Vdash(\sigma)[|\langle \mathbf{m}, \mathbf{a}\rangle|_{\mathfrak{F}} \neq \sigma]$. Let $q=p$, $c=\langle m,a\rangle$ and $\mathfrak{F}$ be $1=1$.

Subcase III$_2$: otherwise. Thus there is a $q_0 \leqslant p$ and a $\gamma < \kappa_1$ such that

$$q_0 \Vdash |\langle \mathbf{m}, \mathbf{a}\rangle|_{\mathfrak{F}} = \gamma.$$

Suppose (for the sake of a contradiction) that

$$(1) \qquad q_0 \Vdash (x)(\mathsf{E}\sigma)\left[x \in W\left(\mathbf{e}, \mathbf{a}, H_\gamma^{\mathfrak{F}}\right) \rightarrow |x|_{\mathfrak{F}} = \sigma\right].$$

Then by Lemma 16, $q_0 \Vdash (x)(\mathsf{E}\sigma)_{\sigma<\alpha}[\cdots]$ for some $\alpha < \kappa_1$, and consequently $q_0 \Vdash [|b|_{\mathfrak{F}} < \alpha]$, which contradicts the supposition that $p \Vdash (\sigma)[|b|_{\mathfrak{F}} \neq \sigma]$. Thus (1) is false, and so there exist $q \leqslant q_0$ and c such that

$$q \Vdash \left[\mathbf{c} \in W\left(\mathbf{e}, \mathbf{a}, H_\gamma^{\mathfrak{F}}\right) \;\&\; (\sigma)(|\mathbf{c}|_{\mathfrak{F}} \neq \sigma)\right].$$

Let $\mathfrak{F}$ be $|\langle \mathbf{m}, \mathbf{a}\rangle|_{\mathfrak{F}} = \gamma \;\&\; \mathbf{c} \in W(\mathbf{e}, \mathbf{a}, H_\gamma^{\mathfrak{F}})$.

Lemma 18. *Suppose* $p \Vdash (\sigma)[|\mathbf{b}|_{\mathfrak{F}} \neq \sigma]$. *Then there exist a* q *and a bounded sentence* $\mathfrak{F}$ *such that*

(i) $p \geqslant q$ *and* $q \Vdash \mathfrak{F}$; *and*

(ii) *for all* $T \subseteq 2^\omega$, *if* $\mathfrak{F}$ *is a true statement about* T, *then* $b \notin \mathcal{O}^T$.

Proof. Moschovakis (1967) showed that every co-recursively enumerable (in ^{3}E) predicate $P(x)$ is expressible in the form $(\mathsf{E}y)R(x,y)$, where $R(x,y)$ is recursively enumerable (in ^{3}E). The intuition behind R is the same as that behind the desired $\mathfrak{F}$. To say $b \notin \mathcal{O}^T$ is equivalent to saying any attempt to recursively enumerate (in ^{3}E$, T$) b in $\mathcal{O}^T$ must fail because the associated computation tree is not well-founded. $\mathfrak{F}$ specifies an infinite descending path through that tree. The path is obtained by repeated application of Lemma 17.

Let $p_0 = p$ and $b_0 = b$. Assume p_i and b_i satisfy the initial supposition of Lemma 17; let $\langle p_{i+1}, b_{i+1}, \mathfrak{F}_{i+1}\rangle$ bear the same relation to $\langle p_i, b_i\rangle$ that $\langle q, c, \mathfrak{F}\rangle$ bears to $\langle p, b\rangle$ in Lemma 17.

Let q be $\bigcup\{q_i | i<\omega\}$, and let $\mathfrak{F}$ be a bounded sentence logically equivalent to the infinite conjunction of the $\mathfrak{F}_i$'s. (The ordinal rank of $\mathfrak{F}$ is the supremum of the ordinal ranks of the $\mathfrak{F}_i$'s, and $\mathfrak{F}$ has a real parameter d such that $(d)_i$ is a Gödel number of $\mathfrak{F}_i$ for all $i>0$.) Clearly $q \Vdash \mathfrak{F}$.

Fix $T \subseteq 2^\omega$. Suppose (for the sake of a contradiction) that $\mathfrak{F}$ is a true statement about T and that $b \in \mathcal{O}^T$. By property (iii) of Lemma 17, $b_i \in \mathcal{O}^T$ for all i, and $|b_i|_T$ $(i<\omega)$ is an infinite descending sequence of ordinals.

Lemma 19. *For each p there exists a T that satisfies p and is generic on all bounded $\mathfrak{F}$'s.*

Proof. Let $\mathfrak{F}_a$ be the bounded sentence whose Gödel number is a. A function P will be defined so that for each a, if a is the Gödel number of a bounded sentence, then either $P(a) \Vdash \mathfrak{F}_a$ or $P(a) \Vdash \sim\mathfrak{F}_a$. The characteristic function of T will be $\bigcup \{P(a) | a \in 2^\omega\}$.

P is defined by transfinite recursion on $\prec$, a well-ordering of 2^ω that is recursive in ${}^3\mathrm{E}$ and recursively regular. Fix a. Let $P_0(a)$ be the union of the given p and the $P(b)$'s for all $b \prec a$. Suppose $a = \ulcorner \mathfrak{F}_a \urcorner$ and $P_0(a)$ is a forcing condition. (If the supposition is false, let $P(a)$ be $P_0(a)$.) Let $P(a)$ be the least q such that $q \leqslant P_0(a)$ and either $q \Vdash \mathfrak{F}_a$ or $q \Vdash \sim\mathfrak{F}_a$. ("least" means least with respect to $\prec$ with the understanding that forcing conditions are coded by reals as in Proposition 13.) By Sublemma 14, $P(a)$ can be found effectively from $P_0(a)$ and a.

The definition of P is successful if $P_0(a)$ is a forcing condition for all a. This last is proved by induction. Fix a, and assume $P_0(b)$ is a forcing condition for all $b \prec a$. Let D be the collection of all $b \prec a$ such that b is the Gödel number of a bounded sentence. Observe that $P \upharpoonright a$ is recursive in ${}^3\mathrm{E}, D, a$ because D decides if $P(b)$ (for each $b \prec a$) is to be $P_0(b)$ or an extension of $P_0(b)$ that forces $\mathfrak{F}_b$ or $\sim\mathfrak{F}_b$. Thus the definition by recursion of $P \upharpoonright a$ becomes an effective transfinite recursion relative to ${}^3\mathrm{E}, D, a$. By Lemma 5, D is recursive in ${}^3\mathrm{E}, c, a$ for some real c, and so $P_0(a)$ is recursive in ${}^3\mathrm{E}, c, a$. $P_0(a)$ is a union of compatible forcing conditions, hence it is the characteristic function of a set of reals. It need only be shown that the domain of $P_0(a)$ is bounded above (with respect to $\prec$) by some real d. For each $b \prec a$, let $Q(b)$ be the least upper bound of the domain of $P(b)$. Clearly $Q \upharpoonright a$ is recursive in ${}^3\mathrm{E}, c, a$. The recursive regularity of $\prec$ implies that the range of $Q \upharpoonright a$, and hence the domain of $P_0(a)$, is bounded above by some d.

Proof of Theorem 9. Fix e and c. For the sake of a contradiction, assume that for all $S \subseteq 2^\omega$: $\{e\}({}^3\mathrm{E}, c, S)$ is defined if and only if $S \leqslant {}^3\mathrm{E}, a$ for some a. Let b be $\langle e_0, c \rangle$. Then $S \leqslant {}^3\mathrm{E}, a$ for some a if and only if $b \in \mathcal{O}^S$. By Lemma 18 there is a forcing condition p such that either (B) or (C) holds.

(B) $p \Vdash |\mathbf{b}|_{\mathfrak{T}} = \gamma$ for some $\gamma < \kappa_1$.

(C) $p \Vdash \mathfrak{F}$, where $\mathfrak{F}$ is a bounded sentence with the following property. For all $T \subseteq 2^\omega$, if $\mathfrak{F}$ is a true statement about T, then $b \notin \mathcal{O}^T$.

By Lemma 19 there exists a T_1 that satisfies p and is generic on all bounded sentences. Let S be any set recursive in ${}^3\mathrm{E}, a$ for some a. Clearly $\varnothing \Vdash \mathfrak{T} \neq \mathbf{S}$, where $\varnothing$ is the null forcing condition. It follows that T_1 is not recursive in ${}^3\mathrm{E}, a$ for any a. Hence $b \notin \mathcal{O}^{T_1}$, and so (C) must hold.

Let $\sigma_0<\kappa_1$ be a limit ordinal greater than the ordinal rank of $\mathcal{F}$ (mentioned in (C)), and greater than the ordinal needed to compute $\prec$ recursively from ^{3}E. Repeat the construction of Lemma 19 from the vantage point of σ_0 to obtain a T_2 that satisfies p, is generic on all bounded sentences of ordinal rank less than σ_0, and is first order definable over H_{σ_0}. Thus $T_2 \leqslant {}^3\mathrm{E}, a$ for any a such that $a\in\mathcal{O}$ and $|a|=\sigma_0$, and so $b\in\mathcal{O}^{T_2}$. But (C) implies $b\notin\mathcal{O}^{T_2}$.

Indexicality is one of two notions needed for the absolute formulation of Post's problem. The other is ordinal recursiveness. Let $R\subseteq 2^{\omega}$. R is said to be *ordinal recursive* in ^{3}E if there exist an integer e and H-sets $H^1,\ldots,H^m$ such that

$$R=\{e\}({}^3\mathrm{E},H^1,\ldots,H^m).$$

By an H-set is meant any set in the hierarchy $\{H_\sigma|\sigma<\kappa_1\}$. A typical ordinal recursive set is a set first order definable over some H_σ by means of a formula without real parameters. Suppose S is recursive in $^3\mathrm{E}, b$. The real b plays a double role. First $b_0\in\mathcal{O}^T$ and $H_{|b_0|_T}$ is the universe over which S is defined. Second b_1 is the real parameter needed in the definition of S. Thus every set recursive in $^3\mathrm{E}, b$ for some real b is ordinal recursive if and only if every real is ordinal recursive.

Proposition 20. *Suppose there exist a well-ordering W of 2^{ω} such that W is recursive in ^{3}E. Then every set recursive in $^3\mathrm{E}, b$ for some $b\in 2^{\omega}$ is ordinal recursive in ^{3}E.*

Proof. Let γ be the height of W. Let f be the unique function that maps $\{H_\sigma|\sigma<\gamma\}$ in an order-preserving fashion onto W. Thus $f(H_\sigma)=b$ if and only if the height of b in W is σ. Since γ is recursive in ^{3}E, the function f^{-1} must also be recursive in ^{3}E. Hence b is recursive in $^3\mathrm{E}, f^{-1}(b)$ for each b, and so each real is ordinal recursive in ^{3}E.

The next lemma clarifies the connection between H-sets and ordinal recursive sets. Two sets of reals are said to have the same ^{3}E-*degree* if each is recursive in the other and ^{3}E.

Lemma 21. (i) *Let H^1 and H^2 be H-sets. Then there exists an H-set H^3 such that $H^1\times H^2$ has the same ^{3}E-degree as H^3.*

(ii) *R is ordinal recursive in ^{3}E if and only if R is first order definable over some H_σ by means of a formula whose only parameters are integers.*

Proof. (ii) follows from (i). The proof of (i) is an effective transfinite recursion that defines functions f and g with the following properties:

$H_\sigma \times H_\tau$ is first order definable over $H_{f(\sigma,\tau)}$ via the formula (with integer parameters only) whose Gödel number is $g(\sigma,\tau)$. Fix $\langle\gamma,\delta\rangle$ and assume f and g are already defined for all $\langle\sigma,\tau\rangle$'s below $\langle\gamma,\delta\rangle$. Define $f(\gamma,\delta)$ to be the strict least upper bound of $f(\sigma,\tau)$ for all $\langle\sigma,\tau\rangle$ less than $\langle\gamma,\delta\rangle$. The definition of $g(\gamma,\delta)$ has two cases. First suppose there is a $\rho<f(\gamma,\delta)$ such that $H_\gamma \times H_\delta$ is first order definable over H_ρ. Then $g(\gamma,\delta)$ is the Gödel number of a formula that defines (over $H_{f(\gamma,\delta)}$) the least such ρ, call it ρ_0, and also defines $H_\gamma \times H_\delta$ over H_{ρ_0}. Now suppose no such ρ exists. Then $\langle H_\gamma, H_\delta\rangle$ is the least pair of H-sets whose product is not first order definable over H_ρ for any $\rho<f(\gamma,\delta)$. This last sentence is the desired definition of $H_\gamma \times H_\delta$ over $H_{f(\gamma,\delta)}$.

It follows from Theorem 9 and Proposition 20 that the class of all sets ordinal recursive in ^{3}E need not be recursively enumerable in ^{3}E. That shortcoming is eliminated by the next definition, which combines the notions of ordinal recursiveness and indexicality. R is said to be *effectively ordinal recursive* in ^{3}E if there exist H-sets $H^1,\ldots,H^m$ such that R is recursive in $^3\mathrm{E}, H^1,\ldots,H^m$, and each H^i is recursive in $^3\mathrm{E}, R$.

Lemma 22. *R is effectively ordinal recursive if and only if R has the same ^{3}E-degree as some H-set.*

Proof. A consequence of Lemma 21(i).

A set is said to be H^* if it has the same ^{3}E-degree as an H-set. The absolute formulation of Post's problem for ^{3}E resembles Normann's remedy above save that H^*-sets stand in place of arbitrary sets recursive in ^{3}E and a real.

Absolute Post Problem for ^{3}E: Do there exist recursively enumerable (in ^{3}E and some fixed H^*-set) classes A and B of H^*-sets such that for each H^*-set R,

$$A \not\leq {}^3\mathrm{E}, B, R \quad \text{and} \quad B \not\leq {}^3\mathrm{E}, A, R$$

on the class of all H^*-sets?

The class of all H^*-sets is recursively enumerable in ^{3}E. The natural enumeration of the H^*-sets parallels the development of the ordinals less than κ_1 by iteration of the ^{3}E jump. It follows there is a useful well-ordering of the objects allowed to occur as members of recursively enumerable classes even when there is no well-ordering of 2^ω recursive in ^{3}E. And if such a well-ordering of 2^ω does exist, then Proposition 20 implies that every real is an H^*-set, and consequently that the incomparability of A and B asked for in the absolute problem is the most possible. Thus parity is maintained between the objects put into recursively enumerable classes and the objects that index reduction procedures without any weakening of incomparability requirements. The axioms of set theory leave open whether or not every real is ordinal recursive in ^{3}E, hence a positive solution to the absolute Post problem for ^{3}E leaves open how much incomparability is achieved, but nonetheless renders that incomparability as definite as the ordinals.

A last word about the choice of H^*-sets as the objects to be enumerated. An H^*-set has all the advantages of indexicality and ordinal recursiveness. The principal argument in favor of indexicality was given just before Proposition 8, and the one for ordinal recursiveness in the previous paragraph. But there is still another point to consider. Why H^* rather than H? The H-sets are of course indexical and ordinal recursive, and in addition they form a recursive (in ^{3}E) class. Why go to H^*? One answer is suggested by Lemma 22, namely that the H^*-sets constitute the largest class of objects suitable for the absolute study of Post's problem relative to ^{3}E. A second answer, perhaps more practical than the first, is that the H^*-sets have useful closure properties exemplified by the statement and proof of Lemma 21(i).

In order to state a morally certain conjecture concerning the absolute Post problem, it is necessary to define two notions of recursively enumerable projectum relative to ^{3}E.

The *greater projectum* is denoted by ρ. It is the least $\sigma \leqslant \kappa_1$ such that there exists an H-set H^1 with the following property: every H-set is recursive in $^3\mathrm{E}, H^1, H_\tau$ for some $\tau < \sigma$.

The *lesser projectum* is denoted by η. It is the least $\sigma \leqslant \kappa_1$ such that there exists an H-set H^1 with the following property: there is a recursively enumerable (in $^3\mathrm{E}, H^1$) class of H-sets contained in $\{H_\tau | \tau < \sigma\}$, but not recursive in $^3\mathrm{E}, H$ for any H-set H.

Proposition 23. (i) $\eta \leqslant \rho$.

(ii) *Suppose there exists a well-ordering of 2^ω recursive in $^3\mathrm{E}, b$ for some real b. Then the least height attained by any such well-ordering is equal to ρ.*

(iii) *Suppose there exists a recursively regular well-ordering of 2^ω recursive in $^3\mathrm{E}, b$ for some real b. Then $\eta = \rho$.*

Proof. (iii) follows from Lemma 5.

Inexorable limitations of space and time compel this paper to end with a mixture of questions, conjectures and halftruths that will be studied elsewhere.

(1) Suppose $\eta = \rho$. At this writing it appears very likely that the absolute Post problem has an affirmative answer.[5] NORMANN's work (1975, 1977)

[5]Yes, it does. Cf. SACKS (1980).

corresponds to the case when ρ is less than κ_1 and is a recursively regular cardinal. A quite different state of affairs ensues when ρ is singular; that case is handled with the aid of κ_r^b, the greatest reflecting ordinal relative to $\langle {}^3\mathrm{E}, b\rangle$, pinned down by HARRINGTON (1973) in his Ph.D. thesis. The supposition $\eta = \rho$ is analogous to a fundamental truth of α-recursion theory: an α-recursively enumerable subset of an ordinal less than α^*, the Σ_1 projectum of α, is α-finite. A closely related supposition, termed adequacy, is made by FENSTAD (1979).

(2) Does there exist a generic extension of L in which the absolute Post problem has a negative answer? in which every subset of 2^ω recursive in ${}^3\mathrm{E}$ and a real is indexical? The more appealing conjecture, in both cases, is yes.

(3) All the results of this paper, save for Proposition 5, lift to ${}^n\mathrm{E}$ when $n > 3$. Some changes in definitions and proof are necessary. They are designed to take into account the difference between the Gandy and Grilliot selection principles, a difference manifest only if $n > 3$.

(4) Is there an absolute version of Theorem 9? One that does not assume some well-ordering of the reals is recursive in ${}^3\mathrm{E}$? It might be possible to show that the class of all sets ordinal recursive in ${}^3\mathrm{E}$ is not recursively enumerable in ${}^3\mathrm{E}, R$, for any ordinal recursive set R, by assuming that $\rho < \kappa_1$ and ρ is recursively regular.

References

FENSTAD, J. E.
[1979] General Recursion Theory (Springer-Verlag, Berlin), to appear.

FRIEDBERG, R.
[1957] Two recursively enumerable sets of incomparable degrees of insolvability, *Proc. Nat. Acad. Sci.*, **43**, 236–238.

HARRINGTON, L.
[1973] Contributions to recursion theory in higher types, Ph.D. Thesis, M.I.T.

KLEENE, S. C.
[1959] Recursive functionals and quantifiers of finite type, *Trans. Am. Math. Soc.*, **91**, 1–52.
[1963] Recursive functionals and quantifiers of finite type, II, *Trans. Am. Math. Soc.*, **108**, 106–142.

KREISEL, G.
[1961] Set theoretic problems suggested by the notion of potential totality, in: Infinitistic Methods (Pergamon Press, Oxford), pp. 103–140.

MOSCHOVAKIS, Y. N.
[1967] Hyperanalytic predicates, *Trans. Am. Math. Soc.*, **138**, 249–282.

MUCHNIK, A. A.

[1956] On the unsolvability of the problem of reducibility in the theory of algorithms, *Dokl. Akad. Nauk. USSR*, **108**, 194–197.

NORMANN, D.

[1975] Degrees of functionals, Preprint Series in Mathematics No. 22, Oslo.

[1977] A note on reflection, preprint, Oslo.

POST, E.

[1944] Recursively enumerable sets of positive integers and their decision problem, *Bull. Am. Math. Soc.*, **50**, 284–316.

SACKS, G. E.

[1966] Post's problem, admissible ordinals and regularity, *Trans. Am. Math. Soc.* **124**, 1–23.

[1977] RE sets higher up, in: Logic, Foundations of Mathematics and Computability Theory (D. Reidel Publ. Co., Dordrecht).

SIMPSON, S.

[1974] Post's problem for admissible sets, in Generalized Recursion Theory (North-Holland, Amsterdam).

SPECTOR, C.

[1955] Recursive wellorderings, *J. Symbolic Logic*, **20**, 151–163.

J. Barwise, H. J. Keisler and K. Kunen, eds., *The Kleene Symposium*
©North-Holland Publishing Company (1980) 223–265

Lambda Calculus: Some Models, Some Philosophy

Dana Scott
Oxford University, Oxford, Great Britain

Dedicated to Professor S.C. Kleene on the occasion of his 70th birthday

0. Introduction

In this essay yet another attempt at an exposition of why the λ-calculus has models is made. The λ-calculus was one of the first areas of research of Professor Kleene, an area in which the experience he gained was surely beneficial in his later development of recursive function theory. In what transpires below, the dialogue will be found to involve Professor Curry rather more than Kleene, since the former has written more extensively on the foundational aspects of combinatory logic. Nevertheless the early works of Church, Curry, Kleene, and Rosser were very closely integrated, and the contributions of Kleene were essential. Thus, the topic does not seem inappropriate to the occasion.

Section 1 provides a very short historical summary, and it will be found that there is considerable overlap with CURRY (1979), which is also in this volume. An earlier version of Professor Curry's paper was in any case the incentive to write the present paper, and the reader should consult Curry's contribution for further references and philosophical remarks.

In Section 2 there is a review of the theory of functions and relations as sets leading up to the important notion of a *continuous set mapping*. In Section 3 the problem of the self-application of a function to itself as an argument is discussed from a new angle, and it is shown that—under the reduction of continuous set mappings to multi-relations—a coherent set-theoretical definition is indeed possible. The model (essentially due to PLOTKIN (1972)) of the basic laws of λ-calculus thus results.

Section 4 relates self-application to recursion by the proof of David Park's Theorem to the effect that the least fixed-point operator and the "paradoxical" combinator are the same in a wide class of well-behaved models. The connection thus engendered to recursion theory (r.e. sets) is outlined, and the section concludes with some remarks on recent results about ill-behaved models and on induction principles.

Section 5 returns to the theme of type theory, and a construction of an (η)-model with fewer type distinctions is presented, There is a brief discussion of how to introduce more type distinctions into models *via* equivalence relations, a topic deserving further study. Finally, Section 6 takes up various points of philosophical disagreement with Professor Curry which can be discussed in the light of the construction presented here. There are many questions remaining, some of which are touched upon. An appendix exhibits a moderately strong axiomatic theory suggested by the models that may help the reader see the difference between the originally proposed calculus and the outlook developed by the author.

1. Some historical background

Priority for the invention of the type-free calculus of functions (*pace* Frege) goes to Moses Schönfinkel in a lecture at Göttingen in December 1920. This talk was written up by H. Behmann and published as SCHÖNFINKEL (1924); a translation with a useful introduction by W.V. Quine is to be found in van HEIJENOORT (1967, pp. 355–366). Sometime around 1926–1927 as a graduate student, Haskell Curry in an analysis of "the process of substitution" independently discovered the combinators; but then in 1927 in "a literature search" he came across the Schönfinkel paper. Full credit is apportioned by him to Schönfinkel in his thesis (CURRY, 1930). According to CURRY (1968), Alonzo Church prepared a manuscript in 1928 on a system with λ-abstraction, and the publication, CHURCH (1932), indicates that it was work done as a National Research Fellow 1928–1929. In this paper on p. 352 and in nearly the same words in CHURCH (1941, pp. 3–4) we find Schönfinkel only mentioned in connection with the reduction of multiple-argument functions to monadic functions. Of course, it is fair to say that Schönfinkel was concerned merely with a kind of *definitional reduction* of primitives, and he proposed no *postulates* from which properties of these general functions could be derived. Such postulates were the contribution of Curry and Church.

Unfortunately Church, to my knowledge, has never explained as fully as Curry has how he was led to his theory. He must have been strongly influenced by Frege (via Russell), and he hoped to solve the paradoxes—not through the theory of types, but by the rejection of the law of the excluded middle. In CHURCH (1932, p. 347), it is stated that such combinations as occur in the Russell Paradox (namely, (λx **not** $(x(x))$) (λx. **not** $(x(x))$), which converts to its own negation) simply *fail to have a truth value*. Thus, we do not have here an intuitionistic theory, but a failure of excluded middle because functions are only *partially defined*. Alas, in

KLEENE and ROSSER (1935) it was shown that Church's system (which was employed by Kleene in KLEENE (1934) (written in 1933) and his thesis, KLEENE (1935) (accepted in September, 1933)) is *inconsistent*. The proof was later very much simplified by Curry and can be found in CURRY and FEYS (1958, pp. 258–260). It applies to various systems proposed by Curry, also, but not to his thesis, which is just the "equational" theory of combinators. This is essentially the system of CHURCH (1941), and the "system of symbolic logic" in that monograph is condensed to a very few pages (§21, pp. 68–71). The consistency of these systems is very forcibly demonstrated by the well-known theorem of CHURCH and ROSSER (1936).

The connections between the systems of Curry and Church were spelled out in his thesis by Barkley Rosser in ROSSER (1935) (written in 1933), who emphasized particularly the elimination of variables. This theme was also taken up by Quine in QUINE (1936) and again in QUINE (1971). It seems very strange to me that in his description of the method of Schönfinkel, QUINE (1971) does not mention the problem of consistency. He says (pp. 8–9).

> "Schönfinkel was the first to reduce analysis to algebra. He was the first to analyze the variable, by showing how to translate it contextually into constant terms. But his treatment is less pure than one could wish; it analyzes the variable only in combination with a function theory that is in effect general set theory."

But an inconsistent set theory it is as soon as we try to give axioms of the usual sort! All later attempts have to make strong restrictions to get anything like a workable set theory. None of these systems are, in my own opinion, particularly natural or beautiful; I do not attempt to catalogue them at this point, therefore—though some further comments are given in the last section of the paper. At this stage it seems fair to say that the model theory of these systems is certainly not very well developed. References and less negative discussion can be found in CURRY (1979) and SELDIN (1980).

For the equational systems proposed by Curry and Church the consistency proof *via* the Church–Rosser Theorem is **not** a great comfort to my mind. As with many proof-theoretic arguments, the result is very sensitive to the exact formulation of the rules. Thus, in a note, KLOP (1977) (see also KLOP (1979)), it is shown that if we extend the usual system by, say, surjective pairing functions, then the Church–Rosser Theorem *no longer holds*. Such an extension is so "natural" in the original style of Schönfinkel, that this result looks very unfortunate. Of course, someone may formulate some modified syntactical property of reduction that will imply that not all terms are interconvertible, but I do not think any such argument has yet been published. There is, however, a consistency proof by *models* (cf. SCOTT (1977), which Klop does not seem to know about) which allows certain extensions: namely, by the properties of anything that exists in the model. And this brings us to the problem of models and their place in the discussion.

Historically my first model for the λ-calculus was discovered in 1969 and details were provided in Scott (1972) (written in 1971). What I have called the "graph model" was found by Gorden Plotkin and is given in Plotkin (1972). It was rediscovered by me in 1973, and a simplified version—with proper credit to Plotkin—is found with rather full proofs in Scott (1976), which also contains many historical remarks and many references. Motivation for these discoveries is, it is hoped, very fully exposed in Scott (1977) and Scott (1977a). The two kinds of models are not unrelated, but we shall not be able to go into details here. The graph model was in effect introduced *twice* earlier in recursion theory quite explicitly but at the time *not* identified as a model (see the discussion of *enumeration operators* in Scott (1976, pp. 575 ff.)). This failure is an interesting case history in the psychology of discovery. On this score we find in Curry (1979) the following closing passage:

> "The history of combinatory logic shows that progress can result from the interaction of different philosophies. One who, as I do, takes an empirical view of mathematics and logic, in the sense that our intuitions are capable of evolution, and who prefers constructive methods, would never discover the models which Scott proposed. On the other hand, it is doubtful if anyone, with what seems to be Scott's philosophy, would have discovered combinatory logic. Both of these approaches have added to the depth of our understanding, and their interaction has produced more than either would have done."

The last sentence is very kind, but I am afraid I cannot agree with all of what was said before that. In the first place, the models *are* constructive. In the second place, as an intellectual exercise—even though the combinators are some 12 years older than I am (I was born in 1932) and we cannot change the past—I think I can defend "my philosophy" in a way that will show how they and their laws (in the form of the graph model) could have been discovered by the extension of known elementary ideas—ideas certainly easily understandable in 1928 when Curry and Church were at work on their first systems.

As a sidelight on the question of discovery, I only recently noticed this passage in Whitehead and Russell (1910, p.280):

> "Again, let us denote by '$\times n$' the relation of m to $m \times n$; then if we denote by 'NC' the class of all cardinal numbers, $\times n$"NC will denote all numbers that result from multiplying a cardinal number by n, i.e., all multiples of n. Thus, e.g., $\times 2$"NC will be the class of all even numbers."

These remarks were included merely for the sake of illustration of what happens when we take the image of a class under a function; however, if anyone had been thinking of a general theory of functions, he would have noticed here how a binary function was reduced to one-place functions $(m \times n = \times(n)(m))$. But no one as far as I know *did* think of that rather simple point before Schönfinkel, who was instead motivated by the idea of having something like a Sheffer-stroke operator for predicate calculus. Did he read *Principia*? We cannot hope to know at this distance in time (for a little more information on Schönfinkel's life, see the review KLINE (1951)); and since he did not write his paper himself, we cannot know what references to the literature he would have made.

2. Some thoughts on functions

What functions are there? Often a *representation* helps in seeing in simpler terms why certain functions exist; we do not need to claim, however, that the representation embodies the essence of the function concept—it is just used to establish possibility. For example, power series can show that a function like e^x is well-defined, differentiable, and that it satisfies its differential equation. (Other properties like being one–one onto the positive reals or satisfying $e^{x+y} = e^x \times e^y$, though elementary, are perhaps not instantly recognizable from this definition.) But not all functions have power-series expansions. It may even be the case that the difficulties about convergence led rather directly in the 19th century to a more abstract notion of function and, step by step, to the "logical monsters" deplored by Poincaré.

From particular functions one goes on to functions in general—and to *spaces* of functions. Who first suggested that a function could be regarded as a set of ordered pairs? By 1914 both Wiener and Hausdorff were doing just that (for relations as well as functions and with pairs reduced to sets as well), but I do not find any earlier references in their works or in some other books I consulted. It is not a very important historical question for our discussion, however, since it is certain that by 1914 functions as abstract objects were well understood. I stress this because I wish to argue that even a set-theoretical discovery of λ-calculus would have been not only possible but even well motivated at an early state. That it was not discovered this way is no argument against my thesis, because many things remain undiscovered even long after all the essential ingredients are available.

In the discussion below we employ the common functional notation $y = F(x)$ to indicate that the *point* x is mapped to the *point* y under the function F. For simplicity, we suppose that both x and y belong to the *same* set A. For *subsets* $X, Y \subseteq A$, as also write $Y = F(X)$ to mean that Y is the *image* of the set X under the mapping F. (See formula (1) below.) In the *Principia* the notation was $F'x$ and $F''X$, respectively, because the image of a set was regarded as the *plural* of the image of a point.

Formally we define set images by:

(1) $$F(X)=\{y|\exists x\in X.\, y=F(x)\}.$$

It may seem unrigorous not to distinguish notationally between $F(x)$ and $F(X)$, but since we are not at this point worrying about a *theory* of sets, we simply imagine A as fixed for the time being. Further, without loss of generality, we could stipulate that A and its power set are *disjoint*; hence, there need be no ambiguity between $x\in A$ and $X\subseteq A$.

Eq. (1) is a particular case of the *image under a relation*. Thus, if R is a binary relation between elements of A, we can define:

(2) $$R(X)=\{y|\exists x\in X.\, yRx\},$$

where, as always, the variables have their obvious ranges: $x,y\in A$ and $X\subseteq A$. (This is meant to correspond to *Principia*, Definition *37.10; and it is in any case standard.) If we like, we can think of R as a *multi-valued* function; that is, even when $X=\{x\}$, the image $R(\{x\})$ can be "plural". For single-valued functions, it is obvious that

$$y=F(x) \quad \text{iff} \quad \{y\}=F(\{x\}).$$

Now even for relations R, the correspondence $X\mapsto R(X)$ *is* a function on the power set of A (with values in the same set of sets). We should ask: "What kind of a function is this?" The answer is well-known (though I could not find it in the nearly endless list of formulae in the *Principia*). Such functions are *distributive* in the sense that they distribute over all unions of sets. Since every set is the union of its singleton subsets, this comes down to:

(3) $$R(X)=\bigcup\{R(\{x\})|x\in X\}.$$

Any correspondence satisfying (3) is distributive. If a **mapping** on sets satisfies (3), then we can define a **relation** between points by:

(4) $$yRx \quad \text{iff} \quad y\in R(\{x\}).$$

Then in view of (3), the *defined* meaning of $R(X)$ in (2) is the same as the *given* meaning of $R(X)$ as a set mapping. We can summarize:

Proposition. *There is a one–one correspondence of interdefinability between binary (point) relations and distributive set mappings provided by formulae (2) and (4).*

Even if this simple theorem is not in *Principia*, the authors (as well as Wiener, Jourdain, Hardy, etc.) would have certainly understood it at once.

It is time now for a *new* idea. Let us recall first how relations are reduced to sets of ordered pairs. We use the notation $\langle x,y\rangle$ for pairs and $A^2 = A \times A$ for the cartesian product. A relation is just a subset $R \subseteq A^2$ and yRx means the same as $\langle y,x\rangle \in R$. The question is: "What should we say about n-ary relations?" Using the notation $\langle x_0, x_1, \ldots, x_{n-1}\rangle$ for n-tuples and A^n for the n-fold Cartesian product, we define the set of *all* finite sequences by:

(5) $$A^* = \bigcup_{n=0}^{\infty} A^n.$$

(The case $n=0$ has $A^0 = \{\langle\rangle\}$, where $\langle\rangle$ is the *empty tuple*.)

An *n-ary relation* is a subset $S \subseteq A^n$. It seems reasonable to call a subset $P \subseteq A^*$ a *multi-relation*. The new idea has to do with how to generalize formula (2) for the image. This is one such way:

(6) $$P(X) = \{y \mid \exists n \exists x_1, \ldots, x_n \in X.\ \ \langle y, x_1, \ldots, x_n\rangle \in P\}.$$

In case $P \subseteq A^2$ (and we agree that the various A^n are pairwise disjoint), then (6) reduces to (2) exactly—provided we substitute P for R in (2). In case $P \subseteq A^{n+1}$, then P is an $(n+1)$-ary relation; formula (6) could be said to define the image of X^n, if we regard P as a multi-valued mapping where $\langle x_1, \ldots, x_n\rangle \mapsto y$. We might wish to define more generally the image of $X_1 \times X_2 \times \cdots \times X_n$, where the n sets are allowed to be distinct, but it is useful enough to stick to the one-place set mapping in (6) for many purposes.

What is achieved by this new definition? In the case of binary relations, we had $y \in R(X)$ just in case there is *one* element $x \in X$ to which y is R-related. In the case of multi-relations, $y \in P(X)$ requires a *finite subset* $\{x_1, \ldots, x_n\} \subseteq X$ with y P-related to $\langle x_1, \ldots, x_n\rangle$. Clearly then the mapping $X \mapsto P(X)$ need *not* be distributive, because we might have $y \in P(\{x_1, \ldots, x_n\})$ but $y \notin P(\{x_i\})$ for $1 \leqslant i \leqslant n$. But nevertheless the mapping does have a special property which generalizes (3):

(7) $$P(X) = \bigcup \{P(E) \mid E \subseteq X, E \text{ finite}\}.$$

We call such functions from sets to sets *continuous*, because a "finite approximation" $E^1 \subseteq P(X)$ is already determined as being true by a finite approximation $E \subseteq X$ where $E^1 \subseteq P(E)$. In fact, the biconditional

$$\forall X \forall E^1 \left[E^1 \subseteq P(X) \leftrightarrow \exists E \subseteq X.\ E^1 \subseteq P(E)\right]$$

is equivalent to the validity of (7) for all $X \subseteq A$, provided we restrict the ranges of the variables E^1 and E to *finite* subsets of A. As another characterization of continuous set mappings, we might mention that they are exactly the ones that distribute over *directed* unions of sets (equivalently: over unions of *chains* of sets). But (7) as a definition will be sufficient for our purposes.

Just as with binary relations, if we have a *given* continuous set mapping P, then we can define a multi-relation P by:

(8) $$\langle y, x_1, \ldots, x_n \rangle \in P \quad \text{iff} \quad y \in P(\{x_1, \ldots, x_n\}).$$

As before, in view of (7), the $P(X)$ defined by (6) will be equal to the given $P(X)$.

A small point of difference: every $P \subseteq A^*$ determines a continuous set mapping, and every continuous map is obtained in this way. However, *different* P's may determine the *same* map. (Thus, whether $\langle\,\rangle \in P$ is of no moment as empty sequences do not figure in (6); for greater regularity and for a later application we will assume that (8) *allows* $\langle\,\rangle \in P$.) There is not, therefore, a one–one correspondence between continuous set mappings and subsets $P \subseteq A^*$. In order to regain the uniqueness we had with binary relations, we need to require that the subset $P \subseteq A^*$ satisfy a certain "fullness" condition:

(9) $$\langle\,\rangle \in P \text{ and whenever } \langle y, x_1, \ldots, x_n \rangle \in P$$
$$\text{and } \{x_1, \ldots, x_n\} \subseteq \{y_1, \ldots, y_k\}, \quad \text{then } \langle y, y_1, \ldots, y_k \rangle \in P.$$

Condition (9) makes P *maximal* among the subsets of A^* determining the desired continuous set mapping.

There is still the question: "Why are continuous set mappings *interesting*?" The answer is quite pleasant. In TARSKI (1930) (written in 1928), condition (7) is given **exactly** is his Axiom 4 on the abstract properties of the *consequence relation*. Of course, we are *not* assuming here the special properties of a closure operator, where additionally

$$X \subseteq P(X) = P(P(X)),$$

since these are not even true for relational images (unless the relation is special). But closure operators can provide fine examples of continuous mappings. Aside from the set of logical consequences of a set of sentences, other familiar algebraic examples of continuous closure operators would be, say, to let $P(X)$ be the *subgroup* of A generated by X in the case that the set A carries the structure of a group. Then P would satisfy the above extra conditions as a set mapping.

It does not seem at all far fetched to suggest that in 1928 a proposed Master's thesis could have had the topic of investigating the theory of *arbitrary* continuous set mappings. I can see in my mind's eye exactly the kind of paper that would have been submitted to *Fundamenta Mathematicae*. Here is a summary of results and definitions:

(i) *A continuous set function is defined as one satisfying formula (7) above; an alternative definition states that the function preserves directed unions of sets.*

(ii) *Continuity in several variables is defined as continuity in each variable separately.*

(iii) *As a side remark it is pointed out that* **P***A, the power set of A, has a topology with the sets of the form* $\{X \subseteq A \mid E \subseteq X\}$, *where* $E \subseteq A$ *is finite, as a basis for the open sets; continuity as defined in* (i) *and* (ii) *is proved to be the same as topological continuity on* **P***A and on the product space* $(\mathbf{P}A)^n$. *This remark would not actually be needed in the sequel, but it helps show why the concepts are natural.*

(iv) *The composition of continuous functions (in any number of variables) is continuous; this of course follows from* (iii), *but a direct argument is very elementary.*

(v) *The notion of* A^* *and a multi-relation is defined.*

(vi) *The image of a set under a multi-relation is defined and it is noted that the mapping is continuous.*

(vii) *It is proved that every continuous function is obtainable from a multi-relation, and in fact there is a maximal such (cf. formula (8)).*

(viii) *The maximal multi-relations in* (vii) *are proved to be exactly the "full" relations in the sense of formula (9).*

As it stands, up to this point, the paper seems rather thin and even might not have been accepted for publication. In the next section we turn to some additional ideas that could have been added to it to give it somewhat greater depth.

By the way, a simple observation as to why continuity is better than distributivity (the binary-relation case) concerns functions of several variables. Suppose $F(X, Y)$ is distributive separately in X and in Y. Then the composition (with identity functions) resulting in $F(X,X)$ is continuous *but not necessarily distributive*; further compositions like $F(F(X,X), F(X,X))$ remain continuous but are even further from being distributive. The reason is that the condition $z \in F(X,X)$ though equivalent to $\exists x \in X \exists y \in X.\ z \in F(\{x\},\{y\})$, is not necessarily equivalent to $\exists x \in X.\ z \in F(\{x\},\{x\})$. By taking the point-wise union of continuous functions we get other continuous functions that are even less distributive. This remark shows that a study of set mappings might have suggested the notion of continuity independently of the well-known examples of "algebraic" closure operations.

3. Some reflections on self-application and combinators

As in Scott (1973) I recall a passage from the introduction to Church (1941):

> "In particular it is not excluded that one of the elements of the range of arguments of a function f should be the function f itself. This possibility has been frequently

denied, and indeed, if a function is defined as a correspondence between two previously given ranges, the reason for the denial is clear. Here, however, we regard the operation or rule of correspondence, which constitutes the function, as being first given, and the range of arguments then determined as consisting of the things to which the operation is applicable. This is a departure which is natural in passing from consideration of functions in a special domain to the consideration of functions in general, and it finds support in consistency theorems which will be proved below."

The philosophical stance assumed by Church does not fit so very well to the λ-**K**-calculus where *every* application has to be taken as meaningful. (An interesting discussion of the differences between the λ-**I**-calculus and the λ-**K**-calculus will be found in Barendregt (1980).) If we think instead of partial recursive functions, where rules are programs and where programs have code numbers, then, following Kleene, an application $\{e\}(n)$ may or may not be "meaningful" (that is, the Turing machine need not halt). There are many cases, however, where $\{e\}(e)$ *does* converge, and such self-applications are quite meaningful. The drawback (if it is one) is that the method of code numbers takes functions in *intension*. It may well happen that $\{n\}=\{m\}$ as partial recursive functions, but $\{e\}(n)\neq\{e\}(m)$. Many of the calculi considered by Church and Curry take the function concept to be *extensional*, and it is rather more difficult in such a context to be precise about the meaning of "rule of correspondence". It is really not the least bit fair of Church in the above passage to invoke the consistency proof *via* the Church–Rosser Theorem, since this gives no *intuitive* justification of the choice of reduction rules which are given in advance.

There are nevertheless many theories of functions in extension where self-application is no trouble whatsoever. To have such a model $\mathfrak{M}$, just let $\mathfrak{M}=\mathbb{R}$, the real numbers. Define application as *addition*, so that $x(y)=x+y$. This is the model for a "theory" of *translations*. In $\mathfrak{M}$, the axiom that has $\forall z.\ x(z)=y(z)$ always implying $x=y$ is trivially valid. The only difficulty with this model is that so *few* functions from $\mathfrak{M}$ into $\mathfrak{M}$ are represented by elements of $\mathfrak{M}$ ("few", despite the fact that there are a continuum number of distinct translations). For example, even though $x(x)$ is always meaningful, this is a function **not** represented in the model; that is, there does not exist an element $a\in\mathfrak{M}$ with $a(x)=x(x)$ for all $x\in\mathfrak{M}$. But $x(x)=2x$ is surely a "harmless" function. Why not throw in *all* linear functions $x\mapsto\alpha+\beta x$ and not just the translations $x\mapsto\alpha+x$? The difficulty now is that the linear functions require *two* parameters α and β. There is no "nice" function of the type $[\cdot,\cdot]:\mathfrak{M}\times\mathfrak{M}\to\mathfrak{M}$ where we could then define application so that $[\alpha,\beta](x)=\alpha+\beta x$. The point is that we would want $x(y)$ to be a linear polynomial in each variable separately. But then $x(x)$ is no longer linear but rather is quadratic. (In any case, the project is doomed as soon as $x\mapsto 1+x$ is allowed; this function has *no* fixed point, but in a full theory of combinators *all* functions have fixed points.) The rub comes when we try to combine the definability of $x(y)$ along with what Curry calls *combinatory completeness*. This point is perhaps not made clear by Church, and his few words of justification do not seem to me to be sufficient.

Is there, then, a model that combines the definability of application (and self-application) along with a flexible theory of functions? But hold, we are not allowed to ask this question because we have put ourselves back to 1928 (or earlier) in a state ignorant of the work of Curry and Church. It is required that we *discover* the λ-calculus—the combinators *and* their laws—on our own.

So let us consider again the thoughts we had on functions in the last section. We left off after having written a first chapter of a Master's thesis about continuous set functions and their representation by multi-relations $P \subseteq A^*$. We defined there in formula (6) an application $P(X)$ for all $X \subseteq A$. This seems fairly powerful. The idea of set- and function-abstraction was certainly in the air at that time (Curry and Church did not start in a vacuum), thus the idea of a *notation* (say, like that of Russell, or Peano, or Frege) for *abstraction* could have occurred to our Master's candidate. I will not try to construct an "original" notation, but will use Church's:

(1) $$\lambda X.\ \tau[X] = \{\langle\rangle\} \cup \{\langle y, x_1, \ldots, x_n\rangle \mid y \in \tau[\{x_1, \ldots, x_n\}]\}$$

Here, $\tau[X]$ is an expression with possibly a free set variable X and $\tau[\{x_1,\ldots,x_n\}]$ is short for the substitution of the set-term $\{x_1,\ldots,x_n\}$ for the variable X in τ. Bound variables and substitution were well understood in 1928 even if Curry is right that the substitution process is messy. Formula (1) is a set-theoretical construction (of subsets of A^*), so there is no unhistorical "anticipation" of Church.

It is without difficulty that we imagine the recognition of the following two laws:

(α) $$\lambda X.\ \tau[X] = \lambda Y.\ \tau[Y],$$

provided of course that Y is not free in $\tau[X]$ and that substitution resolves clashes between free and bound variables. Similarly, there is a principle of extensionality:

(ξ) If, for all $X \subseteq A$, we have $\tau[X] = \sigma[X]$,
then $\lambda X.\ \tau[X] = \lambda X.\ \sigma[X]$.

Perhaps it is also useful to mention a stronger version of this law which takes special advantage of the fact that abstracts are sets:

(ξ*) If, for all $X \subseteq A$, we have $\tau[X] \subseteq \sigma[X]$,
then $\lambda X.\ \tau[X] \subseteq \lambda X.\ \sigma[X]$.

These laws are justified by the usual behaviour of bound variables and very, very elementary extensionality properties of set formation in (1).

The next law would, admittedly, require a little more imagination. The point is that we have already seen that there is a one–one correspondence between certain multi-relations and continuous set mappings. What would require a grasp of formal manipulation would be the insight of how to *express* this in the above notation. But, since class abstraction was already known, we can hope our hypothetical Master's candidate would write:

$$(\beta) \qquad (\lambda X.\ \tau[X])(Y)=\tau[Y],$$

provided the mapping $x \mapsto \tau[X]$ is a **continuous** set mapping on subsets of A. This law expresses exactly the fact, *already noted*, that if we use (1) above to define a multi-relation from a continuous map, then under the definition of application we achieve *the same* mapping by applying the multi-relation to any suitable argument $Y \subseteq A$.

The aspect to emphasize at once is that all this discussion is almost purely notational: (α), (β), (ξ) are just *notational variants of facts already known*. There is no hint of tricks of self-application yet, and we need to ask: "How could a coherent notion of self-application be based on such set-theoretical definitions?"

Look again at the definition given in (1) above. There is a plain distinction of **type** involved, because while $X \subseteq A$ is intended for *arguments*, the *results* of what is defined are multi-relations

$$\lambda X.\ \tau[X] \subseteq A^*.$$

Would anyone (let alone our Master's candidate) ever be tempted to confuse these types (that is, to confuse subsets of A and subsets of A^*)?

Obviously, I want the answer to be "yes", but can I motivate it? Let us stop to reason that a confusion of the kind requested would require that subsets of A^* be at the same time subsets of A. If the set A were *closed under the formation of finite sequences*, then we would have $A^* \subseteq A$ and, trivially, $X \subseteq A^*$ would imply $X \subseteq A$.

Are there such sets $A^* \subseteq A$? Perhaps Russell and Wiener might have balked at first, because Wiener's definition of pairs would—strictly applied—have resulted in sets of *infinite type*. But Kuratowski, Hausdorff, Zermelo or Von Neumann would have had no trouble with the suggestion. For Russell, it would have been simple enough to *axiomatize* the finite-sequence-forming operation and to point out how many models there are where we can regard $A^* \subseteq A$. It would even be easy to go further and eliminate *all* type distinctions previously found by getting $A^* = A$. (In Zermelo's theory, consider the *least* set closed under the formation of finite sequences—"generated" from $\varnothing$, so to speak.)

Having had the idea of closing up A under sequences, we remark that the self-application $P(P)$ makes perfectly good sense, because now $P \subseteq A^* \subseteq A$ and $P(P)$ is defined. Even better, from eq. (6) of the last section, we see that $X(Y)$ *always* makes sense for all $X, Y \subseteq A$. Note that

$$X(Y) = (X \cap A^*)(Y),$$

but, since $A^* \subseteq A$, we lose no continuous functions this way. Also it is *obvious* from the definition that $X(Y)$ is continuous in *both* X and Y (the operation is even distributive in X). This is the main insight: the uniformly "type-free" definition of $X(Y)$ is permitted for a very elementary set-theoretical reason. But will we be motivated to go further?

Once a person has an operation he naturally tries to *iterate* it. A particularly simple case of iteration is something like $P(X)(Y)(Z)$. Take P fixed. The set P is a multi-relation applied to X. But the result (possibly better regarded as restricted to A^*) is a multi-relation we can apply to Y; and so on. Once we say to ourselves: "Sets determine multi-relations; the values are sets; these values determine multi-relations in turn", there is no reason to stop.

Now $P(X)(Y)(Z)$ is a three-place function, continuous in all its arguments (recall what we already remarked about composition). We should ask: "What three-place continuous functions do we get in this way?" Surely this is a natural question. The answer is "all".

Theorem. *Assume* $A^* \subseteq A$. *Let* $X_0, X_1, \ldots, X_{n-1} \mapsto \tau[X_0, X_1, \ldots, X_{n-1}]$ *define a continuous set mapping of n-variables. Let*

$$P = \lambda X_0 \lambda X_1 \cdots \lambda X_{n-1}.\ \tau[X_0, \ldots, X_{n-1}].$$

Then P is a multi-relation such that

$$P(X_0)(X_1) \cdots (X_{n-1}) = \tau[X_0, X_1, \ldots, X_{n-1}]$$

for all $X_0, X_1, \ldots, X_{n-1} \subseteq A$.

This result is the n-ary generalization of the principle (β) already mentioned. Recall that—given Chapter 1—(β) followed by *definition*. The above theorem would follow at once from *iteration* of principle (β) provided we establish (the true)

Lemma. *If* $X_0, X_1, \ldots, X_n \mapsto \tau[X_0, X_1, \ldots, X_n]$ *is a continuous set mapping of* $(n+1)$ *variables, then*

$$X_0, X_1, \ldots, X_{n-1} \mapsto \lambda X_n.\ \tau[X_0, X_1, \ldots, X_n]$$

is a continuous set mapping of n-variables.

The proof of the lemma is an elementary consequence of formula (1) of this section (where $\tau[X]$ is replaced by $\tau[X_0, X_1, \ldots, X_n]$ and X is replaced by X_n). One only has to take a free set variable, replace it by a directed union, and then bring the union to the outside of the set abstraction by valid set-theoretical principles.

The point of the theorem is that iterated *application* is the "inverse" of iterated *abstraction*—provided we have a lemma to the effect that abstraction preserves continuity.

Now the Master's thesis is taking on more substance; however, at this point self-application has still played no major role. Rather, we have motivated and implemented in the present theory Schönfinkel's idea that n-ary functions can be reduced to monadic functions (at least in this context of continuous set functions). The question that remains is: "Where would the combinators have come from?"

It is not fair to answer that Schönfinkel's paper was published before 1928 and that our Master's candidate may have read it. Let us try to get him to discover the combinators himself. Recall that under the assumption $A^* \subseteq A$ we can regard application as a binary set operation:

$$X(Y) = \{y \mid \exists n \exists y_1, \ldots, y_n \in Y.\ \langle y, y_1, \ldots, y_n \rangle \in X\}. \tag{2}$$

(This is just (6) of the last section.) Even in the notation of *Principia* it would have been readily seen that this is continuous in both variables—even if we did not cheat and use the three dots. Compositions of continuous functions are continuous; λ-abstractions of continuous functions are continuous; these constructs can be iterated as much as we please; hence, consider the iterated combination

$$\mathbf{B} = \lambda F \lambda G \lambda X.\ F(G(X)).$$

By the Theorem, $\mathbf{B}(F)(G)$ represents the *composition* of F and G; that is, $\mathbf{B}(F)(G)$ is a multi-relation representing the function represented by the composition of the functions represented by the multi-relations F and G. (This talk of representation is clearly becoming too complex.) Therefore, **B** itself represents the "general idea" of composition. This is probably the first non-trivial combinator to be discovered. As is common, we shall use in the sequel the more readable infix notation:

$$F \circ G = \mathbf{B}(F)(G).$$

Admittedly, this discussion all seems totally trivial *after* the work of Curry and Church, but there are two remarks I might make. In the first place, even today the combinators are not all that well-known (though studies in Computer Science have given them a new lease on life). Thus there is a contemporary problem of motivating them when, say, trying to explain the idea to other mathematical but non logically-trained colleagues. The interpretation proposed here is, I feel, a rather direct way of showing that a non-trivial combinatory algebra is possible (and it is definitely easier than the non-extensional algebra of Gödel numbers under $\{e\}(n)$ as application). Having realized this, the second remark may have greater weight: I believe that in 1928 someone could have reasoned: "If I *can* iterate certain continuity-preserving operations, then I *should* iterate them and try to find out what they give." Many mathematical discoveries have been made on less clear grounds. It strikes me as a suitably "empirical" approach.

So ends Chapter 2 of the hypothetical Master's thesis: the assumption $A^* \subseteq A$ has been utilized more for iteration of concepts than for self-application. What has been shown is that a full theory of continuous functions (as exactly represented by multi-relations) is achieved in such a way that a notation of application and abstraction is completely meaningful in any degree of intermixing (iteration). Moreover, certain obvious laws (such as (α), (β), (ξ), (ξ^*)) hold in what has now become a "type-free" manner. These laws are enough to justify many laws of combinators (e.g. the associativity of **B**). The construction has provided a model for a coherent (though not as yet very formal) theory of application and abstraction—a theory which, as we shall see, is open to many extensions.

It should be stressed that the theory of continuous functions for which these laws are valid is one where the notion of function (as a set mapping) is replaced by the notion of a representing multi-relation. The map $X \mapsto \tau[X]$ (if continuous) is represented perfectly by the corresponding multi-relation $\lambda X.\ \tau[X]$ of formula (1): we could call this multi-relation the *graph* of the map. But the reader must heed the fact that only *continuous* set mappings have graphs of this kind.

Looking back at the early writings on λ-calculus such a point of view may seem very limited, and what has to be done is to illustrate the *scope* of such a theory of continuous functions. In any case in what follows we are simply going to identify continuous functions with their graphs and use this idea as our principal notion of function. Note, however, one advantage of our approach over the formal, purely axiomatic induction of λ-calculus by Curry and Church: in this context the combinators are seen as *special* continuous functions—there are many more such functions than merely those defined in some restricted notation. In this way the combinators (in combination with other operators) are given a greater range of applicability over just the "logical" uses intended by their discoverers.

(Returning from 1928 for a moment to 1972, it should be remarked that Plotkin defined the model by forming not A^* but the closure of A under ordered pairs of finite subsets of the set being constructed. The two closures are of the same order of complexity, but the author feels that the use of A^* makes the definitions look more elementary.)

4. Some notes on iteration and computability

In the previous section self-application (better: confusion of types) was made palatable (or at least: coherent) through the multi-relational representation of continuous set mappings and the closure condition $A^* \subseteq A$. It should be pointed out again, however, that very little advantage was taken of the possibility of self-application, as the "natural" combinators do not emphasize self-application. Since Church was fundamentally concerned with the paradoxes, he considered the question from the start, but his monograph CHURCH (1941) does not on the surface exploit the freedom of self-application very much.

Whether our Master's candidate would have grasped the significance of diagonal applications is not so compelling an hypothesis. But, nevertheless $X(X)$ is at once seen to be a continuous set function; perhaps the step to looking at

$$(\lambda X.\ X(X))(\lambda X.\ X(X))$$

would have required a certain curiosity. Certainly, the paradoxes were widely discussed. So even the kind of formula that enters in their derivations is not all that weird:

$$(\lambda X.\ F(X(X)))(\lambda X.\ F(X(X))).$$

(Here, as we have already remarked, **not** should replace F for the sake of paradox.) If we regard the above as a continuous map $F \mapsto \mathbf{Y}(F)$(that is, $\mathbf{Y}$ is λF of the above expression), then—as in the derivation of the Russell Paradox—we have by a use of (β)

$$\mathbf{Y}(F) = F(\mathbf{Y}(F)).$$

(Surely this is the way Curry discovered the so-called Paradoxical Combinator; note his derivation of the Russell Paradox in CURRY (1930, p. 511).)

Perhaps by now our Master's student is working on his Doctor's thesis. If he were the kind of person to submit papers to *Fundamenta*, he would have surely read KNASTER (1928) written in 1927. This note contains the general Fixed-Point Lemma for Monotone Set Mappings, a joint result of Knaster and Tarski. By a monotone map, we understand a function such that

(1) $\quad X \subseteq Y$ always implies $F(X) \subseteq F(Y)$.

Here $X, Y \subseteq A$ and the values of the function have the same type: $F(X) \subseteq A$. The following theorem specializes from complete lattices the statement of TARSKI (1955), a paper written in 1953 and reporting on results of 1939 and earlier.

Theorem. *Every monotone set mapping $F: \mathbf{P}A \to \mathbf{P}A$ has a least fixed point, and indeed the set of all fixed points, $\{X \subseteq A \mid X = F(X)\}$, forms a complete lattice under $\subseteq$.*

Of course, continuous set mappings are monotone (*Hint*: consider the directed family of sets $\{X, Y\}$, where $X \subseteq Y$), and so the theorem applies to that case. We need not detail the whole and quite elementary proof of the theorem here, but a few words about it might be helpful. Let us take the problem of proving the existence of the *least* fixed point. Consider the family of sets "closed" under F, namely $\{X \subseteq A \mid F(X) \subseteq X\}$. Even though F is not a closure operation and is not assumed to be any better than monotone, we can still easily prove that this family of "closed" sets is closed under arbitrary intersections; in particular, the intersection of the *whole* family belongs to the family itself (that this intersection exists comes from the fact that the set A is itself "closed"). Call this intersection X_0. We have $F(X_0) \subseteq X_0$, and then by monotonicity $F(F(X_0)) \subseteq F(X_0)$. This proves that $F(X_0)$ is "closed" also; whence, $X_0 \subseteq F(X_0)$ because X_0 is already known to be the least one in that family. Obviously, then, X_0 is not only a fixed point but also the least one in the family of fixed points (since every fixed point is "closed").

In the continuous case, the least fixed point of F can be "found" by a simple infinite iteration; calling the least fixed-point operator **fix** we can write:

$$\text{(2)} \qquad \mathbf{fix}(F) = \bigcup_{n=0}^{\infty} F^n(\emptyset).$$

On the right, $\emptyset$ is the empty set. The proof by continuity that the operator does indeed give the least fixed point is *very* elementary, for one has only to note that on the right the union is directed. (In the non-continuous case, the iteration has to be continued into the transfinite.)

In KLEENE (1952), the nub of the proof is embedded in his proof of the First Recursion Theorem (pp. 348 ff), where the verification that partial recursive *functionals* are continuous is carried back to first principles. But note that Kleene is proving two things: the least fixed point exists *and* it is computable. (Strictly speaking, he is dealing with partial functions not subsets; a discussion of the obvious connection is given in SCOTT (1976), which also relates closely to the considerations of the present paper.) These facts ought to be separated, since the existence part is more general and more elementary; in fact, we can prove existence long before we have introduced tools to make computability (a kind of definability) precise. Keep in mind, however, that this remark does not eliminate any of the problems of showing that *particular* functions are continuous; all we have at hand so far from the last section is a method for generating a large number of continuous—and quite complex—functions from other continuous functions.

Now look back at the definition of **fix** in (2) above. For each integer n, the map $F \mapsto F^n(\varnothing)$ is continuous by the composition principle. (Now we are, as usual, identifying functions with their graphs as mutli-relations.) Any point-wise union of continuous functions is continuous. *Hence*, the union map $F \mapsto \mathbf{fix}(F)$ is continuous: **fix** must have a graph. What is it?

By now we should probably give up the rhetorical device of arguing that all this could have been done in 1928. I believe I have at least argued that the *discovery* of the combinators and their elementary laws could have been given a set-theoretical (and, for my taste, *natural*) grounding in 1928. On the other hand, whether their applications and uses would have been so quickly recognized is not clear, since the theory of recursive functions took some time and somewhat different motivation in order to get started. (Mainly, Gödel and the incompleteness theorems were required.) Note, however, that the reason that Knaster and Tarski introduced the fixed-point method was to produce the *recursions* needed in set-theoretical arguments like the Schröder–Bernstein Theorem and many generalizations. Perhaps, then, the step to a general theory of recursion could have come forward along these lines—but such conjectures are more or less pointless.

So we return to the question of relating **fix** to the "pure" combinators. Since $\mathbf{fix}(F)$ is always the least fixed point of the mapping $X \mapsto F(X)$, and since $\mathbf{Y}(F)$ is *some* fixed point, then $\mathbf{fix}(F) \subseteq \mathbf{Y}(F)$ for all F. It was a most useful discovery of David Park (1969) that $\mathbf{fix} = \mathbf{Y}$ was actually possible. (The proof of Park was given for a slightly different kind of interpretation of λ-calculus, but I had no trouble in recasting the idea for a similar proof in Scott (1976, p. 569 f).) In order to give the proof here, some analysis of A^* *within* A is required.

Let $\prec$ be a binary relation on A. We say $\prec$ is a *well-founded relation*, if whenever $X \subseteq A$ is non-empty, then $x \in X$ exists with no $y \prec x$ also satisfying $y \in X$. (The element x is "minimal" in X with respect to $\prec$.) We say that A^* is *progressive* in A (with respect to $\prec$) if whenever we have $x_0, x_1, \ldots, x_n \in A$, then $x_i \prec \langle x_0, x_1, \ldots, x_n \rangle$ for $1 \leqslant i \leqslant n$. (Tuples are always worse than their (later) terms.) The essential point of the well-founded, progressive case is that every element of A^* can be well-pictured as a *tree*:

each node, if an element of A^*, has as its immediate descendants the terms of the sequence of *positive* index. If you come to a sequence of length less than two or come to an element of $A \backslash A^*$, then you stop. What the assumption amounts to is that this tree is always *finite*.

Theorem. *If* $\prec$ *is a well-founded relation on* A *and if* $A^* \subseteq A$ *is progressive in* A, *then the paradoxical combinator* **Y** *is the least fixed-point operator.*

Proof. Let F be a multi-relation representing a continuous set function. Let $B = F(B)$ be *any* fixed point of F. For the moment write $\Delta = \lambda X.\ X(X)$. Let $U = \lambda X.\ F(X(X))$, and recall that $\mathbf{Y}(F) = \Delta(U)$. We know $\Delta(U)$ is a fixed point of F, but we must show that $\Delta(U) \subseteq B$.

Let us write $\langle x_0, x_1, \ldots, x_n \rangle \ll U$ provided $\{x_1, \ldots, x_n\} \subseteq U$. By continuity

$$\Delta(U) = \bigcup \{\Delta(\{x_1, \ldots, x_n\}) | \langle x_0, x_1, \ldots, x_{n-1} \rangle \ll U\}.$$

Thus, we need only show that

$$\langle x_0, x_1, \ldots, x_n \rangle \ll U \text{ always implies } \Delta(\{x_1, \ldots, x_n\}) \subseteq B.$$

So suppose *not*. Let S be the set of tuples $\ll U$ for which the inclusion *fails*. Since $\prec$ is well-founded, let $\langle x_0, x_1, \ldots, x_n \rangle$ be a minimal element of S. We seek a contradiction.

To this end, let y_0 be any element of $\Delta(\{x_0, x_1, \ldots, x_n\}) \backslash B$. Because

$$y_0 \in \{x_1, \ldots, x_n\}(\{x_1, \ldots, x_n\}),$$

we see from formula (2) of the last section that there are elements $y_1, \ldots, y_m$ where

$$\{y_1, \ldots, y_m\} \subseteq \{x_1, \ldots, x_n\}, \quad \text{and} \quad \langle y_0, y_1, \ldots, y_m \rangle \in \{x_1, \ldots, x_n\}.$$

But then, by assumption,

$$\langle y_0, y_1, \ldots, y_m \rangle \prec \langle x_0, x_1, \ldots, x_n \rangle.$$

Since $\langle y_0, y_1, \ldots, y_m \rangle \ll U$, we know then that $\Delta(\{y_1, \ldots, y_m\}) \subseteq B$. But

$$\langle y_0, y_1, \ldots, y_m \rangle \in U = \lambda X.\ F(X(X)),$$

so by definition $y_0 \in F(\Delta(\{y_1, \ldots, y_m\}))$. By monotonicity

$$F(\Delta(\{y_1, \ldots, y_m\})) \subseteq F(B) = B,$$

so $y_0 \in B$, which is impossible!

Actually we need not invoke an arbitrary $\prec$ on A for the above result, because there is always a *least* relation for which A^* is progressive in A. We can call the whole structure $A^* \subseteq A$ a *well-founded model for tuples* provided this least relation is well-founded.

But what does this have to do with computability and Kleene's First Recursion Theorem? We need some definitions. A structure $A^* \subseteq A$ will be said to be *computable with respect to an enumeration* $A = \{a_n | n \in \mathbb{N}\}$ (where $\mathbb{N} = \{0, 1, 2, \ldots\}$) provided that we can show uniformly that the relationship

$$\langle a_{n_0}, a_{n_1}, \ldots, a_{n_{k-1}} \rangle = a_{n_k}$$

is recursively enumerable in $n_0, n_1, \ldots, n_k$.

In the same vein we say a *subset* $U \subseteq A$ is *computable* if $\{n \in \mathbb{N} | a_n \in U\}$ is recursively enumerable. A *continuous set function* $X_0, X_1, \ldots, X_{m-1} \mapsto \tau[X_0, X_1, \ldots, X_{m-1}]$ is *computable* provided its graph $\lambda X_0 \lambda X_1 \cdots \lambda X_{m-1}.\ \tau[X_0, X_1, \ldots, X_{m-1}]$ is computable.

Theorem. (i) *In a computable structure, the computable functions contain all the computable constants and are closed under composition and* λ*-abstraction;*

(ii) *Application is a computable function (of two arguments), and the computable functions are closed under point-wise application;*

(iii) *If* $G(X_0, X_1, \ldots, X_n)$ *is computable function, then so is the least solution to the equation*

$$F(X_1, \ldots, X_n) = G(F, X_1, \ldots, X_n);$$

in fact, **fix** *is a computable function;*

(iv) *Provided* $A^* \subseteq A$ *is well-founded,* **fix** *can be proved computable by explicit definition as the* **Y**-*combinator.*

Proof. Though the theorem has been stated for *functions* (of several variables)—which we could think of as defined by *terms* (with the appropriate number of free variables) built up from constant symbols for computable subsets of A by application and λ-abstraction—here is a case where the argument is easier by combinators. We reduce all λ-definitions to the well-known combinators **S** and **K**, which we have to prove are computable subsets of A. Then we need only check that if U and V are computable sets, then so is $U(V)$. This will take care of parts (i) and (ii). Recursion in part (ii) still requires that we show **fix** to be computable; however, in part (iv) in the well-founded case, there is nothing to prove, because **Y** is λ-definable by our previous theorem.

Turning now to the details we recall

$$\mathbf{K} = \lambda X \lambda Y.\ X.$$

By the definition of λ-abstraction this comes out:

$$\begin{aligned}\mathbf{K} &= \{\langle\rangle\} \cup \{\langle x_0, x_1, \ldots, x_n \rangle | x_0 \in \lambda Y.\ \{x_1, \ldots, x_n\}\} \\ &= \{\langle\rangle\} \cup \{\langle\langle\rangle, x_1, \ldots, x_n \rangle | x_1, \ldots, x_n \in A\} \\ &\quad \cup \{\langle\langle y_0, y_1, \ldots, y_m \rangle, x_1, \ldots, x_n \rangle | y_0 \in \{x_1, \ldots, x_n\}\}.\end{aligned}$$

This means that $a \in \mathbf{K}$ is equivalent to

$$a = \langle\rangle \vee \exists n \exists x_1,\ldots,x_n \in A.\ a = \langle\langle\rangle, x_1,\ldots,x_n\rangle \vee$$
$$\exists n,m \exists x_1,\ldots,x_n \in A \exists y_0,\ldots,y_m \in A \exists z \in A.\ a = \langle z, x_1,\ldots,x_n\rangle \wedge z$$
$$= \langle y_0, y_1,\ldots,y_m\rangle \wedge [y_0 = x_1 \vee \cdots \vee y_0 = x_n]].$$

This is wholly existential on r.e. predicates (r.e. with respect to the enumeration), so **K** is computable.

In order to be able to state the formula for **S** it is convenient to make some definitions:

$$A_{[0]} = \{\langle\rangle\};$$
$$A_{[n+1]} = \{\langle u_0, u_1,\ldots,u_m\rangle \mid u_0 \in A_{[n]} \wedge u_1,\ldots,u_m \in A\}.$$

We recall that

$$\mathbf{S} = \lambda F \lambda G \lambda X.\ F(X)(G(X)).$$

(**S** is of course the combinator for point-wise application of two functions.) Now we can write:

$$\mathbf{S} = A_{[0]} \cup A_{[1]} \cup A_{[2]} \cup$$
$$\{\langle\langle\langle x_0, x_1,\ldots,x_k\rangle, g_1,\ldots,g_m\rangle, f_1,\ldots,f_n\rangle \mid$$
$$x_0 \in \{f_1,\ldots,f_n\}(\{x_1,\ldots,x_k\})(\{g_1,\ldots,g_m\}(\{x_1,\ldots,x_k\}))\}.$$

To show that $a \in \mathbf{S}$ is characterizable by an existential predicate, the main difficulty is in showing

$$x_0 \in \{f_1,\ldots,f_n\}(\{x_1,\ldots,x_k\})(\{g_1,\ldots,g_m\}(\{x_1,\ldots,x_k\}))$$

is existential. The first step is to make this equivalent with

$$\exists p \exists t_1,\ldots,t_p \in \{g_1,\ldots,g_m\}(\{x_1,\ldots,x_k\}).$$
$$\langle x_0, t_1,\ldots,t_p\rangle \in \{f_1,\ldots,f_n\}(\{x_1,\ldots,x_k\}).$$

It then takes two more applications of the definition of application to get down to the elements of the finite sets. What is new in this calculation over the one for **K** is that the clause of the form $t_0,\ldots,t_p \in U$ is really a *finite conjunction* of variable length. A more formal (and less dotty) notation for finite sequences would give us existential quantifiers mixed with *bounded* universal quantifiers (finite conjunctions) which would be enough to show that the predicate is r.e.

We use the same kind of argument to show the closure of the computable sets under application. The case of **fix** is similar, since $a \in \mathbf{fix}$ means

$$\exists n,m \exists f_0, f_1,\ldots,f_m \in A.\ a = \langle f_0, f_1,\ldots,f_m\rangle \wedge f_0 \in \{f_1,\ldots,f_m\}^n(\varnothing).$$

The iterated application has to be shown to be uniformly r.e. in the parameters (including n) with respect to the underlying enumeration of A. It should be clear how to do this. The reason why it is sufficient for (iv) merely to consider **fix** is that we can write:

$$F = \mathbf{fix}(\lambda H \lambda X_1 \cdots \lambda X_n.\ G(H, X_1, \ldots, X_n));$$

and, since G is computable, everything on the right is also.

An interesting corollary to this theorem is the observation that the computable subsets of any computable structure form a model of the λ-calculus. This is one rather strong reason why this approach to model building can be considered "constructive."

What is the difference between the ***Fixed-Point Theorem*** and the ***First Recursion Theorem***? The former only involves the proof that fixed points *exist* and that the least fixed-point operator is *continuous*. (The second part of this statement is essential for *iterated* recursive definitions: things introduced by fixed points can be employed for further introductions by fixed points—the reader will get the (fixed) point here if he recalls that parameters have to be allowed.) The latter statement further requires a proof that (assuming we start with computable things) the results are computable. The Second Recursion Theorem (due to Kleene) would demonstrate that not only can we effectively Gödel number all these definitions, but in a recursion for a function F the right-hand side can employ the Gödel number of the very function being defined (that is, there is a function whose recursive definition calls for its own Gödel number). We do not discuss the details here for the present model, but a quite neat version was given in SCOTT (1976, Section 3).

We should stop to consider whether suitable structures exist. The minimal one where A equals the closure of $\varnothing$ under finite sequences is computable and well-founded. It does not make any difference here what kind of sequence we consider (what our model is), because *all* theories of sequences must satisfy

$$\langle x_0, \ldots, x_{n-1} \rangle = \langle y_0, \ldots, y_{m-1} \rangle \leftrightarrow n = m \wedge \forall i < n.\ x_i = y_i,$$

and, when we start with "nothing," all theories lead to the same minimal model $A = A^*$ (up to isomorphism). The same would go for a model generated by *atoms* (i.e., elements which are non-sequences). This kind of structure is the minimal solution of

$$A = B \cup A^*,$$

where we require that $B \cap A^* = \varnothing$. All such models are well-founded structures determined up to isomorphism by the cardinality of B. Provided that B is countable, there are many enumerations of A making the structure computable. Provided also that we strengthen the definition of well-foundedness to include $x_0 \prec \langle x_0, x_1, \ldots, x_n \rangle$, then *all* strongly well-founded structures are of this form.

In an enumeration making $A^* \subseteq A$ computable it is not necessarily the case that $B = A \setminus A^*$ is computable (r.e. with respect to the enumeration). Similarly, the inequality predicate $x \neq y$ need not be computable; perhaps these stronger conditions ought to be imposed in the definition. In a different direction perhaps we should only consider computable those predicates r.e. with respect to *all* enumerations of A making $A^* \subseteq A$ (or whatever) computable.

So much for remarks on "standard" models. A "non-standard" model (but one with only *finite* sequences) is formed by making up a "peculiar" one–one correspondence between A and A^*. Call it $\pi : A^* \leftrightarrow A$. The theory of tuples then uses $\pi(\langle x_0, \ldots, x_{n-1} \rangle) \in A$ in place of $\langle x_0, \ldots, x_{n-1} \rangle$ (This is what is done for $A = \mathbb{N}$ when we use a Gödel numbering of finite sequences.) We could also do a similar theory with atoms, but the extra generality is not needed here. A different way of introducing atoms into A is suggested in the next section.

It is a quite remarkable discovery of BAETEN and BOERBOOM (1978) that such non-standard models completely change the behavior of the resulting λ-calculus. (Earlier, J. Owlett when a graduate student at Oxford had found that the connection between **fix** and **Y** could be ruined by a non-standard tupling.) The result of Baeten and Boerboom (proved for the graph model of SCOTT (1976) but clearly transferable to the present context) can be stated as follows (A is assumed denumerable):

Theorem. (i) *Let $X \subseteq A$ be arbitrary. There is a choice of $\pi : A^* \leftrightarrow A$ so that under this notion of tuple $\Delta(\Delta) = X$ (instead of $\varnothing$ as in the well-founded case).*

(ii) *Given any closed λ-term τ, there is a choice of $\pi : A^* \leftrightarrow A$ so that $\Delta(\Delta) = \tau$.*

The proof of (ii) is by a fairly straight-forward forcing construction, where the forcing conditions are certain finite partial functions $P \subseteq \pi$ (for $\pi : A^* \leftrightarrow A$). It would be worthwhile to check whether the proof works with terms involving **fix** as a primitive combinator and not defined as **Y**. The resulting tupling is non-recursive, by the way.

Having provided a somewhat detailed analysis of iteration in these models and a glimpse of how recursive definitions enter *via* combinators, we ought to conclude this section with some remarks on how *properties* of the recursively defined functions should be proved.

In Section 6 of CURRY (1979) we find a useful, but brief review of combinatory arithmetic and Kleene's early work on the λ-definability of recursive functions (see the Curry paper for the explicit references). Perhaps we should remark that the word "definability" is not quite properly used in this regard in view of later work on the connection between recursion theory and formalized calculi: it would be better to say that Kleene established the *numeralwise representability* of partial recursive functions in the (pure) λ-calculus. The reason for making this verbal distinction is that the "numerals" are each taken *separately* (Curry calls them the combinators $\mathbf{Z}_n$), and there is no predicate in the theory for the *class* of integers. Therefore, even though we can see the results of any *one* calculation, there is no way to formulate—in the theory—a proof by mathematical induction in order to establish *general* facts about the integers. (The variable n is outside the system, for example.) This strikes me as something of a drawback, but of course Curry was striving for the weaker, more basic, more ultimate foundational systems he wanted to see *common* to all formalized theories. Not all theories, obviously, should be as strong as first-order arithmetic. Be that as it may, there is still a question of just how (or where) we are to do our inductions.

Instead of introducing by one scheme or the other the integers into our present system, we will fix attention for the sake of illustration on the combinator **flx** as the embodiment of the iteration concept. For us this is reasonable because we always have continuous set functions at the back of the mind, and the least-fixed-point construction is quite fundamental in this context. We have already seen how to use **flx** in definitions; what remains is to see how it comes into inductions.

In a highly schematic way, we could consider *directed-complete predicates* of subsets of A. Such a predicate $\mathcal{P}(X)$ has the property of being **closed** under directed unions of sets. Now the least fixed point is a directed union; thus, it is certainly valid in our model that the following holds for all directed-complete predicates:

$$(\iota\uparrow) \qquad \mathcal{P}(\varnothing)\wedge\forall X[\mathcal{P}(X)\rightarrow\mathcal{P}(F(X))]\rightarrow\mathcal{P}(\mathbf{flx}(F)).$$

The only trouble with $(\iota\uparrow)$, the principle of directed-complete induction, is that it would require some machinery for the introduction of the predicates. If we want a more elementary principle of the nature of (ξ) or of (η), then we need to take a form of the predicate expressible in our previous notation. Here is one example:

$$(\iota^*) \qquad P(\varnothing)\subseteq Q(\varnothing)\wedge\forall X[P(X)\subseteq Q(X)\rightarrow P(F(X))\subseteq Q(F(X))] \\ \rightarrow P(\mathbf{flx}(F))\subseteq Q(\mathbf{flx}(F)).$$

It is simple to verify that $P(X) \subseteq Q(X)$ is a directed-complete predicate. Quite a lot can be done with this form of induction, though not everything. An interesting example of something statable by combinators but requiring induction to prove was given in SCOTT (1976, p. 534):

$$\mathbf{fix}(\lambda F \lambda X.\ G(X)(F(X))) = \lambda X.\ \mathbf{fix}(G(X)).$$

This equation can easily be rewritten in combinators: $\mathbf{B}(\mathbf{fix}(\mathbf{S})) = \mathbf{B}(\mathbf{fix})$. Other examples and references can be found in BARENDREGT (1977, pp. 1121 and 1126).

As special cases of (ι^*), we remark that the implication

$$F(A) \subseteq A \rightarrow \mathbf{fix}(F) \subseteq A$$

follows at once by the substitutions $P = \lambda X.\ X$ and $Q = \lambda X.\ A$. In the appendix to this paper we collect together some axioms suggested by our model construction (cf. also SCOTT (1973)). The reader should note two things: in the first place, this is a first-order theory not just an equational theory. (One reason for this is the fact that it seems impossible to regard the *unrestricted* quantifiers as combinators—they are not continuous as operators.) Secondly, what we propose is a very weak theory, because it could be interpreted within the theory of r.e. sets and very likely proved consistent in ordinary first-order arithmetic. In any case it is formulated in as close a form to the original view of combinatory logic as we can come under the plan of modelling functions by continuous set mappings.

5. Some aspects of type theory

In this section of the paper we shall restrict attention to the models where $A^* = A$. As we pointed out before, not only do such sets exist, but this equation eliminates completely the distinction in type between subsets $X \subseteq A^*$ and $X \subseteq A$. Every set in our "universe" is at the same time a set of sequences, and conversely. We saw this was helpful as regards the discussion of continuous set functions; however, this initial elimination of one type distinction hardly eliminates all type distinctions—there is another important one close at hand.

Is there a difference between functions and arguments? Certainly in *use*, but the model presented above has shown that functions may be incorporated as objects among the arguments. That was how we justified self-application. But, with reference to condition (9) of Section 2, it will be seen that (even when $A^* = A$) only *certain* subsets of A^* are used to represent functions; in a precise sense in this model there are "fewer" functions than arguments. Thus, despite our being able to give a useful meaning to $X(Y)$

for all $X, Y \subseteq A$, the distinction between function and argument remains on the level of the object of the model.

We can easily give symbolic form to the distinction by employing in a well-known way the λ-notation. What does $\lambda X.\ P(X)$ represent (in the model) for arbitrary $P \subseteq A$? *Answer*: the arbitrary continuous function. Condition (9) referred to above is equivalent to the satisfaction of this equation:

$$(\eta) \qquad P = \lambda X.\ P(X)$$

Often suggested as a universally valid *law* of λ-calculus, it is often wrongly called the axiom of extensionality—law (ξ) is correctly the extensionality principle for the λ-calculus. CURRY (1979) speaks of "strict" extensionality, which is fair enough. The strictness consists of the requirement that every object *uniquely* represents a function. As is well-known, we could replace (the universal generalization of) (η) by the biconditional:

$$(\varphi) \qquad P = Q \leftrightarrow \forall X.\ P(X) = Q(X).$$

By a very exact analogy with the axiom of set theory, whereby two sets with the same *elements* are equal, we can read (φ) as saying that two functions with the same *values* are equal. The rub is that in general we do *not* known whether P and Q are always the chosen representatives of functions. In the case of (ξ), the two λ-abstracts are by primary intent the representatives of the functions in question, and so we say they are equal in an extensional theory. (φ) above says too much, for, just as in set theory, we can imagine a universe where some objects are not functions (some objects are not sets—atoms, for instance). I prefer to call (φ) a principle of functionality, meaning that every object is (uniquely) to be regarded as a function. This should not restrict the use of the word "functionality" for other uses—for example functionality *relative* to certain mapping properties of the kind we shall discuss below.

Indeed, law (η) always *fails* in the kind of model constructed above, because, for $P = \varnothing$, it is clear from formula (1) of Section 3 that $\lambda X.\ \varnothing(X) \neq \varnothing$ for a trivial reason. (It would do no good to leave out the element $\langle\rangle \in \lambda X.\ \tau[X]$. Consider $R = \{\langle x, x\rangle \mid x \in A\}$, then $R(X) = X$ for all X but $R \neq \lambda X.\ X$ according to the actual definition.) What *is* true in all these models—and this is the reason I have carried around a seemingly superfluous empty sequence—is a somewhat weaker law:

$$(\eta^-) \qquad P \subseteq \lambda X.\ P(X).$$

Owing to our assumption that $A^* = A$, arbitrary subsets of A do satisfy (η^-). For if $a \in P$, then either $a = \langle\rangle$, in which case the element belongs to $\lambda X.\ P(X)$ by definition of abstraction, or else we have $a = \langle y_0, y_1, \ldots, y_n\rangle$, in which case $y_0 \in P(\{y_1, \ldots, y_n\})$, by definition of application, and hence again the element $a \in \lambda X.\ P(X)$.

Interesting as this is, it does not at once answer our question about the distinction between argument and function: this model still makes the distinction, but we want to know whether there is *some* (non-trivial) model in which the law (η) holds. I gave an answer in 1969 with my first model construction by a method that has often been given the unfortunate name "Scottery". (An integrated presentation is planned for Barendregt (1980) and a very thorough discussion is contained in Plotkin and Smyth (1978), where the process is given a categorical formulation incorporating suggestions of several other people. Another presentation together with the connections with the topological and lattice-theoretical aspects of continuous lattices will appear in Gierz et al. (1980).) A direct construction (without inverse limits) was mentioned in Scott (1976, p. 549 ff), but people have not enjoyed very much reading it there; thus, let me explain once more using the models of this paper how easy it is without trying to put the approach in a wider context. Essentially the same proof is given in Plotkin (1972), but the details (by "retracts") as presented here are very much simpler.

Principle (η^-) can be stated purely in terms of combinators and inclusions between them. We have, in fact,

$$\lambda X.\ X \subseteq \lambda F \lambda X.\ F(X),$$

and this is just the start of a sequence of such containments. Define recursively:

(1) $\qquad D_0 = \lambda X.\ X$

(2) $\qquad D_{n+1} = \lambda F \lambda X.\ D_n(F(D_n(X)))$

We can prove the:

Lemma. *For all integers n,*
(i) $D_n \subseteq D_{n+1}$;
(ii) $D_n \circ D_n = D_n$.

Proof. For $n=0$, both (i) and (ii) are clear from what we have already said. Thus assume the case of n and pass to $n+1$.

We can write by (2) above:

$$D_{n+1}(F) = D_n \circ F \circ D_n,$$
$$D_{n+2}(F) = D_{n+1} \circ F \circ D_{n+1}.$$

Hence if $D_n \subseteq D_{n+1}$, then $D_{n+1} \subseteq D_{n+2}$ follows by monotonicity and (ξ).

Also we see

$$D_{n+1}\circ D_{n+1}=\lambda F.\ D_{n+1}(D_{n+1}(F))$$
$$=\lambda F.\ D_n\circ D_n\circ F\circ D_n\circ D_n.$$

Thus if $D_n\circ D_n=D_n$, then $D_{n+1}\circ D_{n+1}=D_{n+1}$.

Now define $D_\infty=\bigcup\{D_n|n=0,1,2,\ldots\}$. We have:

Theorem. *$D_\infty=\lambda F.\ D_\infty\circ F\circ D_\infty=D_\infty\circ D_\infty$, consequently the fixed points of D_∞ are closed under application and the following form of λ-abstraction:*

$$\lambda_\infty X.\ \tau[X]=\lambda X.\ D_\infty(\tau[D_\infty(X)]).$$

Moreover, as a model, the fixed points of D_∞ satisfy (α), (β), (ξ) and (η).

Proof. Both of the first two equations follow by continuity in view of (i) of the lemma. The first comes from (2) and the second from (ii) of the lemma. If $X=D_\infty(X)$ and $Y=D_\infty(Y)$, then

$$X(Y)=D_\infty(X)(Y)$$
$$=(D_\infty\circ X\circ D_\infty)(Y)$$
$$=D_\infty(X(D_\infty(Y)))$$
$$=D_\infty(X(Y)).$$

A simple calculation also shows that

$$D_\infty(\lambda_\infty X.\ \tau[X])=\lambda_\infty X.\ \tau[X],$$

in view of the equations already proved.

If τ is a term built up by application and λ_∞, then we can leave off the first D_∞ in the formula defining λ_∞ *provided* we assume all free variables have values in the model. It is then easy to check (β), again provided free variables are restricted to the model. The reason that (ξ) holds is that if we assume $\tau[X]=\sigma[X]$ holds for all $X=D_\infty(X)$, then $\tau[D_\infty(X)]=\sigma[D_\infty(X)]$ holds for *all* X. We then employ (ξ) (unrestricted) and the definition of λ_∞. Finally (η) in the model is just a restatement of the first equation of the theorem.

The idea of the theorem is this: the first map D_0 does nothing; the second map D_1 turns everything into a function; the third map D_1 turns everything into a functional in the following sense: given F, it is changed into a mapping which takes its argument, turns that into a function, performs F on the result, and finally converts the answer into a function. In general, D_{n+1} makes everything into an $(n+1)$st-order functional by performing suitable conversions on arguments and values with the help of D_n. How far can this go on? The answer is: indefinitely! The limit functional D_∞ works arbitrarily deeply on arguments and values, but owing to nice continuity properties of the construction it satisfies the neat fixed-point equations of the theorem. Note, however, that this would not work out so well if we did not have the $n=0$ case of the inclusion (i) of the lemma.

It should also be noted that the method of proof involves a fixed point —but apparently not one that can be stated in pure λ-calculus. Thus D_∞ is the least fixed point of the following equation:

$$D_\infty = \lambda X.\ X \cup \lambda F.\ D_\infty \circ F \circ D_\infty.$$

We must take care that the theorem is not trivial. In the minimal model for sequence theory, where $A^* = A$ and A^* is generated from "nothing," the least fixed point of D_∞ is analyzed as follows. Indeed by the above the least fixed point is just $D_\infty(\varnothing)$ and we can see

$$D_0(\varnothing) = \varnothing$$

and

$$D_{n+1}(\varnothing) = \lambda X_n.\ D_n(\varnothing);$$

because for all Y it is true that $\varnothing(Y) = \varnothing$. This means that

$$D_\infty(\varnothing) = \lambda X_0.\ \varnothing \cup \lambda X_1 \lambda X_0.\ \varnothing \cup \cdots \cup \lambda X_n \cdots \lambda X_1 \lambda X_0.\ \varnothing \cup \cdots.$$

Trivial as this seems, a strict application of our definitions reveals that $D_\infty(\varnothing) = \bigcup \{A_{[n]} \mid n = 0,1,2,\dots\}$. (See in this regard the calculation of the combinators **K** and **S** in the last section.) Note that every time a new λ comes in a new factor of $A_{[n]}$ goes into the union. But in the minimal model the union of all the $A_{[n]}$ is just A^*; so the only fixed point of D_∞ is the maximal one, A itself. (The situation here is different from the proof mentioned in Scott (1976).)

To make a repair (and I did not notice this problem until I started to write up the paper) we must find a non-minimal model; it will, however, turn out to be well-founded though not strongly so. Let A be the closure of the *one* element set $\{\star\}$ under finite sequences. Now here we find that $A \setminus A^* = \{\star\}$ because the element $\star$ should be regarded as a *non-sequence*. (There are many ways in set theory to find such elements.) This model is of course well-founded as we have already remarked. We are next going to take a *quotient* by the least equivalence relation $\equiv$ where

$$\star \equiv \langle\star\rangle \equiv \langle\langle\star\rangle\rangle \equiv \langle\langle\langle\star\rangle\rangle\rangle \equiv \langle\langle\langle\langle\star\rangle\rangle\rangle\rangle$$
$$\equiv \langle\langle\langle\langle\langle\star\rangle\rangle\rangle\rangle\rangle \equiv \langle\langle\langle\langle\langle\langle\star\rangle\rangle\rangle\rangle\rangle\rangle \equiv \cdots,$$

and where $\langle x_0, x_1, \dots, x_{n-1}\rangle \equiv \langle y_0, y_1, \dots, y_{n-1}\rangle$ if $x_i \equiv y_i$ for all $i < n$. The model that results is $A/\equiv$, which we can think of as the same as $A^*/\equiv$ or $(A/\equiv)^*$ by a slight shift of meaning of the *-operator.

This model would *not* be well-founded in the stronger sense where $x_0 \prec \langle x_0, x_1, \ldots, x_{n-1}\rangle$, but it *is* well-founded in the sense used in Section 4 for the proof of the **Y**-Theorem. (The quotient could be regarded as resulting from a repetitious replacement of $\langle\star\rangle$ by $\star$ in a given sequence until no occurrence of $\langle\star\rangle$ remains.) In this model $\star\notin A_{[n]}$, and so the *least* fixed point of D_∞ is **not** the *greatest* fixed point, which is still $A=\lambda X.\ A$. Thus, the fixed-point set of D_∞ has at least two elements, and in this way we have found a non-trivial model for (η). In such a model a further type distinction has been eliminated because *all* elements can be regarded as (unique representatives) of functions (continuous functions). But whether it is really *profitable* to eliminate such distinctions is another question. Note that we could have adjoined as many distinct $\star$-elements as we wished. These elements act just like atoms; thus, the (η)-model would contain something as complex as the space of all continuous set mappings on the (infinite) set of atoms.

Let us therefore turn to the opposite question of how—given a nice λ-calculus model—it is possible to *introduce* type distinctions. There is a point in this, because the distinctions allow us to sort out differences between elements according to natural properties. The advantage of starting with a λ-calculus model is that the whole of the discussion can be built on *one* notation for function abstraction. (An ordinary type theory has, strictly speaking, different application and abstraction notions at all types.) The price for *one* notation for functions is *several* notations for equivalence relations for representing the different types, but this is not so bad since the different types are different in any case.

A considerable amount of detail has already been given in SCOTT (1975) and SCOTT (1976, Section 7). Without making the formulation too heavy, we can describe here briefly how the method works; a deeper investigation would require some familiarity with the theory of continuous lattices and their subspaces. Some further very interesting uses of the idea can be found in PLOTKIN (1973).

Types, for many purposes, can be identified with equivalence relations on (subsets of) our model. Indeed, let $\mathfrak{S}\subseteq \mathbf{P}A\times\mathbf{P}A$ be a transitive and symmetric relation. The set of self-related elements, $\{X\subseteq A \mid X\,\mathfrak{S}\,X\}$, may be regarded as the subspace of the model in question, and this is the subspace of which we are interested in the quotient modulo $\mathfrak{S}$. (We shall often write $X:\mathfrak{S}$ as short for $X\,\mathfrak{S}\,X$.) Though this is our interest, we shall not actually take the quotient, for it is easier to work with the representatives of the equivalence classes directly.

For example, let $\mathfrak{E}$ and $\mathfrak{F}$ be two such equivalence relations. We define the equivalence corresponding to the function space, call it $[\mathfrak{E} \to \mathfrak{F}]$, by the formula:

(3) $P[\mathfrak{E} \to \mathfrak{F}]Q$ iff whenever $X \mathfrak{E} Y$, then $P(X) \mathfrak{F} Q(Y)$.

That $[\mathfrak{E} \to \mathfrak{F}]$ is an equivalence relation is clear. Note that P is always equivalent to $\lambda X.\ P(X)$, thus we can regard the equivalence classes as consisting of *functions*. Note, too, that the construction can be iterated—in this way we pass to a notion of *higher-type* function. The reason for stressing equivalence relations rather than classes is that our functions are meant to be *extensional*, in the sense that equivalent arguments should get equivalent values. In words it is easy to read (3): two functions are equivalent if they do equivalent things to equivalent arguments. Keep in mind, however, that (3) has further import depending on how demanding the given equivalence relations are. The point is that (3) implies that if an argument lies in the first subspace, then the value *must* lie in the second subspace—the function is well-defined, therefore.

This plan for defining types *via* equivalence relations has many features of a theory of functionality of the kind advocated by Curry; however, our types are not "obs", that is to say elements of the model. The equivalence relations are constructs *over* the model not elements *of* the model. One approach to having obs represent types (better: classes) was taken in SCOTT (1975), but then a transfinite truth definition is needed in seeing which classes the obs define. This may not be a bad thing, but it is less elementary than we care to be at the moment. There would be no trouble, by the way, in having a theory of equivalence relations (rather than classes) done in the form of the 1975 paper.

Thus there are many approaches to the sorting out of the elements, and still many questions about the nature of possible subsets. In particular, the question of which λ-expressions have types and which types completely determine λ-expressions seems rather basic. For instance, the common combinators are very well behaved as regards type:

$$\mathbf{K}: \mathfrak{E} \to [\mathfrak{F} \to \mathfrak{E}],$$

$$\mathbf{S}: [\mathfrak{E} \to [\mathfrak{F} \to \mathfrak{G}]] \to [[\mathfrak{E} \to \mathfrak{F}] \to [\mathfrak{E} \to \mathfrak{G}]].$$

$$\mathbf{B}: [\mathfrak{F} \to \mathfrak{G}] \to [[\mathfrak{E} \to \mathfrak{F}] \to [\mathfrak{E} \to \mathfrak{G}]].$$

Some further details are given in the cited references, but it seems fair to say that the study of this idea has hardly begun. Here, for example, is a question. The combinator **fix** is very important, but does it have a special character as regards functionality? We are tempted to write:

$$\mathbf{Y}: [\mathfrak{E} \to \mathfrak{E}] \to \mathfrak{E}.$$

This is not true in general, since we did not put any closure conditions on our equivalence relations (say, closure under directed unions). We should then ask: "Which are the best closure conditions?" as well as the previous question: "How do we prove that a combinator has no functionality?"

6. Some conclusions and some questions

We have spoken at great length about functions and their properties in this essay. In Section 11 of CURRY (1979), Professor Curry gives the well-known reasons why sets can be reduced to functions, and he then continues:

> "Thus, it is simpler to define a set in terms of a function than vice versa (for a similar idea cf. the set theory of VON NEUMANN (1928)); but the idea is repugnant to many mathematicians, and probably to Scott. This has been a great handicap and source of misunderstanding."

May I disassociate myself from these sufferers of repugnance? I feel I understand rather well the logical interrelationships between sets and functions. I would be very happy indeed to reduce sets to functions *if there were any good theory to do this in*. In my opinion there **does not** exist at the present time such a theory—owing to our troubles with the paradoxes. The theory of Von Neumann, for example, turned out to be easier to state as a set/class theory rather than a function theory. What is needed for a workable set theory (regardless of what sets *are*) is a strong comprehension axiom. As far as I can see, the Curry programme has not as yet produced a straight-forward theory that is anywhere near as workable as the standard Zermelo–Fraenkel system (or the system augmented by classes). However, there is a rather fundamental point about the contrast between *extensional* and *intensional* theories of sets and functions, which is hardly touched on in the literature on combinatory logic. For an interesting, and very likely workable intensional theory of functions, see FEFERMAN (1980) and the related papers cited therein. As regards the question of which comes first: the function or the set, it is not a question of repugnance or prejudice on my part that causes me to formulate constructions within set theory but a problem of helplessness. And I can pinpoint rather narrowly where I think the trouble lies.

For the sake of argument think of a set as a truth-valued function. (I know this over simplifies Curry's approach, but a more subtle view is not needed for the point I will make.) Instead of $X \in A$ we will write $A(X)$; to *assert* $X \in A$ means to assert $A(X) = \mathbf{true}$. On the other hand, to assert $X \notin A$ means to assert $A(X) = \mathbf{false}$. The domain of the function A is "everything"—but now the rub starts. The Russell Paradox shows that the domain cannot really be everything if we were to allow full comprehension. There is no way around Cantor's Theorem that *there are more functions than there are arguments and values*. (And, from what I can understand, the introduction of some concept of "proposition" to replace the two, very separate truth values does not seem to help.)

As is common knowledge to logicians, the way to make $A(X)$ always meaningful is to restrict by some manner the total possible range of functions. The choice of restriction for the Zermelo – Fraenkel set theory is to make the number of X for which $A(X)=$ **true** *very small* compared to the total number of things (at least insofar as these comparisons of number are expressible in the system). The very feature that makes this view of sets easy to grapple with is that we *do not* need regard sets as functions! The "half" of A consisting of the set $\{X|A(X)=$ **true**$\}$ is enough to determine the whole of A; the other, larger "half" $\{X|A(X)=$ **false**$\}$ is completely determined by the first half. The reason why a half loaf is better than a whole is that—in building up sets—we can regard the first, positive half as **FIXED** long before the rest of the elements that would have to enter into the negative half ever come into view. (This idea of "earlier" and "later" sets is made quite precise in the theory of the *rank* of a set.) There is a certain advantage to regarding the universe of sets as being "open ended" (at the top end, at least) even though we have accepted certain laws as pertaining to *all* sets—no matter how "late" these sets come in. The consequence of this view (which, for all I know, may very well be repugnant to Curry) is that the domain of $A(X)$ as a function is not very well determined on the negative side: our usual set theory is not symmetric in its use of **true** and **false**.

Now the system *New Foundations* of QUINE (1944) (see QUINE (1953) or (1963) for the history of his system) was supposed to restore the **true**–**false** symmetry by a different kind of restriction on the comprehension axiom. One would hope that Quine's theory would give at once a theoretical basis for a theory of combinators. But it does not—at least if one construes the word "function" the way Quine does as a set of ordered pairs. The comprehension terms needed for the combinators are simply not "stratified" (in Quine's well-known terminology). And why not? Because functions are **binary** relations and *New Foundations* is **not** a suitable system for a general "type-free" theory of binary relations.

This is a, perhaps, not much remarked fact, but it is very easy to explain. Quine's theory *looks* type free, but—sadly—this is only an illusion. When we restrict attention to *one-place* predicates determined by stratified formulae, then it is very true that Quine lets us be *ambiguous* as to type. We can give the free variable any type we want and then start counting up and down from there. ("Negative" types are permitted, if we like.) When, however, we come to *two-place* predicates, then the story is quite different. There are *two* free variables to cope with now. We can slide the type indices up and down the scale, but in general we can never alter the numerical value of the *difference* between the types of the two variables. In other words, though Quine was successful in banning type distinctions for sets (one-place predicates), he still is faced with *infinitely many* type distinctions for binary relations. Thus, for example, the relations $=$ and $\in$ of equality and membership are of essentially different type in Quine's theory: the type difference is 0 in the first case, 1 in the second.

The reader has surely remarked by now that the theory of *continuous* functions employed in this paper could have been carried out equally well in *New Foundations*. In fact, there is a definition of ordered pair in *New Foundations* so that the formula $z=\langle x,y\rangle$ is stratified with the three variables all of the *same* type; moreover, all objects of the theory are ordered pairs, i.e. $\mathrm{V}=\mathrm{V}\times\mathrm{V}$. Though I have not carried out the details, I do not think it would be difficult to change the definition a little so that we could say $\mathrm{V}=\mathrm{V}^*$. With this understanding, all the basic definitions of this paper would go through, since the defining formulae we have used here are stratified; in particular $Z=X(Y)$ is stratified with all variables of the same type. This is nice, but why does it not settle the question of the relationship of set theory and combinatory logic?

The answer lies in the word "continuous." In order to have the function-theoretic comprehension principle (β) by our approach, we had to make the restriction that $X\mapsto\tau[X]$ defined a continuous set mapping. The kind of comprehension terms needed for set theory (particularly those with quantifiers) *are just not continuous*. We seem to have the choice:

lots of sets but no combinators

or

lots of combinators but few sets.

By "set" here we mean the characteristic functions of a set represented by a function of the theory. This is no *proof* that there is no good mixture, but there does seem to be some evidence that the two notions of function in the two kinds of theories are not quite the same. Combinators in the model of this paper behave more like the *classes* of the Von Neumann–Bernays–Gödel theory. People have tried to make classes self-applied, but a "canonical" theory has not been found that has won general favour. Just as we could carry out the construction in Quine's system, we could have worked within VBG class theory and spoken about continuous class functions. There is a chance that this might lead to some axiomatic niceties and produce a blend of Curry and Church with Von Neumann–

Bernays–Gödel, but the author is not so certain the effort is worth the trouble. (Such a study might be worth a Master's thesis, however. The candidate should recognize that there are *degrees* of continuity, and that in this paper we have only employed *finitary continuity*. In a full class theory there are transfinite notions of continuity that would probably be more useful.)

Aside from the question of what to do next—if anything—one might ask: "What is special about the combinatory logic of *New Foundations*?" But, as we have no models for Quine's theory, there might not be much to say. (The models of JENSEN (1969) for the theory with atoms could, however, give something new.) A more interesting question might come out of the Von Neumann–Bernays–Gödel set/class theory.

I believe I can now also make clearer my attitude toward *type theory* that Curry discusses in Section 9 of his paper. Professor Curry recalls the harsh tone of SCOTT (1969), written just a few weeks before I discovered the first model construction. The paper was therefore never published, and I recanted on some of my remarks. What I especially reputiated was my feeling (at one time held very strongly) that *combinatory logic did not make good mathematical sense at all*—for instance, in not linking up with the ordinary theory of functions: continuous real-valued functions come to mind. This is not an issue of set-theoretic foundations vs. function-theoretic foundations. It is just a question of having some interesting mathematics. Well, on that score the situation has changed: at least I now know how to embed *every* topological space (along with its function space) into a model of combinatory logic (see SCOTT (1972)).

But, really, nothing has changed in my view since 1969 as regards type theory. I assert: ***it is impossible to eliminate from logic and mathematics ALL type distinctions***. As has been illustrated above, *certain* types can be "confused" and then objects of other types can be "forgotten", but no magic so far has ever made a set A and its powerset $\mathbf{P}A$ *equal*. Some types are distinct whether anyone chooses to discuss the difference or not. I certainly did not mean to say in 1969 that we need exactly Russell's theory, but I did mean—and still mean—that the kind of type difference that Russell recognized will always be present *somewhere* in a theory of logical objects. Whether the flexibility of combinatory logic will soften the pain of living with these (necessary) distinctions remains to be fully demonstrated in my opinion.

Some comment is also required on Curry's remarks on *conceptualization*. We read (Section 10):

> "While it is true that concocting formalisms entirely without regard to interpretation is probably fruitless, yet it is not necessary that there be "conceptualization" in terms of current mathematical intuitions. In fact, mathematical intuition is a result of evolution. Mathematicians depend on their intuitions a great deal; let us hope they always will. But the mathematical intuitions of today are not the same as those of a thousand years ago. Combinatory logic may not have had a conceptualization in what seems Scott's sense; but it did have an interpretation by which it was motivated. The formation of functions from other functions by substitution does form a structure, and this structure it analyzed and formalized. For progress we need the freedom to let our intuitions develop further; this included the possibility of formalizing in new ways."

First "evolution": Though geometry has evolved over 2000 years and the attitude toward the concept of number has radically altered, *still we can sense the continuity of ideas*. The Pythagorean Theorem is still true in Euclidean geometry and the old proofs still stand, even though the Greeks might not have been happy with a proof by analytic geometry. What I always found disturbing about combinatory logic was what seemed to me to be a *complete lack* of conceptual continuity. There were no functions known to anyone else that had the extensive properties of the combinators *and allowed self-application*. I agree that people might wish to have such functions, but very early on the contradiction found by Kleene and Rosser showed there was trouble. What I cannot understand is why there was not more discussion of the question of *how* the notion of function that was supposed to be behind the theory was to be made even midly harmonious with the "classical" notion of function. The literature on combinatory logic seems to me to be somehow silent on this point. Perhaps the reason was that the hope of "solving" the paradoxes remained alive for a long time —and may still be alive. Perhaps the reason was that many people gave up working in the theory. Whatever the reason, I do not think I am reading the record unfairly.

Next "substitution": This is not the place to discuss the well-taken criticisms of the complexity of substitution in the formulation of rules in formal theories, nor do we have time to discuss the pros and cons of real and apparent variables and a logic without variables. The question I have about basic motivation concerns the "structure" Curry mentions in the quote given above. What structure??? I agree that we can regard Group Theory as an analysis of the structure of bijective functions under composition, Boolean Algebra as an analysis of sets under inclusion, Banach Space Theory as an analysis of functions under convergence of infinite series, etc. etc. But Combinatory Logic? It just does not seem to me to be a sound step in *analysis* to say: "We now permit our functions to be self-applied." Just like that. Clearly, after seeing so many *analogous* composition operations of different types, we would dearly wish to put them all into one big **B**; but the step to **B(B)(B)**, though it may be a small step for Curry, does seem like a big step for the rest of us—especially in the shadow of the paradoxes.

Some acknowledgments

The author is most grateful both to the University of Oxford for granting him a sabbatical year and to the Guggenheim Foundation for a Fellowship. Special thanks are due to the Xerox Palo Alto Research Corporation and the Computer Science Laboratory for their particular generosity in supplying such good working conditions and remarkable text-editing facilities which made the writing of the paper possible. The excellent typing of Melinda Maggiani has been an essential part of putting together the final draft. The many occasions I have had to lecture on the topic of this paper since the Kleene Symposium have also been most useful. I should also like to record here by warm thanks to Professor Kleene for many kindnesses and much help over the years.

Appendix: Some axioms

Throughout the paper we have alluded to various laws of combinators and λ-expressions without being very systematic; thus, it would seem helpful to collect together what is essential by way of formal properties. In the following list, we have tried to follow Curry's notation for the names of laws as closely as possible. However, the theory of this paper takes $\subseteq$ rather than $=$ a primitive and defines the latter. When laws are strengthened by the use of $\subseteq$ we have added an asterisk; in the one case of a weakening we have added a minus to the name.

(o^*) $\varnothing \subseteq X$,

(ρ^*) $X \subseteq X$,

(τ^*) $X \subseteq Y \wedge Y \subseteq Z \rightarrow X \subseteq Z$,

(σ^*) $X = Y \leftrightarrow X \subseteq Y \wedge Y \subseteq X$,

(μ^*) $X \subseteq Y \rightarrow Z(X) \subseteq Z(Y)$,

(ν^*) $X \subseteq Y \rightarrow X(Z) \subseteq Y(Z)$,

(α) $\lambda X.\ \tau[X] = \lambda Y.\ \tau[Y]$,

(β) $(\lambda X.\ \tau[X])(Y)=\tau[Y]$,

(ξ^*) $\forall X.\ \tau[X]\subseteq\sigma[X]\rightarrow\lambda X.\ \tau[X]\subseteq\lambda X.\ \sigma[X]$,

(η^-) $P\subseteq\lambda X.\ P(X)$,

(π^*) $F(\mathbf{fix}(F))\subseteq\mathbf{fix}(F)$,

(ι^*) $P(\varnothing)\subseteq Q(\varnothing)\wedge\forall X[P(X)\subseteq Q(X)\rightarrow P(F(X))\subseteq Q(F(X))]$
$\rightarrow P(\mathbf{fix}(F))\subseteq Q(\mathbf{fix}(F))$.

(δ^*) $\mathbf{fix}=\lambda F.\ (\lambda X.\ F(X(X)))(\lambda X.\ F(X(X)))$.

The last law has a special character, and the reader might wish to leave it off in view of the large number of models in which it fails. It should also be kept in mind that Curry also formulates his laws as **rules**; we on the other hand in speaking of models have been thinking in terms of first-order theories and the usual notion of truth in models. Nevertheless, our models obviously give interpretations of (some of) Curry's systems. We have also not had time to discuss it here, but the above system is more general than the models of this paper in that we have not formulated principles corresponding to the fact that $\mathbf{P}A$ is an atomic Boolean algebra—indeed there are any number of interesting models of the above which **do not** form Boolean algebras under $\subseteq$. We have not had time here either to investigate other primitives corresponding to the way in which $A=A^*$ was built from sequences. Finally, we should remark that it is known that the above system is weak, because, with the introduction of **S** and **K** and with a definition of λ, the whole system is *finitely axiomatizable*.

Notes added in Proof (February 1980)

I am much obliged to Professors Church, Curry, and Seldin who wrote me comments and corrections to the original manuscripts. In particular Professor Church wrote briefly to the editors on 2 June 1979 as follows:

> To the best of my recollection I did not become acquainted with Frege in any detail until somewhat later than the period about which Scott is writing, say 1935 or 1936. No guarantee for this, it is just a recollection of something never accurately recorded. But I was attracted to Frege because he does give priority to functions over sets, and his system can be made consistent (presumptively) by imposing a simple type theory. To this I would now add that no doubt such a system can be given as much set-theoretic strength as desired by adjoining strong axioms of infinity.

On 1 May 1979 Professor Curry wrote me a long letter explaining his attitudes toward various of the points I had brought up. I hope to take account of these remarks in future publications. In the meantime, however, it seems useful to quote two technical remarks from his letter bearing directly on the details:

> Concerning p. 225, my derivation of the inconsistency in Curry and Feys (1958), pp. 258-260 (which came originally from JSL 7, pp. 115-117 (1942) is not exactly a simplification of the proof of Kleene and Rosser. I assumed the existence of the combinator **K**, whereas they did not, so that the result is, at least superficially, weaker. In his thesis (1968), pp. 19f, Bunder showed that, if **K** is present, the Kleene–Rosser theorem follows from my assumption. A simplified derivation of the actual Kleene–Rosser theorem has not yet, to my knowledge, been given.
>
> Concerning p. 238, this is not exactly the way I discovered **Y**. I am not sure just how the discovery was made, or when I adopted the letter "Y" for it. To settle this would require prolonged search in my cellar; which is hardly worth while. I think that the treatment in Curry and Feys (1958, §5G) is fairly close to the original approach. This is essentially as follows: If F is the Russell function and N is negation, then
>
> $$Fx = N(xx).$$
>
> But
>
> $$N(xx) = \mathbf{B}Nxx = \mathbf{W}(\mathbf{B}N) = \mathbf{BWB}N.$$
>
> We thus get the paradox by taking
>
> $$F = \mathbf{B}N\mathbf{B}N.$$
>
> To get Y we just express this FF as a function of N. There are various ways of doing this. One way given in Curry and Feys (1958, §5G), is
>
> $$\mathbf{Y} = \mathbf{WS}(\mathbf{BWB}).$$
>
> However I do not think I used **S** in my earliest work; I did not appreciate its potentialities until later. Another possibility is
>
> $$\mathbf{Y} = \mathbf{W}(\mathbf{B}(\mathbf{C}(\mathbf{BWB}))(\mathbf{BWB})).$$
>
> Your treatment may be essentially isomorphic, so to speak, to this, but it seems strange to me. (At a seminar at Harvard about 1926 Whitehead cited Suzanne Langer for the functional (or predicate) form of the Russell paradox; but I think I once saw it in Russell's *Principles of Mathematics*.)

References

Barendregt, H. P.

[1977] The type-free lambda calculus, *Handbook of Mathematical Logic*, edited by J. Barwise (North-Holland, Amsterdam), pp. 1091–1132.

[1980] *The Lambda Calculus: Its Syntax and Semantics* (North -Holland, Amsterdam), to appear.

BAETEN, J. and B. BOERBOOM

[1978] Ω can be anything it shouldn't be, Mathematisch Instituut, Utrecht, Preprint no. 84, 9 pp.

CHURCH, A.

[1932/33] A set of postulates for the foundation of logic. *Ann. Math.*, **33**, (1932), 346–366. Second paper, *ibid.*, **34** (1933), 839–864.

[1941] *The Calculi of Lambda Conversion* (Princeton University Press) second edition, 1951, reprinted 1963 by University Microfilms, Ann Arbor, MI, 82 pp.

CHURCH, A. and J. B. ROSSER

[1936] Some properties of conversion, *Trans. Am. Math. Soc.*, **39**, 472–482.

CURRY, H. B.

[1929] An analysis of logical substitution. *Am. J. Math.*, **51**, 363–384.

[1930] Grundlagen der kombinatorischen Logik, *Am. J. Math.*, **52**, 509–536 and 789–834.

[1967] Logic, combinatory, in: *Encyclopedia of Philosophy*, *Vol. 4* (Macmillan Co. and The Free Press, New York), pp. 504–509.

[1968a] Combinatory logic, in: *Contemporary Philosophy*, *a Survey*, edited by R. Klibansky (La Nuova Italia, Editrice, Firenze), pp. 295–307.

[1968b] Recent advances in combinatory logic, *Bull. Soc. Math. Belg.*, **20**, 233–298.

[1980] Some philosophical aspects of combinatory logic, in: *The Kleene Symposium*, edited by J. Barwise, H. J. Keisler and K. Künen (North-Holland, Amsterdam).

CURRY, H. B. and R. FEYS

[1958] *Combinatory Logic*, *Vol. I* (North-Holland, Amsterdam).

CURRY, H. B., J. R. HINDLEY and J. P. SELDIN

[1972] *Combinatory Logic*, *Vol. II* (North-Holland, Amsterdam).

FEFERMAN, S.

[1980] Constructive theories of functions and classes, in: *Logic Colloquium* '78, edited by D. van Dalen (North-Holland, Amsterdam), pp. 159–224.

GIERZ, G., K. H. HOFMANN, K. KEIMEL, J. D. LAWSON, M. MISLOVE and D. S. SCOTT

[1980] *A Compendium of Continuous Lattices* (Springer-Verlag, Berlin), to appear.

Jensen, R. B.
[1969] On the consistency of a (slight?) modification of Quine's new foundations, *Synthese*, **19**, 250–263.

Kleene, S. C.
[1934] Proof by cases in formal logic, *Ann. Math.*, **35**, 524–544.
[1935] A theory of positive integers in formal logic, *Am. J. Math.*, **57**, 153–173 and 219–244.
[1936] λ-definability and recursiveness, *Duke Math. J.*, **2**, 340–353.
[1952] *Introduction to Metamathematics* (North-Holland, Amsterdam, and P. Noordhof, Groningen).

Kleene, S. C. and J. B. Rosser
[1935] The inconsistency of certain formal logics, *Ann. Math.*, **36**, 630–636.

Kline, G. L.
[1951] Review of: Foundations of mathematics and mathematical logic, by S. A. Anovskaa, *J. Symbolic Logic*, **16**, 46–48.

Klop, J. W.
[1977] A counterexample to the Church–Rosser property of λ-calculus $+\mathbf{D}MM \to M$, Mimeographed note dated November 1977, Mathematisch Instituut, Utrecht, 2 pp.
[1979] A counterexample to the Church–Rosser property of λ-calculus with surjective pairing, Mathematisch Instituut, Utrecht, Preprint No. 102, 14 pp.

Knaster, B.
[1928] Un théorème sur les fonctions d' ensembles, *Ann. Soc. Polon. Math.*, **6**, 133–134.

Park, D. M. R.
[1970] The *Y* combinator in Scott's lambda-calculus models, Univ. of Warwick, Unpublished notes.

Plotkin, G. D.
[1972] A set-theoretical definition of application, Memo. MIP-R-95, School of Artificial Intelligence, Univ. of Edinburgh, 32 pp.
[1973] Lambda definability and logical relations, Memo. SAI-RM-4, School of Artificial Intelligence, Univ. of Edinburgh, 20 pp.

Plotkin, G. D. and M. B. Smyth
[1978] The category-theoretic solution of recursive domain equations, Memo. DAI-RR-60, Department of Artificial Intelligence, Univ. of Edinburgh, 42 pp.

Quine, W. V. O.
[1936] A reinterpretation of Schönfinkel's logical operators, *Bull.*

Am. Math. Soc., **42**, 87–89.

[1944] New foundations for mathematical logic, *Am. Math. Monthly*, **44**, 70–80.

[1953] *From a Logical Point of View* (Harvard Univ. Press, 1953, second ed., 1961, reprinted, Harper and Row, 1963), 184 pp.

[1963] *Set Theory and Its Logic* (Harvard Univ. Press, 1963, second ed. 1969), 361 pp.

[1971] Algebraic logic and predicate functors, in: *Logic and Art: Essays in Honor of Nelson Goodman*, edited by R. Rudner and Scheffler (Boobs Merrill), pp. 214–238.

ROSSER, J. B.

[1935] A mathematical logic without variables, *Ann. Math.*, **36**, 127–150, and *Duke Math. J.*, **1**, 328–355.

SCHÖNFINKEL, M.

[1924] Über die Bausteine der mathematischen Logik, *Math. Ann.*, **92**, 305–316.

SCOTT, D.

[1969] A type-theoretical alternative to CUCH, ISWIM, OWHY, unpublished.

[1972] Continuous lattices, in: *Proc. 1971 Dalhousie Conference, Lecture Notes in Mathematics, Vol. 274* (Springer-Verlag, New York), pp. 97–136.

[1973] Models for various type-free calculi, in: *Proc. IVth Int. Congr. for Logic, Methodology and the Philosophy of Science, Bucharest*, edited by P. Suppes et al. (North-Holland, Amsterdam), pp. 157–187.

[1975] Combinators and classes, in: *λ-Calculus and Computer Science*, edited by C. Böhm, Lecture Notes in Computer Science, vol. 37 (Springer-Verlag, Berlin), pp. 1–26.

[1976] Data types as lattices, *SIAM J. Comput.* **5**, 522–587.

[1977] Logic and programming languages, *Comm. ACM*, **20**, 634–641.

[1977a] An appreciation of Christopher Strachey and his work, in: *Foreward to Denotational Semantics* by J. E. Stoy (MIT Press), pp. XV–XXX.

SELDIN, J. P.

[1980] Curry's program, to appear.

TARSKI, A.

[1930] On some fundamental concepts of metamathematics, in translation by J. H. Woodger, in: *Logic, Semantics, Metamathematics* (Cambridge Univ. Press, 1956), pp. 30–37.

[1955] A lattice-theoretical fixpoint theorem and its applications, *Pac. J. Math.*, **5**, 285–309.

VAN HEIJENOORT, J.

[1967] *From Frege to Gödel: A Source Book* (Harvard Univ. Press).

VON NEUMANN, J.

[1928] Die Axiomatisierung der Mengenlehre, *Math. Zeits.*, **27**, 669–752.

WHITEHEAD, A. N. and B. RUSSELL

[1910] *Principia Mathematica, 3 Vols.* (University Press, Cambridge, 1910–1913; second edition, 1925–1927).

This page intentionally left blank

J. Barwise, H. J. Keisler and K. Kunen, eds., *The Kleene Symposium*

The Hierarchy Based on the Jump Operator*

Stephen G. Simpson

Department of Mathematics, The Pennsylvania State University, University Park, PA 16802, U.S.A.

Dedicated to Professor S. C. Kleene on the occasion of his 70th birthday

Abstract: The problem of iterating the jump operator into the transfinite is discussed from several points of view.

0. Introduction

This paper is essentially the text of the author's invited address to the Kleene Symposium on June 21, 1978. We dedicate this paper to Professor Stephen C. Kleene on the occasion of his seventieth birthday.

Unless otherwise specified, by a *set* we mean a set of natural numbers. The *jump operator* is a well-known canonical method of passing from a set X to a more complicated set X^*, defined by

$$m \in X^* \leftrightarrow \{m\}(X,m) \simeq 1$$

(see KLEENE and POST (1954)). The purpose of this paper is to report on the work that has been done on the problem of iterating the jump operator into the transfinite. As with all hierarchies, the goal of such an iteration is to classify sets in increasing levels of complexity. It turns out that transfinite iterates of the jump operator correspond closely to the known set-theoretical hierarchies (GÖDEL, 1939; JENSEN, 1972; SILVER, 1971b; DODD and JENSEN, 1976; MITCHELL, 1974). At the same time these iterates lead to interesting questions concerning degrees of unsolvability (SPECTOR, 1955; BOOLOS and PUTNAM, 1968; JOCKUSCH and SIMPSON, 1976; HODES, 1977). Thus is forged an ineluctable bond between recursion theory and set theory.

*Preparation of this paper was partially supported by NSF grant MCS 77–13935.

1. Finite iterates of the jump operator

These are defined by setting $X^0 = X$ and $X^{n+1} = (X^n)^*$ for all n. In other words, the finite iterates of the jump operator

$$j:\text{sets}\rightarrow\text{sets}$$

are defined by composition, $j^n = j\circ\cdots\circ j$ (n times). Note that the nth jump operator j^n still carries sets to sets. However, degrees of unsolvability, as opposed to sets, are relevant as can be seen from the following theorem of KLEENE (1943) and POST (1948).

Theorem 1.1. *Let X be a set. Then $X \leqslant_T \varnothing^n$ if and only if X is Δ^0_{n+1}.*

In other words, the degree of unsolvability of X is less than or equal to that of the nth jump of the empty set, if and only if X is expressible in both $n+1$ quantifier forms in the arithmetical hierarchy. This theorem illustrates the naturalness of the jump operator inasmuch as one application of the jump operator corresponds to one numerical quantifier.

2. Iteration through the constructive ordinals

In this section we define a streamlined variant of Kleene's notation system $\mathcal{O}$ (KLEENE (1938)) and use it to iterate the jump operator through the constructive ordinals.

Let e be an index of a recursive binary relation $\leqslant_e$ on the natural numbers. The *field* of $\leqslant_e$ is defined to be the set of all x such that $x \leqslant_e x$. We write $x <_e y$ to mean that $x \leqslant_e y$ and $x \neq y$. We define $\mathcal{O}$ to be the set of all e such that $<_e$ is a well-ordering of the field of $\leqslant_e$. If $e \in \mathcal{O}$ we write $|e|$ for the order type of $<_e$ and refer to e as a *notation* for the ordinal $|e|$. An ordinal is said to be *constructive* or *recursive* if it has at least one notation in $\mathcal{O}$. The least nonconstructive ordinal is denoted ω_1.

Let ρ be a fixed recursive function of two variables such that

$$x \leqslant_{\rho(e,z)} y \leftrightarrow x \leqslant_e y <_e z$$

for all x,y,z,e. Thus $|\rho(e,z)|$ runs through the ordinals less than $|e|$ as z runs through the field of $\leqslant_e$. For $e \in \mathcal{O}$ we define

$$H_e = \{2^z 3^m : z \in \text{field}(\leqslant_e) \text{ and } m \in H^*_{\rho(e,z)}\}.$$

For example, if the field of $\leqslant_e$ is the empty set, then $H_e = \varnothing$. If the field of $\leqslant_e$ is a one element set, then $H_e = \{2^z 3^m : m \in \varnothing^*\}$ where z is the unique element of the field. It is easy to check that if $|e|$ is a finite ordinal n, then H_e has the same degree of unsolvability as $\varnothing^n$. More generally, for all

$e \in \mathcal{O}, H_e$ can be described as the result of "iterating the jump operator $|e|$ times" along the well-ordering $<_e$. The H-sets were introduced by DAVIS (1950) and MOSTOWSKI (1951).

It is tempting to define the $|e|$th iterate of the jump operator (applied to $\varnothing$) to be the set H_e. Unfortunately, H_e depends on the notation e and not just on the ordinal $|e|$. It can even be shown that there exist e and e' in $\mathcal{O}$ such that $|e| = |e'|$ but H_e and $H_{e'}$ are not recursively isomorphic. Moreover, we know from proof theory that the problem of choosing a "natural" or "canonical" notation for an arbitrary constructive ordinal is far from trival.

Thus we find a serious obstacle to a satisfactory definition of $\varnothing^{\alpha}$, the αth jump, for arbitrary $\alpha < \omega_1$. The following remarkable theorem of SPECTOR (1955) overcomes this obstacle by employing the concept of degrees of unsolvability.

Theorem 2.1. *If* $|e| = |e'|$, *then* $H_e \equiv_T H_{e'}$, *i.e.* H_e *and* $H_{e'}$ *have the same degree of unsolvability.*

Thus we have a well-defined mapping from the constructive ordinals into the degrees of unsolvability,

$$D = \text{degrees} = \text{sets} / \equiv_T,$$

defined by $\mathbf{0}^{\alpha} = \text{degree}(H_e)$ where $|e| = \alpha$.

Similarly, by the well-known procedure of relativization, we may define $\mathbf{d}^{\alpha}$ for all degrees $\mathbf{d}$ and ordinals $\alpha < \omega_1^{\mathbf{d}}$ where $\omega_1^{\mathbf{d}}$ is the least ordinal not recursive in $\mathbf{d}$. In particular we obtain transfinite iterates of the jump operator

$$j^{\alpha} : D \to D$$

for all $\alpha < \omega_1$.

The above definition of $\mathbf{0}^{\alpha}$, given by Spector's theorem, is satisfying in that $\mathbf{0}^{n} = \text{degree}(\varnothing^{n})$ for all finite n, and $\mathbf{0}^{\alpha+1} = \text{jump}(\mathbf{0}^{\alpha})$ whenever this makes sense. However, Spector's theorem leaves open the exact nature of the dependence of $\mathbf{0}^{\lambda}$ on $\{\mathbf{0}^{\alpha} : \alpha < \lambda\}$ when λ is a limit ordinal. Clearly $\mathbf{0}^{\lambda}$ is an upper bound of $\{\mathbf{0}^{\alpha} : \alpha < \lambda\}$, but can we say more? This question will be discussed in a more general setting from two different points of view in Sections 4 and 5 respectively.

3. The degree of Kleene's $\mathcal{O}$

The class of hyperarithmetical sets is defined by

$$\text{HYP} = \{X : X \leqslant_T H_e \text{ for some } e \in \mathcal{O}\}.$$

Thus a set is hyperarithmetical if and only if its degree of unsolvability is $\leq \mathbf{0}^{\alpha}$ for some $\alpha<\omega_1$. The following theorems of KLEENE (1955, 1959a), SPECTOR (1960), and GANDY (1960) are relevant to the problem of pushing the iteration of the jump operator beyond ω_1.

Theorem 3.1. *Kleene's* $\mathcal{O}$ *is a complete* Π^1_1 *set.*

Theorem 3.2. $\mathrm{HYP}=\Delta^1_1$; *i.e. a set is hyperarithmetical if and only if it is expressible in both one-quantifier forms in the analytical hierarchy.*

Theorem 3.3. *Kleene's* $\mathcal{O}$ *is* Σ^1_1 *over* HYP, *i.e. expressible in existential set quantifier form where the set quantifier ranges over* HYP.

These theorems suggest that Kleene's $\mathcal{O}$ should be in some sense the "next natural set" after HYP. This suggestion is confirmed by the following corollary which is expressed in terms of degrees of unsolvability.

Corollary 3.4. *Let X be a set.*
(i) *X is Δ^1_1 over* HYP *if and only if $X\in$* HYP
(ii) *X is Δ^1_2 over* HYP *if and only if $X\leq_T \mathcal{O}$.*
(iii) *More generally, X is Δ^1_{2+n} over* HYP *if and only if $X\leq_T$ the* nth *jump of $\mathcal{O}$.*

Thus we are led to define $\mathbf{0}^{\omega_1}=\text{degree}(\mathcal{O})$ and more generally $\mathbf{0}^{\omega_1+n}=\text{degree}(\mathcal{O}^n)$.

4. Master codes in the constructible hierarchy

The ideas which were introduced above suffice to define $\mathbf{0}^{\alpha}$ for all ordinals $\alpha<\omega_\omega=$ the limit of the first ω admissible ordinals. Further extensions of the jump hierarchy may be defined by other methods. For instance, the ramified analytical hierarchy and the corresponding degrees $\mathbf{0}^{\alpha}, \alpha<\beta_0$, have been discussed by BOYD ET AL. (1969) and JOCKUSCH and SIMPSON (1976). Let us now pass over these piecemeal results and look at a truly far-reaching extension, the constructible hierarchy of GÖDEL (1939).

For our purposes it is convenient to define the constructible hierarchy as follows:

$$L_0=\mathrm{HF}=\{\text{hereditarily finite sets}\};$$

$$L_{\alpha+1}=\{\text{subsets of } L_\alpha \text{ first order definable over } L_\alpha \text{ with parameters}\};$$

$$L_\lambda=\bigcup\{L_\alpha:\alpha<\lambda\} \text{ for limit ordinals } \lambda.$$

Recall now that by a *set* we mean a subset of ω, the set of natural numbers. A set X is said to be $\Sigma_n(L_\alpha)$ if it is first-order definable over L_α by a Σ_n formula with parameters from L_α (cf. JENSEN (1972)). Thus $X \in L_{\alpha+1}$ if and only if X is $\Sigma_n(L_\alpha)$ for some n. A set X is said to be $\Delta_n(L_\alpha)$ if both X and $\omega - X$ are $\Sigma_n(L_\alpha)$.

We say that a set A is a $\Delta_n(L_\alpha)$ *master code* if

$$\forall X(X \leqslant_T A \leftrightarrow X \text{ is } \Delta_n(L_\alpha)).$$

Thus a $\Delta_n(L_\alpha)$ master code is simply a $\Delta_n(L_\alpha)$ set whose degree of unsolvability is maximum among the degrees of all such sets. The concept of a master code is due to JENSEN (1967, 1972).

By a *Jensen degree* let us mean the degree of a $\Delta_n(L_\alpha)$ master code for some positive integer n and ordinal α. Clearly the Jensen degrees are well-ordered in the natural ordering of degrees of unsolvability. It can also be shown that if A is a $\Delta_n(L_\alpha)$ master code, then the jump of A is a $\Delta_{n+1}(L_\alpha)$ master code. Thus the successor operation on the well-ordering of Jensen degrees is given by the jump operator.

The following theorem of JENSEN (1967) shows that there are no unnecessary gaps in the hierarchy of Jensen degrees.

Theorem 4.1. *Suppose that there exists a set X which is* $\Delta_n(L_\alpha)$ *but not an element of* L_α. *Then there exists a* $\Delta_n(L_\alpha)$ *master code.*

Proof (sketch). We know that X and $\omega - X$ are definable over L by Σ_n formulas with parameters. By the uniformization theorem (JENSEN, 1972) we can choose a canonical collection of Σ_n Skolem functions for these formulas and all their subformulas. Let M be the submodel of L_α generated from the parameters by the Skolem functions. The condensation lemma tells us that there is an ordinal $\beta \leqslant \alpha$ such that L_β is isomorphic to M. By construction X is $\Delta_n(M)$, hence $\Delta_n(L_\beta)$. Hence $\beta = \alpha$ since otherwise we would have $X \in L_\alpha$. Hence the inverse image of M under the Skolem functions yields a subset A of ω which encodes L_α. It can then be shown that A is a $\Delta_n(L_\alpha)$ master code.

An argument of the above type also appeared in a paper of BOOLOS and PUTNAM (1968).

An easy consequence of Jensen's theorem is that the Jensen degrees are well-ordered in order type $\aleph_1^L$, the least constructibly uncountable ordinal. We are therefore justified in defining $\mathbf{0}^\alpha$ for all $\alpha < \aleph_1^L$ to be the αth Jensen degree. It can be shown that this definition of $\mathbf{0}^\alpha$ agrees with the definitions in Sections 1, 2 and 3 above for the α's considered there. Thus we have iterated the jump operator through all the constructibly countable ordinals.

5. Degree theoretic characterizations

In this section we discuss the Jensen degrees $\mathbf{0}^\alpha, \alpha < \aleph_1^L$ which were defined above and reexamine them from an algebraic point of view. Specifically, we consider the algebraic structure

$$\mathfrak{D} = \langle D, \cup, \leqslant, \mathbf{0}, j \rangle$$

where D is the set of all degrees of unsolvability, $\mathbf{0}$ is the least element of D, and j is the jump operator. This is an upper semilattice (with admittedly some extra structure) and it is natural to ask whether the degrees $\mathbf{0}^\alpha, \alpha < \aleph_1^L$ can be characterized in lattice theoretic terms. We may begin by defining

$$\mathbf{0}^0 = \mathbf{0}$$

and

$$\mathbf{0}^{\alpha+1} = \text{jump}(\mathbf{0}^\alpha)$$

for successor ordinals $\alpha + 1$, but what is the lattice theoretic aspect of the dependence of $\mathbf{0}^\lambda$ on $\{\mathbf{0}^\alpha : \alpha < \lambda\}$ for limit ordinals $\lambda < \aleph_1^L$? If we define the ideal

$$I_\lambda = \{\mathbf{d} \in D: \quad (\exists \alpha < \lambda) \quad \mathbf{d} \leqslant \mathbf{0}^\alpha\},$$

then clearly $\mathbf{0}^\lambda$ is an upper bound for I_λ, but we would like to characterize $\mathbf{0}^\lambda$ somehow as the "least natural" upper bound for I_λ. This is accomplished by the following theorem.

Theorem 5.1. *Let λ be a limit ordinal less than $\aleph_1^L$. Define ν_λ to be the least ordinal ν such that the set of all degrees of the form $(\mathbf{a} \cup \mathbf{b})^\nu$ with*

$$I_\lambda = \{\mathbf{d} \in D: \mathbf{d} \leqslant \mathbf{a} \text{ and } \mathbf{d} \leqslant \mathbf{b}\} \tag{$*$}$$

has a least element. Then ν_λ exist and $\mathbf{0}^\lambda$ is that least element. Moreover, there exist degrees $\mathbf{a}$ and $\mathbf{b}$ such that $(*)$ holds and $\mathbf{0}^\lambda = (\mathbf{a} \cup \mathbf{b})^{\nu_\lambda}$ and ν_λ is recursive in $\mathbf{a} \cup \mathbf{b}$.

For $\lambda < \omega_1$ the above theorem is essentially due ENDERTON and PUTNAM (1970) and SACKS (1971). In this case it turns out that $\nu_\lambda = 2$ and an even more perspicuous characterization of $\mathbf{0}^\lambda$ is possible.

For $\lambda < \beta_0$ the above theorem is due to JOCKUSCH and SIMPSON (1976). Here β_0 is the least ordinal β such that L_β is a model of ZF minus the power set axiom, and ν_λ can assume any finite value $n \geqslant 2$.

The full Theorem 5.1 appears in the thesis of HODES (1977) who acknowledges some help from Abramson in the proof of the "moreover" clause. The main tool in Hodes' proof is a notion of forcing due to SACKS (1971) in which a condition is a recursively pointed perfect tree P whose degree lies in I_λ. The recursive pointedness means that P is recursive in each of its infinite branches. The "moreover" clause is obtained with the aid of another notion of forcing due to STEEL (1978).

The noteworthy point about Theorem 5.1 is that the next Jensen degree, $\mathbf{0}^\lambda$, is defined in purely algebraic terms from the ideal generated by the previous Jensen degrees. At first glance, the given definition of $\mathbf{0}^\lambda$ may appear circular in that it uses the concept of νth jump where ν is essentially arbitrary. Indeed, it often happens that $\nu_\lambda > \lambda$. But, the "moreover" clause tells us that the result of the given definition remains unchanged if we restrict attention to $\mathbf{a}, \mathbf{b}, \nu$ such that ν is recursive in $\mathbf{a} \cup \mathbf{b}$. For such $\mathbf{a}, \mathbf{b}, \nu$ the definition of $(\mathbf{a} \cup \mathbf{b})^\nu$ is unproblematical (cf. Section 2 above). Thus we really do have a definition of $\mathbf{0}^\lambda$ in terms of simpler concepts.

6. Open problems

The purpose of this section is to discuss several problems which are suggested by results stated in previous sections.

An obvious problem, suggested by Sections 4 and 5, is to extend the degree theoretic hierarchy $\mathbf{0}^\alpha$ to ordinals $\alpha \geqslant \aleph_1^L$. Clearly some unusual hypothesis is called for here since if $V = L$, then the degrees $\mathbf{0}^\alpha, \alpha < \aleph_1^L$ have no upper bound. If we assume the existence of, say, a Ramsey cardinal, then the work of DODD and JENSEN (1976) establishes the existence of a hierarchy of master codes beyond $\aleph_1^L$. The first such master code is of course Silver's remarkable set $0^\#$ (SILVER, 1971a). It therefore seems reasonable to define $\mathbf{0}^{\aleph_1^L}$ to be the degree of $0^\#$. Similarly, we may define $\mathbf{0}^\alpha$ for $\alpha < \aleph_1^K$ to be the degree of the αth master code in the Dodd–Jensen core model K (DODD and JENSEN, 1976). This much is clear. What is not clear is how to characterize these degrees algebraically from below in the style of Theorem 5.1.

We therefore pose the following test problem: find a natural algebraic characterization of the degree of $0^\#$ within the degree-theoretic structure $\mathfrak{D}$. A fact which may be relevant here is that there exists a Σ_1^1 set of degrees of unsolvability whose determinacy is provably equivalent to the existence of $0^\#$. This result is due to HARRINGTON (1979).

In order to be completely honest with the reader, we must now pause to point out that the problem just stated is highly speculative because the existence of $0^\#$ is not firmly established. Indeed, several prominent set theorists have seriously attempted to refute the existence of $0^\#$. So far it is known that

(i) $0^\#$ does not exist in the models of set theory considered by GÖDEL (1939) and COHEN (1966);

(ii) the consistency of the existence of $0^\#$ with set theory cannot be proved in set theory, even if large cardinal axioms of the kinds considered by LÉVY (1971) and SILVER (1970) are assumed;

(iii) the hypothesis of the nonexistence of $0^{\#}$ is an attractive one with far-reaching consequences, e.g. the solution of the singular cardinals problem (DEVLIN and JENSEN, 1975).

However, the nonexistence of $0^{\#}$ has not yet been proved and indeed may be unprovable.

We now turn to another set of problems. It is known from JOCKUSCH and SIMPSON (1976) that many specific Jensen degrees have natural algebraic characterizations within $\mathfrak{D}$. For instance, let α_n be the least ordinal α such that L_α is a model of the Δ_n^1 comprehension axiom of second order arithmetic. Thus $\alpha_0 = \omega, \alpha_1 = \omega_1, \alpha_2 = \omega_1^{E_1}$, and in general α_{3+n} = the first Σ_{2+n} admissible ordinal greater than ω. The results of JOCKUSCH and SIMPSON (1976) show that the degrees $\mathbf{0}^{\alpha_n}$ as well as $\mathbf{0}^{\beta_0}$ have simple algebraic characterizations. Question: Can we do the same for some of the other specific Jensen degrees which arise from the theories of recursion in higher types (KLEENE, 1959b, 1963) and nonmonotonic inductive definitions (RICHTER and ACZEL, 1974)? A good test case here is the Jensen degree $\mathbf{0}^\sigma$ where $\sigma = \omega_1^{E_1^{\#}}$ = the least Σ_1^1 reflecting ordinal (RICHTER and ACZEL, 1974) = the least non-Gandy ordinal (ABRAMSON and SACKS, 1976). This degree $\mathbf{0}^\sigma$ can also be characterized as the largest degree of a set which is Σ_1^1 inductively definable, i.e. recursively enumerable in $E_1^{\#}$ (HINMAN, 1978, Theorem VI.6.14). Does $\mathbf{0}^\sigma$ have a natural algebraic characterization within $\mathfrak{D}$?

It is perhaps worth remarking that $\mathbf{0}^\sigma$ and most other specific Jensen degrees (and also the degree of $0^{\#}$ if it exists) are already known to be first-order definable in $\mathfrak{D}$. This follows from the general definability theorem of SIMPSON (1977, Theorem 3.12). However, the first-order definitions of $\mathbf{0}^\sigma$ which are known at this writing look extremely artificial from the algebraic and degree-theoretic viewpoint. What we lack is a degree-theoretically natural characterization of $\mathbf{0}^\sigma$.

References

ABRAMSON, F., and G. E. SACKS
[1976] Uncountable Gandy ordinals, *J. London Math. Soc. (2)*, **14**, 387–392.

BOOLOS, G., and H. PUTNAM
[1968] Degrees of unsolvability of constructible sets of integers, *J. Symbolic Logic*, **33**, 497–513.

BOYD R., G. HENSEL and H. PUTNAM
[1969] A recursion theoretic characterization of the ramified analytical hierarchy, *Trans. Am. Math. Soc.*, **141**, 37–62.

COHEN, P. J.
[1966] *Set Theory and the Continuum Hypothesis* (Benjamin).
DAVIS, M.
[1950] On the theory of recursive unsolvability, Ph.D. thesis, Princeton, N.J.
DEVLIN, K., and R. B. JENSEN
[1975] Marginalia to a theorem of Silver, in: *Logic Conference, Kiel 1974, Lecture Notes in Mathematics Vol. 499* (Springer-Verlag, Berlin), pp. 115–142.
DODD, T., and R. B. JENSEN
[1976] The core model, unpublished notes.
ENDERTON, H., and PUTNAM, H.
[1970] A note on the hyperarithmetical hierarchy, *J. Symbolic Logic*, **35**, 429–430.
GANDY, R. O.
[1960] Proof of Mostowski's conjecture, *Bull. Acad. Polon. Sci.*, **8**, 571–575.
GÖDEL, K.
[1939] Consistency proof for the generalized continuum hypothesis, *Proc. Nat. Acad. Sci. U.S.A.*, **25**, 220–224.
HARRINGTON, L.
[1978] Analytic determinacy and $0^{\#}$, *J. Symbolic Logic*, **43**, 685–693.
HINMAN, P. G.
[1978] *Recursion Theoretic Hierarchies* (Springer-Verlag, Berlin).
HODES, H.
[1977] Jumping through the transfinite: a study of Turing degree hierarchies, Ph.D. thesis, Harvard University.
JENSEN, R. B.
[1967] Stufen der konstruktiblen Hierarchie, Habilitationsschrift, Bonn.
[1972] Fine structure of the constructible hierarchy, *Ann. Math. Logic*, **4**, 229–308.
JOCKUSCH, C. G., JR., AND S. G. SIMPSON
[1976] A degree theoretic definition of the ramified analytical hierarchy, *Ann. Math. Logic*, **10**, 1–32.
KLEENE, S. C.
[1938] On notation for ordinal numbers, *J. Sympolic Logic*, **3**, 150–155.
[1943] Recursive predicates and quantifiers, *Trans. Am. Math. Soc.*, **53**, 41–73.
[1955] On the forms of the predicates in the theory of constructive ordinals (second paper), *Am. J. Math.*, **77**, 405–428.
[1959a] Quantification of number theoretic functions, *Compositio Math.*, **14**, 23–40.

[1959b] Recursive functionals and quantifiers of finite types, I, *Trans. Am. Math. Soc.*, **91**, 1–52.

[1963] Recursive functionals and quantifiers of finite types, II, *Trans. Am. Math. Soc.*, **108**, 106–142.

KLEENE, S. C., and E. L. POST

[1954] The upper semilattice of degrees of recursive unsolvability, *Ann. Math.*, **59**, 379–407.

LÉVY, A.

[1971] The sizes of the indescribable cardinals, in: *Axiomatic Set Theory, Proc. Symp. Pure Math. 13, Part I* (Am. Math. Soc., Providence, RI), pp. 205–218.

MITCHELL, W.

[1974] Sets constructible from sequences of ultrafilters, *J. Symbolic Logic*, **39**, 57–66.

MOSTOWSKI, A.

[1951] A classification of logical systems, *Studia Phil.*, **4**, 237–274.

POST, E. L.

[1948] Degrees of recursive unsolvability (preliminary report), *Bull. Am. Math. Soc.*, **54**, 641–642.

RICHTER, W., and P. ACZEL

[1974] Inductive definitions and reflecting properties of admissible ordinals, in: *Generalized Recursion Theory* (North-Holland, Amsterdam), pp. 301–381.

SACKS, G. E.

[1971] Forcing with perfect closed sets, in: *Axiomatic Set Theory, Proc. Symp. Pure Math. 13, Part I* (Am. Math. Soc., Providence, RI), pp. 331–355.

SILVER, J. H.

[1971a] Some applications of model theory in set theory, *Ann. Math. Logic*, **3**, 45–110.

[1970] A large cardinal in the constructible universe, *Fund. Math.*, **69**, 93–100.

[1971b] Measurable cardinals and Δ^1_3 well orderings, *Ann. Math.*, **94**, 414–446.

SIMPSON, S. G.

[1977] First order theory of the degrees of recursive unsolvability, *Ann. Math.*, **105**, 121–139.

SPECTOR, C.

[1955] Recursive well orderings, *J. Symbolic Logic*, **20**, 151–163.

[1960] Hyperarithmetical quantifiers, *Fund. Math.*, **48**, 313–320.

STEEL, J.

[1978] Forcing with tagged trees, *Ann. Math. Logic*, **15**, 55–74.

J. Barwise, H. J. Keisler and K. Kunen, eds., *The Kleene Symposium*
©North-Holland Publishing Company (1980) 277–316

Extended Bar Induction of Type Zero

A.S. Troelstra
Department of Mathematics, University of Amsterdam, Amsterdam, The Netherlands

Dedicated to Professor S. C. Kleene on the occasion of his 70th birthday

Abstract: The schema of extended bar induction of type zero is obtained by generalizing bar induction to trees of sequences with elements from a subset of the natural numbers. It is shown that intuitionistic elementary analysis with extended bar induction of type zero added is proof-theoretically equivalent to the (classical or intuitionistic) theory of finitely iterated inductive definitions with positive matrix.

1. Introduction

1.1. In this paper, we discuss and study an extension of the schema BI (or BI_M) of bar induction of type 0, familiar from formalizations of intuitionistic analysis. Most of our notation will be taken from Kreisel and Troelstra (1970), with the following exceptions: the notation $g(n,x)$ for the xth element of the finite sequence (coded by) n is replaced by the more familiar notation $(n)_x$, and j is used for a certain standard pairing function from $N \times N$ onto N, with inverses j_1, j_2. This introduction and the next section presuppose familiarity with the *results* of Kreisel and Troelstra (1970); in the remainder of the paper we also presuppose familiarity with the details of the elimination of choice sequences and realizability for $\mathbf{IDB}_1$, as treated in Kreisel and Troelstra (1970).

1.2. The schema EBI_0

We shall first introduce the schema of extended bar induction of type 0. Let Ax be a predicate of natural numbers, $\forall x\ Ax$ closed; let Pn be a predicate of (finite sequences of) natural numbers. We use $t \in A$ as an alternative notation for At, and put

$$n \in A^{<\omega} \equiv_{\mathrm{def}} \forall x < \mathrm{lth}(n)\ ((n)_x \in A),$$

$$\alpha \in A^{\omega} \equiv_{\mathrm{def}} \forall x\,(\alpha x \in A),$$

$$\mathrm{Bar}(A,P) \equiv_{\mathrm{def}} \forall \alpha \in A^{\omega}\, \exists x P(\bar{\alpha}x) \quad (\text{“}P \textit{ bars } A^{\omega}\text{”}),$$

$$\mathrm{Mon}(A,P) \equiv_{\mathrm{def}} \forall n \in A^{<\omega}\, \forall y \in A\ (Pn \to P(n * \hat{y})) \quad (\text{“}P \textit{ is monotone on } A^{<\omega}\text{”}),$$

$$\mathrm{Ind}(A,P) \equiv_{\mathrm{def}} \forall n \in A^{<\omega}\, (\forall y \in A\ P(n * \hat{y}) \to Pn) \quad (\text{“}P \textit{ is inductive on } A^{<\omega}\text{”}).$$

EBI_0 now becomes the union of all instances of $\mathrm{EBI}_0(A)$, where $\mathrm{EBI}_0(A)$ is

$\mathrm{EBI}_0(A)$ $\quad \mathrm{Bar}(A,P)\wedge\mathrm{Mon}(A,P)\wedge\mathrm{Ind}(A,P)\to P\langle\rangle.$

1.3. Remark. BI is usually stated as a schema with two predicates P, Q:

$$\mathrm{BI} \qquad [\forall\alpha\,\exists x\,P(\bar\alpha x)\wedge\forall n\,\forall y\,(Pn\to P(n*\hat y))\wedge\forall n\,(Pn\to Qn)\wedge$$
$$\wedge\forall n\,(\forall y\,Q(n*\hat y)\to Qn)]\to Q\langle\rangle.$$

However, as observed by Grayson (FOURMAN and HYLAND, 1979), BI is really equivalent to the special form BI′ where P and Q are the same. To see this, one has only to observe that BI for P and Q follows from BI′ applied to the single predicate $P'n\equiv\forall m\,Q(n*m)$. Thus for $Ax\equiv[0=0]$, $\mathrm{EBI}_0(A)$ is equivalent to BI.

EBI_0 is an induction principle for trees A^ω, the elements of the sequences in A^ω restricted to an arbitrary set of natural numbers $\{x\colon Ax\}$; BI refers to the tree N^ω.

1.4. The schema EBI_1

A similar principle EBI_1, referring to sequences with sequences of a certain set $\{\alpha\colon A\alpha\}$ as elements has been considered in the literature in several variants. A variant precisely analogous to EBI_0 may be stated as follows. Let

$$(\beta)_x\equiv_{\mathrm{def}}\lambda y.\beta j(x,y),$$
$$\beta\in B^{<\omega}\equiv_{\mathrm{def}}\exists n\,[(\beta)_0=\lambda x.n\wedge\forall y\,(0<y\leqslant n\to(\beta)_y\in B)\wedge$$
$$\wedge\forall y>n\,((\beta)_y=\lambda x.0)],$$
$$\beta\in B^\omega\equiv_{\mathrm{def}}\forall x\,((\beta)_x\in B).$$

$B^{<\omega}$ codes finite sequences of sequences in B. We may extend the notation $*$ (concatenation) and $\langle\beta\rangle\equiv\hat\beta$ in an obvious way to finite sequences of sequences; and we shall write, for any β, $\bar\beta^1x$ for $\langle(\beta)_0,(\beta)_1,\ldots,(\beta)_{x-1}\rangle$. Furthermore let

$$\mathrm{Bar}^1(B,P)\equiv_{\mathrm{def}}\forall\beta\in B^\omega\,\exists x\,P(\bar\beta^1x),$$
$$\mathrm{Dec}^1(B,P)\equiv_{\mathrm{def}}\forall\beta\in B^{<\omega}\,(P\beta\vee\neg P\beta),$$
$$\mathrm{Mon}^1(B,P)\equiv_{\mathrm{def}}\forall\beta\in B^{<\omega}\,\forall\gamma\in B\,(P\beta\to P(\beta*\hat\gamma)),$$
$$\mathrm{Ind}^1(B,P)\equiv_{\mathrm{def}}\forall\beta\in B^{<\omega}\,(\forall\gamma\in B\,(P(\beta*\hat\gamma))\to P\beta).$$

Then extended bar induction of type 1, with monotonicity, is given by

$\mathrm{EBI}_1(B)$ $\quad \mathrm{Bar}^1(B,P)\wedge\mathrm{Mon}^1(B,P)\wedge\mathrm{Ind}^1(B,P)\to P\langle\rangle.$

In the earlier literature, the variant considered most often is extended bar induction of type 1 with decidability:

$$\mathrm{EBI}_{1,D}(B) \quad \mathrm{Bar}^1(B,P) \wedge \mathrm{Dec}^1(B,P) \wedge (P \subseteq Q) \wedge \mathrm{Ind}^1(B,Q) \rightarrow Q\langle\,\rangle.$$

(Comments on the earlier literature on EBI_0, EBI_1 and its variants are to be found in Subsection 1.9.)

The principal interest of EBI_1 is in the fact that it can be used to show that various models for the theory of functionals of finite type satisfy bar recursion in all finite types (cf. Subsection 1.9); and as was first shown by SPECTOR (1962; more recent treatments in HOWARD (1968), LUCKHARDT (1973), DILLER and VOGEL (1975)), bar recursion in all finite types suffices to give a Dialectica interpretation of classical analysis in a quantifier-free theory of finite types, while on the other hand these various models for bar recursion are definable in the language of analysis, and EBI_1 is valid in classical analysis. Thus $\mathbf{EL}+\mathrm{EBI}_1$ has exactly the proof-theoretic strength of classical analysis.

1.5. Notation. We may generalize $\mathrm{EBI}_0(A)$ (and similarly $\mathrm{EBI}_1(A)$ and several variants to be introduced later) by permitting $\forall x Ax$ (respectively $\forall \alpha A\alpha$) to contain parameters for natural numbers, lawlike functions, and choice sequences; these generalizations are indicated by respectively

$$\text{n-EBI}_0(A), \quad \text{l-EBI}_0(A), \quad \text{c-EBI}_0(A),$$

and similarly for n-EBI_0, l-$\mathrm{EBI}_1(B)$, etc., etc.

1.6. Principal results

The main result of this paper is that EBI_0 holds an intermediate position between BI and EBI_1:

Result 1. *Let* **EL** *be elementary intuitionistic analysis* (*with* $\alpha, \beta, \gamma, \delta$ *as function variables*); $\mathbf{EL}+$n-EBI_0 *is proof-theoretically equivalent to the* (*intuitionistic or classical*) *theory of finitely iterated inductive definitions.*

At the same time, our proof of this result settles half of another open problem, which was suggested to us by G. Kreisel: are EBI_0, EBI_1 consistent with Church's thesis (CT)? Kreisel's question of course refers to a context (such as the system **CS** of KREISEL and TROELSTRA (1970)) where both constructive function variables and choice variables are present, and CT is taken to refer to constructive (lawlike) functions. (It is obviously meaningless to ask whether we can assume CT for the range of the choiće variables, since even BI_0 contradicts CT in this sense.) In KREISEL and TROELSTRA (1970) it has already been shown that BI is compatible with CT for lawlike sequences, in the following precise sense: the "lawlike part" $\mathbf{IDB}_1$ of **CS** is consistent with CT, and $\mathbf{BI} \subseteq \mathbf{CS}$.

Our method of proof yields a solution for EBI_0:

Result 2. *For a certain theory* **CS*** *containing* **CS** *and* EBI_0, *it is consistent to assume* CT.

We do not know whether CT is also compatible with $\mathbf{CS}^* + \mathrm{EBI}_1$ or $\mathbf{CS} + \mathrm{EBI}_1$.

The interest of the first result is twofold. First of all, it shows that the use of EBI_1 in proving various structures to satisfy bar recursion of higher type is essential: EBI_1 cannot be replaced by EBI_0. Note also that this contrasts with the fact that $\mathrm{BI} \leftrightarrow \mathrm{BI}_1 (\equiv \mathrm{EBI}_1(B)$ for $B\alpha \equiv (0=0))$ with the help of continuity axioms (cf. Howard and Kreisel (1966, §7)); the difference in strength between EBI_1 and EBI_0 on the other hand reveals a genuine difference between (bar induction over) type 0 and type 1 objects.

Secondly, $\mathbf{EL} + \mathrm{EBI}_0$ presents us with a natural-looking example of a theory with the strength of the theory of the finitely iterated inductive definitions.

1.7. Extending EBI_0 to arbitrary trees: the schemata EBI_0', EBI_0''.

The following schema EBI_0' of extended bar induction of type 0 over arbitrary trees is at first sight more general than EBI_0.

Let An be a predicate of natural numbers, $\forall n An$ closed, and put

$$\mathrm{Tree}(A) \equiv_{\mathrm{def}} \forall n \forall m (A(n*m) \to An) \wedge A\langle\,\rangle \wedge$$
$$\wedge \exists a \forall n (An \to A(n*\langle an \rangle)),$$

$$\alpha \in \bar{A}^{\omega} \equiv_{\mathrm{def}} \forall x A(\bar{\alpha}x),$$

$$\mathrm{Bar}'(A,P) \equiv_{\mathrm{def}} \forall \alpha \in \bar{A}^{\omega} \exists x P(\bar{\alpha}x),$$

$$\mathrm{Mon}'(A,P) \equiv_{\mathrm{def}} \forall n \forall x (n*\hat{x} \in A \wedge Pn \to P(n*\hat{x})),$$

$$\mathrm{Ind}'(A,P) \equiv_{\mathrm{def}} \forall n \in A\, (\forall x (A(n*\hat{x}) \to P(n*\hat{x})) \to Pn),$$

then we introduce the schema

$\mathrm{EBI}_0'(A)$ $\qquad \mathrm{Tree}(A) \wedge \mathrm{Bar}'(A,P) \wedge \mathrm{Mon}'(A,P) \wedge \mathrm{Ind}'(A,P) \to P\langle\,\rangle.$

As already remarked in Subsection 1.5, we may use n-EBI_0', l-EBI_0' etc. for EBI_0' generalized to cases where $\forall x Ax$ contains parameters.

$\mathrm{EBI}_0(A)$ coincides with $\mathrm{EBI}_0'(A^{<\omega})$ if $\exists x Ax$, hence by Subsection 2.3 EBI_0' generalizes to EBI_0. As we shall see in Section 2, n-EBI_0' is derivable from EBI_0, and l-EBI_0', c-EBI_0' from l-EBI_0, c-EBI_0 respectively; so EBI_0' is only apparently stronger than EBI_0.

A further generalization $\mathrm{EBI}_0''(A)$ is obtained by replacement of Tree(A) by the weaker (and more natural)

$$\mathrm{Tree}'(A) \equiv_{\mathrm{def}} \forall nm (A(n*m) \to An) \wedge A\langle\,\rangle \wedge$$
$$\wedge \forall n (An \to \exists x A(n*\hat{x})).$$

It is an open problem whether EBI_0'' is derivable from EBI_0; proof-theoretically it is not stronger than EBI_0 (Subsection 5.9).

1.8. EBI_0 **with parameters**

As we shall see in the next section, n-EBI_0 is derivable from EBI_0, and via result 2 in Subsection 1.6 we see that l-EBI_0 is equivalent in proof-theoretic strength to n-EBI_0, hence to EBI_0. On the other hand, the strength of **EL** + c-EBI_0, **CS** + c-EBI_0, **CS*** + c-EBI_0 (**CS*** as in Subsection 1.6) presents an open problem. **EL** + c-EBI_0 is at least consistent, since it is a subsystem of classical analysis; relative to **CS** or **CS***, not even the *consistency* of c-EBI_0 is obvious.

In this subsection we wish to discuss the plausibility for $\mathrm{EBI}_0(A)$ where A contains various sorts of function parameters in somewhat greater detail. Let us first remind the reader, for the sake of comparison, of another instance of a schema the validity or plausibility of which is very much dependent on the sort of function parameters present: Markov's schema (TROELSTRA, 1975; LUCKHARDT, 1977).

To some extent similar considerations may be applied to EBI_0. First of all, though BI_0 seems unobjectionable for lawless sequences, EBI_0 becomes blatantly false, even for A which do not contain extra parameters, e.g. $Ax \equiv x \neq 0$, since $\forall \alpha \neg \forall x A(\alpha x)$, i.e. A^ω is empty w.r.t. lawless sequences, making $\mathrm{Bar}(A,P)$ true for all P. The reason is simply that lawless sequences do not admit an overall restriction on their range.

More interesting is the case where the variables α range over **CS**-type choice sequences, but where A may contain an additional *lawless* parameter. In FOURMAN and HYLAND (1979) a sheaf model is presented in which a certain instance of $\mathrm{EBI}_0(A)$ is false; if we transcribe their example, slightly modified into the language of intuitionistic analysis, the result is a counterexample to $\mathrm{EBI}_0(A)$ for an A containing a lawless parameter. Take

$$A_\varepsilon(x) \equiv (\varepsilon x = 0 \vee x = 0)$$

(ε lawless; note that A is even decidable!), and let

$$Pn \equiv \exists i < \mathrm{lth}(n) \exists j < \mathrm{lth}(n)\ ((n)_i = (n)_j),$$

and observe that $\alpha \in A^\omega \equiv \forall x(\varepsilon(\alpha x) = 0 \vee \alpha x = 0)$ is equivalent (by the axiom of open data) to

$$\exists n(\varepsilon \in n \wedge \forall \eta \in n \forall x(\eta(\alpha x) = 0 \vee \alpha x = 0))$$

which implies

(1) $\qquad \exists n \forall x(\alpha x \leqslant \mathrm{lth}(n)).$

Observe that we have applied the axiom of open data here to a predicate not in the language of the formal theory **LS**; to justify this intuitively, we must assume ε to range over a universe of lawless sequences not entering into the construction of any element of the universe over which α ranges.

By (1), A^ω contains only *bounded* sequences, therefore automatically $\forall\alpha\in A^\omega \exists x P(\bar{\alpha}x)$. To establish $\mathrm{Ind}(A,P)$, assume $n\in A^{<\omega}$, $\forall x\in A$ $P(n*\hat{x})$; let $m=\max\{(n)_i: i<\mathrm{lth}(n)\}$, and assume first Ax for some $x>m$; then $P(n*\hat{x})$ indeed implies Pn, since x differs from all $(n)_i$, $i<\mathrm{lth}(n)$.

If there is no $x>m$ such that Ax, it would follow that $\forall x>m$ $(\varepsilon x\neq 0)$, which is false for lawless ε; therefore $\neg\neg\exists x>m\,(Ax)$, hence $\neg\neg Pn$; but since P is decidable, Pn follows. $\mathrm{Mon}(A,P)$ is obvious; and now application of $\mathrm{EBI}_0(A)$ would yield $P\langle\,\rangle$, which is clearly false.

1.9. Earlier work on extended bar induction

Extended bar induction is first mentioned by KREISEL (1963, notably pages 0.16, 0.65, 0.71–72); Kreisel observes that EBI_1 can be used to show that the (extensional) hereditary continuous functionals (Kleene's countable functionals) satisfy bar recursion in all finite types. (*Remark*: on page 0.65 the notation refers to EBI_0, but on page 0.71 specific instances of EBI_1 are stated; by a (trivial) oversight no decidability or monotonicity for P is mentioned on page 0.71.) The result was stated again in KREISEL (1967, p. 249). He does not give a detailed proof; a proof is presented in TROELSTRA (1973, 2.9.10), showing that the appeal to the axiom of choice at higher types, mentioned by Kreisel on page 0.72 is not needed. Cf. also ERSHOV (1974, 1977).

LUCKHARDT (1973, Chapter XIV; see especially p.144) discusses in the context of a theory of finite types forms of extended bar induction with decidability, called $(\mathrm{aBI})^\sigma_\mathrm{D}$, $(\mathrm{hsBI})^\sigma_\mathrm{D}$ and $(\mathrm{hBI})^\sigma_\mathrm{D}$. Here $(\mathrm{aBI})^\sigma_\mathrm{D}$ refers to arbitrary trees of sequences of objects of type σ, and in our notation corresponds with $\mathrm{EBI}''_{\sigma,\mathrm{D}}$; the schema $(\mathrm{hsBI})^\sigma_\mathrm{D}$ on the other hand corresponds with $\mathrm{EBI}_{\sigma,\mathrm{D}}$, while $(\mathrm{hBI})^\sigma_\mathrm{D}$ is an intermediate form where the set of successors of a node (initial segment) depends on the *length* of the node only. Corresponding rules are also discussed, and LUCKHARDT (1973, p.136) explicitly defends extended bar induction as an intuitionistically acceptable principle, extending BI in a natural way. $\mathrm{EBI}_{1,\mathrm{D}}$ is used by Luckhardt to establish bar recursion for a certain generalized term model. We note in passing that in KREISEL (1963), LUCKHARDT (1973), and TROELSTRA (1973) it is always assumed that in $\mathrm{EBI}_0(A)$ or $\mathrm{EBI}_1(A)$ the set $\{x: Ax\}$ or $\{\alpha: A\alpha\}$ contains at least one element; but this condition may be dropped (see Subsection 2.3 below).

SCARPELLINI (1972) uses a variant of EBI_1 to establish for a certain model defined in terms of limit spaces that it satisfies bar recursion; more recently, HYLAND (1975) has shown that Scarpellini's model is essentially equivalent to the model of the continuous functionals (the proof is classical).

Scarpellini uses c-$\mathbf{EBI}_1$ in a form with two predicates P and Q (as BI in its traditional formulation, see Subsection 1.3) and the conclusion strengthened to $\forall\beta\in B^{<\omega}\,Q\beta$. In fact, he primarily uses rules EBIR, EBIR_0 with the main implication replaced by metamathematical$\Rightarrow$. EBIR is the following rule corresponding to the schema $\mathbf{EBI}_{1,D}$:

$$\text{EBIR}\qquad \mathrm{Bar}^1(B,P)\wedge\mathrm{Dec}^1(B,P)\wedge\mathrm{Mon}^1(B,P)\wedge(P\subseteq Q)\wedge$$
$$\wedge\mathrm{Ind}^1(B,Q)\Rightarrow\forall\beta\in B^{<\omega}\,Q\beta.$$

In EBIR_0, B is restricted to predicates of the form $B\beta\equiv B_0\beta\wedge B_1\alpha_1\wedge\cdots\wedge B_s\alpha_s$ where $\forall\beta\,B_0\beta$, $\forall\alpha_i\,B_i\alpha_i$ are closed. The schema $\mathbf{EBI}_1$ can in fact easily be seen to be derivable from EBIR (cf. LUCKHARDT (1973, p.147); or Subsection 2.4 of this paper); this is not clear if EBIR is replaced by EBIR_0 (cf. SCARPELLINI (1972, p.382 under (C))).

In KREISEL (1963, p.0.16) it is also stated that Howard verified that extended bar induction can be Dialectica interpreted by means of bar recursion in all finite types. A proof for $\mathbf{EBI}_{1,D}$ is to be found in LUCKHARDT (1973, p.150, VI and VII). More precisely, Luckhardt's results imply that (in the notation of TROELSTRA (1973, 1.6.12–15, 3.5.10))

$$\mathbf{HA}^{\omega}+\text{EXT-R}'+\mathrm{M}^{\omega}+\mathrm{IP}_0^{\omega}+\mathrm{AC}+\mathrm{BI}_{\mathrm{M}}^{\sigma}\vdash\mathrm{EBI}_{\sigma,\mathrm{D}},$$

where $\mathrm{BI}_{\mathrm{M}}^{\sigma}=\mathrm{BI}^{\sigma}$ is bar induction over objects of type σ, and EXT-R$'$ is the weak rule of extensionality as used in HOWARD (1968). From this the compatibility with certain continuity axioms (such as WC-N from KREISEL and TROELSTRA (1970)) can be obtained, interpreting $\mathbf{HA}^{\omega}$ in the model ICF of the intensional continuous functionals (cf. TROELSTRA (1973, 3.5.16)). This also follows from the fact that $\mathbf{EL}+\mathrm{GC}$ (generalized continuity, including "Brouwer's principle for functions")+c-EBI_1 can be functional-realized in $\mathbf{EL}+$c-EBI_1 which is a subsystem of classical analysis (cf. Subsection 6.3); thus a conjecture of SCARPELLINI (1972, p.382, (E)) is verified.

In contrast, we should mention here the result of SCARPELLINI (1973), where it is shown that a certain system based on intuitionistic logic and containing $\mathrm{BI}_{\mathrm{D}}^{\sigma}$ (i.e. with decidability instead of monotonicity), extensionality and some continuity axioms, is weaker than full classical analysis.

1.10. Description of the contents of the rest of the paper: outline of the proofs of the main results

In order to find an upper bound for the proof-theoretic strength of $\mathbf{EL}+\mathrm{EBI}_0$, we proceed as follows. We first attempt to embed $\mathbf{EL}+\mathrm{EBI}_0$ in a theory $\mathbf{CS}^*$ extending $\mathbf{CS}$ of KREISEL and TROELSTRA (1970) such that the elimination theorem extends as well. $\mathbf{CS}^*$ contains certain generalizations of the continuity axioms and the axiom of analytic data, and contains a choice-free part $\mathbf{IDB}^*$ extending $\mathbf{IDB}_1$. However, the extension $\mathbf{CS}^*$ of

CS which suggests itself in a natural way does not contain (at least not obviously) all instances of EBI_0, but only those instances $EBI_0(A)$ where Ax is restricted to a certain syntactically defined class (the *almost negative* formulae without free or bound choice variables). However, it becomes then possible to show afterwards that addition of full 1-EBI_0 to **CS*** does not change its proof-theoretic strength.

This is seen as follows. One observes that **IDB*** is sound for realizability by numbers (r-realizability), if we let the variables for constructive functions (=lawlike functions) range over the total recursive functions. As in the case of **HA**, we obtain a characterization theorem:

$$(1)\qquad \begin{cases} \mathbf{IDB}^* + ECT_0 \vdash A \Leftrightarrow \mathbf{IDB}^* \vdash \exists x(x\,\mathrm{r}\,A) \\ \mathbf{IDB}^* + ECT_0 \vdash A \leftrightarrow \exists y(y\,\mathrm{r}\,A) \end{cases}$$

(cf. Troelstra (1973, 3.2.30)); A is supposed to be closed w.r.t. constructive function variables). ECT_0, "extended Church's thesis" contains CT, and $y\,\mathrm{r}A$ is almost negative for all A. Now consider an instance of $EBI_0(A)$, A closed w.r.t. constructive function variables. By the elimination theorem, $\mathbf{CS}^* \vdash Ax \leftrightarrow \tau(Ax)$, $\tau(Ax) \equiv A'x$ a formula of $\mathfrak{L}(\mathbf{IDB}^*)$. By the characterization theorem (1), addition of ECT_0 to **IDB*** results in $A'x \leftrightarrow \exists y(y\,\mathrm{r}A'x)$, so $EBI_0(A)$ follows from $EBI_0(\exists y\, B(x,y))$ where $B(x,y) \equiv y\,\mathrm{r}A'x$; but Proposition 2.2 shows that $EBI_0(\exists y\, B(x,y))$ follows from $EBI_0(C)$ where $Cz \equiv B(j_1z, j_2z)$.

The case where A contains parameters for constructive functions reduces to the previous case by ECT_0, as already remarked in Subsection 1.8. Thus it remains to establish the proof-theoretic strength of **IDB***. This is done by comparing **IDB*** on the one hand with $\mathbf{ID}^i_{<\omega}$ (the intuitionistic theory of finitely iterated positive inductive definitions) and $\mathbf{ID}^i_{<\omega}(\mathfrak{O})$ (the intuitionistic theory of inductively defined tree classes of finite order) on the other hand.

$\mathbf{ID}^i_{<\omega} + ECT_0$ is reduced to $\mathbf{ID}^i_{<\omega}$ by realizability, and contains $\mathbf{IDB}^* + ECT_0$. $\mathbf{ID}^i_{<\omega}(\mathfrak{O})$ is contained in $\mathbf{IDB}^* + ECT_0$.

In order to bound the proof-theoretic strength of $\mathbf{EL} + EBI_0$ from below, one shows that a theory equivalent to **IDB*** is properly contained in $\mathbf{EL} + EBI_0$.

All this put together yields the first result in Subsection 1.6; the second result is a corollary.

As to the contents of the paper, Section 2 discusses several logical relationships such as the derivation of n-EBI'_0 from EBI_0; Section 3 is devoted to the description of **IDB***, comparison between **IDB***, $\mathbf{ID}^i_{<\omega}$, $\mathbf{ID}^i_{<\omega}(\mathfrak{O})$, and realizability for these systems; in Section 4, **CS*** is introduced and the elimination theorem proved; Section 5 reaps the harvest and states the main results in a precise form. Finally, in Section 6, we have collected some supplementary discussion and certain results obtained as byproducts of the developments in Sections 3–5.

1.11. Notational conventions. We shall use two disjoint sets of function variables: lawlike (function) variables (indicated by syntactic variables a,b,c,d) and choice (sequence) variables (syntactic variables $\alpha,\beta,\gamma,\delta$). "Lawlike" and "choice" refer to the fact that in familiar systems such as given in KREISEL and TROELSTRA (1970) the intended interpretation of the respective ranges is indeed the species (set) of lawlike functions and the set of choice sequences respectively. **EL**, elementary analysis, contains only *choice* variables, and a quantifier-free axiom of choice; $\mathbf{EL}_1$ is **EL** with AC_{01} added.

For a formal system **H**, $\mathfrak{L}(\mathbf{H})$ indicates the language of **H**. The system $\mathbf{ID}^i_{<\omega}$ contains all inductively defined sets P_A with matrix $A(P,x)$ which is classically (weakly) positive in P. These theories with their classical counterparts are studied in detail in the recent SIEG (1977) or POHLERS (1977).

2. Logical relationships, reductions

2.1. The present section is devoted to proving logical relationships and reductions between variants and special forms of EBI_0 and EBI_1, some of which will be essentially used in establishing our main results, whereas others are intended to provide background information only. Very simple, but essential is the following proposition (to which we already appealed in our outline, in Subsection 1.10).

2.2. Proposition. (i) *Let* $Bx \equiv \exists y A(j(x,y))$; *then* $EBI_0(B)$ *is derivable from* $EBI_0(A)$.

(ii) *Let* $Bn \equiv \exists y A(n,y)$, *then* $EBI_0''(B)$ *is derivable from* $EBI_0''(A')$ *for an* A' *of the form* $\forall m \leqslant t(n) A(m,t'(n,m))$.

(iii) *Similar assertions hold for* EBI_1: *e.g.* $EBI_1(B')$ *is derivable from* $EBI_1(A')$ *if* $B'\alpha \equiv \exists\beta A'(j(\alpha,\beta))$.

Proof. (i) Assume $\mathrm{Bar}(B,P)$, $\mathrm{Mon}(B,P)$, $\mathrm{Ind}(B,P)$ and put

$$Qn \equiv P(k_1^2 n)$$

where k_1^2, k_2^2 are primitive recursive functions satisfying

$$k_i^2\langle\rangle = 0, \qquad k_i^2(n * \hat{x}) = k_i^2(n) * \langle j_i(x)\rangle \qquad (i=1,2).$$

Now it is easy to verify that $\mathrm{Bar}(A,Q)$, $\mathrm{Mon}(A,Q)$, $\mathrm{Ind}(A,Q)$; by $EBI_0(A)$, $Q\langle\rangle$ and hence $P\langle\rangle$.

(ii) Let $\mathrm{Bar}'(B,P)$, $\mathrm{Mon}'(B,P)$, $\mathrm{Ind}'(B,P)$, $\mathrm{Tree}'(B)$ and put

$$A'n \equiv \forall m \leqslant k_1^2 n\, A\big(m, (k_2^2 n)_{\mathrm{lth}(m) \dot{-} 1}\big),$$

$$Qn \equiv P\big(k_1^2 n\big).$$

Then $\mathrm{Bar}'(A',Q)$, $\mathrm{Mon}'(A',Q)$, $\mathrm{Ind}'(A',Q)$, $\mathrm{Tree}'(A')$; thus $Q\langle\rangle$, hence $P\langle\rangle$.

2.3. Reduction of EBI_0 to inhabited predicates

The following proposition will often be used tacitly in the sequel; it permits us to restrict attention to $\mathrm{EBI}_0(A)$ with $\exists x(x \in A)$, with the advantage that restricted quantifiers $\forall x \in A$, $\exists x \in A$ may now be assumed to range over inhabited domains, so all the usual laws of intuitionistic predicate logic apply. We show

Proposition. (i) *To each predicate of natural numbers Ax we can find another predicate By, such that $\exists y\, By$ holds and as a formula has the same parameters as $\exists x Ax$, and such that* $\mathrm{EBI}_0(B)$ *implies* $\mathrm{EBI}_0(A)$.

(ii) *Similarly for* EBI_1 *instead of* EBI_0.

Proof. Assume the premises of $\mathrm{EBI}_0(A)$:

$$\mathrm{Bar}(A,P), \quad \mathrm{Mon}(A,P), \quad \mathrm{Ind}(A,P),$$

and put

$$Bx \equiv [x=0 \vee (x>0 \wedge A(x \dot{-} 1))].$$

For all α we put

$$\alpha^* \equiv \lambda x.(\alpha x \dot{-} 1),$$

and finally

$$Q(\bar{\alpha}x) \equiv [\exists y<x\ (\alpha y=0) \vee (\forall y<x\ (\alpha y \neq 0) \wedge P(\bar{\alpha}^* x))].$$

We first establish

$$\mathrm{Bar}(B,Q), \quad \text{i.e. } \forall \alpha \in B^\omega \exists x\, Q(\bar{\alpha}x).$$

To see this, take any $\alpha \in B^\omega$; if $\alpha 0=0$, we are done, for then $Q(\bar{\alpha}1)$ holds. If $\alpha 0=z_0+1$, we know that Az_0; define a continuous Γ such that

$$\begin{cases} \alpha x \neq 0 \rightarrow (\Gamma\alpha)x = \alpha x \dot{-} 1, \\ \alpha x = 0 \rightarrow (\Gamma\alpha)x = z_0, \end{cases}$$

then obviously $\forall \alpha \in B^\omega\ (\Gamma\alpha \in A^\omega)$. Since $\mathrm{Bar}(A,P)$, there is an x such that $P((\overline{\Gamma\alpha})x)$. If $\exists y<x\ (\alpha y=0)$, we have $Q(\bar{\alpha}x)$; if not, then also $Q(\bar{\alpha}x)$ since $(\overline{\Gamma\alpha})x = \bar{\alpha}^* x$ in this case.

It is straightforward to verify

$$\mathrm{Mon}(B,Q) \equiv \forall n \in B^{<\omega} \forall y \in B\ (Qn \rightarrow Q(n * \hat{y}))$$

(split into cases according to whether $\exists z<\mathrm{lth}(n)\ ((n)_z=0)$ or not, and whether $y=0$ or not); similarly one verifies $\mathrm{Ind}(B,Q)$. $\mathrm{EBI}_0(B)$ then yields $Q\langle\rangle$, and since $\exists y<\mathrm{lth}(0)\ ((0)_y=0)$ is false, we have $\forall y<\mathrm{lth}(0)\ ((0)_y \neq 0) \wedge P\langle\rangle$, hence $P\langle\rangle$. This establishes (i). Part (ii) is proved similarly.

Remark. The same trick can be applied to the *rules* corresponding to EBI_0 and EBI_1.

2.4. Proposition. EBI_0, EBI_1, EBI'_0, EBI''_0 *can be derived from the corresponding rules.*

(In LUCKHARDT (1973, p. 147) established for various forms of $\mathrm{EBI}_{\sigma,D}$.)

Proof. Consider an instance of $\mathrm{EBI}_0(A)$

(1) $\quad \mathrm{Bar}(A,P)\wedge\mathrm{Mon}(A,P)\wedge\mathrm{Ind}(A,P)\to P\langle\rangle$

and put

$$Bx\equiv C\wedge Ax, \qquad Qn\equiv C\to Pn$$

where

$$C\equiv\mathrm{Bar}(A,P)\wedge\mathrm{Mon}(A,P)\wedge\mathrm{Ind}(A,P).$$

Observe that now $\mathrm{Bar}(B,Q)$ is derivable: if $\alpha\in B^\omega$, then C holds, so $P(\bar{\alpha}x)$ for some x, and thus $C\to P(\bar{\alpha}x)$, i.e. $Q(\bar{\alpha}x)$. $\mathrm{Mon}(B,Q)$ is immediate, and $\mathrm{Ind}(B,Q)$ is derivable as well. Rule-EBI_0 then yields $Q\langle\rangle$, hence $C\to P\langle\rangle$, i.e. (1).

Remark. Note that an instance of $\mathrm{EBI}_0(A)$ *without* parameter in $\forall x\,Ax$ may be reduced to an instance of Rule-$\mathrm{EBI}_0(B)$ for a B such that $\forall x\,Bx$ *does* contain parameters.

2.5. Reduction of n-EBI'_0 to EBI_0

Lemma. *Assume* $\mathrm{Tree}(A)$, $\mathrm{Tree}(B)$ *to be derivable, and moreover let* ξ *be a definable function (possibly with parameters) such that*

(1) $\quad \xi\langle\rangle=\langle\rangle,$

(2) $\quad \forall nx\,\exists y\,(\xi(n*\hat{x})=(\xi n)*\hat{y}),$

(3) $\quad Bn\to A(\xi n),$

(4) $\quad Bn\wedge A(\xi n*\hat{x})\to\exists y\,(\xi(n*\hat{y})=\xi n*\hat{x}\wedge B(n*\hat{y})),$

are all derivable. Then, using only axioms and rules of **EL**, $\mathrm{EBI}'_0(A)$ *is derivable from* $\mathrm{EBI}'_0(B)$.

Proof. Note that ξ induces a mapping ξ' from $\bar{B}^\omega$ onto $\bar{A}^\omega$ given by

$$\xi'\alpha=\lambda x.(\xi\bar{\alpha}(x+1))_x.$$

Assume $\mathrm{EBI}'_0(B)$ and

$$\mathrm{Bar}'(A,P),\quad \mathrm{Mon}'(A,P),\quad \mathrm{Ind}'(A,P),$$

and put $Qn\equiv P(\xi n)$. Then it is not difficult to see that $\mathrm{Bar}'(B,Q)$, $\mathrm{Mon}'(B,Q)$, $\mathrm{Ind}'(B,Q)$. As to $\mathrm{Bar}'(B,Q)$, let $\alpha\in\bar{B}^\omega$, then $\forall x\,B(\bar{\alpha}(x+1))$, hence $\forall x\,A(\xi\bar{\alpha}(x+1))$ which is equivalent to $\forall x\,A(\overline{(\xi'\alpha)}(x+1))$; since $\xi'\alpha\in\bar{A}^\omega$ if $\alpha\in\bar{B}^\omega$ (by (3), (2)), it follows that for some x $P(\overline{\xi'\alpha}(x))$ and thus if $x=0$, $P\langle\rangle$, and therefore also $Q\langle\rangle$; if $x>0$, $Q(\bar{\alpha}x)$ follows.

$\mathrm{Mon}'(B,Q)$ and $\mathrm{Ind}'(B,Q)$ are left to the reader. Application of $\mathrm{EBI}'_0(B)$ yields $Q\langle\,\rangle$, hence $P\langle\,\rangle$.

Theorem. $\text{n-EBI}'_0(A)$ *is derivable from* $\mathrm{EBI}_0(C)$ *for suitable* C.

Proof. The proof is carried out in two steps.

(A) We reduce $\text{n-EBI}'_0(A)$ to $\text{n-EBI}_0(A)$, recalling that

$$\text{n-EBI}_0(A) = \text{n-EBI}'_0(A^{<\omega});$$

and thus with $Bn \equiv A^{<\omega}n$ we have

$$B\langle n_0,\dots,n_u\rangle \equiv An_0 \wedge An_1 \wedge \dots \wedge An_u.$$

We define ξ by $\xi\langle\,\rangle = \langle\,\rangle$,

$$\xi\langle n_0,\dots,n_u\rangle = \begin{cases} n_u & \textit{if } \mathrm{lth}(n_u)=u+1 \wedge \xi\langle n_0,\dots,n_{u-1}\rangle \prec n_u, \\ \xi\langle n_0,\dots,n_{u-1}\rangle * \langle a\xi\langle n_0,\dots,n_{u-1}\rangle\rangle & \textit{otherwise.} \end{cases}$$

Note: $\mathrm{lth}(\xi n) = \mathrm{lth}(n)$ for all n.

(1) holds for ξ by definition, (2) is obvious in combination with the preceding remark. $Bn \to A(\xi n)$ follows by induction on $\mathrm{lth}(n)$, in combination with $\mathrm{Tree}(A)$.

Now let $B\langle n_0,\dots,n_u\rangle, A(\xi n * \hat{x}), n = \langle n_0,\dots,n_u\rangle$. Take $\xi n * \hat{x}$ for y; then $B\langle n_0,\dots,n_u,y\rangle$ by definition, and $\xi\langle n_0,\dots,n_u,y\rangle = y = \xi n * \hat{x}$ by the first clause in the definition of ξ. This establishes (4).

(B) We reduce $\text{n-EBI}_0(B)$ to $\mathrm{EBI}_0(C)$ for arbitrary B and suitable C. Let $B(x) \equiv B(x,y) \equiv B_y(x)$, $\forall xy\, B(x,y)$ closed. Take

$$Cz \equiv B(j_1z, j_2z); \qquad B_1n \equiv B^{<\omega}n, \quad C_1n \equiv C^{<\omega}n.$$

We define ξ such that $\xi\langle\,\rangle = \langle\,\rangle$ and

$$\xi(n*\hat{z}) = \begin{cases} \xi n * \langle j_1 z\rangle & \textit{if } j_2 z = y, \\ \xi n * \langle z_0\rangle & \textit{if } j_2 z \neq y, \end{cases}$$

where z_0 is fixed such that $B_y(z_0)$. Now (1)–(4) are readily verified with B_1, C_1 for B, A respectively.

Remark. This method of reduction does not work for EBI_1: the analogues of the ξ used in the proof above would not be effectively defined.

2.6. Proof-theoretic reduction of EBI''_0

For the proof-theoretic reduction of EBI''_0 to $\text{l-EBI}'_0$ we need the following

Lemma. *Assume the following instance* APC(A) *of the axiom of partial choice*

$$\forall n\left[An \rightarrow \exists y\, A(n*\hat{y})\right] \rightarrow$$

$$\exists a \forall n\left[An \rightarrow \exists! y\,(a(n,y)\neq 0) \wedge \forall y\,(a(n,y)\neq 0 \rightarrow A(n*\langle a(n,y) \dot{-} 1\rangle))\right]$$

(A *fixed*); *then* $\mathrm{EBI}_0''(A)$ *is derivable from* 1-EBI_0' (B) for suitable B.

Proof. Assume Tree$'(A)$, Bar$'(A,P)$, Mon$'(A,P)$, Ind$'(A,P)$. From Tree$'(A)$, together with the available instance APC(A) we find a function a such that for all n

$$An \rightarrow \exists! y\,(a(n,y)\neq 0) \wedge \forall y\,(a(n,y)\neq 0 \rightarrow A(n*\langle a(n,y) \dot{-} 1\rangle)).$$

Also

$$A(n*m) \rightarrow An, \quad A\langle\,\rangle.$$

We now define:

$$Bn \equiv n = \langle\,\rangle \vee \left[n \text{ is of the form } n_1^{x_1+1} * n_2^{x_2+1} * n_3^{x_3+1} * \cdots * n_p^{x_p+1} \wedge\right.$$

$$\wedge \forall i\,(0 < i \leqslant p \rightarrow \mathrm{lth}(n_i) = i \wedge An_i) \wedge$$

$$\wedge \forall i\,(0 < i < p \rightarrow a(n_i, x_i) \neq 0 \wedge (n_i \prec n_{i+1})) \wedge$$

$$\left.\wedge \forall y < x_p \quad (a(n_p, y) = 0)\right].$$

Note that $a(n_p, x_p) = 0$ is permitted. We have Tree(B), since obviously $B\langle\,\rangle$ and $B(n*m) \rightarrow Bn$; and given an $n = n_1^{x_1+1} * \cdots * n_p^{x_p+1}$ as indicated, we either have $a(n_p, x_p) = 0$, in which case $Bn \rightarrow B(n*\hat{n}_p)$, or $a(n_p, x_p) \neq 0$, in which case $Bn \rightarrow B(n*\langle n_p * \langle a(n_p, x_p) \dot{-} 1\rangle\rangle)$; $B\langle\,\rangle \rightarrow B\langle n_1\rangle$ for a suitable n_1. We define a function ψ, "the compression function" by

$$m = \langle\,\rangle \rightarrow \psi m = \langle\,\rangle,$$

$$n = n_1^{x_1+1} * n_2^{x_2+1} * \cdots * n_p^{x_p+1} \wedge \forall i\,(0 < i < p \rightarrow n_i \prec n_{i+1}) \rightarrow \psi n = n_p.$$

Now we put

$$Qn \equiv P\psi n.$$

We have to verify Bar$'(B,Q)$, Mon$'(B,Q)$, Ind$'(B,Q)$.

Let $\alpha \in B^\omega$. Each $\bar{\alpha}x$ will be of the form $n_1^{x_1+1} * \cdots * n_p^{x_p+1}$, $\mathrm{lth}(n_i) = i$; p will increase indefinitely with increasing x, as follows from the construction. Hence there is a unique $\beta \in A^\omega$ with $\bar{\beta}p = n_p$; and thus we always find $\bar{\alpha}x$ with $P(\psi\bar{\alpha}x)$, and therefore $Q(\bar{\alpha}x)$. This establishes Bar$'(B,Q)$.

Now consider n, $n*\hat{x}$ such that $Bn, B(n*\hat{x}), Qn$; $Qn \equiv P\psi n, \psi(n*\hat{x}) \succcurlyeq \psi n$, therefore $Q(n*\hat{x})$. Thus $\mathrm{Mon}'(B,Q)$.

Finally, let Bn, and let

$$(1) \qquad \forall x(B(n*\hat{x}) \to P(\psi(n*\hat{x}))).$$

We must distinguish two cases. Let again $n = n_1^{x_1+1}* \cdots *n_p^{x_p+1}$, and assume $B(n*\hat{x})$. Then either $a(n_p, x_p) = 0$, and in this case $x = n_p$ whenever $B(n*\hat{x})$; or $a(n_p, x_p) \neq 0$, and then, whenever $B(n*\hat{x})$, x is of the form $n_p*\hat{y}$ and $A(n_p*\hat{y})$.

In the first case, $P\psi(n*\hat{n}_p)$ follows by (1), and thus $P\psi n$. In the second case, we have $P(n_p*\hat{y})$ for all y such that $A(n_p*\hat{y})$, hence Pn_p, i.e. $P(\psi n)$ (with help of $\mathrm{Ind}'(A,P)$). And if $n = \langle\rangle$, $B(\langle\langle y\rangle\rangle)$ holds for precisely those y such that $A\langle y\rangle$, and hence $P(\langle y\rangle)$ for all such y by (1), and once again by $\mathrm{Ind}'(A,P)$ we have $P\langle\rangle$. Thus, if Bn, then

$$\forall x(B(n*\hat{x}) \to P(\psi(n*\hat{x}))) \to P\psi n,$$

i.e. we have established $\mathrm{Ind}'(B,Q)$. Application of $1\text{-}\mathrm{EBI}_0'(B)$ yields $Q\langle\rangle$, i.e. $P\langle\rangle$; hence we have shown $\mathrm{EBI}_0''(A)$.

3. The system IDB*

3.1. The language of IDB*

Let us use $\mathfrak{L}$ for the language of **EL** as described in KREISEL and TROELSTRA (1970). This is identical with the language of **EL** as used in this paper (cf. Subsection 1.11) except that in KREISEL and TROELSTRA (1970) constructive function variables are used instead of choice variables. $\mathfrak{L}$ contains numerical variables and variables for constructive functions (a, b, c, d, possibly with sub- or superscripts). Let $\mathfrak{L}[P_1, \ldots, P_n]$ be $\mathfrak{L}$ with some predicate letters $P_1, \ldots, P_n$ for unary predicates of constructive functions added.

We now inductively define a set $\mathfrak{C}$ of constants $K_{[A]}$, and a notion of *depth* $d(A)$, as follows.

(i) Let $A(x)$ be an almost negative formula of $\mathfrak{L}$, not containing parameters besides x. $d(A) = 0$, and $[A]$ is a basic type of depth 0; we introduce constants $K_{[A]}$ for unary predicates of constructive functions, for each $[A]$ with $d(A) = 0$.

(ii) Let $A(x, P_1, \ldots, P_n)$ be any almost negative formula of $\mathfrak{L}[P_1, \ldots, P_n]$ (i.e. built from prime formulae of the form $[t = s]$ or $P_i t$ $(1 \leqslant i \leqslant n)$ or $\exists x[t = s]$ by means of $\to, \forall, \wedge$). Let $K_{[A_1]}, \ldots, K_{[A_n]}$ be a set of predicate constants introduced before, and let $\max\{d(A_1), \ldots, d(A_n)\} = p$. If $Bx \equiv A(x, K_{[A_1]}, \ldots, K_{[A_n]})$, then $d(B) = p + 1$, and $[B]$ is a type of depth $p + 1$, and we add a unary predicate constant $K_{[B]}$ for constructive functions to $\mathfrak{C}$.

We shall use $\sigma,\sigma_1,\sigma_2,\ldots,\tau,\tau_1,\ldots,\rho,\rho_1,\ldots$ to range over types. For the type $[x=x]$ of depth 0 we write N. If $Ax \equiv A_1(j_1x)\wedge A_2(j_2x)$, we identify $[A]$ with the cartesian product $[A_1]\times[A_2]$ of types. We note that our set of types is indeed closed under $\times$.

We shall include $K_{[A]}$ only in the language $\mathfrak{L}^*$ of **IDB*** provided we have first established $\exists x Ax$. (This means no essential loss of generality, cf. Propostion 2.3.)

We may (conservatively) also add a special sort of variables ($e^\sigma, f^\sigma, e_1^\sigma$ etc.) for elements of K_σ; this in fact makes K_σ redundant as a primitive since it may be rendered by $K_\sigma a \equiv_{\mathrm{def}} \exists e^\sigma(a=e^\sigma)$.

It is also convenient to have separate variables $n^\sigma, m^\sigma, m_1^\sigma$ etc. ranging over elements of $A^{<\omega}$, and variables $a^\sigma, b^\sigma,\ldots$ ranging over A^ω, where $[A]=\sigma$.

3.2. Axiomatization of IDB*

The axiomatization is completely similar to $\mathbf{IDB}_1$ except for two points.

First of all, the axiom schema AC-NF is replaced by the more general *axiom of partial choice* APC:

APC $\quad \forall x\left[Bx\to\exists a\, C(x,a)\right]\to$

$$\exists bc\forall x\left[Bx\to\exists! y\,(b(x,y)=0)\wedge\forall y\,(b(x,y)=0\to C(x,(c)_{x,y}))\right],$$

where B is almost negative and

$$(c)_{x,y}\equiv\lambda z.c(x,y,z).$$

Secondly, now we not only have the axioms corresponding to the generalized inductive definition of K, but more generally, if $K_{[A]}$ is a constant of **IDB***, and we put

$$A^*(a,Q)\equiv\exists y\,(a=\lambda x.y+1)\vee(a0=0\wedge\forall y\in A\ (\lambda n.a(\hat{y}*n)\in Q))$$

(Q any formula of **IDB***), we shall assume the following axioms for $K_{[A]}$:

$\mathrm{K}_{[A]}1 \quad a=\lambda x.y+1\to a\in K_{[A]}$,

$\mathrm{K}_{[A]}2 \quad a0=0\wedge\forall y\in A\ (\lambda n.a(\hat{y}*n)\in K_{[A]})\to a\in K_{[A]}$,

which may be expressed more concisely as

$\mathrm{K}_{[A]}1\text{–}2 \quad A^*(a,K_{[A]})\to a\in K_{[A]}$,

and the induction principle for $K_{[A]}$ which takes the form

$\mathrm{K}_{[A]}3 \quad \forall a\left[A^*(a,Q)\to Qa\right]\to\forall a\left[K_{[A]}a\to Qa\right]$,

Q any formula of **IDB***.

Intuitively, the elements of $K_{[A]}$ are (inductively defined) neighbourhood functions representing continuous functionals on trees A^{ω}.

Observe that $K_{[N]} \equiv K$ as introduced in $\mathbf{IDB}_1$. In $\mathbf{IDB}_1$, K serves a double purpose: its elements code continuous functionals both of type $N^N \to N$ and of type $N^N \to N^N$. Similarly for $\sigma = [A]$, $\tau = [B]$ we want to code continuous functionals of types $A^{\omega} \to N$ and $A^{\omega} \to B^{\omega}$ (one could also ask for $A^{\omega} \to B$, but this turns out not to be needed). K_σ cannot serve a double purpose as K did, but if we let K_σ itself in the obvious way code continuous functionals of type $A^{\omega} \to N$, we can use the explicitly defined $K_{\sigma,\tau}$:

$$a \in K_{\sigma,\tau} \equiv_{\mathrm{def}} (a0 = 0) \wedge \forall x\, (\lambda n.a(\hat{x} * n) \in K_\sigma) \wedge$$
$$\wedge \forall n\, (an \neq 0 \to an \dot{-} 1 \in B)$$

to code continuous functionals of type $A^{\omega} \to B^{\omega}$. Note that $K_{\sigma,\tau} \subseteq K_{\sigma,N} \subseteq K_\sigma$ (trivially).

3.3. Closure conditions on the sets $K_\sigma, K_{\sigma,\tau}$

The treatment is similar to the treatment of closure conditions on K (Kreisel and Troelstra, 1970, 3.2.), but we use the occasion to give a slightly more economical and perspicuous treatment.

Definition. $a \in M^\sigma \equiv \forall n^\sigma m\, (an \neq 0 \to an = a(n * m))$,

$$b \in K_\sigma^a \equiv \forall n^\sigma\, (an \neq 0 \to \lambda m.b(\langle an \dot{-} 1 \rangle * m) \in K_\sigma),$$
$$b \in K_{\sigma,\tau}^a \equiv \forall n^\sigma\, (an \neq 0 \to \lambda m.b(\langle an \dot{-} 1 \rangle * m) \in K_{\sigma,\tau}).$$

The definition of $a//b$ in Kreisel and Troelstra (1970, 3.2.8) should read

$$a//b \equiv \lambda n.\mathrm{sg}(\mathrm{tl}\, n) \cdot \mathrm{sg}(a(\mathrm{tl}\, n)) \cdot b(\langle h(a, \mathrm{tl}\, n) \dot{-} 1 \rangle * n).$$

We collect everything in a single lemma:

Lemma. (i) *Induction over unsecured sequences. Writing* $y \in \sigma$ *for* Ay, *if* $\sigma \equiv [A]$:

$$\forall e^\sigma \left[\forall n^\sigma (en \neq 0 \to Qn) \wedge \forall n^\sigma (\forall y \in \sigma\, Q(n * \hat{y}) \to Qn) \to Q\langle\,\rangle\right],$$

for all Q *in the language of* $\mathbf{IDB}^*$.

(ii) *Majorizing: for* $b \in K_\sigma$

$$\forall a \in M^\sigma\, (\forall n\, (bn \neq 0 \to an \neq 0) \to a \in K_\sigma).$$

(iii) $K_\sigma a \to K_\sigma(\lambda n.\mathrm{sg}(an))$,

$$K_\sigma a \to K_\sigma(\lambda n.Sy \cdot \mathrm{sg}(an)).$$

(iv) $K_\sigma a \to K_\sigma(\lambda m.a(n^\sigma * m))$, *for all* n^σ.

(v) $K_\sigma(\lambda n.\mathrm{sg}(n))$,

$$K_\sigma(\lambda n.\mathrm{sg}(\mathrm{lth}(n) \dot{-} x)),\ \textit{for all } x.$$

(vi) $K_\sigma a_0 \wedge \cdots \wedge K_\sigma a_p \to K_\sigma(\lambda n. a_0 n \cdot a_1 n \cdot \cdots \cdot a_p n)$.
(vii) $K_\sigma a \wedge K_\sigma^a b \to K_\sigma(\lambda n.\mathrm{sg}(an)\cdot b(\langle an \dot{-} 1\rangle * n))$.
(viii) *Let h be defined as in* KREISEL *and* TROELSTRA (1970, 3.2.3). *Then*

$\forall a \in K_\sigma\ (\lambda n.h(a,n) \in K_\sigma)$,

$\forall a \in K_{\sigma,\tau}\ (\lambda n.h(a,n) \in K_{\sigma,N})$.

(ix) *Let /, // be defined as in* KREISEL *and* TROELSTRA (1970), *then*

$\forall a \in K_\sigma \forall b \in K_\sigma^{\lambda n.h(a,n)}\ (a/b \in K_\sigma)$,

$\forall a \in K_\sigma \forall b \in K_{\sigma,\tau}^{\lambda n.h(a,n)}\ (a//b \in K_{\sigma,\tau})$.

(x) *Let* $a \in K_\sigma$, $b \in K_{\sigma,\tau}$, *and let c satisfy*

$$\begin{cases} an^\sigma \neq 0 \wedge \forall m < n^\sigma(am = 0) \to c(n*m') = b(\langle n^\sigma\rangle * m'), \\ an^\sigma = 0 \to cn = 0. \end{cases}$$

Then $c \in K_\sigma$.

(xi) *Let k be defined as in* KREISEL *and* TROELSTRA (1970, 3.2.6): $k_n = \lambda m.k(n,m)$. $k_n \in K_{\sigma,N}$ *for all* σ, *and if* $n = n^\sigma$: $k_n \in K_{\sigma,\sigma}$ *for all* σ.
(xii) *Let* $a;b$, $a:b$ *be defined as in* KREISEL *and* TROELSTRA (1970, 3.2.11), *then*

$\forall a \in K_\tau \forall b \in K_{\sigma,\tau}(a;b \in K_\sigma)$,

$\forall a \in K_{\tau,\rho} \forall b \in K_{\sigma,\tau}(a:b \in K_{\sigma,\rho})$.

Proof. (i) Completely similar to KREISEL and TROELSTRA (1970, 3.2.1).

(ii) Induction over K_σ w.r.t. b (i.e. an application of $K_\sigma 3$).

(iii) Immediate with (ii).

(iv) Induction over K_σ w.r.t. a.

(v) First half directly (single application of $K_\sigma 2$, plus $K_\sigma 1$); second half by ordinary induction w.r.t. x.

(vi) As for KREISEL and TROELSTRA (1970, 3.2.2(iv)).

(vii) We apply $K_\sigma 3$ w.r.t. a to:

$$\forall b \in K_\sigma^a(\lambda n.\mathrm{sg}(an)\cdot b(\langle an \dot{-} 1\rangle * n) \in K_\sigma).$$

This is immediate for $a = \lambda n.Sy$. So let now

$$(1) \qquad \begin{cases} a0 = 0, \\ \forall x \in \sigma \forall b' \in K_\sigma^{\lambda n.a(\hat{x}*n)}\ (\lambda n.\mathrm{sg}a(\hat{x}*n)\cdot b'(\langle a(\hat{x}*n) \dot{-} 1\rangle * n) \in K_\sigma), \end{cases}$$

$$(2) \qquad b \in K_\sigma^a.$$

Therefore, if $x = an \dot{-} 1$, $an \neq 0$ for some n, then

$$\forall y \in \sigma(\lambda m.b(\hat{x}*\hat{y}*m) \in K_\sigma),$$

and thus for all $y \in \sigma$ the function b_y^* given by

$$b_y^* 0 = 0, \qquad b_y^*(\hat{x}*m) = b(\hat{x}*\hat{y}*m)$$

is an element of K_σ^a, hence of $K_\sigma^{\lambda n.a(\hat{z}*n)}$ for all $z\in\sigma$. Thus

$$\lambda n.\mathrm{sg}(a(\hat{x}*n))\cdot b_x^*(\langle a(\hat{x}*n)\dot{-}1\rangle*n)=$$

$$\lambda n.\mathrm{sg}(a(\hat{x}*n))\cdot b(\langle a(\hat{x}*n)\dot{-}1\rangle*\hat{x}*n)\in K_\sigma$$

(induction hypothesis (1)). Since $\mathrm{sg}(a0)\cdot b(\langle a0\dot{-}1\rangle*n)=0$, we are done.

(viii) Immediate by (ii).

(ix) Easy with (vii) and (viii).

(x) Use (i) applied to

$$Qn\equiv\lambda m.c(n*m)\in K_\sigma.$$

(xi) $k_n0=0$, $k_n(\hat{x}*m)\neq 0\leftrightarrow\mathrm{sg}(\mathrm{lth}(m)\dot{-}x)\neq 0$, and thus by (ii) and (v) $\lambda m.k_n(\hat{x}*n)\in K_\sigma$; then apply $K_\sigma 2$.

(xii) As in KREISEL and TROELSTRA (1970, 3.2.12) except that the appeal to the unnatural 3.2.2(x) now can be replaced by an application of (x).

3.4. The system IDB** and the recursive versions of K_σ

Definition. Let **IDB**** be the formal system obtained by replacing the axiom of partial choice APC in the description of **IDB*** by

QF-AC $\qquad \forall x\exists y A(x,y)\to\exists a\forall x A(x,ax)\quad$ (*A quantifier-free*).

Definition. Let σ be any type in **IDB*** or **IDB****. We define the restriction K_σ^{r} of K_σ to its recursive elements:

$$a\in K_\sigma^{\mathrm{r}}\equiv_{\mathrm{def}}\exists x(\{x\}=a\wedge K_\sigma a),$$

and the corresponding set of Gödel numbers

$$x\in K_\sigma^{\mathrm{R}}\equiv_{\mathrm{def}}\{x\}\in K_\sigma^{\mathrm{r}}\equiv\{x\}\in K_\sigma.$$

Finally we put, for $\sigma=[A]$

$$A_{\mathrm{PR}}(Q,y)\equiv_{\mathrm{def}}\exists x\forall z(\{y\}(z)=x+1)\vee$$
$$(\{y\}(0)=0\wedge\forall z\in\sigma(\Lambda n.\{y\}(\hat{z}*n)\in Q)),$$

$$A_{\mathrm{R}}(Q,y)\equiv_{\mathrm{def}}\forall u\exists v\,T(y,u,v)\wedge A_{\mathrm{PR}}(Q,y).$$

We can prove in **IDB****:

(1) $\qquad A_{\mathrm{R}}(K_\sigma^{\mathrm{R}},y)\to K_\sigma^{\mathrm{R}}y,$

(2) $\qquad \forall y(A_{\mathrm{R}}(Q,y)\to Qy)\to\forall y(K_\sigma^{\mathrm{R}}y\to Qy)\qquad$ for all Q in $\mathfrak{L}$(**IDB****).

Conversely, let **H** be the system obtained from **IDB**** by replacing all K_σ by K_σ^{R}, and adopting (1) and (2); then on assumption of Church's thesis in the form

CT $\qquad \forall a\exists x\forall y\exists z(Txyz\wedge ay=Uz)$

we can explicitly define K_σ by

$$K_\sigma a\equiv\exists x(a=\{x\}\wedge K_\sigma^{\mathrm{R}}x)$$

and prove the axioms for K_σ. The proofs are completely routine. K_σ^{R} consists of Gödel numbers of *total* recursive functions. A corresponding predicate K_σ^{PR} of partial recursive functions is characterized by

$$(3) \qquad A_{\mathrm{PR}}(K_\sigma^{\mathrm{PR}},y)\to K_\sigma^{\mathrm{PR}}y,$$

$$(4) \qquad \forall y\,(A_{\mathrm{PR}}(Q,y)\to Qy)\to\forall y\,(K_\sigma^{\mathrm{PR}}y\to Qy)$$

for all Q in $\mathfrak{L}$(**IDB****).

K_σ^{R} and K_σ^{PR} can be explicitly defined in terms of each other. On the one hand, K_σ^{R} is simply the restriction of K_σ^{PR} to total elements, and the axioms (1), (2) become provable from (3), (4); on the other hand, K_σ^{PR} may be explicitly defined from K_σ^{R}, and (3), (4) proved as follows. We put

$$y_m\equiv\Lambda x.\{y\}(m*x),$$

$$T'(y,m,u)\equiv\forall n\preccurlyeq m\,\exists v\leqslant u\;T(y_n,0,v)$$

and we associate to each partial recursive $\{y\}$ a total $\{y^*\}$ given by

$$\{y^*\}(n)=\begin{cases} z+1 & \text{if } \exists m\preccurlyeq n\,\exists u\leqslant \mathrm{lth}(n)\,\big[\,T'(y,m,u)\wedge \\ & \quad\wedge\forall v\leqslant u\,(T(y_m,0,v)\to Uv=z+1)\wedge \\ & \quad\wedge\forall m'\prec m\,\forall v'\leqslant u\,(T(y_{m'},0,v')\to Uv'=0)\big], \\ 0 & \textit{otherwise}. \end{cases}$$

We define "*y' majorizes y*" and "*y is σ-consistent*" by

$$\mathrm{maj}(y',y)\equiv\forall n^\sigma(\{y\}(n)\neq 0\to$$
$$\exists m\preccurlyeq n\,(\{y\}(m)=\{y'\}(n)\wedge\forall m'\prec m\,(\{y\}(m')=0))$$
$$\mathrm{cons}_\sigma(y)\equiv\forall n^\sigma m(\{y\}(n)\neq 0\to\{y\}(n*m)=\{y\}(n)).$$

It is easily seen that for each y,y^* majorizes y, is N-consistent and total. Now we have the following

Theorem. *If we put*

$$y\in K_\sigma^{\mathrm{PR}}\equiv y^*\in K_\sigma^{\mathrm{R}}\wedge\mathrm{cons}_\sigma(y),$$

then the axioms (3), (4) *become provable from the axioms for* K_σ^{R}.

Proof. Verification of the closure conditions is straightforward. In order to establish (4) for some Q, apply the schema (2) to

$$Q^*y\equiv\forall y'(\mathrm{cons}_\sigma(y')\wedge\mathrm{maj}(y,y')\to Qy')\wedge\forall u\,\exists v\,Tyuv,$$

which yields $\forall y\,(K_\sigma^{\mathrm{R}}y\to Q^*y)$, hence for any $y\in K_\sigma^{\mathrm{PR}}$ we have $K_\sigma^{\mathrm{R}}y^*$, i.e. Q^*y^*, and therefore Qy.

Remark. The preceding method is a slightly more involved variant of the well-known delaying technique, as used for example in transforming recursive neighbourhood functions into primitive recursive ones (see e.g. Kreisel and Troelstra (1970, p. 298)).

3.5. Consistency of IDB* with Church's thesis

This is demonstrated by extending Kleene's realizability for numbers to **IDB***; this requires a relatively minor extension of the corresponding treatment of $\mathbf{IDB}_1$ (Kreisel and Troelstra, 1970, 3.7.1), which may be described as follows. Let B be any sentence of **IDB***, and let B^{r} be obtained from B by relativization of the function quantifiers to total recursive functions; i.e. $\forall a$, $\exists a$ are replaced by $\forall x \in R$, $\exists x \in R$ where $x \in R \equiv \forall y \exists z\, Txyz$, and occurrences of a in prime formulae are replaced by $\{x\}$; prime formulae $K_\sigma(\{x\})$ may be read as $K_\sigma^{\mathrm{R}}x$.

$y \mathrm{r} B$ is now defined essentially as $y \mathrm{r} B^{\mathrm{r}}$ for arithmetic, with an extra stipulation

$$y \mathrm{r} K_\sigma^{\mathrm{R}} x \equiv K_\sigma^{\mathrm{R}} x.$$

(Cf. the treatment in Kreisel and Troelstra (1970, 3.7) and Troelstra (1973, 3.2.29).)

Realizability for *formulae* is defined relative to an assignment of (Gödel numbers of) total recursive functions to constructive function variables free in the formula. E.g. $B(x,a)$ under the assignment $a \mapsto \{y\}$ (y code of a total recursive function) is realized by z iff $z \mathrm{r} B(x,\{y\})$.

Note that almost negative formulae may be defined as before: constructed from prime formulae (i.e. $t=s$ or $K_\sigma \varphi$) and formulae $\exists x(t=s)$ with help of $\forall, \wedge, \rightarrow$. As in the case of realizability for **HA**, $x \mathrm{r} A$ (A not containing free constructive function variables) is almost negative, and

$$\mathbf{IDB^{**}} + \mathrm{ECT}_0 \vdash A \leftrightarrow \exists x (x \mathrm{r} A)$$

for all sentences A. In fact, more is true: for formulae A

$$\mathbf{IDB^{**}} + \mathrm{ECT}_0 \vdash \forall (A \leftrightarrow \exists x (x \mathrm{r} A))$$

where $\forall$ indicates the universal closure.

In order to obtain

$$\mathbf{IDB^{*}} + \mathrm{ECT}_0 \vdash A \Leftrightarrow \mathbf{IDB^{**}} \vdash \exists x (x \mathrm{r} A)$$

for all sentences A, we have to verify the realizability of the principles of generalized inductive definition $\mathrm{K}_\sigma 1$–3 and the axiom of partial choice APC.

To facilitate the verification, we use the following observation. Let Ax be almost negative. On the one hand, we may treat quantification over $\{x: Ax\}$ as a new primitive, a restricted quantifier $\forall x \in A$ with realizability clause

(1) $$v \mathrm{r} \forall x \in A\ (Bx) \equiv \forall x \in A\ (!\{v\}(x) \wedge \{v\}(x) \mathrm{r} Bx);$$

on the other hand, realizability in this sense is equivalent to realizability in the usual sense applied to the un-abbreviated form $\forall x(Ax \rightarrow Bx)$

(2) $$u\,\mathrm{r}\,\forall x(Ax \rightarrow Bx) \equiv$$
$$\forall x\left[!\{u\}(x) \wedge \forall y(y\,\mathrm{r}\,Ax \rightarrow !\{\{u\}(x)\}(y) \wedge \{\{u\}(x)\}(y)\,\mathrm{r}\,Bx)\right].$$

The equivalence of the two definitions follows by observing that for any u satisfying (2), $\Lambda v.\{u\}(x, \psi_A)$ satisfies (1), and conversely, if v satisfies (1), then $\Lambda x \Lambda y.\{v\}(x)$ satisfies (2). Here ψ_A is the partial recursive function such that (TROELSTRA, 1973, 3.2.11)

$$\exists y(y\,\mathrm{r}\,Ax) \leftrightarrow !\psi_A(x) \wedge \psi_A(x)\,\mathrm{r}\,Ax.$$

Definition (1) is often much easier to manage; under this definition, quantification over $\{x: Ax\}$ behaves just like quantification over N w.r.t. realizability; and thus the verification of the realizability of $\mathrm{K}_\sigma 1$–3 which reduces to the realizability of the induction axioms for K_σ^{R} (or K_σ^{PR}) becomes completely similar to the verification of the realizability of K1–3 in KREISEL and TROELSTRA (1970, 3.7).

We also have to verify the realizability of APC. This is easily done by showing it to be derivable from ECT_0. Summing up, we have

Theorem.

$$\mathbf{IDB^{**}} + \mathrm{ECT}_0 \vdash A \leftrightarrow \exists x(x\,\mathrm{r}\,A),$$

$$\mathbf{IDB^{*}} + \mathrm{ECT}_0 \vdash A \Leftrightarrow \mathbf{IDB^{**}} \vdash \exists x(x\,\mathrm{r}\,A)$$

for all sentences A *of* **IDB***.

3.6. Embedding $\mathbf{ID}^{\mathrm{i}}_{<\omega}(\mathcal{O})$ in IDB*

It will be immediately clear that on assumption of ECT_0, **IDB*** is properly contained in the intuitionistic theory of finitely iterated positive inductive definitions $\mathbf{ID}^{\mathrm{i}}_{<\omega}$ ($\mathbf{ID}^{\mathrm{i}}_{<}$ in the notation of SIEG (1977)); for a precise description of this theory we may refer to SIEG (1977).

On the other hand, the theory $\mathbf{ID}^{\mathrm{i}}_{<\omega}(\mathcal{O})$ (or $\mathbf{ID}^{\mathrm{i}}_{<}(\mathcal{O})$) of (inductively defined) finite tree classes is properly contained in **IDB***, as we shall now show.

Put $\mathcal{O}_0 = N$. For the nth tree class $\mathcal{O}_n$ we have the following formula $B_n(Q, x)$ expressing the closure conditions:

$$B_n(Q,x) \equiv x = 0 \vee$$
$$\exists x_0 x_1\left[x = \langle x_0, x_1, n\rangle \wedge x_1 < n \wedge \forall y \in \mathcal{O}_{x_1}\, Q(\{x_0\}(y))\right],$$

such that $\mathcal{O}_n$ satisfies

(1) $$\begin{cases} B_n(\mathcal{O}_n, x) \rightarrow x \in \mathcal{O}_n, \\ \forall x\left[B_n(Q,x) \rightarrow Qx\right] \rightarrow \forall x\left[\mathcal{O}_n x \rightarrow Qx\right]. \end{cases}$$

Note that B_n contains the constants $\mathcal{O}_1,\ldots,\mathcal{O}_{n-1}$. Let us assume the constants $\mathcal{O}_1,\ldots,\mathcal{O}_{n-1}$ to have been defined explicitly in **IDB***; we are now going to construct an explicit definition of $\mathcal{O}_n$. Put

$$Ax \equiv (j_1 x < n) \wedge j_2 x \in \mathcal{O}_{j_1 x}.$$

Let $[A] = \sigma$, and consider K_σ^{PR}. This predicate is characterized by the axioms (3) and (4) of Subsection 3.4. Intuitively it will be clear that K_σ^{PR} permits us to code $\mathcal{O}_n$ as follows: let $\{y\}_n = \lambda z.\{y\}(n * z)$, then "$y \in K_\sigma^{PR}$ represents a tree of $\mathcal{O}_n$", denoted by $y \in T_n$, is given by

$$\begin{aligned} y \in T_n \equiv_{\text{def}} y \in K_\sigma^{PR} \wedge \forall m^\sigma \big[\{y\}_m(0) = 1 \vee (\{y\}_m(0) = 0 \wedge \\ \wedge \exists i < n \forall x \in A \; ((j_1 x = i \to \{y\}_m(\hat{x}) < 2) \wedge \\ \wedge (j_1 x \neq i \to \{y\}_m(\hat{x}) = 2))) \big]. \end{aligned}$$

If $\{y\}_m(0) = 0$, m governs an $\mathcal{O}_n$-tree; only the successors $m * \hat{x}$ of m with $\{y\}_m(\hat{x}) \neq 2$ count as successors of the tree coded by $\{y\}_m$. Now we may take for our explicit definition (the reader should allow himself some time):

$$\begin{aligned} x \in \mathcal{O}_n \equiv_{\text{def}} \exists z \, \exists y \in T_n \; (\{z\}(0) = x \wedge \{y\}_0(0) < 2 \wedge \\ \forall m^\sigma \big[(\{y\}_m(0) = 1 \to \{z\}(m) = 0) \wedge \\ (\{y\}_m(0) = 0 \to \\ \forall u \in A \big[\{y\}_m(\hat{u}) < 2 \to \exists z_0 ((\{z\}(m) = \langle z_0, j_1 u, n \rangle) \wedge \\ \forall v \in A \; \big[j_1 v = j_1 u \to \{z\}(m * \hat{v}) = \{z_0\}(j_2 v) \big] \big) \big] \big) \big]). \end{aligned}$$

It is now straightforward but tedious to check that the present explicit definition indeed satisfies the axioms (1) for $\mathcal{O}_n$.

3.7. Realizability for $\mathbf{ID}^{\mathrm{i}}_{<\omega}(\mathcal{O})$

This may be treated on lines very similar to the treatment for **IDB***. We obtain a characterization theorem exactly as for **IDB***; the least trivial part of the proof is the verification of the realizability of the induction principles within the theory considered.

For $\mathcal{O}_n$ we have to show

$$(1) \qquad \forall x (B_n(Q, x) \to Qx) \to \forall y (\mathcal{O}_n y \to Qy)$$

to be realizable. Assume

$$u \, \mathrm{r} \, \forall x (B_n(Q, x) \to Qx),$$

i.e.

$$v \, \mathrm{r} \, B_n(Q, x) \to \{u\}(x, v) \, \mathrm{r} \, Qx.$$

$v \, \mathrm{r} \, B_n(Q, x)$ means

$$(2)\qquad \begin{cases} j_1 v = 0 \to j_2 v = 0 \wedge x = 0, \\ j_1 v \neq 0 \to j_2 v \mathrm{r} \exists x_0 x_1 \{x = \langle x_0, x_1, n \rangle \wedge x_1 < n \wedge \\ \qquad\qquad\qquad\qquad \forall y \in \mathcal{O}_{x_1} Q(\{x_0\}(y))\}. \end{cases}$$

Let $\psi(y, w)$ be a partial recursive function (which can be read off from (2)) such that:

$$y = \langle y_0, y_1, n \rangle \wedge y_1 < n \wedge w \, \mathrm{r} \, \forall x \in \mathcal{O}_{y_1} Q(\{y\}(x)) \to$$
$$\to !\psi(y, w) \wedge \psi(y, w) \, \mathrm{r} \, B_n(Q, y).$$

Now let $\{z\}$ be a partial recursive function such that

$$y = 0 \to \{z\}(w, u, y) \simeq \{u\}(0, j(0, 0)),$$
$$y \neq 0 \wedge y = \langle y_0, y_1, n \rangle \wedge y_1 < n \to$$
$$\{z\}(w, u, y) \simeq \{u\}(y, \psi(y, \Lambda x.\{w\}(u, \{y_0\}(x))))$$

and thus by the recursion theorem there is a z_0 such that

$$y = 0 \to \{z_0\}(u, y) \simeq \{u\}(0, j(0, 0)),$$
$$y \neq 0 \wedge y = \langle y_0, y_1, n \rangle \wedge y_1 < n \to$$
$$\{z_0\}(u, y) \simeq \{u\}(y, \psi(y, \Lambda x.\{z_0\}(u, \{y_0\}(x)))).$$

Then one proves by induction over $\mathcal{O}_n$:

$$\forall y \in \mathcal{O}_n \, (!\{z_0\}(u, y) \wedge \{z_0\}(u, y) \, \mathrm{r} \, Qy),$$

and thus z_0 realizes instances of (1).

3.8. Theorem. Γ_0 *is the class of formulae defined in* TROELSTRA (1973, 3.6.3), *i.e.* Γ_0 *satisfies* (a) *prime formulae are in* Γ_0, (b) Γ_0 *is closed under* $\wedge$, $\vee$, $\forall$, $\exists$, (c) *A almost negative and* $B \in \Gamma_0 \Rightarrow A \to B \in \Gamma_0$. *Let* $\Gamma_0^{\mathrm{ar}} = \Gamma_0 \cap \mathcal{L}(\mathbf{HA})$. *Then*

$$\mathbf{ID}^{\mathrm{i}}_{<\omega},\quad \mathbf{IDB}^* + \mathrm{ECT}_0,\quad \mathbf{IDB}^*,\quad \mathbf{ID}^{\mathrm{i}}_{<\omega}(\mathcal{O}) + \mathrm{ECT}_0,\quad \mathbf{ID}^{\mathrm{i}}_{<\omega}(\mathcal{O});$$
$$\mathbf{IDB}^{**},\quad \mathbf{IDB}^{**} + \mathrm{ECT}_0$$

all have the same sentences in Γ_0^{ar}.

Proof. $\mathbf{IDB}^* + \mathrm{ECT}_0$ can be alternatively axiomatized as $\mathbf{ID}^* + \mathrm{ECT}_0$, where $\mathbf{ID}^*$ is the inductive theory corresponding to the classes $K^{\mathrm{R}}_{[A]}$. It is obvious from our treatment of realizability that

$$\mathbf{ID}^*,\quad \mathbf{ID}^* + \mathrm{ECT}_0,\quad \mathbf{IDB}^* + \mathrm{ECT}_0,\quad \mathbf{IDB}^*,\quad \mathbf{IDB}^{**}$$

all coincide w.r.t. Γ_0^{ar}.

Also

(i) $\mathbf{ID}^{\mathrm{i}}_{<\omega}(\mathcal{O})$ is explicitly definable in $\mathbf{ID}^*$;

(ii) $\mathbf{ID}^{\mathrm{i}}_{<\omega}(\mathcal{O})$, $\mathbf{ID}^{\mathrm{i}}_{<\omega}(\mathcal{O})+\mathrm{ECT}_0$ coincide w.r.t. Γ_0^{ar};

(iii) $\mathbf{ID}^{\mathrm{i}}_{<\omega}(\mathcal{O})$ and $\mathbf{ID}^{\mathrm{i}}_{<\omega}$ coincide w.r.t. arithmetical sentences, hence a fortiori w.r.t. Γ_0^{ar} (SIEG, 1977);

(iv) $\mathbf{ID}^* \subseteq \mathbf{ID}^{\mathrm{i}}_{<\omega}$ (immediate);

combining these facts leads immediately to the statement of the theorem.

4. CS* and the elimination theorem

4.1. The language of CS*

CS* contains **IDB***. For each type σ of **IDB*** we add choice variables of type σ: $\alpha^\sigma,\beta^\sigma,\gamma^\sigma,\ldots$. For $\sigma=N$ we often drop the superscript.

The formation rules for terms and functors extend the corresponding rules for **CS**. In any case we include the additional rules:

(i) choice variables of type σ are σ-functors;

(ii) if φ is a functor of $K_{\sigma,\tau}$ and ψ is a σ-functor, then $\varphi|\psi$ is a τ-functor;

(iii) if φ is a functor of $K_{\sigma,\tau}$ or K_σ and ψ is a σ-functor, then $\varphi(\psi)$ is a term;

(iv) if φ is a σ-functor, t a term, then φt is a term.

Furthermore, the closure conditions on $K_\sigma, K_{\sigma,\tau}$ permit (already in **IDB***) the introduction of further term-formation rules when the need arises; e.g. $e\in K_{\sigma,\tau}$, $f\in K_{\rho,\sigma}\Rightarrow e:f$ is a functor of $K_{\rho,\tau}$. We shall use $e^{\sigma,\tau}$, $f^{\sigma,\tau}$ for variables ranging over elements of $K_{\sigma,\tau}$.

4.2. Axioms of CS*

CS* is an extension of **IDB*** $\cup$ **CS**, based on many-sorted intuitionistic predicate logic, with the following additional axiom schemata:

(I) $$\forall\alpha^\sigma(A\alpha\to B\alpha)\leftrightarrow\forall e^{\sigma,\sigma}(\forall\alpha^\sigma A(e|\alpha)\to\forall\alpha^\sigma B(e|\alpha))$$

(generalized analytic data, in fact derivable from analytic data),

(II) $$\forall\alpha^\sigma\exists\xi A(\alpha,\xi)\to\exists e^\sigma\forall n^\sigma(en\neq 0\to\exists\xi\forall\alpha^\sigma A(n|\alpha,\xi))$$

(continuity with respect to lawlike objects; ξ here stands for any lawlike sort of variables such as $x,u,a^\sigma,e^\sigma,e^{\sigma,\tau}$), and

(III) $$\forall\alpha^\sigma\exists\beta^\tau A(\alpha,\beta)\to\exists e^{\sigma,\tau}\forall\alpha^\sigma A(\alpha,e|\alpha)$$

(function continuity).

In addition we need axioms expressing that the variables α^σ range over A^ω:

(IV) $$\forall\alpha[\alpha\in A^\omega\to\exists\beta^\sigma(\alpha=\beta^\sigma)]\qquad(\sigma=[A]),$$

(V) $$\forall\alpha^\sigma\exists\beta[\alpha=\beta\wedge A^\omega\beta]\qquad(\sigma=[A]),$$

and also, not involving choice variables

(VI) $\forall n^\sigma(A^{<\omega}n), \quad \forall m(A^{<\omega}m \to \exists n^\sigma(n=m)) \qquad (\sigma=[A])$.

4.3. The elimination mapping for CS*

We extend the elimination mapping for **CS**, defined in KREISEL and TROELSTRA (1970, 7.1) by extending the auxiliary mapping $\mapsto$ as follows:

(i) $\forall\alpha^\sigma(t[\alpha]=s[\alpha]) \mapsto \forall a^\sigma(t[a]=s[a])$;

(ii) $\forall\alpha^\sigma(A\wedge B) \mapsto \forall\alpha^\sigma A \wedge \forall\alpha^\sigma B$;

(iii) $\forall\alpha^\sigma(A\vee B) \mapsto \forall\alpha^\sigma\exists x((x=0\to A)\wedge(x\neq 0\to B))$;

(iv) $\forall\alpha^\sigma(A\alpha\to B\alpha) \mapsto \forall e^{\sigma,\sigma}(\forall\alpha^\sigma A(e|\alpha)\to\forall\alpha^\sigma B(e|\alpha))$;

(v) $\forall\alpha^\sigma\exists x A(\alpha^\sigma,x) \mapsto \exists e^\sigma\forall n^\sigma(en\neq 0\to\forall\alpha^\sigma A(n|\alpha, en \dot{-} 1))$;

(vi) $\forall\alpha^\sigma\exists\xi A(\alpha^\sigma,\xi) \mapsto \exists e^\sigma\forall n^\sigma(en\neq 0\to\exists\xi\forall\alpha^\sigma A(n|\alpha,\xi))$, *for all lawlike sorts of variables* ξ *(except numerical variables)*;

(vii) $\forall\alpha^\sigma\forall\xi A(\alpha^\sigma,\xi) \mapsto \forall\xi\forall\alpha^\sigma A(\alpha,\xi)$, ξ *a variable of a lawlike sort*;

(viii) $\forall\alpha^\sigma\exists\beta^\tau A(\alpha^\sigma,\beta^\tau) \mapsto \exists e^{(\sigma)\tau}\forall\alpha^\sigma A(\alpha,e|\alpha)$;

(ix) $\forall\alpha^\sigma\forall\beta^\tau A(\alpha,\beta) \mapsto \forall e^{\sigma\times\tau,\sigma}\forall f^{\sigma\times\tau,\sigma}\forall\gamma^{\sigma\times\tau}A(e|\gamma,f|\gamma)$;

(x) $\exists\alpha^\sigma A\alpha^\sigma \mapsto \exists a^\sigma A a^\sigma$.

Remark 1. Clause (vi) is slightly simplified when compared with the corresponding clause in KREISEL and TROELSTRA (1970, 7.1.3. (vii)F,K); the modified clause is basically the same (cf. the next remark) and slightly easier to manage.

Remark 2. Axiom schema II, applied with x for ξ, yields together with APC the *schema*

BC-N$^\sigma$ $\qquad \forall\alpha^\sigma\exists x A(\alpha,x)\to\exists e^\sigma\forall\alpha^\sigma A(\alpha,e(\alpha))$,

and with a for ξ and APC, $\sigma=[N]$:

BC-F $\qquad \forall\alpha\exists a A(\alpha,a)\to\exists e\exists b\forall\alpha A(\alpha,(b)_{e(\alpha)})$.

4.4. The elimination theorem

The following theorem can be established for **CS*** along the lines of the proof (in KREISEL and TROELSTRA (1970, § 7)) of the elimination theorem for **CS**.

Theorem. *For the elimination mapping* τ *defined in Subsection* 4.3 *we have* (*A not containing choice parameters free*)

(i) $\tau(A)\equiv A$ *for* A *a formula of* **IDB***,

(ii) **CS***$\vdash\tau(A)\leftrightarrow A$,

(iii) **CS***$\vdash A\leftrightarrow$**IDB***$\vdash\tau(A)$.

Since the proof is largely parallel to the proof of 7.2 and 7.3.8 in KREISEL and TROELSTRA (1970), we shall only discuss the points where the treatment of KREISEL and TROELSTRA (1970) needs to be supplemented.

Lemma A. *Let $t[\alpha^\sigma]$ be a term, $\varphi[\alpha^\sigma]$ a τ-functor. Then there are $e^\sigma, f^{\sigma,\tau}$ such that in* **IDB***

$$\forall a^\sigma(t[a]=e^\sigma(a)), \qquad \forall a^\sigma(\varphi[a^\sigma]=f^{\sigma,\tau}|a^\sigma).$$

Proof. Quite similar to the proof of 7.3.2 in KREISEL and TROELSTRA (1970). For example, let $\varphi[x,\alpha^\sigma]$ be a choice term of type τ; and let $t[\alpha^\sigma]$ be a numerical term. By induction hypothesis $\forall a^\sigma(t[a^\sigma]=e(a^\sigma))$, $\forall a^\sigma\forall x(\varphi[x,a^\sigma]=f^{\sigma,N}_{\langle x\rangle}|a^\sigma)$, where $f_{\langle x\rangle}=\lambda n.f(\hat{x}*n)$. Then $\forall a^\sigma(\varphi(t[a^\sigma],a^\sigma)=\psi|a^\sigma)$, where

$$\psi\equiv\lambda n.\mathrm{sg}(n)\cdot\mathrm{sg}(en)\cdot f^{\sigma,N}(\langle en\dot{-}1\rangle*n)$$

etc. etc.; obviously, $\psi\in K_{\sigma,N}$.

Discussion. We have left unspecified in Subsection 4.1 which functors we actually need in the language. In fact, we can be more economical than in KREISEL and TROELSTRA (1970); e.g. we do not need functors e/f, $e/\!\hat{/}f$, nor do we need the clause "if φ is a K-functor, t is a term, then $\lambda n.\varphi(\hat{t}*n)$ is again a K-functor", thus saving ourselves the trouble of lawlike K-functors containing choice parameters such as e.g. $\lambda n.e(\langle\alpha x\rangle*n)$. All that we do need in this respect can actually be read off from the definition of the translation: "$e|\alpha$", "$n|\alpha$" is really all that matters. The use of $e:f$, $e;f$, e/f etc. in KREISEL and TROELSTRA (1970) may be viewed either as abbreviations, or referring to a proof carried out in a conservative extension. These observations permit us to be economical in the actual verification of the preceding lemma.

Lemma B. *For terms $t[\alpha^\sigma]$, $s[\alpha^\sigma]$ and functors $\varphi[\alpha^\sigma]$, $\psi[\alpha^\sigma]$ we can prove in* **IDB***:

$$\forall a^\sigma(t[a]=s[a])\rightarrow$$
$$[\tau(\forall\alpha^\sigma A(t[\alpha^\sigma],\alpha^\sigma))\leftrightarrow\tau(\forall\alpha^\sigma A(s[\alpha^\sigma],\alpha^\sigma))]$$

for all $A(x,\alpha^\sigma)$ of **CS***, *and similarly*:

$$\forall a^\sigma(\varphi[a]=\psi[a])\rightarrow$$
$$[\tau(\forall\alpha^\sigma B(\varphi[\alpha],\alpha))\leftrightarrow\tau(\forall\alpha^\sigma B(\psi[\alpha],\alpha))]$$

for $B(\xi,\alpha^\sigma)$ of **CS***, *ξ a variable of the same sort as the functors φ,ψ.*

Lemma C. *Let $A\alpha^\sigma$ be any formula of* **CS***. *Then in* **IDB***

$$\forall e^{\sigma,\sigma}(\tau(\forall\alpha^\sigma A\alpha)\rightarrow\tau(\forall\alpha^\sigma A(e|\alpha))).$$

Proof. Cf. KREISEL and TROELSTRA (1970, 7.3.5).

Lemma D. *Let $C(x,\alpha^\sigma)$ be any formula of* **CS***, *then in* **IDB***

$$\forall e^\sigma[\tau(\forall\alpha^\sigma C(e(\alpha),\alpha))\leftrightarrow\forall n^\sigma(en\neq0\rightarrow\tau(\forall\alpha^\sigma C(en\dot{-}1,n|\alpha)))].$$

Proof. Cf. KREISEL and TROELSTRA (1970, 7.3.6).

The implication from left to right is unproblematic; for the converse, we use induction on the logical complexity of C. Let us consider a typical case in the inductive step where the details differ slightly from the details in 7.3.6 of KREISEL and TROELSTRA (1970). Assume for instance that

$$C(x,\alpha)\equiv\exists y\, B(x,\alpha,y),$$

and assume

$$(1)\qquad \forall n^\sigma(e^\sigma n\neq 0\rightarrow\ulcorner\forall\alpha^\sigma\exists y\, B(e^\sigma n\dot{-}1,n|\alpha^\sigma,y)\urcorner)$$

(writing $\ulcorner A\urcorner$ for $\tau(A)$ as in KREISEL and TROELSTRA (1970)). Written in full, (1) becomes

$$\forall n^\sigma(e^\sigma n\neq 0\rightarrow\exists f^\sigma\forall m^\sigma(fm\neq 0\rightarrow\ulcorner\forall\alpha^\sigma B(en\dot{-}1,n:m|\alpha,fm\dot{-}1)\urcorner))$$

Apply APC to find b, $c(n,u,x)$ such that, abbreviating $\lambda x.c(n,u,x)$ as $c_{n,u}$:

$$\forall n^\sigma m^\sigma(e^\sigma n\neq 0\rightarrow\exists! u(b(n,u)=0)\wedge\forall u\exists f^\sigma(b(n,u)=0)\rightarrow(c)_{n,u}=f$$
$$\wedge(fm\neq 0\rightarrow\ulcorner\forall\alpha^\sigma B(en\dot{-}1,n:m|\alpha,fm\dot{-}1)\urcorner)))$$

Without restriction we may assume

$$c_{n,u}m\neq 0\rightarrow c_{n,u}(m*m')=c_{n,u}m.$$

Define f_1 by

$$\begin{cases} f_1 0=0,\\ f_1(\hat{n}*m)=z+1\leftrightarrow\exists y<\mathrm{lth}(m)\ (b(n,y)=0\wedge c(n,y,m)=z+1).\end{cases}$$

We note that $\lambda m.f_1(\hat{n}*m)\in K_\sigma$ for $n=n^\sigma$, $en^\sigma\neq 0$. Thus

$$\forall n^\sigma m^\sigma(e^\sigma n\neq 0\wedge f_1(\hat{n}*m)\neq 0\rightarrow\ulcorner\forall\alpha^\sigma B(en\dot{-}1,n:m|\alpha,f_1(\hat{n}*m)\dot{-}1)$$

Now consider e/f_1, defined as

$$\lambda n.\mathrm{sg}(en)\cdot f_1(\langle h(e,n)\dot{-}1\rangle *n).$$

If $en^\sigma\neq 0$, then $e(h(e,n)\dot{-}1)\neq 0$ and $h(e,n)\dot{-}1\leqslant n^\sigma$, and thus $f_1\in K_\sigma^{\lambda n.h(e,n)}$. Assume

$$(e/f_1)n^\sigma\neq 0.$$

Then

$$en^\sigma\neq 0,\quad (e/f_1)n=f_1(\langle h(e,n)\dot{-}1\rangle *n),$$
$$h(e,n)\dot{-}1\leqslant n,$$

and thus

$$\ulcorner\forall\alpha^\sigma B(en\dot{-}1,(h(e,n)\dot{-}1):n|\alpha,(e/f_1)n\dot{-}1)\urcorner,$$

or

$$\ulcorner\forall\alpha^\sigma B(en\dot{-}1,n|\alpha,\ (e/f_1)n\dot{-}1)\urcorner;$$

with $\forall a(en \dot{-} 1 = e(n|a))$

$$\ulcorner \forall \alpha^\sigma B(e(n|\alpha), n|\alpha, (e/f_1)n \dot{-} 1)\urcorner.$$

Hence

$$\forall n((e/f_1)n \neq 0 \to \ulcorner \forall \alpha^\sigma B(e(n|\alpha), n|\alpha, (e/f_1)n \dot{-} 1)\urcorner)$$

so

$$\exists f^\sigma \forall n^\sigma (fn^\sigma \neq 0 \to \ulcorner \forall \alpha^\sigma B(e(n|\alpha), n|\alpha, fn \dot{-} 1)\urcorner),$$

i.e. $\ulcorner \forall \alpha^\sigma \exists x B(e(\alpha), \alpha, x)\urcorner$ etc.

Proof (of the elimination theorem). Most of the details are completely similar to the proof of 7.3.8 in KREISEL and TROELSTRA (1970). For example, in verifying Q3-N we have to show

$$\ulcorner \forall \alpha^\sigma \forall \beta^\tau [\forall x P(\alpha^\sigma, x) \to P(\alpha^\sigma, t[\alpha^\sigma, \beta^\tau])]\urcorner.$$

That is, we have to show

$$\forall e^\rho \forall f_1 f_2 [\forall x \ulcorner \forall \gamma^{\sigma\times\tau} P(f_1 : e|\gamma, x)\urcorner \to$$

$$\ulcorner \forall \gamma^{\sigma\times\tau} P(f_1 : e|\gamma, t[f_1 : e|\gamma, f_2 : e|\gamma])\urcorner]$$

where $\rho = \sigma\times\tau, \sigma\times\tau, f_1 \in K_{\sigma\times\tau,\sigma}, f_2 \in K_{\sigma\times\tau,\tau}$. Let $e_1 \in K_{\sigma\times\tau}$ be such that

$$\forall c(t[f_1|c, f_2|c] = e_1|c).$$

Assume $(e_1 : e)n^{\sigma\times\tau} \neq 1$, and let

$$\forall x \ulcorner \forall \gamma^{\sigma\times\tau} P(f_1 : e|\gamma, x)\urcorner;$$

then also $\ulcorner \forall \gamma^{\sigma\times\tau} P(f_1 : e|\gamma, (e_1 : e)n \dot{-} 1)\urcorner$, hence by Lemma C

$$\ulcorner \forall \gamma^{\sigma\times\tau} P(f_1 : e : n|\gamma, (e_1 : e)n \dot{-} 1)\urcorner,$$

thus

$$\forall n^{\sigma\times\tau}((e_1 : e)n \neq 1 \to \ulcorner \forall \gamma^{\sigma\times\tau} P(f_1 : e : n|\gamma, (e_1 : e)n \dot{-} 1)\urcorner),$$

therefore by Lemma D

$$\ulcorner \forall \gamma^{\sigma\times\tau} P(f_1 : e|\gamma, (e_1 : e)(\gamma))\urcorner;$$

application of Lemma A, B yields

$$\ulcorner \forall \gamma^{\sigma\times\tau} P(f_1 : e|\gamma, t[f_1 : e|\gamma, f_2 : e|\gamma])\urcorner.$$

The validity of the translations of axiom schemata I, II, III of **CS*** is immediate. The only new axioms to be checked are IV, V. For IV, we have to show $\forall e(\ulcorner \forall \alpha A^\omega(e|\alpha)\urcorner \to \ulcorner \forall \alpha \exists \beta^\sigma (e|\alpha = \beta^\sigma)\urcorner)$ where $[A] = \sigma$. $A^\omega(e|\alpha)$ is in fact $\forall x A(e_{\langle x\rangle}(\alpha))$, Ax in the language of **IDB***, $e_{\langle x\rangle} = \lambda n.e(\hat{x} * n)$. We observe that

$$\ulcorner\forall\alpha\forall x A(e_{\langle x\rangle}(\alpha))\urcorner\leftrightarrow\forall x\ulcorner\forall\alpha A(e_{\langle x\rangle}(\alpha))\urcorner\leftrightarrow$$
$$\forall x\forall n(e_{\langle x\rangle}n\neq 0\rightarrow\ulcorner\forall\alpha A(en\dot{-}1)\urcorner)\leftrightarrow$$
$$\forall x\forall n(e_{\langle x\rangle}n\neq 0\rightarrow A(en\dot{-}1))\leftrightarrow$$
$$\forall x\forall a A(e_{\langle x\rangle}(a))\leftrightarrow\forall a A^{\omega}(e|a).$$

Therefore we have to verify

$$\forall e\left[\forall a A^{\omega}(e|a)\rightarrow\exists f^{N,\sigma}\forall a(e|a=f|a)\right];$$

the premise tells us that we may take $f=e$.
To check V is even easier: we have to show

$$\exists f^{\sigma,N}(\forall a^{\sigma}\left[a=f|a\right]\wedge\ulcorner\forall\alpha^{\sigma}A^{\omega}(f|\alpha)\urcorner),$$

and this is satisfied if we take an f such that $f|a=a$ for all a, since $\ulcorner\forall\alpha^{\sigma}A^{\omega}(f|\alpha)\urcorner\leftrightarrow\forall a^{\sigma}A^{\omega}(f|a)$ (similar to the reasoning above).

5. Principal results

5.1. We are first going to determine the proof-theoretic strength of **CS*** (in Subsections 5.2, 5.3) by combining our results on realizability for **IDB*** and elimination of choice sequences for **CS***. Then we determine the proof-theoretic strength of **EL** + EBI_0, bounding it from above by **CS*** + $\mathrm{EBI}_0+\mathrm{ECT}_0$, and bounding it from below by embedding **IDB**** into **CS*** + EBI_0 (Subsections 5.4–5.9).

5.2. Proposition. *Let Ax be an almost negative formula of* **IDB*** *containing only numerical variables free. Then* **CS*** $\vdash$*n-*$EBI_0(A)$.

Proof. As follows from Subsection 2.3 we may assume **IDB*** $\vdash\exists x Ax$ without loss of generality, and by Subsection 2.5 we may assume $\exists x Ax$ to be closed.

Instances of $\mathrm{EBI}_0(A)$ for P not containing choice parameters can be proved similar to Kreisel and Troelstra (1970, 5.6.1). For P containing choice parameters, we may adapt the argument in Kreisel and Troelstra (1970, 5.7.5).

5.3. Theorem. *Let $\Gamma_0,\Gamma_0^{\mathrm{ar}}$ be as in Theorem* 3.8. *The following theories have the same theorems in* Γ_0^{ar} (ECT_0 *everywhere restricted to* $\mathfrak{L}$(**IDB***)):

(a) $\mathbf{ID}^{\mathrm{i}}_{<\omega}$, $\mathbf{ID}^{\mathrm{i}}_{<\omega}(\mathcal{O})$, **IDB***, **IDB*** + ECT_0;
(b) **CS***, **CS*** + ECT_0;
(c) **CS*** + l-EBI_0, **CS*** + l-EBI_0 + ECT_0.

Proof. For the theories under (a) we have already demonstrated their equivalence w.r.t. Γ_0^{ar} in Section 3. Furthermore **CS*** is conservative over **IDB*** by the elimination theorem of Section 4, hence a fortiori conservative over **IDB*** w.r.t formulae of Γ_0^{ar}. Similarly for **CS*** + ECT_0.

Now consider **CS*** + l-EBI_0 + ECT_0. Let l-$\mathrm{EBI}_0(A)$ be any instance of l-EBI_0. By the elimination theorem, there is an $A'x \equiv \tau(Ax)$ in **IDB*** such that

$$\mathbf{CS}^* \vdash A'x \leftrightarrow Ax.$$

Without loss of generality $\mathbf{IDB}^* \vdash \exists x A'x$. Assume A' to contain a lawlike-function parameter b. Under assumption of ECT_0, $A'(x,b)$ corresponds to

$$A''(x,y) \equiv A'(x,\{y\}) \wedge \forall u \exists v\, Tyuv.$$

(b replaced by an arbitrary total recursive function), and moreover

$$\mathbf{IDB}^* + \mathrm{ECT}_0 \vdash A''(x,y) \leftrightarrow \exists z\,(z\, \mathrm{r}\, A''(x,y)).$$

Thus relative to **CS*** + ECT_0, l-$\mathrm{EBI}_0(A)$ is equivalent to n-$\mathrm{EBI}_0(\exists z\, B(x,y,z))$ where $B(x,y,z) \equiv z\, \mathrm{r}\, A''(x,y)$; and by Proposition 2.2 this is derivable from n-$\mathrm{EBI}_0(C)$ where $C(x,y) \equiv B(j_1x, y, j_2x)$; and n-$\mathrm{EBI}_0(C)$ in turn is derivable in **CS*** (Proposition 5.2). Thus we see that in fact

$$\mathbf{CS}^* + \text{l-EBI}_0 + \mathrm{ECT}_0 = \mathbf{CS}^* + \mathrm{ECT}_0.$$

5.4. Behaviour of almost negative formulae under elimination

Lemma. *Let A be an almost negative formula of* **CS**, *i.e. built from prime formulae and formulae $\exists x[t=s], \exists a[t=s], \exists \alpha[t=s]$ by means of $\wedge, \forall, \rightarrow$. Then $\tau(A)$ is (equivalent to) an almost negative formula of* $\mathbf{IDB}_1$, *provably in* $\mathbf{IDB}_1$.

Proof. Let $e \in K$; $e_{\langle x \rangle} = \lambda n.e(\hat{x} * n)$; we define e^* primitive recursively in e:

$$e^*0 = 0;$$

$$e^*\langle x_0,\ldots,x_p\rangle = \begin{cases} q_0 + 1 & \text{if we can find } r_0, q_0 \text{ such that} \\ & r_0 = \min_r[\exists q \leqslant r(e_{\langle q\rangle}\langle x_0,\ldots,x_r\rangle = 1)] \\ & \quad \text{and } r \leqslant p, \text{ and} \\ & q_0 = \min_q[e_{\langle q\rangle}\langle x_0,\ldots,x_{r_0}\rangle = 1], \\ 0 & \text{otherwise.} \end{cases}$$

Observe that automatically $e^* \in M$, i.e.

$$\forall nm\,(e^*n \neq 0 \rightarrow e^*n = e^*(n*m)),$$

and if $e^* \in K$, then

$$\forall a\,(e_{\langle e^*(a)\rangle}(a) = 0).$$

As a sublemma we need

(1) $\forall ef(\forall a(e_{\langle f(a)\rangle}(a)=0)\to e^*\in K)$.

To establish (1), assume $\forall a(e_{\langle f(a)\rangle}(a)=0)$ and introduce $f'\in K$ with $\forall n(f'n<1)$ by the stipulation $f'n=1\leftrightarrow\exists z(fn=z+1)\wedge e(\langle fn\dot{-}1\rangle * n)=1 \wedge \mathrm{lth}(n)>fn$, i.e.

$$f'n=\mathrm{sg}(fn)\cdot\mathrm{sg}(e\langle fn\dot{-}1\rangle * n)\cdot\mathrm{sg}(\mathrm{lth}(n)\dot{-}fn).$$

We observe that $\forall n(f'n\neq 0\to e^*n\neq 0)$. To see that $f'\in K$, we first show by induction over K w.r.t. f

(2) $\forall x(\lambda n.\mathrm{sg}(fn)\cdot\mathrm{sg}((\mathrm{lth}(n)+x)\dot{-}fn)\in K)$,

then apply respectively the closure conditions (iii) of Subsection 3.3 to conclude that $\lambda n.\mathrm{sg}(en)\in K$, (vii) to conclude that $\lambda n.\mathrm{sg}(fn)\cdot\mathrm{sg}(e\langle \mathrm{fn}\dot{-}1\rangle * n)\in K$, and finally (vi) with (2) to conclude that $f'\in K$. Now $e^*\in K$ by (ii) of Subsection 3.3.

Now we are ready to prove the assertion of the lemma, by induction on the logical complexity of A. To be precise, we show $\ulcorner\forall\alpha B\urcorner$ to be almost negative, using as our induction hypothesis that for all formulae C of complexity less than B, $\ulcorner\forall\alpha C\urcorner$ is almost negative.

We have four basis cases: $\forall\alpha(t[\alpha]=s[\alpha])$, $\forall\alpha\exists x(t[\alpha,x]=s[\alpha,x])$, $\forall\alpha\exists a(t=s)$, $\forall\alpha\exists\beta(t=s)$. The first case is unproblematic, and the third and fourth case can be reduced to the second. In the second case, modulo logical equivalence we may restrict attention to

(3) $\forall\alpha\exists x(t'[\alpha,x]=0)$

where $t'[\alpha,x]=|t[\alpha,x]-s[\alpha,x]|$. The translation of (3) is equivalent to

(4) $\exists e\forall a(t'[a,e(a)]=0)$;

and thus, if φ is a K-functor representing $\lambda x.t'[a,x]$, i.e. $\varphi_{\langle x\rangle}(a)=t'[a,x]$ for all a, (4) is equivalent to

$$\varphi^*\in K;$$

or if we prefer a representation-independent rendering

$$\forall e(\forall ax(e_{\langle x\rangle}(a)=t'[a,x])\to e^*\in K).$$

The induction steps are almost trivial.

5.5. Theorem. $\mathrm{EBI}_0(A)$, *for* $\forall x Ax$ *closed,* A *almost negative in the language of* **CS**, *is derivable in* **CS***.

Proof. An immediate corollary of the preceding lemma.

5.6. Although a digression from the principal aim, we complete the picture of the behaviour of almost negative formulae by inserting the following lemma:

Lemma. (i) *Let A be an almost negative formula of* **CS**, *and assume instead of a generalized inductive definition for K only the axioms:*

$$K=K_1, \quad \textit{where}$$

$$K_1 a \equiv \forall b \exists x \forall c \leqslant b\, (a(\bar{c}x) \neq 0) \wedge a \in M \qquad (M=M^N),$$

together with an axiom

$$K_1 a \rightarrow \forall \alpha \exists x\, (a(\bar{\alpha}x) \neq 0).$$

Then $\tau(A)$ is again equivalent to an almost negative formula, provably in $\mathbf{EL}_1$.

(ii) *As in* (i), *except that $K=K_1$ is now replaced by $K=K_0$, where*

$$K_0 a \equiv a \in M \wedge \forall b \exists x\, (a(\bar{b}x) \neq 0).$$

Then $\tau(A) \leftrightarrow A'$ provably in $\mathbf{EL}_1$, *where A' is obtained by relativizing all choice quantifiers to lawlike quantifiers.*

Proof. For (i), this runs completely parallel to the proof of Lemma 5.4, except that $K_1 a$ is explicitly defined instead of being a prime formula, and is in fact equivalent to the almost negative formula

$$\forall b \exists x \forall n\, (\mathrm{lth}(n)=x \wedge \forall y<x\, ((n)_y \leqslant by) \rightarrow an \neq 0) \wedge a \in M.$$

For (ii), the proof is straightforward by induction on the complexity of A; consider e.g. the implicational case:

$$\ulcorner \forall \alpha\, (B(\alpha) \rightarrow C(\alpha)) \urcorner \leftrightarrow \forall e\, (\ulcorner \forall \alpha\, B(e|\alpha) \urcorner \rightarrow \ulcorner \forall \alpha\, C(e|\alpha) \urcorner)$$
$$\leftrightarrow \forall e\, (\forall a\, B'(e|a) \rightarrow \forall a\, C'(e|a)) \leftrightarrow \forall a\, (B'a \rightarrow C'a)$$

(induction hypothesis for B, C, followed by a specialization of e to elements e' such that $\forall b\, (e'|b=a)$).

5.7. Remark. (i) has been implicitly used in Troelstra (1977A), in (2°) at the end of Section 2. (2°) should have been stated more carefully as: almost negative predicates are transformed by τ into predicates which are in **EL** equivalent to almost negative predicates.

5.8. Lemma. *Let* **S** *be the theory in the language of* **EL**, *axiomatized as* $\mathbf{EL}_1$ + *all instances of* $\mathrm{EBI}_0(A)$ *for almost negative A. Then*

$$\mathbf{IDB}^{**} \subseteq \mathbf{S} \subseteq \mathbf{CS}^*$$

where $\mathbf{IDB}^{**}$ *is now, of course, expressed in the language of* **EL**, *so instead of "lawlike variables" a, b, c, d choice variables $\alpha, \beta, \gamma, \delta$ are used.*

Remark. Obviously, one may add to **S** any axioms which can be expressed in $\mathfrak{L}$(**EL**) and which are derivable in **CS***, such as

$$\forall\alpha\left[A^{\omega}\alpha\rightarrow\exists\beta B(\alpha,\beta)\right]\rightarrow\exists\gamma\forall\alpha\left[A^{\omega}\alpha\rightarrow!\gamma|\alpha\wedge B(\alpha,\gamma|\alpha)\right]$$

for almost negative Ax with $\forall x Ax$ closed.

Proof. First of all we have to show that to each K_σ of **IDB**** there is in **S** an *explicitly definable* K'_σ provably satisfying the axioms for K_σ (but now relative to choice variables).

To see this, let $K_{[A]}a$ ($\sigma=[A]$) be defined by an inductive clause

$$B(Q,K_{\tau_1},\ldots,K_{\tau_n},a)$$

where $K_{\tau_1},\ldots,K_{\tau_n}$ are (obviously) of a degree lower than degree (σ). B is in fact of the form

$$\exists y\forall x(ax=y+1)\vee(a0=0\wedge\forall y\in A\ (\lambda n.a(\hat{y}*n)\in Q))$$

where A contains $K_{\tau_1},\ldots,K_{\tau_n}$. Let A^* be obtained replacing all constructive function variables by corresponding choice variables, and replacing $K_{\tau_1},\ldots,K_{\tau_n}$ by $K'_{\tau_1},\ldots,K'_{\tau_n}$ respectively. We define K'_σ as

$$K'_\sigma\alpha\equiv\forall\beta\in A^{*\omega}\exists x\left(\alpha(\bar{\beta}x)\neq 0\right)\wedge$$
$$\wedge\forall n\in A^{*<\omega}\forall m(\alpha n\neq 0\rightarrow\alpha n=\alpha(n*m)).$$

If A is almost negative, and $K'_{\tau_1},\ldots,K'_{\tau_n}$ have already been constructed as almost negative predicates, it follows that $K'_\sigma\alpha$ itself is almost negative: the prime formulae $K_{\tau_i}(\varphi[a])$ occurring in A are replaced by almost negative formulae $K'_{\tau_i}(\varphi[\alpha])$.

A^* is almost negative and expressed in $\mathfrak{L}$(**EL**)$=\mathfrak{L}$(**S**), and thus $\mathrm{EBI}_0(A^*)$ is available in **S**; and as may be seen from KREISEL and TROELSTRA (1970, proof of 5.6.2), this enables us to prove the generalized-inductive-definition axioms for K'_σ.

On the other hand, **S** $\subseteq$ **CS**, since for Ax almost negative, $\forall x Ax$ closed, $Ax\in\mathfrak{L}$(**EL**), we have

$$\mathbf{CS}^*\vdash\tau(Ax)\leftrightarrow Ax;$$

$\tau(Ax)$ is again almost negative by Lemma 5.4; therefore $\mathrm{EBI}_0(\tau(A))$, hence also $\mathrm{EBI}_0(A)$ is available in **CS***.

5.9. Theorem. $\mathbf{EL}_1$ + n-EBI_0 *has the same proof-theoretic strength as* $\mathbf{ID}^{\mathrm{i}}_{<\omega}$: *and if* CONT_1 *is* C-C′* *of* KREISEL *and* TROELSTRA (1970, 5.7.6), *then the theories*

(a) $\mathbf{ID}^{\mathrm{i}}_{<\omega}$, $\mathbf{ID}^{\mathrm{i}}_{<\omega}(\mathcal{O})$, **IDB***, **CS***,
(b) $\mathbf{EL}_1$ + n-EBI_0 + CONT_1, $\mathbf{EL}_1$ + n-EBI_0,
(c) $\mathbf{EL}_1$ + n-EBI''_0 + CONT_1, $\mathbf{EL}_1$ + n-EBI''_0,

all have the same theorems in Γ^{ar}_0.

Proof. We observe that **IDB**** (formulated with choice variables) is contained in $\mathbf{EL}_1 + \mathrm{EBI}_0$, hence

$$\mathbf{ID}^{\mathrm{i}}_{<\omega} \cap \Gamma_0^{\mathrm{ar}} \subseteq (\mathbf{EL}_1 + \mathrm{EBI}_0) \cap \Gamma_0^{\mathrm{ar}}.$$

On the other hand, $\mathbf{EL}_1 + \text{n-EBI}_0 + \mathrm{CONT}_1 \subseteq \mathbf{CS}^* + \mathrm{EBI}_0 \subseteq \mathbf{CS}^* + \mathrm{EBI}_0 +$ ECT_0 (relative to $\mathcal{L}(\mathbf{IDB}^*)) \subseteq \mathbf{CS}^* + \mathrm{ECT}_0$, so

$$(\mathbf{EL}_1 + \text{n-EBI}_0 + \mathrm{CONT}_1) \cap \Gamma_0^{\mathrm{ar}} \subseteq \mathbf{CS}^* \cap \Gamma_0^{\mathrm{ar}} = \mathbf{ID}^{\mathrm{i}}_{<\omega} \cap \Gamma_0^{\mathrm{ar}},$$

and this yields the statement of the theorem for (a), (b). We have also made use of the equivalence of EBI_0 and n-EBI_0 relative to **EL** (cf. Subsection 2.5). For the systems under (c), the theorem is a consequence of the fact that

$$\mathbf{CS}^* + \mathrm{ECT}_0 + \mathrm{EBI}_0 \supset \mathbf{CS}^* + \text{l-EBI}'_0 \vdash \text{n-EBI}''_0(A)$$

for almost negative A. For **CS*** contains the axiom of partial choice, and thus by Lemma 2.6 $\mathrm{EBI}''_0(A)$ becomes derivable from l-$\mathrm{EBI}'_0(B)$ for suitable B. But then, as before, $\mathrm{EBI}''_0(A)$ becomes derivable for all A on addition of ECT_0 (for **IDB***-formulae).

5.10. Remark. Thus we have obtained a "natural" theory, $\mathbf{EL}_1 + \mathrm{EBI}_0$, which has the same proof-theoretic strength as the (intuitionistic or classical) theory of finitely iterated inductive definitions. The theory is more natural in the following sense: in $\mathbf{ID}^{\mathrm{i}}_{<\omega}$ or $\mathbf{ID}^{\mathrm{i}}_{<\omega}(\mathcal{O})$ there is no reason, once one has already introduced so many constants for inductively defined sets in the language, to stop at finite iterations; whereas the most obvious strengthenings of EBI_0 either require a transition to EBI_1 or addition of new constants to the language.

Or, expressed in yet another way: EBI_0 absorbs all the ω levels of **IDB***.

6. Byproducts and additional comments

6.1. In this section we have collected a miscellany: a proof of the function-realizability of EBI_1, thereby answering a conjecture of Scarpellini (1972) affirmatively; generalizations of results in Troelstra (1974, 1975) concerning instances of generalized continuity (GC) and the fan theorem which are conservative over intuitionistic arithmetic, and finally some open problems.

We start with the realizability of EBI_1.

6.2. Lemma. *For each instance F of* $\mathrm{EBI}_1(A)$, *A almost negative*:

$$\mathbf{EL} + \mathrm{EBI}_{1,\mathrm{D}}(A) \vdash \exists \varepsilon (\varepsilon \mathrm{r}^1 F)$$

(*Notation of* Troelstra (1973, 3.3)).

Proof. It is slightly easier to use a formulation of EBI_1 where $\mathrm{Mon}(A,P)$ is replaced by

$$\mathrm{Mon}^*(A,P) \equiv \forall\alpha\in A^{<\omega}\,\forall\beta\in A^{<\omega}\,(P\alpha\to P(\alpha*\beta)).$$

For convenience we introduce $\sigma,\sigma',\sigma'',\ldots$ as variables ranging over $A^{<\omega}$. Assume A almost negative, and

$$\gamma\,\mathrm{r}^1\,\mathrm{Bar}(A,P)\wedge\mathrm{Mon}^*(A,P)\wedge\mathrm{Ind}(A,P),$$

i.e. for certain $\gamma_0,\gamma_1,\gamma_2$, extracted primitive recursively from γ:

(1) $$\gamma_0\,\mathrm{r}^1\,\forall\alpha\in A^\omega\,\exists x\,P(\bar\alpha^1 x),$$

(2) $$\gamma_1\,\mathrm{r}^1\,\forall\sigma\sigma'(P\sigma\to P(\sigma*\sigma')),$$

(3) $$\gamma_2\,\mathrm{r}^1\,\forall\sigma\big(\forall\beta\in A\;P(\alpha*\hat\beta)\to P\alpha\big).$$

In the rest of the argument we shall frequently make use of the observation made in 3.4 in connection with numerical realizability, namely that $\gamma\,\mathrm{r}^1\,\forall\alpha\in A\,(B\alpha)$ may be treated as $\forall\alpha\in A\,(!\gamma|\alpha\wedge\gamma|\alpha\,\mathrm{r}\,B\alpha)$ for almost negative A. Spelling out (1) we see that

(1′) $$\forall\alpha\in A^\omega\,\big[!\gamma_0|\alpha\wedge j_2(\gamma_0|\alpha)\,\mathrm{r}^1\,P(\bar\alpha^1(j_1(\gamma_0|\alpha)(0)))\big],$$

and similarly we obtain from (2)

(2′) $$\forall\sigma\sigma'\alpha\big(\alpha\,\mathrm{r}^1\,P\sigma\to!\gamma_1|(\sigma,\sigma',\alpha)\wedge\gamma_1|(\sigma,\sigma',\alpha)\,\mathrm{r}^1\,P(\sigma*\sigma')\big).$$

Let $\sigma\mapsto\sigma^*$ be the standard embedding of $\sigma\in A^{<\omega}$ into A^ω (obtained by dropping the length, i.e. $\sigma^*(x,y)=\sigma(x+1,y)$ for $x<\mathrm{lth}(\sigma)$, and putting $\lambda y.\sigma^*(x,y)=\beta_0$ for some fixed $\beta_0\in A$ if $x\geqslant\mathrm{lth}(\sigma)$). Put

$$\varphi(\gamma,\alpha)\equiv j_1(\gamma_0|\alpha)(0)$$

and let $\sigma_\gamma,\sigma'_\gamma$ be such that

(4) $$\varphi(\gamma,\sigma^*)\leqslant\mathrm{lth}(\sigma)\to\sigma=\sigma_\gamma*\sigma'_\gamma\wedge\mathrm{lth}(\sigma_\gamma)=\varphi(\gamma,\sigma^*);$$

we may assume $\sigma_\gamma,\sigma'_\gamma$ to be primitive recursive in σ,γ.
From (1′), (2′), (4) it follows that

(5) $$\varphi(\gamma,\sigma^*)\leqslant\mathrm{lth}(\sigma)\to\varphi'(\gamma,\sigma)\,\mathrm{r}^1\,P\sigma$$

where

$$\varphi'(\gamma,\sigma)\equiv\gamma_1|(\sigma_\gamma,\sigma'_\gamma,j_2(\gamma_0|\sigma^*)).$$

Now define ε such that

$$\varphi(\gamma,\sigma^*)\leqslant\mathrm{lth}(\sigma)\to\varepsilon|(\varepsilon,\gamma,\sigma)\simeq\varphi'(\gamma,\sigma),$$

$$\varphi(\gamma,\sigma^*)>\mathrm{lth}(\sigma)\to\varepsilon|(\varepsilon,\gamma,\sigma)\simeq\gamma_2|(\sigma,\Lambda^1\beta.\varepsilon'|(\gamma,\sigma*\hat\beta)).$$

By the recursion theorem for partial continuous function application (TROELSTRA, 1973, 1.9.16) we find ε_1 such that

$$\varphi(\gamma,\sigma^*) \leqslant \mathrm{lth}(\sigma) \to \varepsilon_1|(\gamma,\sigma) \simeq \varphi'(\gamma,\sigma),$$

$$\varphi(\gamma,\sigma^*) > \mathrm{lth}(\sigma) \to \varepsilon_1|(\gamma,\sigma) \simeq \gamma_2|(\sigma, \Lambda^1\beta.\varepsilon_1|(\gamma,\sigma*\hat{\beta})).$$

Application of $\mathrm{EBI}_{1,\mathrm{D}}(A)$ to

$$P'\sigma \equiv \varphi(\gamma,\sigma^*) \leqslant \mathrm{lth}(\sigma),$$

$$Q'\sigma \equiv !\varepsilon_1|(\gamma,\sigma) \wedge \varepsilon_1|(\gamma,\sigma)\,\mathrm{r}^1\,P\sigma$$

yields $Q'\langle\rangle$, i.e. $\varepsilon_1|(\gamma,\langle\rangle)\mathrm{r}^1 P\langle\rangle$.
Obviously, $\Lambda\gamma.\varepsilon_1|(\gamma,\langle\rangle)$ realizes

$$\mathrm{Bar}(A,P) \wedge \mathrm{Mon}^*(A,P) \wedge \mathrm{Ind}(A,P) \to P\langle\rangle.$$

6.3. Theorem. $\mathbf{EL} + \mathrm{EBI}_1 + \mathrm{GC} \vdash A \Rightarrow \mathbf{EL} + \mathrm{EBI}^{\mathrm{an}}_{1,\mathrm{D}} \vdash \exists\varepsilon(\varepsilon\,\mathrm{r}\,A)$, *where* $\mathrm{EBI}^{\mathrm{an}}_{1,\mathrm{D}}$ *indicates the restriction of* $\mathrm{EBI}_{1,\mathrm{D}}$ *to* $\mathrm{EBI}_{1,\mathrm{D}}(A)$ *for almost negative* A.

Proof. We observe first that if $\mathbf{EL} + \mathrm{EBI}^{\mathrm{an}}_1 + \mathrm{GC} \vdash B$, then

$$\mathbf{EL} + \mathrm{EBI}^{\mathrm{an}}_{1,\mathrm{D}} \vdash \exists\varepsilon(\varepsilon\,\mathrm{r}^1\,B)$$

(here $\mathrm{EBI}^{\mathrm{an}}_1$ is EBI_1 restricted to almost negative tree predicates), by the previous lemma. Also, $\mathrm{EBI}(A)$ for arbitrary A is relative to GC equivalent to $\mathrm{EBI}(\exists\beta(\beta\,\mathrm{r}^1 A\alpha))$, and this is *derivable* from $\mathrm{EBI}(B)$ where $B\gamma \equiv j_1\gamma\,\mathrm{r}^1 A(j_2\gamma)$; cf. Subsection 2.2. Therefore $\mathbf{EL} + \mathrm{EBI}^{\mathrm{an}}_1 + \mathrm{GC} = \mathbf{EL} + \mathrm{EBI}_1 + \mathrm{GC}$.

Remark. For EBI''_1 a similar result holds.

6.4. Instances of generalized continuity conservative over arithmetic

The methods of TROELSTRA (1974, 1977) are easily combined with the developments of this paper to show that also the following instances of generalized continuity GC are conservative over **HA**:

(1) $$\forall\alpha \in A^\omega \exists\beta B(\alpha,\beta) \to \exists\gamma \forall\alpha \in A^\omega\, B(\alpha,\gamma|\alpha),$$

where A^ω is defined as in Subsection 1.2, Ax almost negative, containing only number-variables free. More precisely, we have

Theorem. **EL** + all instances of GC of the form (1) is conservative over **HA**.

Remark. Similarly, the methods of TROELSTRA (1974) may be extended to obtain conservativeness over $\mathbf{EL} = \mathbf{EL}_1 + \mathrm{AC}_{0,1}$ for certain generalizations of the fan theorem to finitely branching trees coded by almost negative formulae.

6.5. Open problems

Some of these we have already encountered:

(i) Is the schema $\mathbf{EBI}_0''$ derivable from $\mathbf{EBI}_0$ (cf. Subsection 1.7) without the use of APC? The method of reducing $\mathbf{EBI}_0'$ to $\mathbf{EBI}_0$ (Subsection 2.5) does not apply to $\mathbf{EBI}_0''$.

(ii) Similar questions may be asked in connection with $\mathbf{EBI}_1$. In this case, if we formulate $\mathbf{EBI}_1'$, $\mathbf{EBI}_1''$ in analogy to $\mathbf{EBI}_0'$, $\mathbf{EBI}_0''$, we do not know whether the arrows in

$$\mathbf{EBI}_1'' \rightarrow \mathbf{EBI}_1' \rightarrow \mathbf{EBI}_1$$

can be reversed (relative to **EL**, say).

(iii) Is c-$\mathbf{EBI}_0$ proof-theoretically stronger than $\mathbf{EBI}_0$? A negative answer for **EL** does not obviously imply a negative answer for **CS*** or **CS**. If the answer is negative, we may still ask whether c-$\mathbf{EBI}_0$ is derivable from $\mathbf{EBI}_0$. Similarly for the schemata $\mathbf{EBI}_0'$, $\mathbf{EBI}_0''$, $\mathbf{EBI}_1$, $\mathbf{EBI}_1'$, $\mathbf{EBI}_1''$ and the corresponding forms with parameters.

As to the possibility of a refutation of c-$\mathbf{EBI}_0$ in a theory such as **CS**, the search can be narrowed down by the following two observations:

(1) In **CS** we have $\forall\alpha\neg\neg\exists a(\alpha=a)$. So if $\mathrm{Bar}(A,P)$, $\mathrm{Mon}(A,P)$, $\mathrm{Ind}(A,P)$, A containing a parameter α, $Pn\equiv P(n,\alpha,\ldots)$ we find with l-$\mathbf{EBI}_0$ under the assumption $\exists a(\alpha=a)$ $P(n,\alpha,\ldots)$, and thus, with $\neg\neg\exists a(\alpha=a)$, $\neg\neg P(\langle\rangle,\alpha,\ldots)$; therefore on assumption of l-$\mathbf{EBI}_0$ we won't find our counterexample in the form $\neg P(\langle\rangle)$; and if P itself does not contain choice parameters, we also cannot expect to obtain a refutation of $\forall\alpha(\mathrm{Bar}(A,P)\wedge\mathrm{Mon}(A,P)\wedge\mathrm{Ind}(A,P)\rightarrow P\langle\rangle)$, this being derivable from l-$\mathbf{EBI}_0$.

(2) Since c-$\mathbf{EBI}_1$ can be shown to be realizable by functions (see Subsection 6.3), the system of KLEENE and VESLEY (1965) + c-$\mathbf{EBI}_1$ is consistent; this shows that our only chance of deriving a contradiction in **CS** + c-$\mathbf{EBI}_0$ or **CS** + c-$\mathbf{EBI}_1$ is to utilize the one axiom in **CS** preventing the identification "lawlike = choice": BC-F, continuity for lawlike functions.

(iv) Is $\mathbf{EBI}_0 + \mathbf{EL}_1$ conservative over $\mathbf{EL} + \{\mathbf{EBI}_0(A)$: *A almost negative*$\}$ for arithmetical sentences? and similarly with **CS*** instead of $\mathbf{EL}_1$?

The preceding questions can easily be multiplied by refining the distinctions.

(v) (G. Kreisel) Which inductively defined sets of $\mathbf{ID}^i_{<\omega}$ are explicitly definable in $\mathbf{EL}_1 + \mathbf{EBI}_0$? We recall that three basic types of explicit definition of sets introduced by monotone generalized inductive definitions are known:

(a) by means of comprehension principles, as the least predicate satisfying certain closure conditions;

(b) from below, defining for a g.i.d. condition $A(P,x)$ a sequence P_A^α (α an ordinal) such that

$$P_A^0 = \varnothing, \quad P_A^\alpha(x) \leftrightarrow (\exists \beta < \alpha) A(P_A^\beta, x).$$

In the classical case there is always a closure ordinal (the least non-recursive ordinal) γ such that $P_A^{\gamma+1} = P_A^\gamma$.

(c) By use of unsecured sequences (via principles of bar induction) this gives certain tree-like g.i.d. sets directly; other g.i.d. sets may be obtained from these by explicit definition (cf. §4 of KREISEL and TROELSTRA (1970)).

As to (c), it will be clear that certain improvements over §4 of KREISEL and TROELSTRA (1970) are possible, at least if we also use APC, the axiom of partial choice; the notion of strictly positive $A(P,a)$ may then be enlarged so as to include closure under the logical operations $\forall x \in B$, B almost negative, containing only numerical variables free. But it is not clear how to deal with $A(P,a)$ where P may have weakly positive occurrences, such as positive occurrences of $\neg\neg P\varphi$ or negative occurrences of $\neg P\varphi$.

However, the exact extent of explicit definability of g.i.d. sets by methods (b) or (c) deserves further study.

Acknowledgements

The author owes a special debt of gratitude to H. Luckhardt, who unselfishly took the considerable trouble of going in detail through a preliminary version of this paper, providing valuable comments and corrections. Luckhardt's writings and conversations renewed my interest in extended bar induction. G. Kreisel provided stimulating correspondence and criticism of an early draft of the introduction.

References

DILLER, J., and H. VOGEL

[1975] Intensionale Funktionalinterpretation der Analysis, in: *Proof Theory Symposium Kiel 1974*, edited by J. Diller and G. H. Müller (Springer, Berlin), pp. 56–72.

ERSHOV, Y. L.

[1974] The model G of the theory BR, *Soviet Math. Dokl.*, **15**, 1158–1161; translation of *Dokl. Akad. Nauk. SSSR*, **217**, 1004–1006.

[1977] Constructions 'by finite', in: *Logic, Foundations of Mathematics, and Computability Theory*. (*Part one of the proceedings of the fifth international Congress of Logic, Methodology and Philosophy of Science, London, Ontario, Canada–1975.*), edited by R. E. Butts and J. Hintikka (Reidel, Dordrecht), pp. 1–9.

FOURMAN, M., and J. M. E. HYLAND

[1979] Sheaf models for analysis, in: *Applications of Sheaves*, edited by M. Fourman, C. Mulvey and D. Scott (Springer, Berlin).

HOWARD, W. A.

[1968] Functional interpretation of bar induction by bar recursion, *Compositio Math.*, **20**, 107–124.

HOWARD, W. A., and G. KREISEL

[1966] Transfinite induction and bar induction of types zero and one, and the rôle of continuity in intuitionistic analysis, *J. Symbolic Logic*, **31**, 325–358.

HYLAND, J. M. E.

[1975] Recursion theory on the countable functionals, Thesis, University of Oxford, Oxford.

KLEENE, S. C., and R. E. VESLEY

[1965] *The Foundations of Intuitionistic Mathematics, Especially in Relation to Recursive Functions* (North-Holland, Amsterdam).

KREISEL, G.

[1963] Introduction to Vol. 2, in: The Stanford Report on the foundations of analysis, Stanford University, Stanford.

[1967] Mathematical logic: what has it done for philosophy?, in: *Bertrand Russell, Philosopher of the Century*, edited by R. Schoenman (Allen and Unwin, London), pp. 201–272.

KREISEL, G., and A. S. TROELSTRA

[1970] Formal systems for some branches of intuitionistic analysis, *Ann. Math. Logic*, **1**, 229–387.

LUCKHARDT, H.

[1973] *Extensional Gödel Functional Interpretation. A Consistency Proof of Classical Analysis* (Springer, Berlin).

[1977] Über das Markov-Prinzip II, *Arch. Math. Logik Grundlagenforsch.*, **18**, 147–157.

POHLERS, W.

[1977] Beweistheorie der iterierten induktiven Definitionen, Habilitationsschrift, Ludwig-Maximilians Universität, München.

SCARPELLINI, B.

[1972] A formally constructive model for bar recursion of higher types, *Z. Math. Logik Grundlagen Math.*, **18**, 321–383.

[1973] On bar induction of higher types for decidable predicates, *Ann. Math. Logic*, **5**, 77–163.

Sieg, W.

[1977] Trees in metamathematics, Thesis, Stanford University, Stanford.

Spector, C.

[1962] Provably recursive functionals of analysis: a consistency proof of analysis by an extension of principles formulated in current intuitionistic mathematics, in: *Proceedings of Symposia in Pure Mathematics V*, edited by J. C. E. Dekker (Am. Math. Soc., Providence, RI), pp. 1–27.

Troelstra, A. S.

[1973] Chapter I–IV, in: *Metamathematical Investigation of Intuitionistic Arithmetic and Analysis*, edited by A. S. Troelstra (Springer, Berlin).

[1974] Note on the fan theorem, *J. Symbolic Logic*, **39**, 584–596.

[1975] Markov's principle and Markov's rule for theories of choice sequences, in: *Proof Theory Symposium Kiel 1974*, edited by J. Diller, and G. H. Müller (Springer, Berlin), pp. 370–383.

[1977] Special instances of generalized continuity which are conservative over intuitionistic arithmetic, *Indag. Math.*, **39**, 55–65.

[1977A] Some models for intuitionistic finite type arithmetic with fan functional, *J. Symbolic Logic*, **42**, 194–202.

J. Barwise, H. J. Keisler and K. Kunen, eds., *The Kleene Symposium*

Intuitionistic Analysis: The Search for Axiomatization and Understanding

Richard Vesley
Department of Mathematics, State University of New York at Buffalo, Buffalo, NY 14214, U.S.A.

Dedicated to Professor S. C. Kleene on the occasion of his 70th birthday

Abstract: This paper is an historical sketch, tracing the development of current axiomatizations of Brouwer's analysis.

Three quarters of a century ago L. E. J. Brouwer defended his doctoral thesis in Amsterdam and arose almost instantly as the leader of the intuitionistic program in the foundations of mathematics. His vigorous attacks on the views of Hilbert and Russell were well-known long before the world began to appreciate the positive aspects of his own program. Over a quarter of a century ago, as Brouwer's long career drew near its close, an effort began to use formal systems and their interpretations to study his mathematical analysis, i.e. the intuitionistic theory of real numbers and sets and functions of them. The aim was to accomplish for analysis, an area to which Brouwer had given much of his attention, what had by that time already been achieved for intuitionistic natural number arithmetic—axiomatization (based on Heyting (1930)), interpretation (Kleene, 1945; Nelson, 1947), and, it was hoped, improved understanding. The man we recognize at this meeting was the first to chart a successful course in this effort. He was for a time an unaccompanied pioneer and then a leader in this enterprise.

Today we enjoy a much clearer view than before of the fruits of Brouwer's piercing intuition, of what these results are and how they can be linked. In these brief remarks, I concentrate on how this has come about. So as not to diffuse the focus, I consider only highlights. I leave unmentioned much of great, perhaps greater, mathematical beauty and interest that has arisen as a consequence, but that is not immediately relevant to understanding intuitionistic analysis (e.g., relations to classical systems or among systems containing principles denied or deliberately not affirmed intuitionistically, unintended interpretations, etc.). I omit also careful definitions and statements of results. These are mostly quite complicated and are readily available in the literature, which already includes several excellent surveys, e.g. TROELSTRA (1969) and VAN DALEN (1973); two have appeared very recently, DUMMETT (1977) and TROELSTRA (1977).

Our review begins with Brouwer's work and is devoted to what we have learned about the specifically Brouwerian notion of choice sequence and the mathematics based on it. A more complete historical account, even within the same subject framework, would have to evaluate the influence on Brouwer of earlier constructivists, particularly the French semi-intuitionists. A more comprehensive subject treatment, even starting with Brouwer, might deal with the several varieties of analysis now entitled in one way or another (e.g., by being based on intuitionistic logic) to be called intuitionistic. Among these Brouwer's analysis possesses a coherence matched only by a few. Worth noting here by contrast are two of these few, both based on the belief that real numbers should be given by laws. The Russian constructive analysis uses Markov algorithms (equivalently, recursive functions) to represent real numbers; it allows but tries to avoid deviations from intuitionistic logic to employ Markov's principle. The theory is exceedingly formal in its development and thus very clear but in this respect far removed from the spirit of Brouwer. The analysis of E. Bishop is much more in this spirit. Bishop requires that real numbers be given by laws, but does not require that laws be represented by recursive functions.

Now Brouwer did study from time to time what he called the "reduced continuum" of lawlike real numbers. But he stated forcefully that it was precisely in going beyond this that intuitionistic analysis had made a new contribution and acquired a unique character. He wrote (BROUWER, 1952, p.142):

> "One might fear that intuitionist mathematics must necessarily be poor and anaemic, and in particular would have no place for analysis. But this fear would have presupposed that infinite sequences generated by the intuitionist...would have to be fundamental sequences, i.e. predeterminate infinite sequences which like classical ones, proceed in such a way that, from the beginning, the *m*th term is fixed for each *m*. Such however is not the case...a much wider field of development which includes analysis, and in several places far exceeds the frontiers of classical mathematics, is opened by the
>
> SECOND ACT OF INTUITIONISM
>
> *which recognizes the possibility of generating new mathematical entities:*
>
> *firstly in the form of infinitely proceeding sequences* $p_1, p_2, \ldots,$ *whose terms are chosen more or less freely from mathematical entities previously acquired; in such a way that the freedom of choice existing perhaps for the first element* p_1 *may be subjected to a lasting restriction at some*

following p_ν and again and again to sharper lasting restrictions or even abolition at further subsequent p_ν's, while all these restricting interventions, as well as the choices of the p_ν's themselves, at any stage may be made to depend on possible future mathematical experiences of the creating subject;

secondly, in the form of mathematical species, i.e. properties supposable for mathematical entities previously acquired...."

A point of this statement is that intuitionistic mathematics involves notions not studied in traditional mathematics. This is especially obvious in analysis, where the real numbers are obtained from the infinitely proceeding sequences (or "choice sequences") mentioned in the quotation. In describing these sequences, Brouwer refers to subjective and temporal conditions in a way that appears beyond conventional mathematical treatment.

A second but related point combines with this first to make formidable the task of understanding Brouwer's analysis. This second point is the intuitionistic rejection of logical or linguistic bases for mathematics. In BROUWER (1952) this constitutes part of the first of his two "Acts of Intuitionism". This view is very explicit from the beginning (BROUWER, 1907, pp.72-73). Evidently it will concern all who aim to axiomatize intuitionistic theories. But it raised a particularly puzzling problem in connection with analysis. Axiomatization for logic and arithmetic turned out to be, or could be seen as, careful selection of proper subsets of classical logical and linguistic rules. But clearly in analysis axiomatization, if possible at all, demanded more than suppressing certain classical truths. For there were statements like the one quoted above in which Brouwer claimed to be moving beyond classical mathematics and also others in which he apparently did affirm classically false propositions (e.g., the uniform continuity theorem of BROUWER (1924)). In this situation it was not of much comfort to be told by the intuitionist that logic and language are not equal to the job of providing a foundation.

Thus, although (or because) Brouwer's analysis arose as the positive part of a foundational critique of classical analysis, and so was built with great sensitivity to its own foundations, it nevertheless presented difficuties to anyone wishing to dissect it by traditional means.

A consequence was that, unaided by any linguistic or logical frame, the reader of Brouwer's papers had to find harmony among such observations as the following. Real numbers, if they were Cauchy sequences, were neither assumed to be given by a law nor to be classical set-theoretic sequences. By contrast, functions on the real number were law-given (BROUWER, 1923, p.3). Allowed as means of proving theorems about real functions were considerations about arbitrary proofs (BROUWER, 1924, pp. 189–90), where these were emphatically not proofs in any formal systems. Certain classically true propositions were deemed false, while other classically false ones were deemed true.

Brouwer's own efforts to elucidate his foundations remain of course an inspiration and a measure by which we test our own subsequent attempts to understand. Yet in themselves they did not bring the understanding that might have been wished. For example, over a period of 30 years (from BROUWER (1924) to BROUWER (1954)), Brouwer repeated with slight change the argument for his "bar theorem". Seen in hindsight, following the work of Kleene, we know that this argument involved the assertion of an induction principle independent of the rest of intuitionistic analysis; on its surface, it involved assertions about the nature of certain mathematical proofs. On either count, one might have expected the argument to evoke a critique, an explication, further use or extension. There was none. (The first use of some of the ideas, but in a non-intuitionistic context, is in KLEENE (1955)).

Aside from Brouwer's own work the earliest attempt to provide a systematization of his ideas in analysis was the "hesitant beginning" (in the words of TROELSTRA (1977, p.132)) of HEYTING (1930). The bar theorem just mentioned was not included; no classical or neutrally plausible interpretations or further studies ensued.

The next noteworthy event, and one with lasting consequences, was the argument of KLEENE (1950). He showed that if lawlike was interpreted as recursive, then one could give a mathematical demonstration supporting Brouwer's assertion that his continuum involved more than the "reduced continuum" of lawlike reals. Prior to this one might have expected Brouwer's analysis to have an unintended model in the recursive real numbers. As early as Brouwer (1912, pp.91–92) it had been said that only real numbers given by laws could be constructed (though cf. bottom p.92–top p.93). Kleene investigated Brouwer's "fan theorem" (BROUWER, 1954, p.15, 1927, p.66). He exhibited an instance of this which cannot hold if the sequences range over only recursive ones. This example has had a longer life and utility than any other metamathematical result in this area. It is used in the two recent books cited above, DUMMETT (1977, pp.73-74) and TROELSTRA (1977, pp.83-84).

In the 1950 paper Kleene also made some preliminary remarks concerning his intention to formalize intuitionistic analysis. In part the objective was to extend the work begun in the realizability interpretation of intuitionistic arithmetic and to tie together the two aspects of constructivity represented by intuitionism and recursive function theory. Such a connection was invited particularly by Brouwer's references to "laws" and by the apparent prominence in intuitionism of the concept of a "spread" (BROUWER, 1954; called "Menge" in BROUWER, 1925). It suffices here to say that a spread is a law accepting or rejecting initial segments of choice sequences, and mapping accepted segments into segments of corresponding sequence outputs. Different spreads represent the real axis, particular finite or infinite closed intervals, etc.

Through the 1950s and especially in the latter part of the decade there grew an interest in the foundations of intuitionistic analysis. In large part this was because of work under way by Kleene and begun by Kreisel. An important additional contribution was the appearance of HEYTING (1956). Although Heyting did not present formalisms or interpretations, his book in its organization and in some detail reinforced especially two of Brouwer's views. First, the fundamental significance of analysis in the intuitionistic program; Heyting developed the theory of real numbers early and that theory influenced the much later formulation of logical principles. Second, the rooting of analysis in the notion of choice sequence. Heyting made it clear that this notion was what distinguished intuitionistic analysis, and not, for instance, either adherence to a particular logical dogma or some special concept of lawlike objects. Lawlike functions, and in particular spreads, were assigned by Heyting to a somewhat subsidiary, though necessary, role.

These same points were to receive emphasis in the formalism of Kleene. Already in 1950 and again in 1957 Kleene outlined a choice of primitives for a system of analysis, which included only variables α, $\beta, \ldots$ to represent choice sequences and say $x, y, \ldots$ to represent natural numbers. Constructive or lawlike functions were not taken as primitive. Reasons for this last decision were later discussed in KLEENE (1967, 1969). Of course, under Church's Thesis, which is not accepted intuitionistically, constructive functions could be represented. More relevant from the viewpoint we are taking here is that this economical language turned out to enable the expression of much of Brouwer's analysis in a reasonably direct and (it seems) undistorted manner. While sometimes criticized for omission, as noted below, it has also been described as a "*tour de force*" (DUMMETT, 1977, p.311). In any event, Kleene's system already in its language emphasized conspicuously the choice sequence concept. As matters turned out, the first point mentioned above, the fundamental significance of analysis in the intuitionist program, was also illustrated in the formalism; for certain classically valid logical principles were formally refuted by examples in analysis. But this latter development was not yet explicit. For in 1950 a satisfactory list of postulates for choice sequences had yet to be distilled.

This is an appropriate place to note that while the above reflects one plausible reading of Brouwer, it is also possible to find support for the, not necessarily incompatible, view that a proper primitive concept for intuitionism should be that of a proof or a construction (KREISEL, 1959, p.109), and that intuitionistic mathematics including analysis should be built on this basis. KREISEL (1960) outlines the beginning of such a development; GOODMAN (1970) shows that intuitionistic logic and arithmetic can be so developed. However, so far as can be seen at this time, this route does not lead to analysis. That is, if one wishes to have choice sequences in a theory of constructions, one must introduce them as primitive.

Still another significant basic concept was introduced in KREISEL (1958), that of lawless sequence (at first, i.e. in 1958, called free or absolutely free choice sequence). It is debatable to what extent this notion was envisioned by Brouwer (cf. TROELSTRA (1977, pp. 131-132) and references cited there). Kreisel's was certainly the first detailed discussion and was not inspired by any Brouwerian formulation (KREISEL, 1967, p.180). In 1958 Kreisel gave some axioms for lawless sequences, including the distinctive continuity axiom: if a (one-place) predicate holds at all in a given spread b of lawless sequences, then it holds for a certain subspread of (in general, infinitely many different) lawless sequences, i.e. all those beginning with a certain specified initial segment. The special case if b is the universal spread of all lawless sequences is

$$A(\alpha)\supset\exists n\forall\beta(\bar{\alpha}(n)=\bar{\beta}(n)\supset A(\beta)),$$

where $\bar{\alpha}(n)$ is the natural number encoding the sequence

$$\langle\alpha(0),\alpha(1),\ldots,\alpha(n-1)\rangle.$$

This axiom schema, while expressing the primary and a very natural aspect of the lawless sequence concept, nevertheless may appear paradoxical taken together with the few other natural axioms for the concept. In 1958 Kreisel conjectured the possibility of a consistency proof by translation into a classical system. This possibility was later verified (KREISEL, 1968) in a translation which in fact reduced the theory of lawless sequences to a neutral theory of constructive functions.

The lawless sequences were not in 1958 intended to provide a foundation for intuitionistic analysis. They enter in this review for two reasons. First, the naturalness and simplicity of their formulation has inspired efforts to use them as such a foundation, efforts which though not completely successful have yielded results that we must acknowledge (MYHILL, 1967; VAN DALEN and TROELSTRA, 1970; cf. TROELSTRA, 1977, Chapter 4). Second, the consistency-eliminability results of Kreisel just cited are a prototype (at least pedagogically; I am not sure about historically) for the analogous Kreisel–Troelstra results for choice sequences proper, to be noted below.

We return to the search for an axiomatization of analysis. We have mentioned the fan theorem; this was used by Brouwer in the proof of his uniform continuity theorem, and the latter was in the intuitionists' analysis a pinnacle. It was presented frequently by Brouwer in writing and lecturing. It was displayed prominently in HEYTING (1956). KLEENE (1957) described his evolving system as including at least an axiom of choice

$$\forall x\exists y A(x,y)\supset\exists\alpha\forall x A(x,\alpha(x))$$

and the fan theorem in the form

$$\forall \alpha_B \exists n A(\alpha,n) \supset$$
$$\exists z \forall \alpha_B \exists n \forall \gamma_B \{\forall x [x<z \supset \gamma(x)=\alpha(x)] \supset A(\gamma,n)\},$$

where $\forall \alpha_B(\;)$ is $\forall \alpha[\forall x(\alpha(x) \leqslant \beta(\bar{\alpha}(x)) \supset (\;)]$. In KREISEL (1958) a similar schema was given for lawless sequences.

There was yet some distance to go before achieving a satisfactory anatomy of at least a large portion of intuitionistic analysis. For Brouwer had obtained the fan theorem as a consequence of his bar theorem. Now the latter came under scrutiny. The challenge was, and is, that to a degree unmatched in any other of his arguments (with one possible exception to be noted below), Brouwer relied here on observations concerning the nature of intuitionistic proofs. Kleene extracted from Brouwer's argument a schema for proof by induction (called subsequently by Kreisel: "bar induction"). In KLEENE and VESLEY(1965) (hereafter: "FIM"), this schema was given in several equivalent forms. One is (*26.3a of FIM, p.54)

$$\forall z(\mathrm{Seq}(z) \supset R(z) \vee \neg R(z)) \& \forall \alpha \exists x R(\bar{\alpha}(x)) \&$$
$$\forall z(\mathrm{Seq}(z) \& R(z) \supset A(z)) \& \forall z(\mathrm{Seq}(z) \& \forall s A(z * \langle s \rangle) \supset A(z))$$
$$\supset A(\langle\rangle)$$

where $\mathrm{Seq}(z)$ expresses that z encodes a sequence, $*$ is concatenation, $\langle\rangle$ is the empty sequence, $\langle s \rangle$ the sequence consisting of just s. Kleene's formulation was, as he later remarked (KLEENE, 1973, p.13), "somewhat delicate"; e.g. the decidability condition on R, or some equivalent, was implicit but not obvious in Brouwer's presentation.

Soon Spector, also studing Brouwer's argument and aided by Kleene's formulations and Kreisel's suggestions, had achieved a generalized schema, one for definition of a functional by recursion ("bar recursion") in higher types. (At lowest type, this schema is validated in the proper context by bar induction.) SPECTOR (1962) presented for the first time in print a bar schema.

Kleene's bar induction postulate is valid both intuitionistically and classically. (The function variables can be taken as choice sequences or as classical number theoretic functions.) The fan theorem is not. Even the fathoming, in part, of Brouwer's deepest argument leaves a gap which must be filled by a non-classically valid principle, something sufficient to enable the derivation of the fan theorem.

Here Kleene formulated "Brouwer's principle for numbers", according to which functionals from choice sequences to the natural numbers were required to be continuous. In FIM, p.73 this principle is

$$\forall \alpha \exists x A(\alpha,x) \supset \exists \tau \forall \alpha \exists y \{\tau(\bar{\alpha}(y))>0 \&$$
$$\forall z(\tau(\bar{\alpha}(z))>0 \supset y=z) \& A(\alpha,\tau(\bar{\alpha}(y)) \dot{-} 1)\}.$$

As stated the principle does more than give continuity; it is also a choice principle giving the existence of a functional, and it further prescribes how such a functional is to be represented by a function from N to N. In fact Kleene postulated as his second novel schema a more general one, "Brouwer's principle for functions", which is like the principle for numbers but with antecedent $\forall\alpha\exists\beta A(\alpha,\beta)$ and correspondingly altered conclusion. The principle for numbers is a consequence. These principles, or their equivalents, have come to be called $\forall\alpha\exists x$-continuity and $\forall\alpha\exists\beta$-continuity. (KLEENE (1959) and KREISEL (1959) independently discussed classically systems of countable (Kleene) or continuous (Kreisel) functionals of higher type with representing functions in $N\to N$.)

So, for FIM two quite unprecedented axiom schemes had been selected (before 1960); one, classically valid, for bar induction, and the other, classically invalid, for Brouwer's principle or continuity. As Kleene showed in FIM, surprisingly it is the classically valid one of these which justifies, in a sense, the claim in Brouwer's "Second Act" statement above that choice sequences are something new. For the bar induction schema forces in the interpretation sequences beyond the lawlike (recursive) ones. Equally striking is the compression into the single schema for Brouwer's principle of all the non-classical features of the system. Thus, as FIM contended in detail, two postulates can carry the weight for much of analysis of Brouwer's claims that with choice sequences he was going beyond classical mathematics, floating free of logico-linguistic foundations and in a position to refute classical truths.

Kreisel's approach to formalizing Brouwer's bar argument and to the non-classical continuity requirement was different. He proposed (by 1963) a system for analysis containing two (kinds of) postulates for choice sequences which are invalid if the function variables range over classical number theoretic functions. Further, his system contained variables $a,b,\ldots$ ranging over constructive functions, as seemed more in keeping with Brouwer. The first of the non-classical schemes relied on the extraction from Brouwer's bar argument of a description of a particular inductively defined class K of constructive functions which could be used to represent continuous functionals. Combining this description with the requirement that any functional from choice sequences to choice sequences should be continuous Kreisel postulated

$$\forall\alpha\exists\beta A(\alpha,\beta)\supset\exists a\in K\forall\alpha A(\alpha,a(\alpha)).$$

The corresponding postulate with β replaced by a variable for a constructive function was also included. The other of Kreisel's non-classical schemes was reminiscent of the lawless sequences scheme above; it asserts that if a predicate A holds of a choice sequence α, then there is a spread a including α such that A holds of all elements of a:

$$A(\alpha)\supset\exists a(\mathrm{Spr}(a)\&\ \alpha\in a\&\forall\beta\in aA(\beta)),$$

where Spr(a) expresses that a is a spread. The Kleene postulates can be derived easily in this Kreisel system.

Both the Kleene system and this Kreisel system were finally described in publication in 1965 (in FIM and KREISEL (1965)). Each was accompanied by justification of the appropriateness of the system. Such a justification could consist of two parts: first, metatheoretic arguments establishing consistency relative to some neutral or classical theory and also supporting the constructivity of the asserted postulates; second, an elaboration of known intuitionistic mathematics within the system. Kleene in FIM introduced his functional realizability interpretation and used it to provide the first; a reasonably extensive collection of known intuitionistic results, including some of the counterexamples to classical logical principles, was developed to provide the second. The Kreisel system including Kleene's could be supported by the identical second justification. For the first Kreisel provided an interpretation of his system into a neutral formal system which did not involve choice sequences but only natural numbers and constructive functions (and the class K). The treatment is parallel to but more complicated than the eliminability result of KREISEL (1968) for lawless sequences (which also depends on K).

Curiously, both these systems for intuitionistic analysis were challenged in respect of their treatments of spreads. Kleene's system was said not to permit proper representation of spreads because of the lack of constructive function variables. Kreisel's system, by contrast, overvalued the spread concept. TROELSTRA (1967) showed that the second above of Kreisel's two non-classical schemes was inconsistent with the identification of constructive with recursive functions. Though intuitionistically Church's Thesis is not accepted, neither is its negation. So Troelstra proposed a change in the axiom, into one he calls the axiom of analytic data; the idea is much the same as Kreisel's but the role previously played by spreads is assumed by ranges of continuous functionals:

$$A(\alpha) \supset \exists a \in K(\exists \beta(\alpha = a(\beta)) \& \forall \beta A(a(\beta)).$$

(Actually, this is not quite accurate, for Troelstra uses a special kind of variable for elements of K.) Under this change the desired elimination theorem can still be obtained, as was verified in KREISEL and TROELSTRA (1970), where much additional material concerning this new system CS also appears.

In TROELSTRA (1971) CS was shown to be a conservative extension of the system of FIM.

In the third edition of his book HEYTING (1971) implicitly acknowledged the intuitionists' interest in the account stemming from Kleene of intuitionist principles. For he added a discussion, prepared with the aid of Troelstra, on bar induction and continuity schemes.

The essence of Kleene's approach is to treat Brouwer's choice sequences as identical with the classical number-theoretic sequences constructively considered, where the guide to "constructively considered" is provided by the realizability interpretation. The essence of the Kreisel–Troelstra approach is to treat choice sequences as figures of speech; sentences containing choice sequence variables can be reduced to equivalent sentences not containing such variables but referring only to constructive functions. Both systems give a limpid description of certain principles for employing choice sequences, apparently including all that Brouwer had occasion to use, save as discussed below.

It is not hard to conjecture how Brouwer might greet either approach as a faithful mirror of his intention. Surely he would not admit either as definitive, since he would insist on the distinction between mathematics on the one hand and logic and language on the other. And presumably this would save him from any worry that, separately but in equal measure, these interpretations appear to threaten if not refute the contention that new entities (choice sequences) are being created which cannot be obtained classically.

But would his insight provide any refutation of these interpretations? In this connection there are important and provocative observations of MYHILL (1967). Just as it appeared that intuitionistic analysis was being tied up into a neat package of formal rules, Myhill arose to insist that Brouwer's views were not being adequately taken into account. He pointed particularly to two aspects of Brouwer's work and to a previously unobserved connection between them. One was Brouwer's use of continuity principles; the other was his use of what are often called "creative subject" arguments. The latter arguments were used occasionally by Brouwer to refute the intuitionistic analogues of certain classically true statements. These arguments could not be fitted into existing intuitionistic formalisms. This had not been felt embarrassing, mostly because what was lost in their absence was simply a few negative results. Nothing positive had been established by such arguments. What Myhill emphasized, and KREISEL (1967) also, was that from such arguments there might be distilled new and useful intuitionistic principles. Kripke had proposed a logical scheme which would provide the "creative subject" results and which would have positive consequences. Myhill made a strong case that Kripke's schema should be part of intuitionistic analysis. This schema, at least in its stronger forms, conflicts with $\forall\alpha\exists\beta$-continuity, a principle of CS and FIM. Myhill contended that this is symptomatic of the other inadequacy of these systems in reflecting Brouwer's thought. What could be justified intuitionistically, and what gives all that is needed for analysis, is a weaker continuity principle ($\forall\alpha\exists!\beta$-continuity), not in conflict with Kripke's schema. Here I shall not go into the deeper criticism made by Myhill, except to say that the issue was raised when and how are choice sequences to be considered as intensionally presented and when as extensionally presented. There is as yet no consensus. Nor is there consensus on which, if any, version of Kripke's schema should be accepted.

If the case against $\forall\alpha\exists\beta$-continuity prevails and the acceptable weaker principle is substituted, the consequence for CS is that the proof of the eliminability result fails, while the consequence for FIM is that a realizable principle is revealed to be intuitionistically invalid. The latter is not unprecedented but is one more concern to add to any doubts about the closeness of the interpretation to Brouwer's intention.

Myhill had a view of choice sequences which he believed settled these questions and which rested on the lawless sequence notion. But this view justified the previously mentioned overly strong axiom involving spreads which succumbed in 1968 to Troelstra.

Since Myhill's work there has been a new or at least strongly reinforced awareness that there may be a variety of different kinds of choice sequences, not only all possessing mathematical interest but several of which may be inherent in Brouwer's writings. For, as any serious reading reveals and as is to be expected, Brouwer's views over half a century seem to have changed and developed. In this connection a valuable service has been done the student recently in Heyting's edition of the first volume of Brouwer's works.

In the past decade by far the most active worker in intuitionistic analysis has continued to be Troelstra, who indefatigably produces interesting and intricate results, within and about analysis and in and for systems weaker and stronger. It would be hopeless to attempt a review of these here.

Presumably, one way in which we shall learn more about intuitionistic analysis is through attempts to use its principles, in the sharp forms now available to us, to achieve (informally) results of various kinds. W. VELDMAN (1976) and H. DE SWART (1976) have used intuitionistic analysis, as projected by Beth, to achieve a completeness theorem for intuitionistic predicate logic. Originally Kreisel used the lawless sequences for results concerning completeness. These two enterprises are elegantly contrasted and developed in TROELSTRA (1977). Informative uses are bound to be made of versions of Kripke's schema. One hopes this will also be true of bar induction, or of principles that may lie behind it.

My theme has been the quest for an understanding of Brouwer's analysis. If some of what I have said smacks more than usual in mathematics of textual translation and interpretation that is because for decades the task has been to catch up to (possibly we should say: to perfect) Brouwer's thought. If we have not a complete understanding yet, we have certainly moved much closer than 28 years ago when Kleene projected his search. I have tried to note the most important mathematical steps in that quest. KREISEL (1971) has said the work that most needs doing now is philosophical. From one viewpoint the obstacles in understanding have not been removed. But they have been bypassed, and most workers today use bypass routes pioneered by Kleene. For it was Brouwer's vision what theory should be attained and Kleene's insight which first found implicit in Brouwer a usable, more conventionally mathematical (axiomatic) road to that theory.

Surely philosophers and mathematicians can find much left to explore in the issues that have been opened. At some indeterminate point, no doubt, we shall pass, or have passed, beyond the limits of Brouwer's own discernment into a clearer view and keener understanding than he possessed, led there by the work of Heyting, Kleene, Kreisel, Troelstra, Myhill and the rest. But this Brouwer knew, and it is entirely in keeping with the intuitionistic spirit. I close by quoting words of Brouwer, spoken originally in tribute to Mannoury, but words that can apply equally to Brouwer himself, or to others like Kleene who have built remarkable mathematics (BROUWER, 1946, p.193):

> "For the undertone of ...[your] argument has not whispered: 'Behold, some new acquisitions for our museum of immovable truths', but something like this: 'Look what I have built for you out of the structural elements of our thinking. These are the harmonies I desired to realize. Surely they merit that desire? This is the scheme of construction which guided me. Behold the harmonies, neither desired nor surmised, which after the completion surprised and delighted me. Behold the visions which the completed edifice suggests to us, whose realization may perhaps be attained by you or me one day.' "

References

BROUWER, L. E. J.

[1907] Over de grondslagen der wiskunde, Thesis, Amsterdam, English translation in (1975), pp. 11–101.

[1912] Intuitionisme en formalisme (Amsterdam), also (1913) in *Bull. Am. Math. Soc.*, **20**, 81–96.

[1923] Begrundung der Funktionlehre unabhangig vom logischen Satz vom ausgeschlossen Dritten, Erster Teil, Stetigkeit, Messbarkeit, Derivierbarkeit, Royal Dutch Acad., 1st Section, No. 2.

[1924] Beweis, dass jede volle Funktion gleichmassig stetig ist, *Proc. Roy. Dutch Acad.*, **27**, 189–193.

[1925] Zur Begründung der intuitionistischen Mathematik I, *Math. Ann.*, **95**, 453–472.

[1927] Uber Definitionsbereiche von Funktionen, *Math. Ann.*, **97**, 60–75.

[1946] Address delivered on Sept. 16, 1946, at the University of Amsterdam by Professor L.E.J. Brouwer on the conferment upon Professor G. Mannoury of the honorary degree of Doctor of Science, in (1975), pp.472–476.

[1952] Historical background, principles and methods of intuitionism, *South African J. Sci.*, **49**, 139–146.

[1954] Points and spaces, *Can. J. Math.*, **6**, 1–17.

[1975] *Collected Works, Vol. 1: Philosophy and Foundations of Mathematics*, edited by A. Heyting (North-Holland, Amsterdam).

De Swart, H.

[1976] Another intuitionistic completeness proof, *J. Symbolic Logic*, **41**, 644–662.

Dummett, M.

[1977] *Elements of Intuitionism* (Oxford Univ. Press, Oxford).

Goodman, N.

[1970] A theory of constructions equivalent to arithmetic, in: *Intuitionism and Proof Theory*, edited by A. Kino, J. Myhill and R. Vesley (North-Holland, Amsterdam), pp.101–120.

Heyting, A.

[1930] Die formalen Regeln der intuitionistischen Logik and die formalen Regeln der intuitionistischen Mathematik, *S.-B. Preuss. Akad. Wiss., Physil-Math. Kl.*, 42–56 and 158–169.

[1956] and [1971] *Intuitionism, and Introduction* (North-Holland, Amsterdam) 1st ed., 1956; 3rd ed., 1971.

Kleene, S. C.

[1945] On the interpretation of intuitionistic number theory, *J. Symbolic Logic*, **10**, 109–124.

[1950] Recursive functions and intuitionistic mathematics, in: *Proceedings of the International Congress of Mathematicians (1950) vol. 1, (1952)*, pp.679–685.

[1955] On the forms of the predicates in the theory of constructive ordinals, *Am. J. Math.*, **77**, 405–428.

[1957] Realizability, in: *Summaries of Talks Presented at the Summer Institute of Symbolic Logic in 1957 at Cornell University, Vol. 1* (Institute for Defense Analysis, Princeton, R.I), pp.100–104; Reprint in *Constructivity in Mathematics*, edited by A. Heyting (North-Holland, Amsterdam, 1959).

[1959] Countable functionals, in: *Constructivity in mathematics*, edited by A. Heyting (North-Holland, Amsterdam), pp.81–100.

[1967] Constructive functions, in: "The foundations of intuitionistic mathematics", in: *Logic, Methodology and Philosophy of Science III, Proceedings of the Third International Congress for LMPS (1967)* (North-Holland, Amsterdam), pp.137–144.

[1969] Formalized recursive functionals and formalized realizability, Memoirs Am. Math. Soc. 89.

[1973] Realizability: a retrospective survey, in: *Cambridge Summer School in Mathematical Logic*, edited by H. Rogers and A. Mathias (Springer, Berlin), pp.95–112.

KLEENE, S. C., and R. E. VESLEY

[1965] FIM *The Foundations of Intuitionistic Mathematics, especially in relation to recursive functions* (North-Holland, Amsterdam).

KREISEL, G.

[1958] A remark on free choice sequences and the topological completeness proofs, *J. Symbolic Logic*, **23**, 369–388.

[1959] Interpretation of analysis by means of constructive functionals of finite types, in: *Constructivity in Mathematics*, edited by A. Heyting (North-Holland, Amsterdam), pp.101–128.

[1960] Foundations of intuitionistic logic, in: *Logic, Methodology and Philosophy of Science I, Proceedings of the First International Congress for LMPS (1960)* (Stanford Univ., Stanford), pp.198–210.

[1965] Mathematical logic, in: *Lectures on Modern Mathematics, Vol. 3,* edited by T.L. Saaty (Wiley, New York), pp.95–195.

[1967] Informal rigour and completeness proofs, in: *Problems in the Philosophy of Mathematics*, edited by I. Lakatos (North-Holland, Amsterdam), pp.138–186.

[1968] Lawless sequences, *Compositio Math.*, **20**, 222–248.

[1971] Perspectives in the philosophy of pure mathematics, in: *Logic, Methodology and Philosophy of Science IV, Proceedings of the Fourth International Congress for LMPS (1971)* (North-Holland, Amsterdam), pp.225–277.

KREISEL, G., and A. S. TROELSTRA

[1970] Formal systems for some branches of intuitionistic analysis, *Ann. Math. Logic*, **1**, 229–387.

MYHILL, J.

[1967] Notes towards an axiomatization of intuitionistic analysis, *Logique et Analyse*, **9**, 280–297.

NELSON, D.

[1947] Recursive functions and intuitionistic number theory, *Trans. Am. Math. Soc.*, **61**, 307–368.

SPECTOR, C.

[1962] Provably recursive functionals of analysis: a consistency proof of analysis by an extension of principles formulated in current intuitionistic mathematics, in: *Recursive Function Theory, Proceedings of Symposia in Pure Math, Vol. 5* (Am. Math. Soc., Providence, RI), pp.1–27.

TROELSTRA, A. S.

[1967] The theory of choice sequences, in: *Logic, Methodology and Philosophy of Science III, Proceedings of the Third International Congress for LMPS* (North-Holland, Amsterdam), pp.201–223.

[1969] *Principles of Intuitionism* (Springer, Berlin).

[1971] An addendum, *Ann. Math. Logic*, **3**, 437–439.

[1977] *Choice sequences* (Oxford Univ. Press, Oxford).

VAN DALEN, D.

[1973] Lectures on intuitionism, in: *Cambridge Summer School in Mathematical Logic*, edited by A. Mathias and H. Rogers (Springer, Berlin), pp.1–89.

VAN DALEN, D., and A. S. TROELSTRA

[1970] Projections of lawless sequences, in: *Intuitionism and Proof Theory*, edited by A. Kino, J. Myhill and R. Vesley (North-Holland, Amsterdam), pp.163–186.

VELDMAN, W.

[1976] An intuitionistic completeness theorem for intuitionistic predicate logic, *J. Symbolic Logic*, **41**, 159–166.

This page intentionally left blank

J. Barwise, H. J. Keisler and K. Kunen, eds., *The Kleene Symposium*

The Busy Beaver Method*

Robert P. Daley

Computer Science Department, University of Pittsburgh, Pittsburgh, PA. 15260, U.S.A.

Dedicated to Professor S. C. Kleene on the occasion of his 70th birthday

Abstract: This paper exploits computational complexity notions to obtain a simpler construction of a solution to Post's problem which is different from previous constructions involving the priority method. Moreover, the set constructed admits a rather explicit description.

1. Introduction

In this paper we show how some of the finite injury and infinite injury priority arguments can be simplified by making explicit use of the primitive notions of axiomatic computational complexity theory. Phrases such as "perform n steps in the enumeration of W_i" certainly bear witness to the fact that many of these complexity notions have been used implicitly from the early days of recursive function theory. However, other complexity notions such as that of an "honest" function are not so apparent, neither explicitly nor implicitly. Accordingly, one of the main factors in our simplification of these diagonalization arguments is the replacement of the characteristic function χ_A of a set A by the function ν_A, which is the next-element function of the set A. Another important factor is the use of busy beaver sets (see Daley (1978)) to provide the basis for the required diagonalizations thereby permitting rather simple and explicit descriptions of the sets constructed. Although the differences between the priority method and our method of construction are subtle, they are nonetheless real and noteworthy. Further applications of this method can be found in Daley (1980).

In preparation for the results which follow we devote the remainder of this section to the requisite definitions and notions as well as some preliminary lemmas. A more comprehensive discussion of many of the notions in this section can be found in Daley (1978).

*This research was supported by NSF Grant MCS 76-00102.

An *acceptable Gödel numbering* (see ROGERS (1967)) is a sequence of functions $\{\phi_i^X\}$ such that

(ϕ1) for every set X of natural numbers, every X-computable partial function is contained in the list $\{\phi_i^X\}$,

(ϕ2) there exists a "universal program" u such that for every set X of natural numbers,

$$(\forall i)(\forall p)\left[\phi_u^X(i,p)=\phi_i^X(p)\right],$$

(ϕ3) there exists a total computable function S (the *S-m-n* function) such that for every set X of natural numbers,

$$(\forall i)(\forall j)(\forall p)\left[\phi_{S(i,j)}^X(p)=\phi_i^X(j,p)\right].$$

We will say that i is a program for ϕ_i^X, and $\phi_i^X(p)\downarrow(\phi_i^X(p)\uparrow)$ will denote the fact that $\phi_i^X(p)$ is defined (undefined). We will abbreviate $\phi_i^\varnothing$ by ϕ_i and use W_i and $\operatorname{dom}\phi_i$ to denote the domain of ϕ_i. Hereafter, X will always denote a set of natural numbers and $X|s=\{r \mid r\in X \text{ and } r\leqslant s\}$.

A *computational complexity measure* for $\{\phi_i^X\}$ (see BLUM (1967a), LYNCH (1976), SYMES (1971)) is a sequence of functions $\{\Phi_i^X\}$ such that

(Φ1) for every set X of natural numbers, $\operatorname{dom}\Phi_i^X=\operatorname{dom}\phi_i^X$,

(Φ2) for every set X of natural numbers, the predicate $\Phi_i^X(p)=q$ is decidable from i, p, q, and X.

A *program size measure* (see BLUM (1967b)) is any total computable function μ such that there exists a total computable function κ such that $\kappa(n)$ is the cardinality of the finite set $\{i \mid \mu(i)=n\}$.

Complexity notions are used in this paper to achieve simplifications in two fundamental ways. The first way, which we now take up, provides for the simplification of the calculations which verify that the sets constructed have the desired properties, i.e., that certain diagonalizing computations are preserved in the limit. To this end we will now restrict the models of computation which we will consider by making additional assumptions about computational complexity measures and program size measures. These assumptions are in no way unnatural or counterintuitive, but are quite reasonable especially if one takes a programming language approach to the description of algorithms.

A *measured universal programming system* (m.u.p.s.) is a system $\Omega=\langle\{\phi_i^X\},\mu,\{\Phi_i^X\}\rangle$ where $\{\phi_i^X\}$ is an acceptable Gödel numbering, μ is a program size measure and $\{\Phi_i^X\}$ is a computational complexity measure for $\{\phi_i^X\}$, such that

(Ω1) $(\exists c)(\forall i)(\forall j)[\mu(S(i,j))\leqslant c\times(\mu(i)+\mu(j))]$,

(Ω2) $(\forall i)[0<\mu(i)\leqslant\log_2 i]$,

(Ω3) $(\forall X)(\forall i)(\forall p)[\Phi_i^X(p)>\mu(i)$ and $\Phi_i^X(p)>\phi_i^X(p)]$,

(Ω4) $(\forall X)(\forall i)(\forall p)(\forall s\geqslant\Phi_i^X(p))[\phi_i^{X|s}(p)=\phi_i^X(p)$ and $\Phi_i^{X|s}(p)=\Phi_i^X(p)]$,

(Ω5) $(\forall X)(\forall i)(\forall p)(\forall s)[\Phi_i^{X|s}\leqslant s\Rightarrow\phi_i^{X|s}(p)=\phi_i^X(p)$ and $\Phi_i^{X|s}(p)=\Phi_i^X(p)]$.

Condition (Ω1) restricts the *S-m-n* function S to be a simple transformation as occurs in ordinary programming languages. Condition (Ω2) is reasonable in any programming language with an alphabet of at least two symbols. Conditions (Ω1) and (Ω2) imply that for any program transformation σ of the form $\sigma(i)=S(k,i)$,

$$(\dagger) \qquad \left(\overset{\infty}{\forall} i\right)[\mu(\sigma(i)) \leqslant i].$$

Condition (Ω3) is reasonable in any programming system where the cost (i.e., the complexity) of a program takes into account the time required to compile that program and the time to print the output.

Conditions (Ω4) and (Ω5) insure that the computational complexity measure takes into account the members of X "used" in a computation, which is an essential notion of relative computations (see SHOENFIELD (1971). Consider a model of computation in which the oracle information is stored on a tape as a sequence of zeros and ones, which specify the characteristic function of the oracle set, and suppose that the computational complexity measure for this model counts the numbers of bits of memory used in the computation. Then (Ω4) says roughly that if $\phi_i^X(p)$ halts, then one can obtain the same results by providing ϕ_i with any sufficiently large ($>\Phi_i^X(p)$) portion of the oracle tape. Similarly, (Ω5) says that if ϕ_i halts on input p having read only s bits of the oracle tape, then no information contained in the remainder of the tape can affect the computation.

For any set X of natural numbers the next element function of X is defined by, $\nu_X(p)=\min\{q \mid q>p \text{ and } q\in X\}$. We use $\deg X$ to denote the degree of unsolvability of the set X.

Lemma 1.1. *$\deg X \leqslant \deg Y \Leftrightarrow (\exists i)[\phi_i^Y=\nu_X]$.*

Let $\Omega=\langle\{\phi_i^X\},\mu,\{\Phi_i^X\}\rangle$ be some m.u.p.s. Let λ be the program transformation such that

$$(\forall i)(\forall j)(\forall p)[\phi_{\lambda(i,j)}(p)=\phi_i(\Phi_j(p))]$$

and define

$$f(0)=0,$$
$$f(1)=\mu(\min\{i \mid (\forall p)[\phi_i(p)=p]\}),$$
$$f(n+1)=\max\{\mu(\lambda(i,j)) \mid \mu(i)\leqslant f(n) \text{ and } \mu(j)\leqslant f(n)\}.$$

Next we construct the program size measure μ^0 by scaling down μ by a factor of f by defining $\mu^0(i)=n \Leftrightarrow f(n-1)<\mu(i)\leqslant f(n)$. Clearly, $\Omega^0=\langle\{\phi_i^X\},\mu^0,\{\Phi_i^X\}\rangle$ is a m.u.p.s, and corresponds to a programming language in which the operation (or subroutine) λ is considered a primitive operation (or library subroutine) and hence whose use incurs a minimal size cost.

The second fundamental use of complexity notions is the construction of extremely fast growing functions which will form the basis of our diagonalizations and which will insure that conflicts between the various required properties do not arise and hence that injuries to these requirements can be avoided. To this end we define the following functions

$$b^0(n)=\max\{\Phi_i()|\phi_i()\downarrow \text{ and } \mu^0(i)\leqslant n\},$$

$$g^0(p)=\begin{cases}\min\{\mu^0(i)|\Phi_i()=p \text{ and } \mu^0(i)\leqslant p\}, & \text{if such an } i \text{ exists,}\\ p, & \text{otherwise.}\end{cases}$$

$$b^0(n,s)=\begin{cases}\max\{\Phi_i()|\Phi_i()\leqslant s \text{ and } \mu^0(i)\leqslant n\}, & \text{if such an } i \text{ exists,}\\ s, & \text{otherwise.}\end{cases}$$

where $\phi_i()$ indicates that program i has no inputs, and define

$$B^0=\{b^0(n)\},$$

$$B_s^0=\{b^0(n,s)|n\leqslant s\},$$

$$A^0=\bar{B}^0, \quad \text{the complement of } B^0.$$

Clearly, the functions g^0 and $b^0(n,s)$ and the set B_s^0 are recursive. The following lemma summarizes the most important properties of the busy beaver set B^0.

Lemma 1.2 (a) $(\forall n)(\forall s)[b^0(n,s)\leqslant b^0(n+1,s)$ *and* $b^0(n,s)\leqslant b^0(n,s+1)]$.

(b) $(\forall n)(\forall s\geqslant b^0(n))[b^0(n,s)=b^0(n)]$.

(c) A^0 *is recursively enumerable.*

(d) $\deg A^0=O'=\deg B^0$.

(e) $(\forall i)(\forall n\geqslant\mu^0(i))[\phi_i(b^0(n))\downarrow\Rightarrow\phi_i(b^0(n))<b^0(n+1)]$.

(f) $(\forall n)[b^0(n)<b^0(n+1)]$.

(g) $(\forall p)[p\in B^0\Leftrightarrow b^0(g^0(p))=p]$.

Proof. (c): $A^0=\operatorname{dom}\phi$ where ϕ is the partial recursive function defined by,

$$\phi(p)=\min\{q|q>p \text{ and } g^0(q)\leqslant g^0(p)\}.$$

(d): The halting problem $H=\{i|\phi_i(i)\downarrow\}$ is Turing reducible to B^0, since

$$\phi_i(i)\downarrow\Leftrightarrow\phi_{S(i,i)}()\downarrow\Leftrightarrow\Phi_{S(i,i)}()\leqslant b^0(\mu^0(S(i,i))).$$

(e): Suppose $n\geqslant\mu^0(i)$ and let j be such that $\mu^0(j)=n$ and $\Phi_j()=b^0(n)$. Then $\mu(j)\leqslant f(n)$ and $\mu(i)\leqslant f(n)$ so that $\mu(\lambda(i,j))\leqslant f(n+1)$ and hence $\mu^0(\lambda(i,j))\leqslant n+1$. If $\phi_i(b^0(n))\downarrow$, then

$$\phi_i(b^0(n))=\phi_{\lambda(i,j)}()<\Phi_{\lambda(i,j)}()\leqslant b^0(n+1).$$

From (b) it is clear that $\lim_{s\to\infty} b^0(n,s)=b^0(n)$ so that $b^0(n)$, though non-computable, can be computed in the limit. We can also demonstrate the recursive enumerability of A^0 by exhibiting an effective procedure for enumerating the members of A^0. Because of its relevance to the subsequent constructions we now give such an enumeration procedure. The procedure uses infinitely many markers $\overline{m}$ which will be placed at certain times on top of positive integers which have been arranged in a list. Marker $\overline{m}$ corresponds to the set of computations $\{\phi_i()|\mu^0(i)\leqslant m\}$ and its final resting place will be $b^0(m)$. The construction proceeds in stages and at stage s the position of marker $\overline{m}$ for $m\leqslant s$ will be $b^0(m,s)$. A marked integer will denote any integer on which a marker has been placed.

Stage s: (1) Place marker $\overline{s}$ on integer s.
(2) Let $m=g^0(s)$.
Move all markers $\overline{m}, \overline{m+1}, \ldots, \overline{s}$ onto integer s. Enumerate all integers $\leqslant s$ which are unmarked.

Letting A_s^0 denote those integers which have been enumerated into A^0 by stage s we see that $A_s^0=\overline{B}_s^0|s$, and that B_s^0 consists of all marked integers at stage s. The recursive enumerability of A^0 follows from the fact that s is the only integer which is unmarked at stage $s-1$ and marked at stage s, but $s\notin A_{s-1}^0$ so that $A_{s-1}^0\subseteq A_s^0$.

The final property of B^0 which will be of use to us concerns the "true stages" in the enumeration of A^0 (see SOARE (1976)).

Lemma 1.3 (a) $(\forall s)[B^0|s=B_s^0 \Leftrightarrow s\in B^0]$.
(b) $(\forall n)[B^0|b^0(n)=B^0_{b^0(n)}]$.

2. Post's problem

In this section we construct a recursively enumerable set which is neither recursive nor complete thereby providing a solution to the well-known problem posed by POST in (1944). In order to provide suitable motivation for our construction and to constrast it with the finite injury priority method we first present a summary of the important points of this method. We refer the reader to SOARE (1976) for a more complete description of the finite injury priority method.

Let C be a complete recursively enumerable set. To solve Post's problem we must construct a recursively enumerable set A such that for each i the following conditions are satisfied:

(R_i^-) $\quad \chi_C\neq\phi_i^A$,

(R_i^+) $\quad \chi_A\neq\phi_i$.

The priority construction proceeds in stages, and by A_s we denote those integers $\leq s$ which have been put into A by stage s, and by $\overline{A_s}$ we denote all those integers $\leq s$ which have not been put into A by stage s. Similarly, by $C_s(W_{j,s})$ we denote those integers $\leq s$ which have been put into $C(W_j)$ by stage s of some enumeration procedure for $C(W_j)$.

The condition (R_i^-) is called a negative requirement, since in the priority construction the satisfaction of this condition may require that certain integers be restrained from being put into the set A in order to preserve oracle computations of the form $\phi_i^{A_s}(p)$ by setting $A|s = A_s$.

The condition (R_i^+) is called a positive requirement, since in its more commonly phrased form W_i infinite $\Rightarrow W_i \cap A \neq \varnothing$, its satisfaction may entail putting integers into the set A.

It is possible for these two conditions to be in direct conflict with one another. For example, (R_i^-) may require that a particular integer p be restrained from A (because for some stage $s \geq p$ and some integer $q \leq s$, $\chi_{C_s}(q) \neq \phi_i^{A_s}(q)$ and $p \in \overline{A_s}$), and (R_j^+) may require that the same integer p be put into A (because for some stage $t > s, p \in W_{j,t}$). This conflict is resolved by giving (R_j^+) precedence over (R_i^-) if and only if $j < i$. More precisely, when the above conflict first arises at stage t, p is placed into A_t if $j < i$, and is restrained from A_t otherwise. If $j < i$ and p is placed into A_t, then the computation $\phi_i^{A_t}(q)$ may no longer agree with $\phi_i^{A_s}(q)$ and therefore a new witness for (R_i^-) may have to be found. In this case condition (R_i^-) is said to be injured since action may have been continually taken from stage s onward to preserve $\phi_i^{A_s}(q)$. Similarly, if $j \geq i$ and p is restrained from A_t, then a new witness for (R_j^+) must be found. Thus the positive and negative requirements are ordered with respect to their priority according to the sequence $R_1^- > R_1^+ > R_2^- > R_2^+ > \cdots$. Because the set A which is constructed is essentially the result of resolving many such conflicts, the description of the set A is extremely complex. Our primary concern is the elimination of such conflicts and a simple description of the set constructed.

Suppressing the adversary relationship between the conditions (R_i^-) and (R_i^+) we see that the essence of the diagonalization is the discovery of witnesses w_i^- and w_i^+ to the conditions (R_i^-) and (R_i^+) respectively. The conflicts which arose above for the priority method can be viewed as a result of poor timing, i.e., the result of starting the search for w_i^+ before w_i^- was found and starting the search for w_{i+1}^- before w_i^+ was found. These conflicts could be avoided, of course, by waiting until one witness was found before beginning the search for the next. However, since in some instances no witnesses can nor need be found, such a construction will be unable to handle all requirements. Thus, what is needed is an a priori threshold function θ such that for each i, $\theta_{i-1}^+ < w_i^- < \theta_i^-$, if w_i^- exists, and

$\theta_i^- < w_i^+ < \theta_i^+$, if w_i^+ exists. The function b^0 will be our threshold function and moreover the values $b^0(i)$ will be the witnesses to the requirements.

We give a procedure which enumerates the members of the recursively enumerable set A. The procedure uses two types of markers: real markers $\overline{m}$ and phantom markers $\dddot{n}$. The real markers are the same as those used in the enumeration procedure for A^0 given in Section 1. Indeed, the position of real marker $\overline{m}$ at stage $s \geqslant m$ is $b^0(m, s)$ and its final resting place is $b^0(m)$. The phantom markers are used to establish $B_s = B|s$ for certain integers s, where $B = \overline{A}$, and there may be several phantom markers $\dddot{n}$ for each integer n. The enumeration procedure is defined in stages.

Stage s: (1) Place real marker $\overline{s}$ on integer s.

(2) Let $m = g^0(s)$.

(a) Move all real markers $\overline{m}, \overline{m+1}, \ldots, \overline{s}$ onto integer s.

(b) Remove all phantom markers $\dddot{m}, \dddot{m+1}, \ldots, \dddot{s}$.

(c) Define B_s to be the set of all marked integers, and A_s to be the set of all unmarked integers $\leqslant s$.

(3) For all integers $n \leqslant s$ such that

(i) $b^0(n, s)$ is not marked by any phantom markers, and

(ii) $s = \min\{t \mid t > b^0(n, s)$ and $\Phi_n^{B_t}(b^0(n, s)) \leqslant t\}$, place a phantom marker $\dddot{n}$ on each marked integer between $b^0(n, s) + 1$ and s (i.e., on each member of $B_s|(b^0(n, s), s]$).

Go to stage $s + 1$.

Notice that only during step (2) can new integers be placed into A. It should be clear that A_s and B_s are recursive set functions. Condition (i) of step (3) is not really necessary, although it will allow for a simpler description of the set constructed. Also, $B^0 \subseteq B$ and $B_s^0 \subseteq B_s$.

Lemma 2.1. *A is recursively enumerable.*

Let τ be the program transformation defined by

$$\phi_{\tau(i)}(p) = \min\{s \mid s > p \text{ and } \Phi_i^{B_s}(p) \leqslant s\},$$

and let

$$m_0 = \max\{m \mid \mu^0(\tau(m)) \geqslant m\},$$

$$n_0 = \min\{n \mid b^0(n) > \max\{\phi_{\tau(m)}(b^0(m)) \mid \phi_{\tau(m)}(b^0(m))\downarrow \text{ and } m \leqslant m_0\}\}.$$

The existence of m_0 follows from (†). Thus we have,

$$\forall n \geqslant n_0[\phi_{\tau(n)}(b^0(n))\downarrow \Rightarrow \phi_{\tau(n)}(b^0(n)) < b^0(n+1)], \tag{2.1}$$

$$\forall n < n_0[\phi_{\tau(n)}(b^0(n))\downarrow \Rightarrow \phi_{\tau(n)}(b^0(n)) < b^0(n_0)]. \tag{2.2}$$

Lemma 2.2. (a) $(\forall n)[B|b^0(n)=B_{b^0(n)}]$.
(b) $(\forall n\geqslant n_0)[\phi_{\tau(n)}(b^0(n))=s\Rightarrow B_s=B|s]$.
(c) $(\forall n\geqslant n_0)[\phi_{\tau(n)}(b^0(n))\uparrow\Rightarrow\nu_B(b^0(n))=b^0(n+1)]$.

Proof. The proof of these three parts is based on the following facts. Phantom markers $\dddot{n}$ ultimately are situated on integers greater than $b^0(n)$ and are removed precisely when $\underline{\bar{n}}$ is moved. Therefore, for $n\geqslant n_0$ the only markers which can be located in the interval $(b^0(n), b^0(n+1))$ are the phantom markers $\dddot{n}$. Also, if $s=\phi_{\tau(n)}(b^0(n))$ and $n\geqslant n_0$ then

$$\phi_{\tau(n)}(b^0(n))=\min\{t\,|\,t>b^0(n,s)\text{ and }\Phi_n^{B_t}(b^0(n,s))\leqslant t\},$$

and therefore phantom markers $\dddot{n}$ are placed on all members of B_s in the interval $(b^0(n), s]$.

Lemma 2.3. $\deg B<\deg B^0$.

Proof. From Lemma 2.1 it follows that $\deg B\leqslant\deg B^0$ and so it suffices to show that $\deg B\neq\deg B^0$. We show that for each program $i, \phi_i^B\neq\nu_{B^0}$. Given i choose a $j>n_0$ such that $\phi_j^X=\phi_i^X$. If $\phi_j^B(b^0(j))\uparrow$, then $\phi_i^B=\phi_j^B\neq\nu_{B^0}$, so that we will assume that $\phi_j^B(b^0(j))\downarrow$. Let $t\geqslant\Phi_j^B(b^0(j))$ be such that $t\in B^0$. By (Ω4) we have $\Phi_j^{B|t}(b^0(j))=\Phi_j^B(b^0(j))$ and by Lemma 2.2(a) we have $B|t=B_t$. Therefore, $\Phi_j^{B_t}(b^0(j))=\Phi_j^B(b^0(j))\leqslant t$, and consequently $\phi_{\tau(j)}(b^0(j))\downarrow$.

Let $s=\phi_{\tau(j)}(b^0(j))$. Since $j>n_0$, from (2.1) and Lemma 2.2(b) it follows respectively that $s<b^0(j+1)$ and $B|s=B_s$. Thus,

$$\Phi_j^{B|s}(b^0(j))=\Phi_j^{B_s}(b^0(j))\leqslant s \tag{2.3}$$

from which it follows by (Ω5) that

$$\Phi_j(b^0(j))=\Phi_j^{B|s}(b^0(j)). \tag{2.4}$$

Combining (2.3) and (2.4) with $\phi_j^B<\Phi_j^B$ from (Ω3) we obtain

$$\phi_j^B(b^0(j))\leqslant s<b^0(j+1)$$

but $\nu_{B^0}(b^0(j))=b^0(j+1)$. Therefore, $\nu_{B^0}\neq\phi_j^B=\phi_i^B$.

Lemma 2.4. $\deg B\neq 0$.

Proof. Given a program i choose a program $j\geqslant\max\{n_0,\mu^0(i)\}$ such that $\phi_j^X(p)\uparrow$ for every set X and every integer p. Clearly, $\phi_{\tau(j)}(p)\uparrow$ for all X and p also, so that from $j\geqslant n_0$ by Lemma 2.2(c) we have $\nu_B(b^0(j))=b^0(j+1)$. Since $j\geqslant\mu^0(i)$, if $\phi_i(b^0(j))\downarrow$, then $\phi_i(b^0(j))<b^0(j+1)$ by Lemma 1.2(f) so that $\phi_i\neq\nu_B$.

Putting all the preceding lemmas together we obtain,

Theorem 2.1 (Friedberg–Muchnik). *There exists a recursively enumerable set A such that* $0 < \deg A < 0'$.

Actually, the set A has an additional property of some interest. From Lemma 2.2 and (Ω5) and the proof of Lemma 2.3 it follows that for $i \geqslant n_0$,

$$\phi_i^B(b^0(i))\downarrow \Leftrightarrow \phi_{\tau(i)}(b^0(i))\downarrow \Leftrightarrow \phi_{\tau(i)}(b^0(i)) < b^0(i+1).$$

Now let γ be the program transformation defined by $\phi^X_{\gamma(i)}(p,q) = \phi_i^X(p)$, and let $n = S(\gamma(i), i)$. Then

$$\phi_i^B(i)\downarrow \Leftrightarrow \phi^B_{S(\gamma(i),i)}(b^0(n))\downarrow \Leftrightarrow \phi_{\tau(n)}(b^0(n)) < b^0(n+1).$$

Therefore, we have $\deg A' = \deg B' = 0'$.

Corollary 2.2. *There exists a recursively enumerable set A such that* $0 < \deg A < 0'$ *and* $\deg A' = 0'$.

An informal description of the set can be obtained by considering the following paradigm which elucidates the concurrent process aspect of the construction. Imagine a universe consisting of robot photographers in which the goal of the nth photographer is to stand on $b^0(n)$ and take a photograph of $b^0(n+1)$ (i.e., the final location of the $(n+1)^{\text{st}}$ photographer). Since b^0 is not computable, the photographers must use approximations to b^0 and periodically bulletins updating the current approximations are issued. The photographers are permitted only one photograph at any given location and must publicly display their photograph after it is taken. When an update bulletin is issued and a photographer is informed that she is standing on the wrong spot, then she destroys the photograph taken (if any) at that location and moves to the new location at which she is entitled to take a new photograph.

From Lemma 1.2(f) it is clear that all but the first few photographers will be frustrated in their goals. The position of the nth photographer at time s is the same as the position of the real marker $\underline{\bar{n}}$ at stages s of the enumeration procedure, viz., $b^0(n,s)$. Also, the phantom markers correspond to the photographic images.

What does the set B look like? There are essentially two cases. Suppose $n \geqslant n_0$ and that the nth photographer never takes a photograph from $b^0(n)$. Since eventually all photographers greater than n will move to $b^0(n+1)$ and have destroyed any of their photographs taken before $b^0(n+1)$, and since no photographer less than n can see beyond $b^0(n)$, no record exists in the universe of any events occurring during the interval $(b^0(n), b^0(n+1))$.

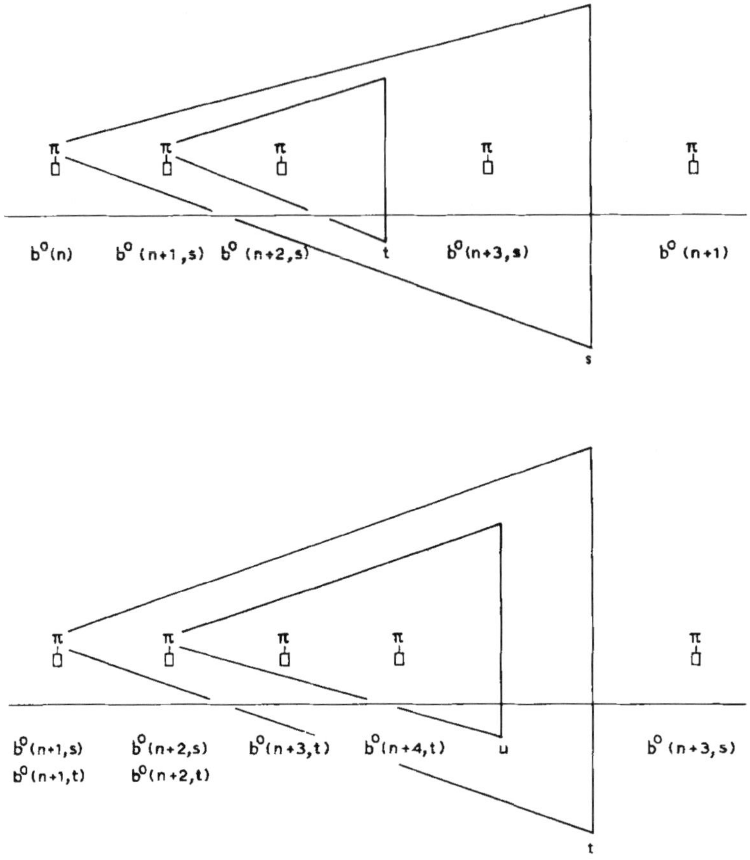

Figure 1. (a) Photograph taken by the nth photographer at time $s=\phi_{\tau(n)}(b^0(n))$; (b) enlargement of photograph taken by the $(n+1)$st photographer at time $t=\phi_{\tau(n+1)}(b^0(n+1,s))$.

Therefore $B^0\cap(b^0(n), b^0(n+1))=\varnothing$, just as in Lemma 2.2(c). Now suppose that the nth photographer does take a photograph from $b^0(n)$ at time s. Then her photograph will contain (primary) images of the photographers greater than n, i.e., will record $b^0(n+1,s),\ldots,b^0(s,s)$. In addition her photograph will contain the photographs of these photographers, which themselves contain (secondary) images of photographers. In this way the nested structure of the set is clearly visible through the various photographic levels of photographers, photographs of photographers, photographs of photographs of photographers, and so on. Fig. 1 depicts this nested structure.

We now contrast our method of construction with the priority method. Have we really managed to eliminate the conflicts and hence the injuries of the priority method? On the surface it appears that we have, but are there any hidden conflicts? Such hidden conflicts can reside only with the calculation of b^0. We therefore investigate the possible interpretation of the movement of $b^0(i)$ as an injury to (R_i^-). We first examine the set of computations $\{\phi_k()\mid \mu^0(k)<i \text{ and } \phi_k()\downarrow\}$ which define $b^0(i)$ for a computation of the form $\phi_{i-1}(b^0(i-1))$, i.e., for a computation for (R_{i-1}^+). Let j be a program such that $\mu^0(j)=i-1$ and $\Phi_j(\,)=b^0(i-1)$. Then since $\phi_{\lambda(i-1,j)}()=\phi_{i-1}(b^0(i-1))$ and $\mu^0(\lambda(i-1,j))<i$, the set of defining computations for $b^0(i)$ contains the computation $\phi_{i-1}(b^0(i-1))$ in the form of $\phi_{\lambda(i-1,j)}()$, and thus $\phi_{i-1}(b^0(i-1))\downarrow$ forces a restart of (R_i^-) just as in the priority method where (R_{i-1}^+) injures (R_i^-). However, to say that (R_i^-) is restarted *because* of (R_{i-1}^+) would be too simplistic a point of view. First of all, it is clear that for many $k\geqslant i$ the event $\phi_k(b^0(i-1))\downarrow$ also forces a restart of (R_i^-), i.e., (R_i^+) takes precedence over (R_i^-) for many $k\geqslant i$. While this might be explained in terms of a more complex ordering among the priorities of the positive and negative requirements, it is also the case that for any $k<i$ the event $\phi_k(b^0(i-1))\downarrow$ forces a restart of (R_i^-). Thus if ϕ_k is total, then ϕ_k causes infinitely many restarts, which is definitely not the case for the finite injury priority method, where each positive requirement can cause only finitely many injuries. Moreover, although (R_i^+) forces a restart of (R_i^-) for any $i\geqslant k, (R_i^-)$ may still obstruct the satisfaction of (R_i^+) unless $\phi_i^B(b^0(i))\uparrow$. In view of the remarks above no ordering of the priorities of the positive and negative requirements can account for such behavior. To further complicate matters there are many events in addition to the positive requirements which can cause a restart of (R_i^-), such as $\phi_i(i)\downarrow$ and in general any $\phi_k(p)\downarrow$ where $\mu^0(S(k,p))<i$. Finally, from Lemma 2.2(b) it is clear that not all the (R_i^-) are eventually satisfied, although this can be remedied by an application of the Recursion Theorem. For these reasons we claim that our method of construction is fundamentally different from the finite injury priority method.

One real advantage of our method is that it allows a top down approach to the construction of such sets, i.e., a construction based on a view of the set at the end of time. We illustrate this as follows. We can incorporate the positive requirements directly into the construction of a solution of Post's problem in a straightforward manner and obtain a recursively enumerable set A^1 such that $B^1=\overline{A}^1$ satisfies,

$$(2.6)\qquad \left(\overset{\infty}{\forall} n\right)\left[\phi_{\tau_1(n)}(b^0(2n))=s\Rightarrow B^1|s=B^1_s\right],$$

$$(2.7)\qquad (\forall n)\left[\nu_{B^1}(b^0(2n+1))=b^0(2n+2)\right],$$

where τ_1 is defined by replacing B in (τ1) by B^1. Thus the even intervals $(b^0(2n), b^0(2n+1))$ of B^0 are used to satisfy the negative requirements, and the odd intervals $(b^0(2n+1), b^0(2n+2))$ of B^0 are used to satisfy the positive requirements. The enumeration procedure for A^1 is obtained from the original enumeration procedure for the set A given in this section by replacing $b^0(n,s)$ by $b^0(2n,s)$ everywhere in step (3). On the other hand by reversing the roles of the odd and even intervals we can obtain a recursively enumerable set A^2 such that $B^2=\overline{A^2}$ satisfies,

$$(2.8)\qquad \left(\overset{\infty}{\forall} n\right)\left[\phi_{\tau_2(n)}(b^0(2n+1))=s \Rightarrow B^2|s=B_s^2\right],$$

$$(2.9)\qquad (\forall n)\left[\nu_{B^2}(b^0(2n))=b^0(2n+1)\right].$$

Clearly, A^1 and A^2 satisfy $O<\deg A^1<O'$ and $O<\deg A^2<O'$. What is remarkable is that A^1 and A^2 represent incomparable degrees. This is easily seen since (2.6) and (2.9) show that

$$\left(\overset{\infty}{\forall} n\right)\left[\phi_n^{B^1}(b^0(2n))<b^0(2n+1)=\nu_{B^2}(b^0(2n))\right],$$

and (2.7) and (2.8) show that

$$\left(\overset{\infty}{\forall} n\right)\left[\phi_n^{B^2}(b^0(2n+1))<b^0(2n+2)=\nu_{B^1}(b^0(2n+1))\right].$$

Of course, Friedberg and Muchnik originally solved Post's problem by constructing a pair of incomparable recursively enumerable sets, but using our method we have constructed two independently defined incomparable recursively enumerable sets, i.e., the definition of A^1 does not involve A^2 and vice versa.

References

Blum, M.

[1967a] A machine independent theory of the complexity of recursive functions, *J. Assoc. Comput. Mach.*, **14**, 322–336.

[1967b] On the size of machines, *Information and Control*, **11**, 257–265.

Daley, R.

[1978] On the simplicity of busy beaver sets, *Z. Math. Logik Grundlagen Math.*, **24**, 207–224.

[1980] Busy beaver sets and the degrees of unsolvability, to appear.

Lynch, N., A. Meyer and M. Fischer

[1976] Relativizations of the theory of computational complexity, *Trans. Am. Math. Soc.*, **220**, 243–287.

Post, E.

[1944] Recursively enumerable sets of positive integers and their decision problems, *Bull. Am. Math. Soc.*, **50**, 284–316.

Rogers, H.

[1967] *Theory of Recursive Functions and Effective Computability* (McGraw Hill, New York).

Schoenfield, J.

[1971] *Degrees of Unsolvability* (North-Holland, Amsterdam).

Soare, R.

[1976] The infinite injury priority method, *J. Symbolic Logic*, **41**, 513–530.

Symes, M.

[1971] The extension of machine-independent computational complexity theory to oracle machine computation and to the computation of finite functions, Ph. D. Dissertation, Dept. of Applied Analysis and Comput. Sci., Univ. of Waterloo.

This page intentionally left blank

J. Barwise, H. J. Keisler and K. Kunen, eds., *The Kleene Symposium*

On Kleene Degrees of Analytic Sets

Karel Hrbacek
Department of Mathematics, City College of The City University of New York, 138th Street and Convent Avenue, New York, NY 10031, U.S.A.

*Stephen G. Simpson**
Department of Mathematics, Pennsylvania State University, University Park, PA 16802, U.S.A.

Dedicated to Professor S. C. Kleene on the occasion of his 70th birthday

Abstract: We show that the following assertion is true in a wide class of models of Zermelo–Fraenkel set theory. There exist $2^{\aleph_0}$ analytic subsets of the Cantor space having pairwise incomparable Kleene degrees. It follows by a lemma of Kuratowski that no two of these sets are Borel isomorphic.

By Cantor space we mean the complete separable metric space 2^ω of all subsets of $\omega = \{0,1,2,\ldots\}$. Subspaces A and B of 2^ω are said to be Borel isomorphic if there exists a Borel isomorphism of A onto B, i.e. a one–one mapping of the points of A onto the points of B which induces a one–one correspondence between the Borel sets of A and the Borel sets of B.

Lemma 1 (KURATOWSKI, 1966, p.436). *Let f be a Borel isomorphism of A onto B. There exist Borel sets $A_1 \supseteq A$ and $B_1 \supseteq B$ such that f can be extended to a Borel isomorphism f_1 of A_1 onto B_1.*

Lemma 1 has been used to study the problem of the existence of non-Borel analytic sets which are pairwise not Borel isomorphic. MAITRA and RYLL-NARDZEWSKI (1970) proved the existence of two such sets, and MAULDIN (1976) proved the existence of three such sets. Both of these results depended on the Axiom of Constructibility, $V = L$.

We say that A is Kleene reducible to B if there exists a partial recursive functional Φ and a parameter $p \in 2^\omega$ such that $\Phi(x,p,B,{}^2\mathrm{E}) = 1$ for all $x \in A$, and $\Phi(x,p,B,{}^2\mathrm{E}) = 0$ for all $x \in 2^\omega - A$; cf. KLEENE (1959, 1963) and SOLOVAY (1971). The substitution theorems of KLEENE (1959) imply that the relation of Kleene reducibility is reflexive and transitive. A Kleene degree is an equivalence class of subsets of 2^ω under the equivalence

*Partially supported by NSF grant MCS 77-13935.

relation of mutual Kleene reducibility. The Kleene degrees have an obvious partial ordering induced by Kleene reducibility.

Lemma 2 (essentially due to KLEENE (1963). (i) *B is a Borel set if and only if B has the same Kleene degree as the empty set.*

(ii) *If B is a Borel set, then A and* $(A-B)\cup(B-A)$ *have the same Kleene degree.*

(iii) *If* $f:2^\omega\to 2^\omega$ *is Borel measurable, then* $f^{-1}(A)$ *is Kleene reducible to A.*

Lemma 2 suggests that the Kleene degree of a set A is a natural measure of the complexity of A, provided Borel sets are taken as having minimum complexity.

There is a well-known analogy between recursive subsets of ω and Borel subsets of 2^ω. Pursuing this analogy, we may say that recursively enumerable sets correspond to co-analytic sets, the enumeration theorem corresponds to the existence of a universal co-analytic set, the "single-valuedness theorem" (ROGERS, 1967, p.71) corresponds to Kondo's theorem, Turing degrees (ROGERS, 1967, p.137) correspond to Kleene degrees, and the jump operation (ROGERS, 1967, p.254) corresponds to the superjump (ACZEL and HINMAN, 1974). Thus we arrive at the following analog of Post's problem. Does there exist an analytic set which is neither a Borel set nor Kleene equivalent to a universal analytic set?

This analog of Post's problem was solved completely by HARRINGTON (1978) and STEEL (1980) who proved that the following three statements are pairwise equivalent. (1) There exists an analytic set which is neither a Borel set nor Borel isomorphic to a universal analytic set. (2) There exists an analytic set which is neither a Borel set nor Kleene equivalent to a universal analytic set. (3) There exists $p\in 2^\omega$ such that $p^{\#}$ does not exist. However, Harrington and Steel left open the status of the Kleene degree analog of the Friedberg–Muchnik theorem, i.e. the conjecture that there exist two analytic sets whose Kleene degrees are incomparable. It seems reasonable to guess that (3) should imply the following stronger conjecture. (4) There exist $2^{\aleph_0}$ analytic sets whose Kleene degrees are pairwise incomparable. HRBACEK, (1978) proved (4) among other things, assuming $V=L$.

The purpose of this note is to prove (4) under an assumption which is stronger than (3) but weaker than $V=L$ and true in a wide class of models of Zermelo–Fraenkel set theory. Let V be the universe, and let L be Gödel's universe of constructible sets. We say that V is a forcing extension of L if there exists a partial ordering $P\in L$ and an L-generic filter $G\subseteq P$ such that $V=L[G]$; cf. JECH (1978, §16). Our result is:

Theorem 1. *Assume that V is a forcing extension of L. To each countable set x of countable ordinals we can associate an analytic set A_x, such that A_x is Kleene reducible to A_y if and only if $x \subseteq y$.*

Corollary. *If V is a forcing extension of L, there exist $2^{\aleph_0}$ analytic sets no two of which are Borel isomorphic.*

To prove the corollary, let F be a family of $2^{\aleph_0}$ subsets of ω which are pairwise incomparable with respect to inclusion. By Theorem 1 the analytic sets $A_x, x \in F$, are pairwise incomparable with respect to Kleene reducibility. It follows by Lemmas 1 and 2 that no two of these sets are Borel isomorphic.

Our proof of Theorem 1 will be based on a characterization of Kleene reducibility in terms of admissible sets. A good source of basic admissibility theory is Barwise (1975). For any set or class B, a formula ψ in the language of set theory with an extra symbol B is said to be $\Delta_0(B)$ if it is built up from atomic formulas $x=y$, $x \in y$, $x \in B$ by means of propositional connectives $\wedge, \vee, \sim, \rightarrow, \leftrightarrow$ and bounded quantifiers $\forall x \in y$, $\exists x \in y$. A formula φ is said to be $\Sigma_1(B)$ if it is of the form $\exists x \psi$ where ψ is $\Delta_0(B)$. A nonempty transitive set M is said to be B-admissible if it is closed under the formation of unordered pairs $\{x,y\}$ and unions $\bigcup x$ and satisfies the separation and bounding principles

$$\forall a \exists b \forall x (x \in b \leftrightarrow x \in a \wedge \psi(x,a))$$

and

$$\forall a (\forall x \exists y \psi(x,y,a) \rightarrow \exists b \forall x \in a \exists y \in b \psi(x,y,a))$$

for all $\Delta_0(B)$ formulas ψ not mentioning b. A subset A of M is said to be $\Delta_1(M,B)$ if there exist $\Sigma_1(B)$ formulas φ_0 and φ_1 and a parameter $p \in M$ such that for all $x \in M$,

(1) $\quad x \in A \leftrightarrow M$ satisfies $\varphi_1(x,p)$

and

(2) $\quad x \in A \leftrightarrow M$ satisfies $\sim\varphi_0(x,p)$.

Lemma 3. *Let A and B be subsets of 2^ω. A is Kleene reducible to B if and only if there exist $\Sigma_1(B)$ formulas φ_0 and φ_1 and a parameter $p \in 2^\omega$ such that for all B-admissible sets M containing ω and p as elements,* (1) *and* (2) *hold for all $x \in 2^\omega \cap M$.*

We omit the straightforward but tedious proof of this lemma. The idea of the proof is that $\varphi_i(x,p)$ expresses the existence of a well-founded computation tree showing that $\Phi(x,p,B,{}^2\mathrm{E})=i$. Some of the details can be found in HRBACEK (1978) and SHOENFIELD (1968).

For any set or class B and transitive set M, we let fodo(M,B) be the set of all subsets of M which are first-order definable over the structure $\langle M,\epsilon,B\cap M\rangle$ allowing parameters from M. We put $L_0[B]=$ the empty set, $L_{\xi+1}[B]=\mathrm{fodo}(L_\xi[B],B)$, $L_\lambda[B]=\bigcup\{L_\xi[B]\colon \xi<\lambda\}$ for limit ordinals λ, and $L[B]=\bigcup\{L_\xi[B]\colon \xi$ is an ordinal$\}$. This is called the constructible hierarchy relative to B.

An ordinal α is called B-admissible if there exists a B-admissible set M such that α is the least ordinal not an element of M. It is well-known that an ordinal α is B-admissible if and only if the transitive set $L_\alpha[B]$ is B-admissible. For any ordinal ξ we denote by ω_ξ^B the ξth ordinal in order of magnitude which is either B-admissible or a limit of B-admissible ordinals. If B is the empty set, then B-admissible sets and ordinals are called admissible, $\Delta_1(M,B)$ is written $\Delta_1(M)$, and ω_ξ^B is written ω_ξ. The ξth infinite initial ordinal is denoted $\aleph_\xi$.

Lemma 4. *Assume that V is a forcing extension of L. Let F be a function from the ordinals into the ordinals such that the graph of F is a primitive recursive relation (cf.* JENSEN *and* KARP *(1971)). Then given $p\in 2^\omega$ we can find arbitrarily large countable ordinals $\sigma<\aleph_1$ such that $\omega_{\sigma+\xi}^p=\omega_{\sigma+\xi}$ for all $\xi\leqslant F(\sigma)$.*

Proof. Let $P\in L$ be a partial ordering and $G\subseteq P$ an L-generic filter such that $V=L[G]$. An easy forcing argument shows that if α is admissible and $P\in L_\alpha$, then α is G-admissible. Given $p\in 2^\omega$ let κ be an ordinal such that $\omega_\kappa=\kappa$, $P\in L_\kappa$, and $p\in L_\kappa[G]$. Then clearly $\omega_{\kappa+\xi}^p=\omega_{\kappa+\xi}$ for all ordinals ξ. Hence, by a Löwenheim–Skolem argument, we can find arbitrarily large countable ordinals σ such that $\omega_{\sigma+\xi}^p=\omega_{\sigma+\xi}$ for all $\xi\leqslant F(\sigma)$.

Let W be the set of all $z\in 2^\omega$ such that the binary relation $\{\langle i,j\rangle\colon 2^i3^j\in z\}$ is a well-ordering of ω. For $z\in W$, the order type of $\{\langle i,j\rangle\colon 2^i3^j\in z\}$ is denoted $|z|$. Let J be a 2-place primitive recursive function with primitive recursive inverses which maps the ordered pairs of ordinals one–one onto the ordinals. If x is a countable set of countable ordinals, let W_x be the set of all $z\in W$ such that $\omega_{J(\xi,\nu)}\leqslant|z|<\omega_{J(\xi,\nu)+1}$ for some ordinals ξ and ν such that $\nu\in x$. The proof of the following lemma is not difficult and we omit it.

Lemma 5. *W_x is coanalytic. If $x\subseteq y$, then W_x is Kleene reducible to W_y.*

Lemma 6. *Assume that V is a forcing extension of L. If x is not a subset of y, then W_x is not Kleene reducible to W_y.*

Proof. Suppose for a contradiction that W_x is Kleene reducible to W_y. Put $A = W_x$ and $B = W_y$ and let φ_0, φ_1 and p be as in Lemma 3. Let $q \in 2^\omega$ be such that p and y are elements of $L_\gamma[q]$ where $\gamma = \omega_1^q$. Let ν be a countable ordinal in $x - y$. By Lemma 4 (applied e.g. to the function $F(\sigma) =$ least $\tau > \sigma$ such that for all $\nu < \sigma$ there exists $\xi < \tau$ such that $\sigma < J(\xi, \nu) < \tau$), we can find a countable ordinal ξ such that $\omega_{J(\xi,\nu)+1}$ is q-admissible. Put $\alpha = \omega_{J(\xi,\nu)+1}$ and $\beta = \omega_{J(\xi,\nu)}$. Then $\gamma \leqslant \alpha$, $\beta < \alpha$, and for each $z \in B$ we have either $|z| < \beta$ or $|z| \geqslant \alpha$. By a theorem of Sacks (1976) we can find $r \in 2^\omega$ such that $q \in L_\alpha[r]$ and $\alpha = \omega_1^r$. Put $M = L_\alpha[r]$. Then M is an admissible set and $|z| < \beta$ for all $z \in B \cap M$. Also $y \in M$ so $B \cap M$ is $\Delta_1(M)$. Hence M is B-admissible. Also $p \in M$ so $A \cap M$ is $\Delta_1(M)$. But A contains all $z \in W$ such that $\beta \leqslant |z| < \alpha$. Hence $W \cap M$ is $\Delta_1(M)$. Hence the hyperjump of r is an element of M. This is impossible since $\alpha = \omega_1^r$ so Lemma 6 is proved.

To prove Theorem 1, put $A_x = 2^\omega - W_x$. Lemma 5 implies that A_x is analytic. Lemma 2 implies that A_x and W_x have the same Kleene degree. Lemmas 5 and 6 imply that W_x is Kleene reducible to W_y if and only if $x \subseteq y$. This completes the proof of Theorem 1.

References

Aczel, P., and P. G. Hinman

[1974] Recursion in the superjump, in: *Generalized Recursion Theory* (North-Holland, Amsterdam), pp. 3–41.

Barwise, K. J.

[1975] *Admissible Sets and Structures* (Springer-Verlag, Berlin).

Harrington, L.

[1978] Analytic determinacy and $0^\#$, *J. Symbolic Logic*, **43**, 685–693.

Hrbacek, K.

[1978] On the complexity of analytic sets, *Z. Math. Logik Grundlagen Math.*, **24**, 419–425.

Jech, T.

[1978] *Set Theory* (Academic Press, New York).

Jensen, R., and C. Karp

[1971] Primitive recursive set functions, in: *Axiomatic Set Theory*, *Part I* (Am. Math. Soc., Providence, RI), pp. 143–176.

Kleene, S. C.

[1959] Recursive functionals and quantifiers of finite type I, *Trans. Am. Math. Soc.*, **91**, 1–52.

[1963] Recursive functionals and quantifiers of finite type II, *Trans. Am. Math. Soc.*, **108**, 106–142.

Kuratowski, K.

[1966] *Topology*, *Vol. I* (Academic Press, New York).

MAITRA, A., and C. RYLL-NARDZEWSKI

[1970] On the existence of two analytic non-Borel sets which are not isomorphic, *Bull. Acad. Polon. Sci.*, **18**, 177–178.

MAULDIN, D. R.

[1976] On non-isomorphic analytic sets, *Proc. Am. Math. Soc.*, **58**, 241–244.

ROGERS JR., H.

[1967] *Theory of Recursive Functions and Effective Computability* (McGraw Hill, New York).

SACKS, G. E.

[1976] Countable admissible ordinals and hyperdegrees, *Advances in Math.*, **20**, 213–262.

SHOENFIELD, J. R.

[1968] A hierarchy based on a type 2 object, *Trans. Am. Math. Soc.*, **134**, 103–108.

SOLOVAY, R. M.

[1971] Determinacy and type 2 recursion (abstract), *J. Symbolic Logic*, **36**, 374.

STEEL, J. R.

[1980] Analytic sets and Borel isomorphisms, *Fund. Math.*, to appear.

J. Barwise, H. J. Keisler and K. Kunen, eds., *The Kleene Symposium*

A Semantics for Kleene's j-expressions

David P. Kierstead

Department of Mathematics, University of Wisconsin at Madison, Madison, WI 53706, U.S.A.

Dedicated to Professor S. C. Kleene on the occasion of his 70th birthday

Abstract: S. C. Kleene (*Generalized Recursion Theory II*, edited by J. E. Fenstad et al. (North-Holland, Amsterdam)) proposed a system for computation in higher types which was based on the syntactic manipulation of formal symbols, called j-expressions. No adequate semantics for these expressions could be based on the classical (total) type structure over $\mathbb{N}$. In particular, not all j-expressions corresponded to objects of type j, and two j-expressions corresponding to the same object could not necessarily be substituted for each other in other j-expressions without altering their meaning. In this paper we provide a semantics which eliminates these problems. The type structure for this semantics is obtained by adding one extra object ⓤ at type 0 and, at type $(j+1)$, allowing all *monotone*, partial functions from type j into $\mathbb{N}$. The original type structure is embeddable into the extended type structure in a manner which preserves Kleene's computations.

1. Introduction

Surely every reader of S. C. Kleene's pioneering papers (Kleene, 1959, 1963) in higher type recursion theory has been puzzled (if not dismayed) by the fact that two important principles fail in this theory. The first is the substitution principle—that if $\phi(\alpha^n, \mathfrak{A})$ and $\theta(\tau^{n-1}, \mathfrak{A})$ are partial recursive, then certainly $\lambda\mathfrak{A}\phi(\lambda\tau^{n-1}\theta(\tau^{n-1}, \mathfrak{A}), \mathfrak{A})$ should be partial recursive. Indeed, this principle seems so natural that one might find it difficult to conceive of its failing. The second principle is the analogue of the First Recursion Theorem from ordinary recursion theory—that if $\mathbf{F}(\zeta; \mathfrak{A})$ is a partial recursive functional, then the minimal solution of the equation $\mathbf{F}(\zeta; \mathfrak{A}) \simeq \zeta(\mathfrak{A})$ should also be partial recursive. The failure of this principle appears even more anomalous when one considers that the analogue of the Second Recursion Theorem does hold.[1]

The problem underlying both of these failures is demonstrated by an example of Kleene (1963).

[1]Kleene (1959, 1963) did obtain restricted versions of these principles.

Example 1.1 (Kleene). Define:

$$\chi(e,x)\simeq\begin{cases}0 & \text{if } \neg T(e,e,x),\\ \text{undefined} & \text{if } T(e,e,x);\end{cases}$$

$$\phi(\sigma^2,e)\simeq\sigma^2(\lambda x\chi(e,x));$$

$$\theta(\alpha)\simeq 0;$$

$$\psi(e)\simeq\phi(\lambda\alpha\theta(\alpha),e).$$

ϕ is partial recursive and $\phi(\sigma^2,e)$ is defined if and only if $\forall x \neg T(e,e,x)$ (since $\lambda x\chi(e,x)$ is total, whence a type-1 object, if and only if $\forall x \neg T(e,e,x)$). Were ψ partial recursive, then ψ would be partial recursive in the sense of ordinary recursion theory (KLEENE, 1959, XVII). This is impossible since $\psi(e)$ is defined if and only if $\forall x \neg T(e,e,x)$, and the latter is not recursively enumerable.

Perhaps one feels cheated by this example. One could easily argue that $\theta(\alpha)$ is defined without using any values of $\alpha(x)$, and therefore $\phi(\lambda\alpha\theta(\alpha),e)$ should be defined for all e. It is easy to construct more complicated θ where it is not so clear for what values of x the corresponding values of $\alpha(x)$ are used by θ in determining $\theta(\alpha)$; but there is still the feeling that if $\lambda x\chi(e,x)$ is (correctly) defined for all of these x, then $\phi(\lambda\alpha\theta(\alpha),e)$ should be defined. Kleene seems to have shared these misgivings and made a suggestion for remedying the situation (KLEENE, 1963, p.111). KLEENE (1978) resurrected this idea. He argued that there was no need to attempt a computation of $\chi(e,x)$ until its value was called for in the course of computing $\theta(\lambda x\chi(e,x))$. Thus we might simply carry $\chi(e,x)$ along as a formal expression until such a value is called for. Of course we might soon find ourselves considering formal expressions within formal expressions ad nauseam, and it would be no small task to keep all of the syntax straight. Having taken this approach, Kleene was able to include analogues of the substitution and first recursion principles as primitive operations and produce an eminently workable theory.

KLEENE (1978) developed a system for computation in higher types which was based entirely on the syntactic manipulation of formal symbols, called "j-expressions". Under a given assignment Ω to the free variables of the language, each 0-expression is either undefined or defined with some natural number as its value; each $(j+1)$-expression B naturally represents a partial function $[\mathrm{B}]_\Omega$ from type j into the natural numbers. $[\mathrm{B}]_\Omega$ need not be total, and thus need not be a type-$(j+1)$ object. In the course of proving certain theorems in Section 3 of KLEENE (1978), it was necessary to know that one j-expression B could be substituted (freely) for another $\bar{\mathrm{B}}$ in a 0-expression E without affecting the value of E under Ω. In Example 1.2 we will present an example which shows that the condition that $[\mathrm{B}]_\Omega=[\bar{\mathrm{B}}]_\Omega$

is insufficient to justify these substitutions—even when $[B]_\Omega$ is total. Kleene postponed the justification of these substitutions pending the development of an appropriate semantics for his system.

In this paper we develop a semantics in which:

(1) every j-expression represents an object, and

(2) semanticly equivalent expressions may be freely substituted for each other.

In Theorem 5.4 we give the condition under which one j-expression may be substituted for another in the system of KLEENE (1978).

We assume familiarity with KLEENE (1978). All unexplained terminology and notation is from that paper. We use the notations $\downarrow$ and $\uparrow$ for "is defined" and "is undefined" respectively. E will always denote a 0-expression and $\Theta=\theta_1,\ldots,\theta_l$ a list of "assumed function" symbols. We use $E\equiv E^*$ to mean that E and E* are the same formal string of symbols.

Example 1.2. Let B be the 1-expression $\lambda a\ \mathrm{cs}(a,0,0)$. ($\mathrm{cs}(a,b,c)\simeq b$ if $a=0$, c if $a>0$.) Let $\bar{B}$ be $\lambda a0$. It is easy to construct a 0-expression A which is undefined under any assignment (see Section 3 of KLEENE (1978)). Let E be the 0-expression B(A). Then, under any Ω, E is undefined, but $\bar{B}(A)=0$. Thus $\bar{B}$ may not be substituted for B in E—even though $[B]_\Omega=[\bar{B}]_\Omega$ for any Ω. The problem is that the "obvious" interpretation $[B]_\Omega$ of B does not take into account whether B "uses" its argument or not.

2. The extended types

We define $\hat{T}^{(j)}$ (the set of *extended type-j objects*, henceforth denoted by *type* $\hat{j}$) by recursion on j.

$$\hat{T}^{(0)}=\{ⓤ,0,1,2,\ldots\}=\mathbb{N}\cup\{ⓤ\}$$

(ⓤ is a new object suggesting "undefined").

Define $\alpha^0\subset\beta^0\leftrightarrow\alpha^0=ⓤ$ or $\alpha^0=\beta^0$.

$\hat{T}^{(j+1)}=$ the set of all monotone partial functions from $\hat{T}^{(j)}$ into $\mathbb{N}$. (α^{j+1} is *monotone* if: whenever $\alpha^{j+1}(\beta^j)\downarrow$ and $\beta^j\subset\gamma^j$, then $\alpha^{j+1}(\gamma^j)\downarrow$ and is equal to $\alpha^{j+1}(\beta^j)$—where "$\subset$" is the relation defined above at type $\hat{0}$ and has the usual, set-theoretical meaning at all higher levels.) Let us write $ⓤ^j$ for the totally undefined type-$\hat{j}$ object. Monotonicity requires (among other things) that if $\alpha^{j+1}(ⓤ^j)=w$, then $\alpha^{j+1}=\lambda\beta^j w$.[2]

[2]That some semantics satisfying conditions (1) and (2) above was necessary, and that this would entail an extension of the type structure along these lines, were suggested by KLEENE (1978 and private communication). KLEENE (1980a) has also pursued this idea and arrived at a semantics significantly different from ours. Our type structure agrees with his at the 0 and 1 levels but differs from there on.

3. Computations

We now take Ω to be an assignment of objects from the extended types to the variables of the language and of monotone, partial functions on these types to the function symbols Θ. We give an inductive definition of $[\mathrm{E}]_\Omega = w$ (E has value $w \in \mathbb{N}$ under Ω). Formally, we define $\mathrm{K} = \{\langle \mathrm{E}, \Omega, w\rangle : [\mathrm{E}]_\Omega = w\}$ by stages K^ξ, indexed by ordinals, taking $\mathrm{K} = \bigcup_\xi \mathrm{K}^\xi$. In Section 3.4 we will show how this may be viewed as a computation procedure. We will use the following notation:

$\Omega[\gamma^j/\chi^j]$ is the assignment obtained from Ω by assigning χ^j to γ^j,

$$\mathrm{K}^{<\xi} = \bigcup_{\delta<\xi} \mathrm{K}^\delta,$$

$[\alpha^j]_\Omega =$ the type-$\hat{j}$ object assigned to α^j by Ω,

$$[\mathrm{E}]_\Omega^{<\xi} = \begin{cases} w & \text{if } \langle \mathrm{E}, \Omega, w\rangle \in \mathrm{K}^{<\xi}, \\ \text{ⓤ} & \text{if there is no such } w, \end{cases}$$

for any $(j+1)$-expression B,

$$[\mathrm{B}]_\Omega^{<\xi} = \lambda\chi^j[\mathrm{B}(\gamma^j)]_{\Omega[\gamma^j/\chi^j]}^{<\xi}.$$

($[\mathrm{B}]_\Omega^{<\xi}$ is the part of $[\mathrm{B}]_\Omega$ built up before stage ξ.)

We will need to know that $[\mathrm{B}]_\Omega^{<\xi} \in \hat{\mathrm{T}}^{(j+1)}$, i.e. that it is monotone, so we prove the following lemma simultaneously with the definition of K.

Lemma 3.1. *If, for each free variable* α^j *of* E, $[\alpha^j]_\Omega \subset [\alpha^j]_{\Omega^*}$, *then* $[\mathrm{E}]_\Omega^{<\xi} \subset [\mathrm{E}]_{\Omega^*}^{<\xi}$. *In particular*, $[\mathrm{B}]_\Omega^{<\xi}$ *is monotone.*

The definition of $\langle \mathrm{E}, \Omega, w\rangle \in \mathrm{K}^\xi$ is by cases, according to the form of E as a 0-expression (see Section 2.2 of Kleene (1978)). In each case, the lemma follows easily from the inductive hypothesis.

Definition 3.2. In each of the nine cases, we specify the condition which must be met in order to have $\langle \mathrm{E}, \Omega, w\rangle \in \mathrm{K}^\xi$.

E1: E is α^0 and $[\alpha^0]_\Omega = w$.

E2: E is $\phi_i(\mathrm{A}_0^0, \ldots, \mathrm{A}_{n_0}^0, \mathrm{A}_1^1, \ldots, \mathrm{A}_{n_r}^r)$ and $\langle \mathrm{E}^*, \Omega, w\rangle \in \mathrm{K}^{<\xi}$ where E^* is the result of making the same substitutions on the right side of the schema introducing ϕ_i as yields E on the left side. Notice that if ϕ_i is introduced via S0 with the "θ_t" actually an η to be phased out later via S11, (say $\phi_i(\eta, \Theta; \mathfrak{A}) \simeq \eta(\mathfrak{A})$ and ϕ_j phases out η by $\phi_j(\Theta; \mathfrak{A}) \simeq \phi_k(\phi_j, \Theta; \mathfrak{A})$), then E^* is $\phi_j(\vec{\mathrm{A}})$, not $\eta(\vec{\mathrm{A}})$.

E4.λ: E is $\{\lambda\alpha^j A\}(B)$ and $\langle S^{\alpha^j}_{B}A|, \Omega, w\rangle \in K^{<\xi}$. ($S^{\alpha^j}_{B}A|$ is the result of substituting B for each free occurrence of α^j in A, after changing bound variables appropriately.)

E4.j: E is $\alpha^j(B)$ and $\hat{\alpha}^j(\beta^{j-1}) = w$, where $\hat{\alpha}^j = [\alpha^j]_\Omega$ and $\beta^{j-1} = [B]^{<\xi}_\Omega$.

E5: E is 0 and $w = 0$.

E6′: E is A′ and $\langle A, \Omega, w_0\rangle \in K^{<\xi}$ where $w_0 + 1 = w$.

E6$\dot{-}$: E is $A \dot{-} 1$ and $\langle A, \Omega, w_0\rangle \in K^{<\xi}$ where $w_0 \dot{-} 1 = w$.

E6cs: E is cs(A, B, C) and either:

(i) $\langle A, \Omega, 0\rangle \in K^{<\xi}$ and $\langle B, \Omega, w\rangle \in K^{<\xi}$, or

(ii) $\langle A, \Omega, n+1\rangle \in K^{<\xi}$ for some n and $\langle C, \Omega, w\rangle \in K^{<\xi}$.

E7: E is θ_t(---) and the condition similar to that of E4.j holds.

This completes the definition of K.

Definition 3.3. Let B be a j-expression and E a 0-expression.

(i) $[B]_\Omega = \bigcup_{\xi} [B]^{<\xi}_\Omega$.

(ii)

$$|E|_\Omega = \begin{cases} \text{the least } \xi \text{ such that } \langle E, \Omega, w\rangle \in K^{\xi} \text{ for some } w & \text{if such a } \xi \text{ exists,} \\ \infty & \text{otherwise.} \end{cases}$$

We will generally use induction on $|E|_\Omega$ in situations where Kleene (in KLEENE (1978)) used "induction on computation trees".

3.4. *Computation trees.* For each E, Ω, we describe the construction of a *computation tree* based on E, Ω. If $[E]_\Omega\downarrow$ the tree will be well-founded (in the sense that all paths are finite) of rank $|E|_\Omega$. Each completed tree will have a *value* associated with it and this value will be $[E]_\Omega$. These trees are essentially the same as those in Section 2.4 of KLEENE (1978), except that we must alter the E1, E4.j and E7 steps (which correspond to the evaluation of a variable) to account for the fact that we are dealing with a different set of objects. We picture our trees as lying on their sides and growing from left to right. The *principal branch* issuing from a vertex is the branch proceeding horizontally to the right. When a tree is completed we will tag the principal branch with the value of the tree. (In order to make our claim about the rank of the tree hold, we must not consider the tag to be a part of the tree.) Each vertex of the tree is occupied by a 0-expression. With each vertex, there is associated an *assignment in force* at that vertex.

Consider a vertex occupied by E with Ω as the assignment in force. By cases on E, we describe the next level of growth beyond this vertex.

E1: E is α^0. If $[\alpha^0]_\Omega = (u)$ there is no growth and the tree cannot be completed. If $[\alpha^0]_\Omega = w$, then the tree is completed by tagging it with the value w.

E2: E is $\phi_i(A_1^0,\ldots,A_{n_0}^0,A_1^1,\ldots,A_{n_r}^r)$. Taking E^* as in the definition of K, we add one more vertex, occupied by E^*, along the principal branch and keep Ω as the assignment in force.

$$E \text{——} E^* \cdots$$

E4.λ: E is $\{\lambda\alpha^j A\}(B)$. We add one more vertex occupied by $S_B^{\alpha^j}A|$ with Ω the assignment in force.

$$\{\lambda\alpha^j A\}(B) \text{——} S_B^{\alpha^j}A| \cdots$$

E4.1: E is $\alpha^1(B)$. Take $\hat{\alpha}^1=[\alpha^1]_\Omega$. *Case* 1: $\hat{\alpha}^1(ⓤ)\downarrow$. The tree is completed by tagging it with the value $\hat{\alpha}^1(ⓤ)$. *Case* 2: otherwise. We place B (under Ω) at the next lower vertex and begin a computation tree for B. If this tree is completed at some stage with value w, and $\hat{\alpha}^1(w)\downarrow$, then the tree for E is completed by tagging it with $\hat{\alpha}^1(w)$. Otherwise the tree for E cannot be completed.

$$\begin{array}{l} \alpha^1 B \text{——} \hat{\alpha}^1(w) \\ \quad \searrow B \cdots w \end{array}$$

E4.j $(j>1)$: E is $\alpha^j(B)$. Take $\hat{\alpha}^j=[\alpha^j]_\Omega$. *Case* 1: $\hat{\alpha}^j(ⓤ^{j-1})\downarrow$. The tree is completed by tagging it with the value $\hat{\alpha}^j(ⓤ^{j-1})$. *Case* 2: $\hat{\alpha}^j(ⓤ^{j-1})\uparrow$. We begin simultaneous computations of $B(\gamma^{j-2})$ under each assignment $\Omega[\gamma^{j-2}/\chi^{j-2}]$ for $\chi^{j-2}\in\hat{T}^{(j-2)}$. As these computations proceed, one step at a time, some of them may become completed with values $w_{\chi^{j-2}}\in\mathbb{N}$. As computations become completed we add them, at lower next vertices, to the computation tree for $\alpha^j(B)$. In this way, we build up, in ξ steps, partial functions $\beta_\xi^{j-1}:\hat{T}^{(j-2)}\to\mathbb{N}$ by taking

$$\beta_\xi^{j-1}(\chi^{j-2})\simeq\begin{cases} w_{\chi^{j-2}} & \text{if the computation of } B(\gamma^{j-2}) \\ & \text{under } \Omega[\gamma^{j-2}/\chi^{j-2}] \\ & \text{was completed in } <\xi \text{ steps,} \\ \uparrow & \text{otherwise.}\end{cases}$$

Notice that β_ξ^{j-1} is just $[B]_\Omega^{<\xi}$. When and if, at some stage ξ, we have $\hat{\alpha}^j(\beta_\xi^{j-1})\downarrow$, the tree for $\alpha^j(B)$ is tagged with the value of $\hat{\alpha}^j(\beta_\xi^{j-1})$.

$$\begin{array}{l} \alpha^j(B) \text{———} \hat{\alpha}^j(\beta_\xi^{j-1}) \\ \qquad\qquad \vdots \\ \qquad \searrow B(\gamma^{j-2})\cdots\Omega[\gamma^{j-2}/\chi_\delta^{j-2}]\cdots w_{\chi_\delta^{j-2}} \\ \qquad\qquad \vdots \\ \qquad \searrow B(\gamma^{j-2})\cdots\Omega[\gamma^{j-2}/\chi_1^{j-2}]\cdots w_{\chi_1^{j-2}} \\ \qquad \searrow B(\gamma^{j-2})\cdots\Omega[\gamma^{j-2}/\chi_0^{j-2}]\cdots w_{\chi_0^{j-2}} \end{array}$$

E5: E is 0. The tree is tagged with 0.

E6′: E is A′. We place A at a lower next vertex and begin computing A under Ω. When and if this computation is completed, say with value w, the tree for A′ is tagged with $w+1$.

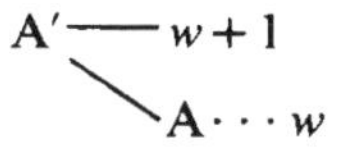

$$A' \underset{\searrow A \cdots w}{\overset{}{\rule{1em}{0.4pt}} w+1}$$

E6$\dotdiv$: E is A $\dotdiv$ 1. Similarly.

$$A \dotdiv 1 \underset{\searrow A \cdots w}{\overset{}{\rule{1em}{0.4pt}} w \dotdiv 1}$$

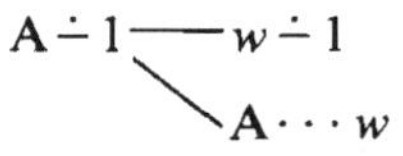

E6cs: E is cs(A, B, C). We compute A (under Ω) at a lower next vertex. When and if this computation is completed, we place either B or C at the next vertex along the principal branch from E according as the value of A was 0 or >0.

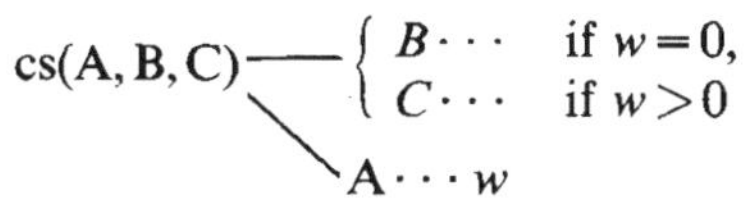

$$\text{cs}(A, B, C) \underset{\searrow A \cdots w}{\overset{}{\rule{1em}{0.4pt}}} \begin{cases} B \cdots & \text{if } w = 0, \\ C \cdots & \text{if } w > 0 \end{cases}$$

E7: E is $\theta_t(A_1^0, \ldots, A_{n_0}^0, A_1^1, \ldots, A_{n_r}^n)$. Similarly to E4.$j$, we build up partial functions $\hat{\alpha}_1^0, \ldots, \hat{\alpha}_{n_r}^r$ until we have sufficiently large arguments for $[\theta_t]_\Omega(\hat{\alpha}_1^0, \ldots, \hat{\alpha}_{n_r}^r)$ to be defined, at which point we tag the tree for E.

This completes the procedure for the construction of computation trees. We shall seldom make actual use of the trees, but they serve well to guide the intuition. If $\langle E, \Omega, w \rangle \in K^\xi$, then the next vertices after E are precisely the expressions whose values were required in the determination that $\langle E, \Omega, w \rangle \in K^\xi$. The entire tree contains precisely the information that was "used" in determining that $\langle E, \Omega, w \rangle \in K$.

3.5. *Discussion.* In this section we attempt to justify aspects of our computations which may appear to violate the spirit of computability.

At the E4.j step we are, in general, obliged to attempt computations of $B(\gamma^{j-2})$ (under assignments $\Omega[\gamma^{j-2}/\chi^{j-2}]$) which cannot be completed. This is a significant deviation from the classical approach where a single uncompletable subcomputation will frustrate the entire computation. The next example will show that, under any semantics satisfying conditions (1) and (2) above, such computations will have to be begun.

If, at step E4.j, we are only allowed to begin computations of $B(\gamma^{j-2})$ under $\Omega^* = \Omega[\gamma^{j-2}/\chi^{j-2}]$ for values of χ^{j-2} such that $[B(\gamma^{j-2})]_{\Omega^*}\downarrow$, and if

$[\alpha^j]_\Omega(ⓤ^{j-1})\uparrow$, there will have to be a χ^{j-2} such that

$$(\forall\beta^{j-1})([\alpha^j]_\Omega(\beta^{j-1})\downarrow\rightarrow\beta^{j-1}(\chi^{j-2})\downarrow).$$

This is because the χ^{j-2} for the first computation of $\mathrm{B}(\gamma^{j-2})$ will have to be chosen with no a priori knowledge of $[\mathrm{B}]_\Omega$ and will have to be in the domain of $[\mathrm{B}]_\Omega$. If conditions (1) and (2) are to hold, the same must be true with any j-expression A in place of α^j. That is, given A,Ω with $[\mathrm{A}]_\Omega(ⓤ^{j-1})\uparrow$ there must be a χ^{j-2} such that

$$(\forall\beta^{j-1})([\mathrm{A}]_\Omega(\beta^{j-1})\downarrow\rightarrow\beta^{j-1}(\chi^{j-2})\downarrow).$$

We now present an example of a 3-expression A for which no such χ^{j-2} may be chosen.[3] There are simpler examples, but they make use of specific derivations (including S11). This example shows that the situation is basic to the definition of a j-expression and independent of the specific schemata allowed.

Example 3.6. Let A be the 3-expression $\lambda\gamma^2[\gamma^2(\lambda a\gamma^2(\lambda b[a]))]$. For each i, let

$$\beta_i^2(\chi^1)\simeq\begin{cases} n & \text{if } \{\langle i,i+n\rangle,\langle i+1,i+n\rangle,\langle i+2,i+n\rangle,\ldots\}\subset\chi^1,\\ i & \text{if } \{\langle i,0\rangle,\langle i+1,1\rangle,\langle i+2,2\rangle,\ldots\}\subset\chi^1,\\ \uparrow & \text{otherwise.}\end{cases}$$

Then $\beta_i^2(\chi^1)\downarrow$ if and only if χ^1 restricted to $\{m\colon m\geqslant i\}$ is either:

(i) $\lambda m[n]$ for some $n\geqslant i$, or

(ii) $\lambda m[m\dot{-}i]$.

Clearly each β_i^2 must be an object in any semantics which satisfies (1) and (2) and has all of the classical type-1 objects. Now, under any Ω, $[\mathrm{A}]_\Omega(\beta_i^2)=i$; but there is no χ^1 such that $(\forall i)\beta_i^2(\chi^1)\downarrow$.

Having abandoned the hope of completing all computations which are initiated (under fixed assignments), there are various ways to handle the E4.j step. One method is to begin under some assignment $\Omega^*=\Omega[\gamma^{j-2}/\chi^{j-2}]$ (e.g. with $\chi^{j-2}=ⓤ^{j-2}$), and then allow χ^{j-2} to be extended as certain information about the course of the computation is discovered. This is essentially the approach which Kleene takes in KLEENE (1930a, b) and we refer the reader to these papers for further details. Our approach is to attempt computations for all possible χ^{j-2} simultaneously (as a generalization of the dovetailing technique from ordinary recursion theory) until we have built up a partial function β_ξ^{j-1} on which α^j is defined. If one

[3]This situation does not occur at types $\hat{1}$ or $\hat{2}$.

views this system as a generalization of recursion relative to a partial function, it is natural to think of α^j as embodied in an oracle which will supply a value $\alpha^j(\beta^{j-1})$ when supplied with an argument β^{j-1} in the domain of α^j, and stand mute otherwise. The usual custom is to consider the entire computation as being frustrated if (the oracle for) α^j is ever questioned with a β^{j-1} not in the domain of α^j. Perhaps this viewpoint is too restrictive. One could view our computation procedure as a process by which α^j is first questioned with β_0^{j-1}. If we immediately receive an answer $\alpha^j(\beta_0^{j-1})$ all is well; but, without worrying about whether we will receive an answer to our first question, we could question α^j with a (possibly transfinite) sequence of arguments $\beta_0^{j-1} \subset \beta_1^{j-1} \subset \cdots \subset \beta_\xi^{j-1} \cdots$ where a value of $\alpha^j(\beta_\xi^{j-1})$ at any one of the arguments would suffice. In this way we would be able to present α^j with arguments on which it was not defined and still not violate the principle that computations should be based only on information provided by α^j, rather than on the lack of such information.

4. Adequacy of the semantics

The next two theorems say that our semantics does what it should, i.e. that it satisfies conditions (1) and (2). We will frequently perform induction on j and ξ simultaneously, assuming the result holds whenever $k<j$ or $k=j$ and $\zeta<\xi$. We refer to this as *induction on* (j,ξ).

Theorem 4.1. *Let* B *be a j-expression. Then* $[\mathrm{B}]_\Omega \in \mathrm{T}^{(j)}$.

Proof. Immediate from Lemma 3.1.

Theorem 4.2. *Let* E *be a* 0*-expression. Let* $\mathrm{D}, \overline{\mathrm{D}}$ *be j-expressions not containing the variable* β^j. *Let* $\mathrm{E_D} \equiv \mathrm{S}_{\mathrm{D}}^{\beta^j}\mathrm{E}|$, $\mathrm{E}_{\overline{\mathrm{D}}} \equiv \mathrm{S}_{\overline{\mathrm{D}}}^{\beta^j}\mathrm{E}|$. *Then, for any* Ω,

$$[\mathrm{D}]_\Omega \begin{Bmatrix} \subseteq \\ = \end{Bmatrix} [\overline{\mathrm{D}}]_\Omega \rightarrow [\mathrm{E_D}]_\Omega \begin{Bmatrix} \subseteq \\ = \end{Bmatrix} [\mathrm{E}_{\overline{\mathrm{D}}}]_\Omega.$$

Before proving Theorem 4.2 we need a rather technical lemma. We will use the lemma to guarantee that if $[\mathrm{A(B)}]_\Omega\downarrow$, say $|\mathrm{A(B)}|_\Omega = \xi$, then we need no more information about $[\mathrm{B}]_\Omega$ than is obtainable in $<\xi$ steps, i.e. that $[\mathrm{A(B)}]_\Omega = [\mathrm{A}(\gamma^j)]_{\Omega[\gamma^j/[\mathrm{B}]_\Omega^{<\xi}]}$. This should be clear, considering that any necessary information about B is computed on the tree for A(B), and thus is computed in $<\xi$ steps. The lemma is stated in a more complicated manner in order to have a workable inductive hypothesis.

Lemma 4.3. *Let* D *be a j-expression not containing* β^j, E *a* 0*-expression and* $E_D \equiv S_D^{\beta^j}E|$. *Suppose* $|E_D|_\Omega = \xi < \infty$. *Let*

$$\zeta = \begin{cases} \xi + 1 & \text{if E is of the form } \beta^j \text{ or } \beta^j(B), \\ \xi & \text{otherwise.} \end{cases}$$

Let $\Omega_0 = \Omega[\beta^j/[D]_\Omega^{<\zeta}]$. *Then* $[E]_{\Omega_0} = [E_D]_{\Omega_0}$ $(= [E_D]_\Omega)$ *and* $|E|_{\Omega_0} \leqslant |E_D|_{\Omega_0}$.

Proof. For the purposes of this lemma, we write $[E]_\Omega = {}^*[\bar{E}]_{\bar{\Omega}}$ to mean $[E]_\Omega = [\bar{E}]_{\bar{\Omega}}$ and $|E|_\Omega \geqslant |\bar{E}|_{\bar{\Omega}}$. We use induction on (j, ξ). Note that, since β^j does not occur in D, $[D]_\Omega = [D]_{\Omega_0}$, $[E_D]_\Omega = [E_D]_{\Omega_0}$, etc. We now take cases on the form of E.

E1: E is α^0. If $\alpha^0 \not\equiv \beta^j$, we are done, so assume $\alpha^0 \equiv \beta^j$ (so $j = 0$). Then $E_D \equiv D$ and $\zeta = \xi + 1$. Clearly $[D]_\Omega = [D]_\Omega^{<\xi+1} = [\beta^0]_{\Omega_0}$ and $|\beta^0|_{\Omega_0} = 0 \leqslant \xi$.

E2: E is $\phi_i(\vec{A})$. Using "*" as in the definition of K (e.g. E* is the next vertex on the tree for E), we have $(E_D)^* \equiv (E^*)_D$. Thus $|(E^*)_D|_\Omega = |(E_D)^*|_\Omega < |E_D|_\Omega = \xi$ so the inductive hypothesis on ξ applies to yield that $[(E_D)^*]_\Omega = {}^*$ $[(E^*)_D]_\Omega = {}^\bullet$ $[E^*]_{\Omega_0}$, whence $[E_D]_\Omega = {}^*$ $[E]_{\Omega_0}$.

Notice that we have used Lemma 3.1 implicitly since, depending on the form of $(E^*)_D$, the inductive hypothesis might yield either:

$$[(E^*)_D]_\Omega = {}^*[E^*]_{\Omega[\beta^j/[D]_\Omega^{\leqslant \xi}]}$$

or

$$[(E^*)_D]_\Omega = {}^*[E^*]_{\Omega[\beta^j/[D]_\Omega^{<\xi-1}]}.$$

But $\xi - 1 < \xi \leqslant \zeta$ so, in either case, we have $[(E^*)_D]_\Omega = {}^*$ $[E^*]_{\Omega_0}$. In future, we will not bother to mention such points.

E4.λ: E is $\{\lambda\alpha^k A\}(B)$. Similarly to case E2.

E4.k: E is $\alpha^k(B)$. First notice that $[B_D]_\Omega^{<\xi} = [B_D]_{\Omega_0}^{<\xi} \subset [B]_{\Omega_0}^{<\xi}$. To see this, we return to the definition of $[B]_{\Omega_0}^{<\xi}$. Consider the case that $k > 1$ so that B_D is a $(k-1)$-expression with $k - 1 > 0$. (The case that $k = 1$ is immediate from the inductive hypothesis on ξ.) Suppose that $[B_D]_{\Omega_0}^{<\xi}(\chi^{k-2})\downarrow$. Let γ^{k-2} be a new variable and let $\Omega_1 = \Omega_0[\gamma^{k-2}/\chi^{k-2}]$. Then

$$\begin{aligned} [B_D]_{\Omega_0}^{<\xi}(\chi^{k-2})\downarrow &\to |B_D(\gamma^{k-2})|_{\Omega_1} < \xi \quad \left(\text{by def. of } [B_D]_{\Omega_0}^{<\xi}\right) \\ &\to \left[B_D(\gamma^{k-2})\right]_{\Omega_1} = {}^* \left[B(\gamma^{k-2})\right]_{\Omega_1} \quad (\text{by ind. hyp. on } \xi) \end{aligned}$$

so $|B(\gamma^{k-2})|_{\Omega_1} < \xi$, i.e. $[B]_{\Omega_0}^{<\xi}(\chi^{k-2})\downarrow$ and $[B]_{\Omega_0}^{<\xi}(\chi^{k-2}) = [B_D]_{\Omega_0}^{<\xi}(\chi^{k-2})$. We now return to E4.$k$.

Case 1: $\alpha^k \not\equiv \beta^j$. Take $\hat{\alpha}^k = [\alpha^k]_\Omega$ $(= [\alpha^k]_{\Omega_0})$. Now $E_D \equiv \alpha^k(B_D)$ so

$$[E_D]_\Omega = \hat{\alpha}^k\left([B_D]_\Omega^{<\xi}\right) = \hat{\alpha}^k\left([B]_{\Omega_0}^{<\xi}\right) = \left[\alpha^k(B)\right]_{\Omega_0}$$

and $|\alpha^k(B)|_\Omega \leqslant \xi$.

Case 2: $\alpha^k \equiv \beta^j$. Then E is $\beta^j(\mathrm{B})$, $\zeta = \xi + 1$, and $\mathrm{E_D} \equiv \mathrm{D(B_D)}$. Let γ^{j-1} be a new variable and let $\Omega_1 = \Omega_0[\gamma^{j-1}/[\mathrm{B_D}]^{<\xi}_{\Omega_0}]$. Then, by the inductive hypothesis on j,

$$[\mathrm{E_D}]_{\Omega_0} =^* [\mathrm{D(B_D)}]_{\Omega_0} =^* [\mathrm{D}(\gamma^{j-1})]_{\Omega_1}.$$

In particular, $|\mathrm{D}(\gamma^{j-1})|_{\Omega_1} \leqslant \xi$ so

$$[\mathrm{E_D}]_{\Omega_0} = [\mathrm{D}]^{<\zeta}_{\Omega_1}([\mathrm{B_D}]^{<\xi}_{\Omega_1}) = [\mathrm{D}]^{<\zeta}_{\Omega_0}([\mathrm{B}]^{<\xi}_{\Omega_0}) = [\beta^j(\mathrm{B})]_{\Omega_0} = [\mathrm{E}]_{\Omega_0}$$

and $|\beta^j(\mathrm{B})|_{\Omega_0} \leqslant \xi$.

E5: trivial.

E6: similarly to E2.

E7: similarly to E4.k, case 1.

Proof of Theorem 4.2. It suffices to consider the "$\subset$" part, from which the "=" part follows by interchanging the roles of D and $\overline{\mathrm{D}}$. Thus we assume that $|\mathrm{E_D}|_\Omega = \xi < \infty$, say $[\mathrm{E_D}]_\Omega = w$, and show that $[\mathrm{E}_{\overline{\mathrm{D}}}]_\Omega = w$. We use induction on (j, ξ), taking cases on the form of E. Most of the cases are routine. We consider two representative cases in detail. Without loss of generality, β^j does not occur bound in E.

E4.λ: E is $\{\lambda\alpha^k\mathrm{A}\}(\mathrm{B})$. By assumption, $\alpha^k \not\equiv \beta^j$ so $\mathrm{E_D} \equiv \{\lambda\alpha^k\mathrm{A_D}\}(\mathrm{B_D})$. Thus, by the inductive hypothesis on ξ, we have

$$\mathrm{E_D} \simeq_\Omega \mathrm{S}^{\alpha^k}_{\mathrm{B_D}}\mathrm{A_D}| \simeq_\Omega \mathrm{S}^{\alpha^k}_{\mathrm{B}_{\overline{\mathrm{D}}}}\mathrm{A}_{\overline{\mathrm{D}}}| \simeq_\Omega \mathrm{E}_{\overline{\mathrm{D}}}.$$

E4.k: E is $\alpha^k(\mathrm{B})$. By the inductive hypothesis on ξ, $[\mathrm{B_D}]^{<\xi}_\Omega \subset [\mathrm{B}_{\overline{\mathrm{D}}}]_\Omega$.

Case 1: $\alpha^k \not\equiv \beta^j$. Let $\hat{\alpha}^k = [\alpha^k]_\Omega$. Then

$$\begin{aligned}[\mathrm{E_D}]_\Omega &= [\alpha^k(\mathrm{B_D})]_\Omega = \hat{\alpha}^k([\mathrm{B_D}]^{<\xi}_\Omega) = \hat{\alpha}^k([\mathrm{B}_{\overline{\mathrm{D}}}]_\Omega) \\ &= [\alpha^k(\mathrm{B}_{\overline{\mathrm{D}}})]_\Omega = [\mathrm{E}_{\overline{\mathrm{D}}}]_\Omega.\end{aligned}$$

Case 2: $\alpha^k \equiv \beta^j$ so $\mathrm{E_D} \equiv \mathrm{D(B_D)}$. Let γ^{j-1} be a new variable. Let $\Omega_0 = \Omega[\gamma^{j-1}/[\mathrm{B_D}]^{<\xi}_\Omega]$ and $\Omega_1 = \Omega[\gamma^{j-1}/[\mathrm{B}_{\overline{\mathrm{D}}}]_\Omega]$. Then

$$\begin{aligned}[\mathrm{E_D}]_\Omega &= [\mathrm{D(B_D)}]_\Omega = [\mathrm{D}(\gamma^{j-1})]_{\Omega_0} \quad \text{(by Lemma 4.3)} \\ &= [\overline{\mathrm{D}}(\gamma^{j-1})]_{\Omega_0} = [\overline{\mathrm{D}}(\gamma^{j-1})]_{\Omega_1} \\ &\qquad \text{(by assumption on } [\overline{\mathrm{D}}]_\Omega \text{ and Lemma 3.1)} \\ &= [\overline{\mathrm{D}}(\mathrm{B}_{\overline{\mathrm{D}}})]_{\Omega_1} \quad \text{(by ind. hyp. on } j) \\ &= [\mathrm{E}_{\overline{\mathrm{D}}}]_\Omega.\end{aligned}$$

5. Embedding Kleene's 1978 theory into the extended theory

KLEENE (1978) considered his computations to be in the context of the classical type structure:

$$\mathrm{T}^{(0)} = \mathbb{N} = \{0, 1, 2, \dots\},$$

$$\mathrm{T}^{(j+1)} = \mathbb{N}^{\mathrm{T}^{(j)}} = \text{the set of all } \textit{total} \text{ functions from } \mathrm{T}^{(j)} \text{ into } \mathbb{N}.$$

In order to transfer some of our results from the extended theory to the unextended theory of Kleene's, we make use of an embedding of that theory into the present one. To avoid confusion, we will (usually) use variables like $\hat{\alpha}^j$ for elements of $\hat{\mathrm{T}}^{(j)}$ and α^j for elements of $\mathrm{T}^{(j)}$. Notice that $\mathrm{T}^{(0)} \subset \hat{\mathrm{T}}^{(0)}$ and $\mathrm{T}^{(1)} \subset \hat{\mathrm{T}}^{(1)}$, but $\mathrm{T}^{(j+2)} \not\subset \hat{\mathrm{T}}^{(j+2)}$. The reason for this is that $\alpha^{j+2} \in \mathrm{T}^{(j+2)}$ is not monotone when considered as a partial function on $\hat{\mathrm{T}}^{(j+1)}$; e.g. $\alpha^2(\lambda\beta^0 0)\downarrow$, $\lambda\beta^0 0 \subsetneqq \lambda\hat{\beta}^0 0$, but the domain of α^2 is $\mathrm{T}^{(1)}$ so $\alpha^2(\lambda\hat{\beta}^0 0)\uparrow$. Of course, α^2 has a minimum monotone extension—this will provide our embedding.

We define $^\circ : \mathrm{T}^{(j)} \to \hat{\mathrm{T}}^{(j)}$ by recursion on j. We will need to prove the following lemma simultaneously with the definition.

Lemma 5.1.

(i) $^\circ$ *is one-to-one,*

(ii) *if* $j \geqslant 1$, *then every* $\mathring{\alpha}^j$ *has the same domain—namely*

$$\{\hat{\beta}^{j-1} \in \hat{\mathrm{T}}^{(j-1)} : (\exists \beta^{j-1})\mathring{\beta}^{j-1} \subset \hat{\beta}^{j-1}\},$$

(iii) *if* $\hat{\alpha}^j \in \hat{\mathrm{T}}^{(j)}$, *then there is at most one* $\alpha^j \in \mathrm{T}^{(j)}$ *such that* $\mathring{\alpha}^j \subset \hat{\alpha}^j$.

Definition 5.2.

$$\mathring{\alpha}^0 = \alpha^0.$$

$$\mathring{\alpha}^{j+1}(\hat{\beta}^j) \simeq \begin{cases} \alpha^{j+1}(\beta^j) & \text{if } \mathring{\beta}^j \subset \hat{\beta}^j \text{ for some } \beta^j \in \mathrm{T}^{(j)}, \\ \uparrow & \text{if there is no such } \beta^j. \end{cases}$$

Notice that part (iii) of the inductive hypothesis assures us that $\mathring{\alpha}^{j+1}$ is well-defined. It is easy to verify the three parts of the lemma ((iii) follows from (i) and (ii)).

If $\mathfrak{A} = \alpha_0^{j_0}, \dots, \alpha_n^{j_n}$, then we write $\mathring{\mathfrak{A}}$ for $\mathring{\alpha}_0^{j_0}, \dots, \mathring{\alpha}_n^{j_n}$. $\mathfrak{A} \subset \mathfrak{B}$ means $\alpha_0^{j_0} \subset \beta_0^{j_0}, \dots, \alpha_n^{j_n} \subset \beta_n^{j_n}$.

For any partial function θ on the unextended types, we may define a monotone, partial function $\mathring{\theta}$ on the extended types by:

$$\mathring{\theta}(\hat{\mathfrak{A}}) \simeq \begin{cases} \theta(\mathfrak{A}) & \text{if } \mathring{\mathfrak{A}} \subset \hat{\mathfrak{A}} \text{ for some } \mathfrak{A}, \\ \uparrow & \text{otherwise.} \end{cases}$$

Given any assignment Ω in the unextended theory, we may define an assignment $\mathring{\Omega}$ in the extended theory by:

$$[\alpha^j]_{\mathring{\Omega}} = ([\alpha^j]_\Omega)^\circ$$
$$[\theta]_{\mathring{\Omega}} = ([\theta]_\Omega)^\circ.$$

Theorem 5.3. *Let* E *be a* 0*-expression and* Ω *an assignment from the unextended theory. Let* $[\mathrm{E}]_\Omega$ *be the result of computing* E *under* Ω *as in* KLEENE (1978). *Then* $[\mathrm{E}]_\Omega \simeq [\mathrm{E}]_{\mathring{\Omega}}$.

Proof. ($\subset$) Suppose $[\mathrm{E}]_\Omega = w$. We show that $[\mathrm{E}]_{\mathring{\Omega}} = w$. The proof is by induction on the computation tree for $[\mathrm{E}]_\Omega$. The only nontrivial cases are E4.j ($j>1$) and E7. We consider only E4.j (E7 being similar).

E4.j ($j>1$): E is $\alpha^j(\mathrm{B})$. Let $\beta^{j-1} = [\mathrm{B}]_\Omega$, $\hat{\beta}^{j-1} = [\mathrm{B}]_{\mathring{\Omega}}$, and $\bar{\alpha}^j = [\alpha^j]_\Omega$. $[\alpha^j(\mathrm{B})]_\Omega\downarrow$ so β^{j-1} is total. Thus a computation of $\mathrm{B}(\gamma^{j-2})$ under each assignment $\Omega[\gamma^{j-2}/\chi^{j-2}]$ occurs on the tree for $[\mathrm{E}]_\Omega$; so the inductive hypothesis yields computations of $\mathrm{B}(\gamma^{j-2})$ under each assignment $\mathring{\Omega}[\gamma^{j-2}/\mathring{\chi}^{j-2}]$. Thus (using Lemma 3.1) $\mathring{\beta}^{j-1} \subset \hat{\beta}^{j-1}$, whence

$$[\mathrm{E}]_{\mathring{\Omega}} \simeq \mathring{\bar{\alpha}}^j(\hat{\beta}^{j-1}) \simeq \bar{\alpha}^j(\beta^{j-1}) \simeq [\mathrm{E}]_\Omega \simeq w.$$

($\supset$) Suppose $[\mathrm{E}]_{\mathring{\Omega}} = w$. We show that $[\mathrm{E}]_\Omega = w$. We use induction on $\xi = |\mathrm{E}|_{\mathring{\Omega}}$. We must be a little careful here. What the inductive hypothesis says is that if $|\mathrm{E}^*|_{\Omega^*} < \xi$ *and* Ω^* is of the form $\mathring{\bar{\Omega}}$ for some assignment $\bar{\Omega}$ from the unextended types, then $[\mathrm{E}^*]_{\Omega^*} \simeq [\mathrm{E}^*]_{\bar{\Omega}}$. Again, we consider only one case.

E4.j *($j>1$):* E is $\alpha^j(\mathrm{B})$. Let $\hat{\beta}^{j-1} = [\mathrm{B}]_{\mathring{\Omega}}^{<\xi}$, and $\bar{\alpha}^j = [\alpha^j]_\Omega$. $[\alpha^j(\mathrm{B})]_{\mathring{\Omega}} = w$ so $\mathring{\bar{\alpha}}^j(\hat{\beta}^{j-1}) = w$. I.e. there is some $\beta^{j-1} \in \mathrm{T}^{(j-1)}$ with $\mathring{\beta}^{j-1} \subset \hat{\beta}^{j-1}$ and $\bar{\alpha}^j(\beta^{j-1}) = w$. Therefore, under any assignment of the form $\mathring{\Omega}[\gamma^{j-2}/\mathring{\chi}^{j-2}]$, $\mathrm{B}(\gamma^{j-2})$ has a computation (in the extended theory) yielding $\beta^{j-1}(\chi^{j-2})$ and terminating in $<\xi$ steps. Now

$$\mathring{\Omega}[\gamma^{j-2}/\mathring{\chi}^{j-2}] = (\Omega[\gamma^{j-2}/\chi^{j-2}])^\circ$$

so, by the inductive hypothesis, under each assignment of the form $\Omega[\gamma^{j-2}/\chi^{j-2}]$, $\mathrm{B}(\gamma^{j-2})$ has a computation yielding $\beta^{j-1}(\chi^{j-2})$. I.e. $\beta^{j-1} = [\mathrm{B}]_\Omega$. Thus

$$[\alpha^j(\mathrm{B})]_\Omega \simeq \bar{\alpha}^j([\mathrm{B}]_\Omega) \simeq \bar{\alpha}^j(\beta^{j-1}) \simeq w.$$

We conclude with our main theorem which gives the criterion under which one j-expression may be substituted for another in the computations of KLEENE (1978). Although technically $[\mathrm{E}]_\Omega$ has no meaning in the unextended theory when $[\mathrm{E}]_\Omega\uparrow$, we will continue to write $[\mathrm{E}]_\Omega \subset [\mathrm{E}^*]_\Omega$ to mean "if $[\mathrm{E}]_\Omega = w$, then $[\mathrm{E}^*]_\Omega = w$".

Theorem 5.4. *Let* $\mathrm{E}, \mathrm{D}, \overline{\mathrm{D}}, \mathrm{E}_{\mathrm{D}}, \mathrm{E}_{\overline{\mathrm{D}}}$ *be as in Theorem* 4.2. *Let* Ω *be an assignment from the unextended types. Then*

$$[\mathrm{D}]_{\dot{\Omega}} \left\{ \begin{matrix} \subseteq \\ = \end{matrix} \right\} [\overline{\mathrm{D}}]_{\dot{\Omega}} \rightarrow [\mathrm{E}_{\mathrm{D}}]_{\Omega} \left\{ \begin{matrix} \subseteq \\ \simeq \end{matrix} \right\} [\mathrm{E}_{\overline{\mathrm{D}}}]_{\Omega}.$$

Proof. As in Theorem 4.2, we need only consider the case of $\subset$. By Theorems 4.2 and 5.3,

$$[\mathrm{E}_{\mathrm{D}}]_{\Omega} \simeq [\mathrm{E}_{\mathrm{D}}]_{\dot{\Omega}} \subset [\mathrm{E}_{\overline{\mathrm{D}}}]_{\dot{\Omega}} \simeq [\mathrm{E}_{\overline{\mathrm{D}}}]_{\Omega}.$$

References

KLEENE, S. C.

[1959] Recursive functionals and quantifiers of finite types, I, *Trans. Am. Math. Soc.*, **91**, 1–52.

[1963] Recursive functionals and quantifiers of finite types, II, *Trans. Am. Math. Soc.*, **108**, 106–142.

[1978] Recursive functionals and quantifiers of finite types revisited, I, in: *Generalized Recursion Theory II*, edited by J. E. Fenstad, R. O. Gandy and G. E. Sacks (North-Holland, Amsterdam), pp. 185–222.

[1980a] Recursive functionals and quantifiers of finite types revisited, II, in: *The Kleene Symposium*, edited by J. Barwise, H. J. Keisler and K. Kunen (North-Holland, Amsterdam).

[1980b] Recursive functionals and quantifiers of finite types revisited, III, to appear.

J. Barwise, H. J. Keisler and K. Kunen, eds., *The Kleene Symposium*

Recursion and Nonmonotone Induction in a Quantifier*

*Phokion G. Kolaitis***

University of California, Los Angeles, CA, U.S.A.

Dedicated to Professor S. C. Kleene on the occasion of his 70th birthday

Abstract: With each (monotone) quantifier Q we associate a new functional $\mathbf{F}_Q^{\hat{}}$ and develop its recursion theory. We establish that if Q is an ω-complete filter, then recursion in $\mathbf{F}_Q^{\hat{}}$ can be characterized in terms of simple and natural nonmonotone inductions involving the quantifier Q.

GRILLIOT (1971) proved that a relation R on an acceptable structure $\mathfrak{A}$ is semirecursive in the Kleene type 2 object $\mathbf{E}$ if and only if it is $\Pi_1^0(\mathfrak{A})$-nonmonotone inductive; moreover, he showed that R is semirecursive in the functional $\mathbf{E}^{\#}$ (i.e. positive elementary inductive on $\mathfrak{A}$) if and only if it is $\Sigma_2^0(\mathfrak{A})$-nonmonotone inductive. The theory of nonmonotone inductive definability was developed further by RICHTER (1971), RICHTER and ACZEL (1974), MOSCHOVAKIS (1974b) and others.

In the mean time, generalized (monotone) quantifiers entered into recursion theory through the contributions of ACZEL (1970, 1975) and MOSCHOVAKIS (EIAS). In the light of this development the question was raised of whether the results of GRILLIOT (1971) extend to generalized quantifiers. This question has also a conceptual aspect requiring the formulation of the right notion of recursion which will capture the properties of nonmonotone inductions in a quantifier Q, since the type 2 object $\mathbf{F}_Q$ and the functional $\mathbf{F}_Q^{\#}$ associated with Q fail in general to do so.

In this paper we attack the above problem by introducing and studying a new functional $\mathbf{F}_Q^{\hat{}}$ which embodies unrestricted quantification with respect to Q, but only deterministic quantification with respect to the dual quantifier $\breve{Q}$. We show that the functional $\mathbf{F}_Q^{\hat{}}$ possesses a smooth recursion theory and it has the additional feature that (in most cases) distinguishes the quantifier Q from its dual, in contrast to the situation with the functionals $\mathbf{F}_Q$ and $\mathbf{F}_Q^{\#}$.

*The results of this paper are contained in Chapters 5 and 6 of the author's Ph.D. dissertation. The author is truly grateful to his advisor Professor Y. N. Moschovakis for his continuous guidance and encouragement, as well as to Professor A. S. Kechris for numerous long conversations in which he offered valuable comments and suggestions on this work.

**Current address: Department of Mathematics, The University of Chicago, Chicago, Il 60637, U.S.A.

The main result of the paper is that if Q is an ω-complete filter, then recursion in $\mathbf{F}_Q^{\wedge}$ can be characterized in terms of simple and natural nonmonotone inductions involving the quantifier Q. The two theorems of GRILLIOT (1971) are instances of this result by simply taking Q to be the universal quantifier $\forall$. What is of added interest here is that this identification of recursion in $\mathbf{F}_Q^{\wedge}$ with nonmonotone inductions in Q depends strongly on the ω-completeness of Q, as we can construct counter-examples even with quantifiers which are filters. This seems to be the first discovery in generalized recursion theory of the new phenomenon, where an interesting recursion theoretic fact depends essentially on the structural and set theoretic properties of the quantifier, and does not hold uniformly for arbitrary ones, as it has until now happened.

1. Preliminaries

This section contains a minimum amount of the necessary background material; detailed definitions of the basic notions can be found in the first few pages of KECHRIS and MOSCHOVAKIS (1977) and KOLAITIS (1978a, b), whose notation we follow here.

1.1. Let $\mathfrak{A}=\langle A,R_1,\ldots,R_n,f_1,\ldots,f_m,c_1,\ldots,c_k\rangle$ be a structure such that ω, $\leqslant_\omega$ are elementary on $\mathfrak{A}$; a *functional* (on A with values in ω) is a partial mapping

$$\Phi: A^n \times \mathcal{P}\mathcal{F}_{k_1} \times \cdots \times \mathcal{P}\mathcal{F}_{k_m} \to \omega$$

which is monotone, i.e. if $f_1 \subseteq g_1,\ldots,f_m \subseteq g_m$ and $\Phi(\bar{x},f_1,\ldots,f_m)=w$, then $\Phi(\bar{x},g_1,\ldots,g_m)=w$ (here $\mathcal{P}\mathcal{F}_k$ is the set of all k-ary partial functions from A into ω).

If $\bar{\Phi}=(\Phi_1,\ldots,\Phi_s)$ is a sequence of functionals on A, then $\mathcal{F}[\bar{\Phi}]$ denotes the smallest class of functionals on A containing $\Phi_1,\ldots,\Phi_s$, the characteristic functions of the relations of the structure and equality, and satisfying certain minimal closure properties (closed under composition, definition by cases, functional substitution etc.). By iterating the operative functionals in $\mathcal{F}[\bar{\Phi}]$, we obtain the notion of a *recursive in* $\bar{\Phi}$ n-ary partial function from A into ω. We define then in the standard way the *semirecursive in* $\bar{\Phi}$ and the *recursive in* $\bar{\Phi}$ relations. The collection of the semirecursive in $\bar{\Phi}$ relations is called the *envelope* of $\bar{\Phi}$ and is denoted by ENV[$\bar{\Phi}$], while the collection of the recursive relations in $\bar{\Phi}$ is the *section* of $\bar{\Phi}$ and is denoted by SEC[$\bar{\Phi}$]. These notions relativize to a finite sequence $\bar{x}=(x_1,\ldots,x_n)$ from A, so that we put ENV[$\bar{\Phi},\bar{x}$] for the collection of the *semirecursive relations in* $\bar{\Phi}$ *from* $\bar{x}$ and SEC[$\bar{\Phi},\bar{x}$] for the collection of the *recursive relations in* $\overline{\Phi}$ *from* $\bar{x}$. Finally, we put

$$\mathbf{ENV}[\overline{\Phi}] = \bigcup \{\mathrm{ENV}[\overline{\Phi},\bar{x}] : \bar{x} \in A^{<\omega}\},$$

$$\mathbf{SEC}[\overline{\Phi}] = \bigcup \{\mathrm{SEC}[\overline{\Phi},\bar{x}] : \bar{x} \in A^{<\omega}\}$$

for the "*boldface*" *semirecursive in* $\overline{\Phi}$ and *recursive in* $\overline{\Phi}$ relations on A.

1.2. In inductive definability a (monotone) quantifier Q on a set A with $\omega \subseteq A$ has been represented by the functionals $\mathbf{F}_Q$ and $\mathbf{F}_Q^{\#}$, where

$$\mathbf{F}_Q(f) = \begin{cases} 0 & \text{if } f \text{ is total and } Qx(f(x)=0), \\ 1 & \text{if } f \text{ is total and } \breve{Q}x(f(x) \neq 0), \\ \text{undefined} & \text{if } f \text{ is not total;} \end{cases}$$

$$\mathbf{F}_Q^{\#}(f) = \begin{cases} 0 & \text{if } Qx(f(x)=0), \\ 1 & \text{if } \breve{Q}x(f(x) \neq 0), \\ \text{undefined} & \text{otherwise} \end{cases}$$

(in general, $f(x) \neq n$ means that $f(x)$ is defined and has value $\neq n$).

If $\exists$ is the existential quantifier on A, then we write $\mathbf{E}$ for the functional $\mathbf{F}_\exists$ and $\mathbf{E}^{\#}$ for the functional $\mathbf{F}_\exists^{\#}$. Notice that the functionals $\mathbf{E}$ and $\mathbf{F}_Q$ are actually type 2 objects in the sense of KLEENE (1959); on the other hand it is well-known that recursion in $\mathbf{E}^{\#}, \mathbf{F}_Q^{\#}$ coincides with positive elementary induction, since a relation R is semirecursive in $\mathbf{E}^{\#}, \mathbf{F}_Q^{\#}$ on a structure $\mathfrak{A}$ if and only if it is ("lightface") Q-inductive on $\mathfrak{A}$.

1.3. Let $\mathfrak{A}$ be an acceptable structure and let Γ be a collection of relations on A; we say that Γ is a ("lightface") *semi-Spector class* on $\mathfrak{A}$ if

(i) Γ contains all the elementary relations on $\mathfrak{A}$ and is closed under $\&, \vee, \forall$ and deterministic $\exists$.

(ii) Γ has the prewell-ordering property and is ω-parametrized. (In general, if Γ is a collection of relations on A and Q is a quantifier on A, we say that Γ is *closed under deterministic* Q if whenever $P(x,\bar{y})$ and $S(x,\bar{y})$ are disjoint relations in Γ, then R is also in Γ, where

$$R(\bar{y}) \Leftrightarrow \forall x(P(x,\bar{y}) \vee S(x,\bar{y})) \& QxP(x,\bar{y}).)$$

A ("lightface") *Spector class* on $\mathfrak{A}$ is a semi-Spector class which is also closed under $\exists$. The notions of a Spector class and a semi-Spector class were introduced by MOSCHOVAKIS (EIAS, 1974b); to avoid confusion we should point out that Moschovakis actually studied the "boldface" versions of these notions, i.e. the classes were assumed to contain all elementary relations defined by arbitrary parameters from A and to satisfy A-parametrization, instead of ω-parametrization.

The next result, due to MOSCHOVAKIS (EIAS, 1974a) and ACZEL (1975), summarizes the properties of recursion in $\mathbf{E}$, $\mathbf{F}_Q$ and positive elementary induction in the quantifier Q (recursion in $\mathbf{E}^{\#}, \mathbf{F}_Q^{\#}$).

1.4 Theorem. *Let $\mathfrak{A}$ be an acceptable structure and let Q be a quantifier on A.*

(i) *The class* $\mathrm{ENV}[\mathbf{E},\mathbf{F}_Q]$ *of the semirecursive relations in* $\mathbf{E},\mathbf{F}_Q$ *is the smallest semi-Spector class on $\mathfrak{A}$ closed under deterministic Q and deterministic $\check{Q}$.*

(ii) *The class* $\mathrm{ENV}[\mathbf{E}^{\#},\mathbf{E}_Q^{\#}]=\mathrm{IND}(\mathfrak{A},Q)$ *of the semirecursive relations in* $\mathbf{E}^{\#},\mathbf{F}_Q^{\#}$ *(Q-inductive relations) is the smallest Spector class on $\mathfrak{A}$ closed under Q and $\check{Q}$.*

1.5. The prewell-ordering property for both $\mathrm{ENV}[\mathbf{E},\mathbf{F}_Q]$ and $\mathrm{ENV}[\mathbf{E}^{\#},\mathbf{F}_Q^{\#}]$ is a consequence of the Stage Comparison Theorem. MOSCHOVAKIS (1976) introduced an abstract notion of *normality* which yields this theorem in the context of functional induction.

Let $\mathfrak{A}$ be a structure such that ω, $\leqslant_\omega$ are elementary on $\mathfrak{A}$ and let $\bar{\Phi}$ be a sequence of functionals on A; we say that the sequence $\bar{\Phi}$ is *strongly normal* if for each functional $\Psi(\bar{x},\bar{f})$ in $\mathfrak{F}[\bar{\Phi}]$, there is a functional $\Delta_\Psi(\bar{x},\bar{f},\bar{\delta})$ in $\mathfrak{F}[\bar{\Phi}]$ taking values 0, 1 only and such that:

(i) if $\Psi(\bar{x},\bar{f}\restriction\{\bar{y}\colon \bar{\delta}(\bar{y})=0\})\downarrow$, then $\Delta_\Psi(\bar{x},\bar{f},\bar{\delta})=0$,

(ii) if $\bar{\delta}$ is total, $\{\bar{y}\colon \bar{\delta}(\bar{y})=0\}\subseteq \mathrm{dom}\bar{f}$ and $\Psi(\bar{x},\bar{f}\restriction\{\bar{y}\colon \bar{\delta}(\bar{y})=0\})\uparrow$, then $\Delta_\Psi(\bar{x},\bar{f},\bar{\delta})=1$.

We also say that such a functional Δ_Ψ is a *normalizing functional* for Ψ.

For any quantifier Q, both sequences $\mathbf{E},\mathbf{F}_Q$ and $\mathbf{E}^{\#},\mathbf{F}_Q^{\#}$ are strongly normal; this fact accounts for many similarities between recursion in $\mathbf{E},\mathbf{F}_Q$ and positive elementary induction in the quantifier Q.

2. Separating a quantifier from its dual by recursion

In this section we introduce a new functional $\mathbf{F}_Q^{\wedge}$ and develop its recursion theory. We show that the functional $\mathbf{F}_Q^{\wedge}$ distinguishes in most cases the quantifier Q from its dual $\check{Q}$, gives rise to interesting semi-Spector and Spector classes, and leads to a natural notion of admissible sets with quantifiers studied recently by BARWISE (1978).

Recursion in the functional $\mathbf{F}_Q^{\wedge}$

2.1. Let A be a set such that $\omega\subseteq A$ and let Q be a quantifier on A. We introduce the following functional $\mathbf{F}_Q^{\wedge}$, which embodies unrestricted quantification with respect to Q, but only deterministic quantification with respect to $\check{Q}$:

$$\mathbf{F}_Q^{\hat{}}(f)=\begin{cases}0 & \text{if } Qx(f(x)=0),\\ 1 & \text{if } f \text{ is total and } \check{Q}x(f(x)\neq 0),\\ \text{undefined} & \text{otherwise.}\end{cases}$$

It is easy to show that $\mathbf{F}_Q$ is recursive in $\mathbf{F}_Q^{\hat{}}$, while $\mathbf{F}_Q^{\hat{}}$ is recursive in $\mathbf{E},\mathbf{F}_Q^{\#}$ (actually $\mathbf{F}_Q\in\mathcal{F}[\mathbf{F}_Q^{\hat{}}]$ and $\mathbf{F}_Q^{\hat{}}\in\mathcal{F}[\mathbf{E},\mathbf{F}_Q^{\#}]$). Therefore the following inclusions hold for the corresponding classes of semirecursive relations:

$$\mathrm{ENV}[\mathbf{E},\mathbf{F}_Q]\subseteq\mathrm{ENV}[\mathbf{E},\mathbf{F}_Q^{\hat{}}]\subseteq\mathrm{ENV}[\mathbf{E},\mathbf{F}_Q^{\#}]$$

and

$$\mathrm{ENV}[\mathbf{E}^{\#},\mathbf{F}_Q]\subseteq\mathrm{ENV}[\mathbf{E}^{\#},\mathbf{F}_Q^{\hat{}}]\subseteq\mathrm{ENV}[\mathbf{E}^{\#},\mathbf{F}_Q^{\#}].$$

2.2. Theorem. *If $\mathfrak{A}=\langle A,R_1,\ldots,R_n,f_1,\ldots,f_m,c_1,\ldots,c_k\rangle$ is a structure such that ω, $<_\omega$ are elementary on $\mathfrak{A}$ and Q is a quantifier on A, then the sequences $\mathbf{E},\mathbf{F}_Q^{\hat{}}$ and $\mathbf{E}^{\#},\mathbf{F}_Q^{\hat{}}$ are strongly normal.*

Proof. It is enough to find normalizing functionals for $\mathbf{E}$ and $\mathbf{F}_Q^{\hat{}}$ which are in the class $\mathcal{F}[\mathbf{E},\mathbf{F}_Q^{\hat{}}]$.

We assign first to $\mathbf{E}$ the functional

$$\Delta_{\mathbf{E}}(f,\delta)=1\dot{-}\mathbf{E}(\lambda x(1\dot{-}\delta(x))),$$

where

$$1\dot{-}a=\begin{cases}1 & \text{if } a=0,\\ 0 & \text{if } a\neq 0 \text{ or } a\notin\omega.\end{cases}$$

In order to construct a normalizing functional for $\mathbf{F}_Q^{\hat{}}$ we consider first the functional

$$\Phi(x,f,\delta)=\begin{cases}1 & \text{if } \delta(x)\neq 0,\\ 1 & \text{if } \delta(x)=0\ \&\ f(x)\neq 0,\\ 0 & \text{if } \delta(x)=0\ \&\ f(x)=0.\end{cases}$$

We put now

$$\Delta_{\mathbf{F}_Q^{\hat{}}}(f,\delta)=\begin{cases}0 & \text{if } \mathbf{F}_Q^{\hat{}}(\lambda x\Phi(x,f,\delta))=0,\\ 0 & \text{if } \mathbf{F}_Q^{\hat{}}(\lambda x\Phi(x,f,\delta))=1\ \&\ \mathbf{E}(\lambda x(1\dot{-}\delta(x)))=1,\\ 1 & \text{if } \mathbf{F}_Q^{\hat{}}(\lambda x\Phi(x,f,\delta))=1\ \&\ \mathbf{E}(\lambda x(1\dot{-}\delta(x)))=0.\end{cases}$$

The functional $\Delta_{\mathbf{F}_Q^{\hat{}}}(f,\delta)$ is clearly in the class $\mathcal{F}[\mathbf{E},\mathbf{F}_Q^{\hat{}}]$ and it is quite straightforward to show that it is a normalizing functional for $\mathbf{F}_Q^{\hat{}}$.

The main properties of recursion in $\mathbf{F}_Q^{\hat{}}$ are given by the next result, which follows from the preceding Theorem 2.2, the Main Lemma of

MOSCHOVAKIS (1974a) and general results in KECHRIS and MOSCHOVAKIS (1977).

2.3. Theorem. *Let $\mathfrak{A}$ be a structure such that ω, $\leqslant_\omega$ are elementary on $\mathfrak{A}$ and let Q be a quantifier on A.*

(i) *The class* $\mathrm{ENV}[\mathbf{E},\mathbf{F}_Q^{\wedge}]$ *of the semirecursive relations in* $\mathbf{E},\mathbf{F}_Q^{\wedge}$ *is closed under* $\&, \bigvee, \forall, Q$, *deterministic* $\exists$, *deterministic* $\breve{Q}$ *and has the prewell-ordering property. A relation $R \subseteq A^n$ is recursive in* $\mathbf{E},\mathbf{F}_Q^{\wedge}$ *if and only if both R and $A^n - R$ are semirecursive in* $\mathbf{E},\mathbf{F}_Q^{\wedge}$.

(ii) *If the structure $\mathfrak{A}$ is acceptable, then* $\mathrm{ENV}[\mathbf{E},\mathbf{F}_Q^{\wedge}]$ *is the smallest semi-Spector class on $\mathfrak{A}$ closed under Q and deterministic $\breve{Q}$. Moreover, in this case* $\mathrm{ENV}[\mathbf{E},\mathbf{F}_Q^{\wedge}]$ *is closed under* $\exists^\omega$ *and*

$$\mathrm{ENV}\left[\mathbf{E},\mathbf{F}_Q^{\wedge}\right]=\mathrm{ENV}\left[\mathbf{E},\mathbf{F}_Q^{\wedge},\mathbf{E}_\omega^{\#}\right],$$

where $\mathbf{E}_\omega^{\#}=\mathbf{E}_{\exists^\omega}^{\#}$ *is the functional embodying existential quantification over ω.*

(iii) *The class* $\mathrm{ENV}[\mathbf{E}^{\#},\mathbf{F}_Q^{\wedge}]$ *of the semirecursive relations in* $\mathbf{E}^{\#},\mathbf{F}_Q^{\wedge}$ *is closed under* $\&, \bigvee, \forall, \exists, Q$, *deterministic* $\breve{Q}$ *and has the prewell-ordering property. A relation $R \subseteq A^n$ is recursive in* $\mathbf{E}^{\#},\mathbf{F}_Q^{\wedge}$ *if and only if both R and $A^n - R$ are semirecursive in* $\mathbf{E}^{\#},\mathbf{F}_Q^{\wedge}$.

(iv) *If the structure $\mathfrak{A}$ is acceptable, then* $\mathrm{ENV}[\mathbf{E}^{\#},\mathbf{F}_Q^{\wedge}]$ *is the smallest Spector class on $\mathfrak{A}$ closed under Q and deterministic $\breve{Q}$.*

The above results show that the theory of recursion in $\mathbf{F}_Q^{\wedge}$ is analogous to recursion in $\mathbf{E},\mathbf{F}_Q$ and positive elementary induction (recursion in $\mathbf{E}^{\#},\mathbf{F}_Q^{\#}$). Many of the similarities between these theories are due to the fact that the functionals involved are strongly normal. In KOLAITIS (1978a, b) we established that normality has also second order consequences (such as the Second Stage Comparison Theorem), which in turn imply a generalized Spector–Gandy type Theorem for recursion in strongly normal, Q-definable functionals. In particular, Theorem 3.6 in KOLAITIS (1978b) and the preceding Theorem 2.2 yield immediately the following normal form for recursion in $\mathbf{F}_Q^{\wedge}$.

2.4. Theorem. *Let $\mathfrak{A}$ be an acceptable structure and let Q be a quantifier on A.*

(i) *A relation R is semirecursive in* $\mathbf{E},\mathbf{F}_Q^{\wedge}$ *if and only if there is a formula $\varphi(Y,\bar{x})$ of the language $\mathfrak{L}^{\mathfrak{A}}(Q)$ of $\mathfrak{A}$ such that*

$$R(\bar{x}) \Leftrightarrow \left(\exists Y \in \mathrm{SEC}\left[\mathbf{E},\mathbf{F}_Q^{\wedge},\bar{x}\right]\right)\varphi(Y,\bar{x}).$$

(ii) *A relation R is semirecursive in* $\mathbf{E}^{\#},\mathbf{F}_Q^{\wedge}$ *if and only if there is a formula $\varphi(Y,\bar{x})$ of the language $\mathfrak{L}^{\mathfrak{A}}(Q)$ of $\mathfrak{A}$ such that*

$$R(\bar{x}) \Leftrightarrow \left(\exists Y \in \mathrm{SEC}\left[\mathbf{E}^{\#},\mathbf{F}_Q^{\wedge},\bar{x}\right]\right)\varphi(Y,\bar{x}).$$

2.5. If $\mathfrak{A}$ is an acceptable structure, then recursion in $\mathbf{E}^{\#}, \mathbf{F}_{\hat{Q}}$ gives also rise to the smallest "boldface" Spector class on $\mathfrak{A}$ closed under Q and deterministic $\check{Q}$, namely the class

$$\mathbf{ENV}\left[\mathbf{E}^{\#}, \mathbf{F}_{\hat{Q}}\right] = \bigcup \left\{\mathrm{ENV}\left[\mathbf{E}^{\#}, \mathbf{F}_{\hat{Q}}, \bar{x}\right] : \bar{x} \in A^{<\omega}\right\}.$$

The Companion Theorem 9E.1 in MOSCHOVAKIS (EIAS) implies that this "boldface" Spector class can be identified with the collection of Σ_1 relations of a certain admissible set. We give here a concrete description of this companion admissible set, using the notion of a *$Q^{\#}, \check{Q}$-admissible set* introduced recently by BARWISE (1978).

Assume that A, M are transitive sets, $A \in M$ and Q is a quantifier on A. The class of $\Delta_0(Q, \check{Q})$ formulas is the smallest collection containing the atomic and negated atomic formulas of the structure $\langle M, \in \rangle$ with parameters from M and closed under $\&, \vee, (\forall x \in y), (\exists x \in y), (Qx \in A), (\check{Q}x \in A)$. A formula is $\Sigma_1(Q, \check{Q})$ if it is of the form $\exists x \varphi$, where φ is $\Delta_0(Q, \check{Q})$.

The set M is *$Q^{\#}, \check{Q}$-admissible* if it is closed under pairing and union, and satisfies the schemata of separation and collection for the $\Delta_0(Q, \check{Q})$ formulas and the following schema of *Q-collection*:

$$(Qx \in A)(\exists y)\varphi(x,y) \Rightarrow (\exists w)(Qx \in A)(\exists y \in w)\varphi(x,y),$$

where φ is $\Delta_0(Q, \check{Q})$.

$Q^{\#}, \check{Q}$-admissibility is an intermediate notion between Q-admissibility and strong Q-admissibility studied by MOSCHOVAKIS (EIAS) (the Q-admissible and strong Q-admissible sets are the $Q, \check{Q}$-admissible and $Q^{\#}, \check{Q}^{\#}$-admissible of BARWISE (1978)). It is now quite straightforward, using the methods of MOSCHOVAKIS (EIAS), to establish the following Companion Theorem for the "boldface" Spector class $\mathbf{ENV}[\mathbf{E}^{\#}, \mathbf{F}_{\hat{Q}}]$.

2.6. Theorem. *Let $\mathfrak{A} = \langle A, \in \restriction A, R_1, \ldots, R_n \rangle$ be an acceptable structure such that A is a transitive set and let Q be a quantifier on A. Put*

$$\mathfrak{A}^{\wedge}(Q) = \bigcap \left\{M : A, R_1, \ldots, R_n \in M \text{ and } M \text{ is } Q^{\#}, \check{Q}\text{-admissible}\right\}.$$

Then $\mathfrak{A}^{\wedge}(Q)$ is $Q^{\#}, \check{Q}$-admissible and its ordinal is the supremum of the prewell-orderings in $\mathbf{SEC}[\mathbf{E}^{\#}, \mathbf{F}_{\hat{Q}}]$. Moreover if $P \subseteq A^n$, then

(i) *P is "boldface" recursive in $\mathbf{E}^{\#}, \mathbf{F}_{\hat{Q}}$ if and only if $P \in \mathfrak{A}^{\wedge}(Q)$.*

(ii) *P is "boldface" semirecursive in $\mathbf{E}^{\#}, \mathbf{F}_{\hat{Q}}$ if and only if P is $\Sigma_1(Q, \check{Q})$ on $\mathfrak{A}^{\wedge}(Q)$.*

The above Theorem 2.6 can be extended to arbitrary acceptable structures by considering admissible sets with urelements, as in BARWISE (1975, 1978). Actually, BARWISE (1978) established that in general if $\mathfrak{A}$ is any structure and Q is a quantifier on A, then there is a smallest $Q^{\#}, \check{Q}$-admissible set (with urelements) above $\mathfrak{A}$. Moreover, he obtained representability characterizations of the $\Sigma_1(Q,\check{Q})$ relations on the smallest $Q^{\#},\check{Q}$-admissible set in terms of infinitary rules of proof, studied first by ENDERTON (1967) and ACZEL (1970, 1972). This last result and the preceding Theorem 2.6 yield immediately representability characterizations of recursion in $\mathbf{E}^{\#},\mathbf{F}_{\check{Q}}^{\wedge}$.

The main examples

We present here some examples which show that in many interesting cases recursion in $\mathbf{F}_Q^{\wedge}$ is different from recursion in $\mathbf{F}_{\check{Q}}^{\wedge}$, in contrast to the situation with the functionals $\mathbf{F}_Q$ and $\mathbf{F}_Q^{\#}$.

2.7. Let $\mathfrak{A}$ be a structure such that $\omega, \leqslant_\omega$ are elementary on $\mathfrak{A}$. If we consider the universal quantifier $\forall$ on $\mathfrak{A}$, then it is easy to see that

$$\mathbf{F}_\forall^{\wedge}(f)=1\dot{-}\mathbf{E}(\lambda x(1\dot{-}f(x)))$$

and therefore $\mathrm{ENV}[\mathbf{E}]=\mathrm{ENV}[\mathbf{F}_\forall^{\wedge}]$. On the other hand $\mathbf{F}_\exists^{\wedge}=\mathbf{E}^{\#}$ and hence recursion in $\mathbf{F}_\forall^{\wedge}$ does not always coincide with recursion in $\mathbf{F}_\exists^{\wedge}$.

Assume now that $\mathfrak{A}$ is acceptable and let $\mathcal{C}=\langle\omega,\cdot\leqslant_\omega,\langle\,\rangle\rangle$ be a *definable coding scheme* on $\mathfrak{A}$, i.e. $\langle\,\rangle : A^{<\omega}\to A$ is a total one to one function such that the corresponding coding and decoding relations and functions are elementary on $\mathfrak{A}$. For each $n\geqslant 2$ let Q_n be the string of n alternations of the universal and the existential quantifier, i.e.

$$Q_n=\underbrace{\forall\exists\cdots}_{n}=\Big\{X\colon\ (\underbrace{\forall x_1\exists x_2\cdots}_{n})(\langle x_1,\dots,x_n\rangle\in X)\Big\}.$$

We can easily show that $\mathfrak{F}[\mathbf{F}_{Q_n}^{\wedge}]=\mathfrak{F}[\mathbf{F}_{\check{Q}_n}^{\wedge}]=\mathfrak{F}[\mathbf{E}^{\#}]$ and hence

$$\mathrm{ENV}\big[\mathbf{F}_{Q_n}^{\wedge}\big]=\mathrm{ENV}\big[\mathbf{F}_{\check{Q}_n}^{\wedge}\big]=\mathrm{ENV}\big[\mathbf{E}^{\#}\big].$$

2.8. If $\mathfrak{A}$ is an acceptable structure, then (relative to a definable coding scheme $\mathcal{C}$) we define the *Suslin quantifier* $\mathcal{S}$ on A, where

$$\mathcal{S}=\{X\subseteq A\colon (\forall x_1\forall x_2\cdots)(\exists n)(\langle x_1,\dots,x_n\rangle\in X)\}.$$

The dual of the Suslin quantifier is the *classical $\mathcal{Q}$-quantifier*, i.e.

$$\mathcal{Q}=\{X\subseteq A\colon (\exists x_1\exists x_2\cdots)(\forall n)(\langle x_1,\dots,x_n\rangle\in X)\}.$$

Another important quantifier in inductive definability is the *open game quantifier G*, where

$$G=\{X\subseteq A\colon (\forall x_1\exists y_1\forall x_2\exists y_2\cdots)(\exists n)(\langle x_1,y_1,\dots,x_n,y_n\rangle\in X)\}.$$

The dual of G is the *closed game quantifier*

$$\check{G}=\{X\subseteq A\colon (\exists x_1\forall y_1\exists x_2\forall y_2\cdots)(\forall n)(\langle x_1,y_1,\dots,x_n,y_n\rangle\in X)\}$$

ACZEL (1975) gave a natural game theoretic interpretation of infinite strings of quantifiers $(Q_1x_1Q_2x_2\cdots)$ where $Q_1, Q_2, \ldots$ is a sequence of quantifiers on A. In particular, with each quantifier Q on A he associated the quantifiers $Q^{\vee}$ and $Q^{\wedge}$, where

$$Q^{\vee} = \{X\colon (Qx_1Qx_2\cdots)(\exists n)(\langle x_1, x_2, \ldots, x_n\rangle \in X)\}$$

and

$$Q^{\wedge} = \{X\colon (Qx_1Qx_2\cdots)(\forall n)(\langle x_1, x_2, \ldots, x_n\rangle \in X)\}.$$

We see for example that $\mathfrak{S} = \forall^{\vee}$, $\mathfrak{A} = \exists^{\wedge}$, $G = (\forall\exists)^{\vee}$ and $\breve{G} = (\exists\forall)^{\wedge}$. ACZEL (1975) established that in general the quantifier $(\breve{Q})^{\wedge}$ is the dual of $Q^{\wedge}$.

The next result is due to MOSCHOVAKIS (EIAS) for the game quantifier and to ACZEL (1975) for arbitrary quantifiers.

2.9. Theorem. *Let $\mathfrak{A}$ be an acceptable structure, Q a quantifier on A and Γ a semi-Spector class on $\mathfrak{A}$. If Γ is closed under the quantifier Q, then Γ is also closed under the quantifier $Q^{\vee}$. In particular, every semi-Spector class on $\mathfrak{A}$ is closed under the Suslin quantifier and every Spector class on $\mathfrak{A}$ is closed under the game quantifier.*

Although every semi-Spector class is closed under the Suslin quantifier, many natural semi-Spector classes are not closed under the classical $\mathfrak{A}$-quantifier. This is a consequence of the following Representation Theorem of MOSCHOVAKIS (1967) (see also Theorem 19.1 in KECHRIS and MOSCHOVAKIS (1977)).

2.10. Theorem. *Let $\mathfrak{A}$ be an acceptable structure and let $\mathbf{F}$ be a type 2 object on A. If $R \in \mathrm{ENV}[\mathbf{E}, \mathbf{F}]$, then there is a relation $P \in \mathrm{ENV}[\mathbf{E}, \mathbf{F}]$ such that*

$$\neg R(\bar{y}) \Leftrightarrow (\exists x_1 \exists x_2 \cdots)(\forall n) P(\langle x_1, \ldots, x_n\rangle, \bar{y}) \Leftrightarrow \mathfrak{A} x P(x, \bar{y}).$$

In particular the semi-Spector class ENV[**E**, **F**] *is not closed under the classical $\mathfrak{A}$-quantifier.*

If Γ and Γ' are semi-Spector classes on the structure $\mathfrak{A}$, then we say that Γ is *strongly contained in* Γ' and we write $\Gamma < \Gamma'$ if every relation in Γ and its complement are in Γ'.

2.11. Proposition. *If $\mathfrak{A}$ is an acceptable structure, then*

$$\begin{aligned}\mathrm{ENV}[\mathbf{E}, \mathbf{F}_{\mathfrak{S}}] = \mathrm{ENV}[\mathbf{E}, \mathbf{F}_{\mathfrak{S}}^{\wedge}] &< \mathrm{ENV}[\mathbf{E}, \mathbf{F}_{\mathfrak{A}}^{\wedge}] \\ &= \mathrm{ENV}[\mathbf{E}^{\#}, \mathbf{F}_{\mathfrak{A}}^{\wedge}] = \mathrm{ENV}[\mathbf{E}^{\#}, \mathbf{F}_{\mathfrak{S}}^{\#}]\end{aligned}$$

Proof. It is clear that $\mathrm{ENV}[\mathbf{E}, \mathbf{F}_{\mathfrak{S}}] \subseteq \mathrm{ENV}[\mathbf{E}, \mathbf{F}_{\mathfrak{S}}^{\wedge}]$. Theorems 1.4 and 2.9 imply that $\mathrm{ENV}[\mathbf{E}, \mathbf{F}_{\mathfrak{S}}]$ is a semi-Spector class closed under $\mathfrak{S}$ and deterministic $\Game$. However, ENV[$\mathbf{E}, \mathbf{F}^{\wedge}_{\mathcal{S}}$] is (by Theorem 2.3) the smallest semi-Spector class closed under $\mathcal{S}$ and deterministic $\Game$, and therefore ENV[$\mathbf{E}, \mathbf{F}_{\mathcal{S}}$] = ENV[$\mathbf{E}, \mathbf{F}^{\wedge}_{\mathcal{S}}$]. A similar argument establishes that ENV [$\mathbf{E}^{\#}, \mathbf{F}^{\wedge}_{\Game}$] = ENV[$\mathbf{E}^{\#}, \mathbf{F}^{\#}_{\mathcal{S}}$]; moreover, $\mathbf{E}^{\#}$ is recursive in $\mathbf{F}^{\wedge}_{\Game}$ and hence ENV[$\mathbf{E}, \mathbf{F}^{\wedge}_{\Game}$] = ENV[$\mathbf{E}^{\#}, \mathbf{F}^{\wedge}_{\Game}$].

We complete the proof of the proposition by showing that ENV[$\mathbf{E}, \mathbf{F}_{\mathcal{S}}$] $<$ ENV[$\mathbf{E}, \mathbf{F}^{\wedge}_{\Game}$]. It is first of all obvious that ENV[$\mathbf{E}, \mathbf{F}_{\mathcal{S}}$] $\subseteq$ ENV[$\mathbf{E}, \mathbf{F}^{\wedge}_{\Game}$]. Assume now that $R \in$ ENV[$\mathbf{E}, \mathbf{F}_{\mathcal{S}}$]; since $\mathbf{F}_{\mathcal{S}}$ is a type 2 object, Theorem 2.10 implies that there is a relation $P \in$ ENV[$\mathbf{E}, \mathbf{F}_{\mathcal{S}}$] such that

$$\neg R(\vec{y}) \Leftrightarrow \Game x P(x, \vec{y}).$$

But also $P \in$ ENV[$\mathbf{E}, \mathbf{F}^{\wedge}_{\Game}$] and this class is closed under the classical $\Game$-quantifier, so that $\neg R \in$ ENV[$\mathbf{E}, \mathbf{F}^{\wedge}_{\Game}$].

We can also show that

$$\begin{aligned} \mathrm{ENV}\left[\mathbf{E}^{\#}, \mathbf{F}_{\mathcal{S}}\right] = \mathrm{ENV}\left[\mathbf{E}^{\#}, \mathbf{F}^{\wedge}_{\mathcal{S}}\right] &\subseteq \mathrm{ENV}\left[\mathbf{E}^{\#}, \mathbf{F}^{\wedge}_{\Game}\right] \\ &= \mathrm{ENV}\left[\mathbf{E}^{\#}, \mathbf{F}^{\#}_{\mathcal{S}}\right]. \end{aligned}$$

However, ENV[$\mathbf{E}^{\#}, \mathbf{F}_{\mathcal{S}}$] is not always strongly contained in ENV[$\mathbf{E}^{\#}, \mathbf{F}^{\wedge}_{\Game}$], since for example on the structure of analysis these two classes coincide with the inductive relations. We should point out also that ENV[$\mathbf{E}, \mathbf{F}_{\mathcal{S}}$] $\subseteq$ ENV[$\mathbf{E}^{\#}, \mathbf{F}_{\mathcal{S}}$] and that these two classes coincide on the structure of arithmetic, but ENV[$\mathbf{E}, \mathbf{F}_{\mathcal{S}}$] $<$ ENV[$\mathbf{E}^{\#}, \mathbf{F}_{\mathcal{S}}$] on the structure of analysis.

2.12. The preceding results about the Suslin and the classical $\Game$-quantifier can be extended to the quantifiers $Q^{\vee}$ and $Q^{\wedge}$. More specifically, using the minimality characterizations of Theorems 1.4 and 2.3, as well as Theorem 2.9 we can establish that for any quantifier Q.

(i) $$\begin{aligned} \mathrm{ENV}\left[\mathbf{E}, \mathbf{F}^{\wedge}_{Q}, \mathbf{F}_{(Q\forall)^{\vee}}\right] &= \mathrm{ENV}\left[\mathbf{E}, \mathbf{F}^{\wedge}_{Q}, \mathbf{F}^{\wedge}_{(Q\forall)^{\vee}}\right] \\ &\subseteq \mathrm{ENV}\left[\mathbf{E}, \mathbf{F}^{\wedge}_{Q}, \mathbf{F}^{\wedge}_{(\check{Q}\exists)^{\wedge}}\right] \\ &= \mathrm{ENV}\left[\mathbf{E}, \mathbf{F}^{\#}_{Q}, \mathbf{F}^{\#}_{(Q\forall)^{\vee}}\right]; \end{aligned}$$

(ii) $$\begin{aligned} \mathrm{ENV}\left[\mathbf{E}^{\#}, \mathbf{F}^{\wedge}_{Q}, \mathbf{F}_{(Q\forall\exists)^{\vee}}\right] &= \mathrm{ENV}\left[\mathbf{E}^{\#}, \mathbf{F}^{\wedge}_{Q}, \mathbf{F}^{\wedge}_{(Q\forall\exists)^{\vee}}\right] \\ &\subseteq \mathrm{ENV}\left[\mathbf{E}^{\#}, \mathbf{F}^{\wedge}_{Q}, \mathbf{F}^{\wedge}_{(\check{Q}\exists\forall)^{\wedge}}\right] \\ &= \mathrm{ENV}\left[\mathbf{E}^{\#}, \mathbf{F}^{\#}_{Q}, \mathbf{F}^{\#}_{(Q\forall\exists)^{\vee}}\right]. \end{aligned}$$

Moschovakis has recently generalized his Representation Theorem 2.10 into the case of recursion in the functional $\mathbf{F}^{\wedge}_{Q}$; from this result it follows that the above inclusions are proper and actually the "smaller" class is strongly contained in the "bigger" one. In particular, by taking Q to be the universal quantifier $\forall$, we obtain from (ii) the following picture for the game quantifier G:

$$\begin{aligned}\mathrm{ENV}[\mathbf{E}^{\#},\mathbf{F}_G] = \mathrm{ENV}[\mathbf{E}^{\#},\mathbf{F}^{\wedge}_{G}] &< \mathrm{ENV}[\mathbf{E}^{\#},\mathbf{F}^{\wedge}_{\check{G}}] \\ &= \mathrm{ENV}[\mathbf{E}^{\#},\mathbf{F}^{\#}_{G}].\end{aligned}$$

We will return to the game quantifier at the end of the paper, giving a different proof of the above result, which sheds more light in the difference between recursion in $\mathbf{F}^{\wedge}_{G}$ and recursion in $\mathbf{F}^{\wedge}_{\check{G}}$.

We conclude this section by pointing out that whenever $\mathrm{ENV}[\mathbf{E}^{\#},\mathbf{F}^{\wedge}_{Q}] \subsetneq \mathrm{ENV}[\mathbf{E}^{\#},\mathbf{F}^{\wedge}_{\check{Q}}]$, then the Companion Theorem 2.6 implies that the smallest $Q^{\#},\check{Q}$-admissible set does not coincide with the smallest $\check{Q}^{\#},Q$-admissible set. In this case the companion of $\mathrm{ENV}[\mathbf{E}^{\#},\mathbf{F}^{\wedge}_{Q}]$ is an example of a $Q^{\#},\check{Q}$-admissible which is not $\check{Q}^{\#},Q$-admissible and therefore it is not strong Q-admissible either.

3. On nonmonotone induction in a quantifier

In this section we establish that if Q is an ω-complete filter, then the classes $\mathrm{ENV}[\mathbf{E},\mathbf{F}^{\wedge}_{Q}]$ and $\mathrm{ENV}[\mathbf{E}^{\#},\mathbf{F}^{\wedge}_{Q}]$ can be characterized in terms of simple nonmonotone inductions in the quantifier Q. This result contains as a special case the main theorems of GRILLIOT (1971) and reinforces the naturalness of the functional $\mathbf{F}^{\wedge}_{Q}$.

We obtain also a similar characterization of nonmonotone inductions in the classical $\mathcal{Q}$-quantifier, which generalizes to arbitrary acceptable structures results of ACZEL (1970) and T. J. GRILLIOT (unpublished data) about the Σ^1_1 relations on the structure of arithmetic.

Background

3.1. A *second order relation* on a set A is a relation with arguments elements and relations on A. If a second order relation φ is of the form $\varphi(x_1,\ldots,x_n,Y)$ and Y is n-ary, then we say that φ is *operative*; in this case φ determines a sequence $\{\varphi^{\xi}\}$ of n-ary relations on A by the equations

$$\varphi^{\xi} = \varphi^{<\xi} \cup \{\bar{x} \colon \varphi(\bar{x},\varphi^{<\xi})\}, \quad \text{where } \varphi^{<\xi} = \bigcup_{\eta<\xi} \varphi^{\eta}.$$

The *fixed point* of φ is the relation $\varphi^{\infty} = \bigcup_{\xi}\varphi^{\xi}$. Assume now that $\mathfrak{D}$ is a collection of second order relations on A; we say that a relation R on A is $\mathfrak{D}$*-inductive* if there is an operative relation φ in $\mathfrak{D}$ and a sequence $\bar{n}$ from ω such that

$$R(\bar{y}) \Leftrightarrow \varphi^{\infty}(\bar{n},\bar{y}).$$

A relation $R \subseteq A^n$ is $\mathfrak{Y}$*-hyperelementary* if both R and $A^n\text{-}R$ are $\mathfrak{Y}$-inductive. We denote these classes of relations by $\mathfrak{Y}$-IND and $\mathfrak{Y}$-HYP.

3.2. If Γ is a class of relations on a set A, then the *dual class* $\check{\Gamma}$ is the collection of the complements of the relations in Γ. The *self-dual class* Δ consists of the relations both in Γ and $\check{\Gamma}$, i.e. $\Delta = \Gamma \cap \check{\Gamma}$. If $\bar{x} \in A^{<\omega}$, then the *relativized class* $\Gamma(\bar{x})$ is the collection

$$\Gamma(\bar{x}) = \{P\colon \text{there is } R \in \Gamma \text{ such that } (\forall \bar{y})(P(\bar{y}) \Leftrightarrow R(\bar{x}, \bar{y}))\}.$$

We define $\check{\Gamma}(\bar{x})$ in a similar way and we put $\Delta(\bar{x}) = \Gamma(\bar{x}) \cap \check{\Gamma}(\bar{x})$.

Assume that Γ is a semi-Spector class on $\mathfrak{A}$ and $\varphi(\bar{x}, Y_1, \ldots, Y_k)$ is a second order relation on A; we say that φ is Γ *on* Δ if for any relations R_i, S_i $(1 \leqslant i \leqslant k)$ in Γ there is a relation P in Γ with the property that: for every $\bar{x}_1, \ldots, \bar{x}_k$ such that

$$(\forall i \leqslant k)(\forall \bar{z}_i)[R_i(\bar{x}_i, \bar{z}_i) \Leftrightarrow \neg S_i(\bar{x}_i, \bar{z}_i)]$$

we have

$$P(\bar{x}, \bar{x}_1, \ldots, \bar{x}_k) \Leftrightarrow \varphi(\bar{x}, \{\bar{z}_1\colon R_1(\bar{x}_1, \bar{z}_1)\}, \ldots, \{\bar{z}_k\colon R_k(\bar{x}_k, \bar{z}_k)\}).$$

We say that φ is Δ *on* Δ if both φ and $\neg\varphi$ are Γ on Δ. Notice that if Q is a quantifier on A and Γ is closed under deterministic Q and deterministic $\check{Q}$, then the graphs of the functionals $F_Q, F_Q^{\#}, F_Q^{\hat{}}$ are all Δ on Δ.

The semi-Spector class Γ is φ*-compact* if for every $\bar{x} \in A^n$, every $R_1, \ldots, R_k \in \Gamma(\bar{x})$ such that $\varphi(\bar{x}, R_1, \ldots, R_k)$ and every $R_1^0, \ldots, R_k^0$ in $\Delta(\bar{x})$ with $R_1^0 \subseteq R_1, \ldots, R_k^0 \subseteq R_k$ there exist $R_1^*, \ldots, R_k^*$ in $\Delta(\bar{x})$ such that $R_1^0 \subseteq R_1^* \subseteq R_1, \ldots, R_k^0 \subseteq R_k^* \subseteq R_k$ and $\varphi(\bar{x}, R_1^*, \ldots, R_k^*)$.

If $\mathfrak{Y}$ is a collection of second order relations on A, then Γ is $\mathfrak{Y}$*-compact* in case Γ is φ-compact for every φ in $\mathfrak{Y}$.

This notion of "compactness" is the "lightface" version of the one introduced by MOSCHOVAKIS (1974b). The next theorem is due to Moschovakis and is an important closure property of semi-Spector classes; a proof of it can be found in KOLAITIS (1978b).

3.3. The nonmonotone first recursion theorem. *Let Γ be a semi-Spector class on an acceptable structure $\mathfrak{A}$ and let $\mathfrak{Y}$ be a collection of second order relations on A containing the relations definable by universal formulas of the language $\mathfrak{L}^{\mathfrak{A}}$ of $\mathfrak{A}$ and closed under &. If Γ is $\mathfrak{Y}$-compact and every relation in $\mathfrak{Y}$ is Δ on Δ, then for every operative second order relation φ in $\mathfrak{Y}$ the fixed point φ^∞ of φ is in Γ.*

Nonmonotone inductions and ω-complete filters

3.4. Let $\mathfrak{A} = \langle A, R_1, \ldots, R_n, f_1, \ldots, f_m, c_1, \ldots, c_k\rangle$ be an acceptable structure and let $\mathcal{C} = \langle \omega, \leqslant_\omega, \langle\,\rangle\rangle$ be a definable coding scheme on $\mathfrak{A}$; we associate with $\mathcal{C}$ the relation "sequence" seq and the functions "length of a sequence" lh, "concatenation" $*$ and q (where $q(x,i)=(x)_i=$ the ith coordinate of x). We consider then the expanded structure

$$\mathfrak{A}(\mathcal{C})=\langle A,R_1,\ldots,R_n,\omega,\leqslant_\omega,\mathrm{seq},\mathrm{lh},q,*,f_1,\ldots,f_m,c_1,\ldots,c_k\rangle.$$

We say that a formula φ of the language $\mathcal{L}^{\mathfrak{A}(\mathcal{C})}$ is *restricted on* $\mathfrak{A}$ (*relative to* $\mathcal{C}$) if all quantifiers in φ occur in one of the following two forms:

$$(\exists x)\left[x\leqslant_\omega t(\bar{y})\&\cdots\right],\qquad(\forall x)\left[x\leqslant_\omega t(\bar{y})\rightarrow\cdots\right],$$

where t is a term of the language $\mathcal{L}^{\mathfrak{A}(\mathcal{C})}$.

A second order relation on A is Σ_k^0 *on* $\mathfrak{A}$ (*relative to* $\mathcal{C}$) or $\Sigma_k^0(\mathfrak{A})$ if it is definable by a formula of the language $\mathcal{L}^{\mathfrak{A}(\mathcal{C})}$ of the form $[(\exists\bar{z}_1)(\forall\bar{z}_2)\cdots(\cdot\bar{z}_k)\varphi(\bar{z}_1,\bar{z}_2,\ldots,\bar{z}_k,\bar{y},\bar{Y})]$, where φ is restricted. In an analogous way we define the notion of a Π_k^0 *second order relation on* $\mathfrak{A}$ (*relative to* $\mathcal{C}$) or $\Pi_k^0(\mathfrak{A})$.

If Q is a quantifier on A, then $Q^{(n)}$ denotes the collection of all second order relations on A definable by a formula of the language $\mathcal{L}^{\mathfrak{A}(\mathcal{C})}(Q)$ of the form:

$$Q_1x_1Q_2x_2\cdots Q_nx_n\varphi\left(x_1,\ldots,x_n,\bar{y},\bar{Y}\right),$$

where $Q_i=\forall$ or $Q_i=Q$ for $1\leqslant i\leqslant n$ and φ is restricted on $\mathfrak{A}$. We also put $Q^{(<\omega)}$ for the union of all these classes of second order relations, i.e. $Q^{(<\omega)}=\bigcup_n Q^{(n)}$.

The next lemma is a key property of the restricted formulas; it can be proved easily by induction on the construction of the restricted formulas.

3.5. Lemma. *Let $\mathfrak{A}$ be an acceptable structure, let $\mathcal{C}$ be a coding scheme on $\mathfrak{A}$ and let $\varphi(v_1,\ldots,v_n,Y)$ be a restricted formula on $\mathfrak{A}$ relative to $\mathcal{C}$ such that Y is a unary relation variable and all negation symbols occurring in φ apply only to atomic subformulas of φ.*

(i) *Let $t_1(\bar{v},\bar{w}),\ldots,t_n(\bar{v},\bar{w})$ be all the terms which occur in φ in the form $t_i\in Y$. For each $\bar{x}\in A^n$ there are finitely many elements $a_1,a_2,\ldots,a_s$ of A of the form $t_i(\bar{x},\bar{j})$, where $\bar{j}$ is a sequence from ω, such that for any unary relations R,S on A if for all $l\leqslant s(a_l\in R\Leftrightarrow a_l\in S)$, then $(\varphi(\bar{x},R)\Leftrightarrow\varphi(\bar{x},S))$.*

(ii) *If $\bar{x}\in A^n$ and $\{S_k\colon k\in\omega\}$ is a sequence of unary relations on A such that $\varphi(\bar{x},S_k)$ and $S_k\subseteq S_{k+1}$ for all $k\in\omega$, then $\varphi(\bar{x},\bigcup_{k\in\omega}S_k)$.*

3.6. A quantifier Q on a set A is a *filter* if it is closed under finite intersections, i.e. whenever X,Y are in Q, then the intersection $X\cap Y$ is also in Q. We say that Q is an *ω-complete filter* if whenever $\{X_n\colon n\in\omega\}$ is a countable sequence of elements in Q, then the intersection $\bigcap_{n\in\omega}X_n$ is also an element of Q.

There are many natural quantifiers in model theory and set theory which are ω-complete filters, for example the dual of the quantifier "there exist uncountably many" or the filter of the closed unbounded sets on a regular cardinal.

3.7. Theorem. *Let $\mathfrak{A}$ be an acceptable structure and let Q be a quantifier on A which is an ω-complete filter. If Γ is a semi-Spector class closed under Q, then Γ is $Q^{(1)}$-compact.*

Proof. Let Γ be a semi-Spector class closed under Q and let φ be a $Q^{(1)}$ formula; let us assume for simplicity that φ is of the form $\varphi(x, Y)$, where Y is a unary relation variable and that there is a restricted formula $\psi(x,z,Y)$ such that all negation symbols occurring in ψ apply only to atomic subformulas of ψ and such that

$$\varphi(x,Y) \Leftrightarrow Qz\psi(x,z,Y).$$

Assume now that $x \in A$, $R \in \Gamma(x)$, $R_0 \in \Delta(x)$, $R_0 \subseteq R \subseteq A$ and $\varphi(x,R)$ holds. We have to show that there is a relation $R^* \in \Delta(x)$ such that $R_0 \subseteq R^* \subseteq R$ and $\varphi(x,R^*)$. Let $U \subseteq A^2$ be a relation in Γ which ω-parametrizes the unary relations in Γ, let $\sigma: U \to \text{Ordinals}$ be a Γ-norm on U and let $e \in \omega$ be such that

$$(\forall y)(R(y) \Leftrightarrow U(e,\langle x,y\rangle)).$$

A standard diagonal argument shows that the set $W_x = \{k \in \omega\colon (k,x) \in U\}$ is in the class $\Gamma(x)$-$\Delta(x)$; moreover, it is well-known that

$$\begin{aligned}&\sup\{\sigma((k,x))\colon k \in W_x\} = \\ &\quad = \sup\{\text{rank}(<)\colon < \text{ is a prewell-ordering in } \Delta(x)\}.\end{aligned}$$

If $(a,b) \in U$, we put

$$R^{<\sigma((a,b))} = \{y \in R\colon (e,\langle x,y\rangle) <^*_\sigma (a,b)\}.$$

We consider now the relation

$$\begin{aligned}P(k,l) \Leftrightarrow {} & (k \in W_x)\&(l \in W_x)\&[(k,x) <^*_\sigma (l,x)] \\ & \&(Qz)(\exists n \in \omega)[((k,x) <^*_\sigma (n,\langle x,z\rangle) <^*_\sigma (l,x)) \\ & \&\psi(x,z,R^{<\sigma((n,\langle x,z\rangle))})].\end{aligned}$$

It is clear that $P \in \Gamma(x)$. Moreover we claim that

$$(\forall x)[k \in W_x \to (\exists l)(l \in W_x \& P(k,l))].$$

To prove this claim notice first of all that if $k \in W_x$, then

$$\begin{aligned}&(Qz)(\exists n \in \omega)[((k,x) <^*_\sigma (n,\langle x,z\rangle))\&((n,\langle x,z\rangle) \in U) \\ &\qquad \&\psi(x,z,R^{<\sigma((n,\langle x,z\rangle))})].\end{aligned}$$

In fact we have that $B=\{z: \psi(x,z,R)\}\in Q$. If $k\in W_x$, then for any $z\in B$ we can find some $n\in\omega$ such that

$$(n,\langle x,z\rangle)\in U, \qquad \sigma((k,x))<\sigma((n,\langle x,z\rangle))$$

and

$$\max_{1\leq i\leq s}\{\sigma((e,\langle x,a_i\rangle)): a_i\in R\}<\sigma((n,\langle x,z\rangle)),$$

where $a_1,a_2,\ldots,a_s$ are the elements of A associated with (x,z) and the restricted formula ψ as in the preceding Lemma 3.5. Since $z\in B$, we have that $\psi(x,z,R)$ and hence part (i) of Lemma 3.5 implies that $\psi(x,z,R^{<\sigma((n,\langle x,z\rangle))})$. We can now complete the proof of the claim by noticing that if there is some $k_0\in W_x$ such that $\neg(\exists l)(l\in W_x\& P(k_0,l))$, then

$$l\notin W_x\Leftrightarrow(Qz)(\exists n\in\omega)\big[((k_0,x)<_\sigma^*(n,\langle x,z\rangle)<_\sigma^*(l,x)) \\ \&\psi(x,z,R^{\sigma((n,\langle x,z\rangle))})\big],$$

so that $W_x\in\Delta(x)$, contradiction. Thus we have established that

$$(\forall k)\big[k\in W_x\rightarrow(\exists l)(l\in W_x\& P(k,l))\big].$$

Since $P\in\Gamma(x)$ and P is a relation on ω, we can find a relation $P^*\subseteq P, P^*\in\Gamma(x)$ such that for each $k\in W_x$ there is exactly one integer $l\in W_x$ such that $P^*(k,l)$. Also since $R_0\in\Delta(x)$ and $W_x\in\Gamma(x)-\Delta(x)$, there is some $k_0\in W_x$ such that $R_0\subseteq R^{<\sigma((k_0,x))}$. Therefore we can find a sequence of integers $\{k_m: m\in\omega\}$ with the property that

$$R_0\subseteq R^{<\sigma((k_0,x))} \quad\text{and}\quad P^*(k_m,k_{m+1}) \quad\text{for all } m\in\omega.$$

If we put

$$B_m=\big\{z: (\exists n\in\omega)((k_m,x)<_\sigma^*(n,\langle x,z\rangle)<_\sigma^*(k_{m+1},x)) \\ \&\psi(x,z,R^{<\sigma((n,\langle x,z\rangle))})\big\},$$

then $B_m\in Q$ for all $m\in\omega$, since $P^*(k_m,k_{m+1})$ is true for all $m\in\omega$. We consider now the relation

$$R^*=\bigcup_{m\in\omega} R^{<\sigma((k_m,x))}$$

and we claim that $R^*\in\Delta(x)$ and $Qz\psi(x,z,R^*)$. It is clear first of all that $R^*\notin\Gamma(x)$, but also

$$y\in R^*\Leftrightarrow(\forall n\geq 2)(\exists u\in\omega)\big[\mathrm{seq}(u)\&\mathrm{lh}(u)=n\&((u)_1=k_0) \\ \&(\forall i)(i<n\rightarrow P^*((u)_i,(u)_{i+1}) \\ \&(((u)_n,x)<_\sigma^*(e,\langle x,y\rangle))\big]$$

and hence $\neg R^*\in\Gamma(x)$. We conclude the proof by showing that

$Qz\psi(x,z,R^*)$. Since Q is an ω-complete filter and $B_m \in Q$ for all $m \in \omega$, we have that $B^* = \bigcap_{m\in\omega} B_m \in Q$. We claim that if $z \in B^*$, then $\psi(x,z,R^*)$ is true. In fact, if we fix a $z \in B^*$, then for any $m \in \omega$ there is an integer n_m such that

$$(k_m, x) <^*_\sigma (n_m, \langle x,z\rangle) <^*_\sigma (k_{m+1}, x) \quad \text{and} \quad \psi(x,z,R^{<\sigma((n_m,\langle x,z\rangle))}).$$

But then

$$R^* = \bigcup_{m\in\omega} R^{<\sigma((k_m,x))} = \bigcup_{m\in\omega} R^{<\sigma((n_m,\langle x,z\rangle))}$$

and by part (ii) of Lemma 3.5 we have that $\psi(x,z,\bigcup_{m\in\omega} R^{<\sigma((n_m,\langle x,z\rangle))})$, hence $\psi(x,z,R^*)$.

3.8. Theorem. *Let $\mathfrak{A}$ be an acceptable structure and let Q be a quantifier on A which is an ω-complete filter. If Γ is a semi-Spector class on $\mathfrak{A}$ which is closed under Q, then Γ is $Q^{(<\omega)}$-compact.*

Proof. The result follows immediately from the preceding Theorem 3.7 and the observation that if Q is an ω-complete filter and the semi-Spector class Γ is closed under Q, then for any $n \in \omega$ the quantifier

$$Q_1 Q_2 \cdots Q_n = \{X\colon (Q_1 x_1)(Q_2 x_2)\cdots(Q_n x_n)(\langle x_1, x_2, \ldots, x_n\rangle \in X)\},$$

where $Q_i = \forall$ or $Q_i = Q$ for $1 \leqslant i \leqslant n$, is also an ω-complete filter and Γ is closed under it.

3.9. Corollary. *Let $\mathfrak{A}$ be an acceptable structure and let Q be an ω-complete filter on A. If Γ is a Spector class on $\mathfrak{A}$ which is closed under Q, then Γ is $\exists^A Q^{(<\omega)}$-compact, where*

$$\exists^A Q^{(<\omega)} = \{\varphi(\bar{x},\bar{Y})\colon \text{there is some } \psi(\bar{x},y,\bar{Y}) \text{ in } Q^{(<\omega)} \text{ such that } \varphi(\bar{x},\bar{Y}) \Leftrightarrow (\exists y)\psi(\bar{x},y,\bar{Y})\}.$$

3.10. Corollary (Essentially GRILLIOT (1971)). *If $\mathfrak{A}$ is an acceptable structure, then*

(i) *every semi-Spector class on $\mathfrak{A}$ is $\Pi^0_1(\mathfrak{A})$-compact,*
(ii) *every Spector class on $\mathfrak{A}$ is $\Sigma^0_2(\mathfrak{A})$-compact.*

3.11. Theorem. *Let $\mathfrak{A}$ be an acceptable structure and let Q be a quantifier on $\mathfrak{A}$ which is an ω-complete filter. Then for any $n \geqslant 2$*

(i) $\mathrm{ENV}[\mathbf{E}, \mathbf{F}_Q^{\smallfrown}] = Q^{(n)} - \mathrm{IND} = Q^{(<\omega)} - \mathrm{IND} =$ *the smallest semi-Spector class on $\mathfrak{A}$ closed under Q and deterministic $\check{Q}$.*

(ii) $\mathrm{ENV}[\mathbf{E}^{\#}, \mathbf{F}_Q^{\smallfrown}] = \exists^A Q^{(n)} - \mathrm{IND} = \exists^A Q^{(<\omega)} - \mathrm{IND} =$ *the smallest Spector class on $\mathfrak{A}$ closed under Q and deterministic $\check{Q}$.*

Proof. (i) It follows from Theorems 2.3 and 3.8 that the semi-Spector class $\mathrm{ENV}[\mathbf{E},\mathbf{F}_Q^{\wedge}]$ is $Q^{(<\omega)}$-compact; it is also clear that every second order relation in $Q^{(<\omega)}$ is Δ on Δ for $\mathrm{ENV}[\mathbf{E},\mathbf{F}_Q^{\wedge}]$ and therefore the Non-monotone First Recursion Theorem 3.3 implies that $Q^{(<\omega)}-\mathrm{IND}\subseteq \mathrm{ENV}[\mathbf{E},\mathbf{F}_Q^{\wedge}]$.

In order to establish that $\mathrm{ENV}[\mathbf{E},\mathbf{F}_Q^{\wedge}]\subseteq Q^{(n)}-\mathrm{IND}$ for any $n\geqslant 2$ we show first that the functions recursive in $\mathbf{E},\mathbf{F}_Q^{\wedge}$ can be generated by a single master functional $\{e\}(x)$ using Kleene schemata. The proof of this fact is similar to the corresponding one for recursion in higher types in KECHRIS and MOSCHOVAKIS (1977). The Kleene schemata for recursion in $\mathbf{E},\mathbf{F}_Q^{\wedge}$ are the same as the ones for recursion in $\mathbf{E}$ together with an additional schema, say S9, for the functional $\mathbf{F}_Q^{\wedge}$.

GRILLIOT (1971) showed that the Kleene schemata for recursion in $\mathbf{E}$ can be simulated by a $\Pi_1^0(\mathfrak{A})$-nonmonotone induction. His argument can be easily generalized to yield that the Kleene schemata for recursion in $\mathbf{E},\mathbf{F}_Q^{\wedge}$ can be simulated by a second order relation which is the disjunction of a $\Pi_1^0(\mathfrak{A})$ relation and a $Q^{(1)}$ relation, hence by a $Q^{(n)}$-nonmonotone induction for any $n\geqslant 2$. More specifically, assume that $\{e\}(x)$ is the master functional for recursion in $\mathbf{E},\mathbf{F}_Q^{\wedge}$ which consists of the schemata S1-S9 (the schemata S1-S8 are the ones for recursion in $\mathbf{E}$ and the schema S9 corresponds to the functional $\mathbf{F}_Q^{\wedge}$). We define an operative second order relation φ such that

$$\{e\}(x)\downarrow\Leftrightarrow\langle 0,e,x\rangle\in\varphi^{\infty},\qquad \{e\}(x)=u\Leftrightarrow\langle 1,e,u,x\rangle\in\varphi^{\infty}$$

and moreover if $\{e\}(x)\downarrow$ and $\{e\}(x)=u$, then $|\langle 0,e,x\rangle|_{\varphi}=|\langle 1,e,u,x\rangle|_{\varphi}$ (where, if $z\in\varphi^{\infty}$, then $|z|_{\varphi}=$ least ξ such that $z\in\varphi^{\xi}$). For the schemata S1-S8 the clauses for φ are as in GRILLIOT (1971), so that these schemata are simulated by a $\Pi_1^0(\mathfrak{A})$ second order relation. Assume now that for the schema S9 we have

$$\{\langle 9,n,e\rangle\}(x)=\mathbf{F}_Q^{\wedge}(\lambda y\{e\}(y,x)).$$

We add now five more clauses in the definition of φ so that for any ordinal ξ

$$\begin{aligned}
&\langle 10,\langle 9,n,e\rangle,x\rangle\in\varphi^{\xi} && \text{if } (\forall y)(\langle 0,e,y,x\rangle\in\varphi^{<\xi}),\\
&\langle 11,\langle 9,n,e\rangle,x\rangle\in\varphi^{\xi} && \text{if } (Qy)(\langle 1,e,0,y,x\rangle\in\varphi^{<\xi}),\\
&\langle 0,\langle 9,n,e\rangle,x\rangle\in\varphi^{\xi} && \text{if } \langle 10,\langle 9,n,e\rangle,x\rangle\in\varphi^{<\xi}\\
& && \quad\vee\langle 11,\langle 9,n,e\rangle,x\rangle\in\varphi^{<\xi},\\
&\langle 1,\langle 9,n,e\rangle,0,x\rangle\in\varphi^{\xi} && \text{if } \langle 11,\langle 9,n,e\rangle,x\rangle\in\varphi^{<\xi},\\
&\langle 1,\langle 9,n,e\rangle,1,x\rangle\in\varphi^{\xi} && \text{if } \langle 10,\langle 9,n,e\rangle,x\rangle\in\varphi^{<\xi}\\
& && \quad\&\langle 11,\langle 9,n,e\rangle,x\rangle\notin\varphi^{<\xi}.
\end{aligned}$$

Each of the first two clauses introduces a "one step delay" and they are inserted in order to insure that

$$|\langle 0,\langle 9,n,e\rangle,x\rangle|_{\varphi}=|\langle 1,\langle 9,n,e\rangle,u,x\rangle|_{\varphi},$$

whenever $\{\langle 9,n,e\rangle\}(x)=u$ (the numbers 10 and 11 are just indices not used before in the definition of φ).

This completes the proof of the fact that the master functional for recursion in $\mathbf{E},\mathbf{F}_Q^{\hat{}}$ can be simulated by a disjunction of a $\Pi_1^0(\mathfrak{A})$ relation and a $Q^{(1)}$ relation and establishes that $\mathrm{ENV}[\mathbf{E},\mathbf{F}_Q^{\hat{}}]\subseteq Q^{(n)}-\mathrm{IND}$ for any $n\geqslant 2$. We should point out that the hypothesis that Q is an ω-complete filter was not used in this argument, so that the inclusion $\mathrm{ENV}[\mathbf{E},\mathbf{F}_Q^{\hat{}}]\subseteq Q^{(n)}-\mathrm{IND}(n\geqslant 2)$ is true for any quantifier Q.

(ii) The second part of the theorem is proved using a similar argument together with Theorem 2.3 and Corollary 3.9.

The main results of GRILLIOT (1971) are instances of the preceding Theorem 3.11, by simply taking Q to be the universal quantifier $\forall$.

3.12. Theorem. *If* $\mathfrak{A}$ *is an acceptable structure, then*

(i) $\mathrm{ENV}[\mathbf{E}]=\Pi_1^0(\mathfrak{A})-\mathrm{IND}=$ *the smallest semi-Spector class on* $\mathfrak{A}$.

(ii) $\mathrm{ENV}[\mathbf{E}^{\#}]=\Sigma_2^0(\mathfrak{A})-\mathrm{IND}=$ *the smallest Spector class on* $\mathfrak{A}$.

3.13. The preceding Theorem 3.11, which identifies recursion in $\mathbf{F}_Q^{\hat{}}$ with nonmonotone inductions in Q, depends essentially on the ω-completeness of Q. Actually, we can construct counterexamples even with quantifiers which are filters. For this we consider the structure of arithmetic $\langle\omega,+,\cdot\rangle$ and the filter

$$\tilde{\mathfrak{S}}=\{Y\subseteq\omega\colon(\forall x_1\forall x_2\cdots)(\exists n)(\forall u)\\ [\mathrm{seq}(u)\&(\langle x_1,\ldots,x_n\rangle\subseteq u)\rightarrow u\in Y]\}.$$

It can be shown that every semi-Spector class on $\langle\omega,+,\cdot\rangle$ is closed under $\tilde{\mathfrak{S}}$ and hence the minimality characterizations of Theorems 1.4 and 2.3 imply that on the structure $\langle\omega,+,\cdot\rangle$

$$\mathrm{ENV}[\mathbf{E},\mathbf{F}_{\tilde{\mathfrak{S}}}]=\mathrm{ENV}[\mathbf{E},\mathbf{F}_{\tilde{\mathfrak{S}}}^{\hat{}}]=\mathrm{ENV}[\mathbf{E}^{\#},\mathbf{F}_{\tilde{\mathfrak{S}}}]=\mathrm{ENV}[\mathbf{E}^{\#},\mathbf{F}_{\tilde{\mathfrak{S}}}^{\hat{}}].$$

Since the Spector class $\mathrm{ENV}[\mathbf{E}^{\#},\mathbf{F}_{\tilde{\mathfrak{S}}}^{\hat{}}]$ is equal to the envelope of the type 2 objects $\mathbf{E},\mathbf{F}_{\tilde{\mathfrak{S}}}$, it follows from a theorem of HARRINGTON, ET AL. (1973) that $\mathrm{ENV}[\mathbf{E}^{\#},\mathbf{F}_{\tilde{\mathfrak{S}}}^{\hat{}}]$ is not Π_2^0-compact (see also HARRINGTON and KECHRIS (1975)). On the other hand, any Π_1^1 second order relation on $\langle\omega,+,\cdot\rangle$ is definable by a $\tilde{\mathfrak{S}}^{(1)}$ formula, so that $\tilde{\mathfrak{S}}^{(1)}-\mathrm{IND}$ is Π_1^1-compact and hence

$$\mathrm{ENV}[\mathbf{E},\mathbf{F}_{\tilde{\mathfrak{S}}}^{\hat{}}]=\mathrm{ENV}[\mathbf{E}^{\#},\mathbf{F}_{\tilde{\mathfrak{S}}}^{\hat{}}]<\tilde{\mathfrak{S}}^{(1)}-\mathrm{IND}.$$

Of course the quantifier $\tilde{\mathfrak{S}}$ is not an ω-complete filter, since for each $n\in\omega$ the set $B_n=\{u\in\omega\colon \text{seq}(u)\&\text{lh}(u)\geqslant n\}$ is an element of $\tilde{\mathfrak{S}}$, but $\bigcap_{n\in\omega}B_n=\varnothing$.

More on the classical $\mathfrak{C}$ and the game quantifier

3.14. Assume that $\mathfrak{A}$ is an acceptable structure and consider the classical $\mathfrak{C}$-quantifier on A (relative to a definable coding scheme $\mathcal{C}$ on $\mathfrak{A}$), i.e.

$$\mathfrak{C}=\{Y\subseteq A\colon (\exists x_1\exists x_2\cdots)(\forall n)(\langle x_1,\ldots,x_n\rangle\in Y)\}.$$

We say that a second order relation $\theta(\bar{x},Y_1,\ldots,Y_k)$ on A is $(\mathfrak{C}\forall)_1$ on $\mathfrak{A}$ if it is definable by a formula of the language $\mathfrak{L}^{\mathfrak{A}(\mathcal{C})}(\mathfrak{C})$ of the form $(\mathfrak{C}y)(\forall z)\varphi(\bar{x},y,z,\bar{Y})$, where φ is restricted on $\mathfrak{A}$ (relative to $\mathcal{C}$). If all the relation variables $Y_1,\ldots,Y_k$ occur in φ positively, then we say that the second order relation $\theta(\bar{x},\bar{Y})$ is *positive* $(\mathfrak{C}\forall)_1$ *on* $\mathfrak{A}$. We write $(\mathfrak{C}\forall)_1-\text{IND}$ and $(\mathfrak{C}\forall)_1^{\text{pos}}-\text{IND}$ for the corresponding classes of inductive relations on A.

3.15. Theorem. *Let $\mathfrak{A}$ be an acceptable structure and let Γ be a Spector class on $\mathfrak{A}$. If Γ is closed under the classical $\mathfrak{C}$-quantifier, then Γ is $(\mathfrak{C}\forall)_1$-compact.*

Proof. (Outline.) Assume that $x\in A$, $R\in\Gamma(x)$, $R_0\in\Delta(x)$, $R_0\subseteq R\subseteq A$ and

$$(\exists x_1\exists x_2\cdots)(\forall k)(\forall y)\psi(x,\langle x_1,\ldots,x_k\rangle,y,R),$$

where ψ is restricted on $\mathfrak{A}$ and all negation symbols in ψ apply only to atomic subformulas of ψ. Let $U\subseteq A^2$ be a relation in Γ which ω-parametrizes the unary relations in Γ, let σ be a Γ-norm on U and let $e\in\omega$ be such that $(\forall w)[R(w)\Leftrightarrow U(e,\langle x,w\rangle)]$. We consider the relation

$$\begin{aligned}P(\langle x_1,n_1,\ldots,x_k,n_k\rangle)\Leftrightarrow(\forall i)\big[&1\leqslant i<k\rightarrow(n_i\in\omega)\\ &\&((n_i,\langle x,x_1,\ldots,x_i\rangle)\\ &<_\sigma^*(n_{i+1},\langle x,x_1,\ldots,x_{i+1}\rangle))\big]\\ &\&\big(R_0\subseteq R^{<\sigma((n_1,\langle x,x_1\rangle))}\big)\&(n_k\in\omega)\\ &\&((n_k,\langle x,x_1,\ldots,x_k\rangle)\in U)\\ &\&(\exists z)\big[((n_{k-1},\langle x,x_1,\ldots,x_{k-1}\rangle)<_\sigma^*z\\ &<_\sigma^*(n_k,\langle x,x_1,\ldots,x_k\rangle))\\ &\&(\forall i\leqslant k)(\forall y)\psi\big(x,\langle x_1,\ldots,x_i\rangle,y,R^{<\sigma(z)}\big)\big].\end{aligned}$$

Using ideas similar to the ones in the proof of Theorem 3.7 and the hypothesis that Γ is closed under the classical $\mathcal{G}$-quantifier, we can show that there is some $m\in\omega$ with $(m,x)\in U$ and such that

$$(\exists x_1\exists n_1\exists x_2\exists n_2\cdots)(\forall k)\big[P(\langle x_1,n_1,\ldots,x_k,n_k\rangle)\&\\ \&((n_k,\langle x,x_1,\ldots,x_k\rangle)<^*_\sigma(m,x))\big].$$

If we fix an $m\in\omega$ and a sequence $n_1,x_1,\ldots,n_k,x_k,\ldots$ as above, then it follows from Lemma 3.5 that

$$(\forall k)(\forall y)\psi\Big(x,\langle x_1,\ldots,x_k\rangle,y,\bigcup_{i\in\omega}R^{<\sigma((n_i,\langle x,x_1,\ldots,x_i\rangle))}\Big).$$

Finally, we put

$$w\in R^*\Leftrightarrow(\exists z)\big[(z\leqslant^*_\sigma(m,x))\&(w\in R^{<\sigma(z)})\&(R_0\subseteq R^{<\sigma(z)})\\ \&(\exists x_1\exists x_2\cdots)(\forall k)(\forall y)\psi(x,\langle x_1,\ldots,x_k\rangle,y,R^{<\sigma(z)})\\ \&(\forall z')(z'<^*_\sigma z\rightarrow\\ (\forall x_1\forall x_2\cdots)(\exists k)(\exists y)\neg\psi(x,\langle x_1,\ldots,x_k\rangle,y,R^{<\sigma(z')}))\big].$$

It is now easy to check that $R^*\in\Delta(x)$, while also it is clear that $R_0\subseteq R^*\subseteq R$ and

$$(\exists x_1\exists x_2\cdots)(\forall k)(\forall y)\psi(x,\langle x_1,\ldots,x_k\rangle,y,R^*).$$

The preceding Theorem 3.15 and an argument analogous to the one in the proof of Theorem 3.11 yield now the following result.

3.16. Theorem. *If $\mathfrak{A}$ is an acceptable structure, then*

$$\mathrm{ENV}[\mathbf{E}^\#,\mathbf{F}^{\wedge}_{\mathcal{G}}]=\mathrm{ENV}[\mathbf{E}^\#,\mathbf{F}^\#_{\mathcal{G}}]=(\mathcal{G}\forall)^{\mathrm{pos}}_1-\mathrm{IND}\\ =(\mathcal{G}\forall)_1-\mathrm{IND}$$

This result can be thought of as a generalization of a theorem of ACZEL (1970) who showed that $\mathrm{ENV}[\mathbf{E}^\#,\mathbf{F}^\#_{\mathcal{G}}]=\Sigma^{1,\mathrm{mon}}_1-\mathrm{IND}$ on the structure of arithmetic $\langle\omega,+,\cdot\rangle$ and a theorem of T. J. GRILLIOT (unpublished data) who proved that $\Sigma^{1,\mathrm{mon}}_1-\mathrm{IND}=\Sigma^1_1-\mathrm{IND}$ on $\langle\omega,+,\cdot\rangle$ (here $\Sigma^{1,\mathrm{mon}}_1-\mathrm{IND}$ is the class of relations which are inductively definable by monotone Σ^1_1 second order relations).

Another generalization of the results of Aczel and Grilliot is the following theorem of HARRINGTON and MOSCHOVAKIS (1974) about the game quantifier.

3.17. Theorem. *Let $\mathfrak{A}$ be an acceptable structure and let G be the game quantifier on $\mathfrak{A}$*

(i) *If Γ is a Spector class on $\mathfrak{A}$ closed under the quantifier $\breve{G}$, then Γ is $(\breve{G})^{(1)}$-compact.*

(ii) $\mathrm{ENV}[\mathbf{E}^{\#}, \mathbf{F}^{\#}_{G}] = (\breve{G})^{(1)} - \mathrm{IND}$.

We conclude with a result which shows the difference between recursion in $\mathbf{F}_{\hat{G}}$ and recursion in $\mathbf{F}_{\hat{\breve{G}}}$.

3.18. Theorem. *If $\mathfrak{A}$ is an acceptable structure, then*

$$\begin{aligned} \mathrm{ENV}\left[\mathbf{E}^{\#}, \mathbf{F}_{G}\right] = \mathrm{ENV}\left[\mathbf{E}^{\#}, \mathbf{F}_{\hat{G}}\right] &< G^{(1)} - \mathrm{IND} < (\breve{G})^{(1)} - \mathrm{IND} \\ &= \mathrm{ENV}\left[\mathbf{E}^{\#}, \mathbf{F}_{\hat{\breve{G}}}\right] = \mathrm{ENV}\left[\mathbf{E}^{\#}, \mathbf{F}^{\#}_{G}\right]. \end{aligned}$$

Proof. From the minimality characterizations of the various classes of relations and the preceding Theorem 3.17 it follows that it is enough to show that $\mathrm{ENV}[\mathbf{E}^{\#}, \mathbf{F}_{\hat{G}}] < G^{(1)} - \mathrm{IND} < (\breve{G})^{(1)} - \mathrm{IND}$.

The proof of Theorem 3.11 shows also that $\mathrm{ENV}[\mathbf{E}^{\#}, \mathbf{F}_{\hat{G}}] \subseteq G^{(1)} - \mathrm{IND}$. On the other hand, since $\mathrm{ENV}[\mathbf{E}^{\#}, \mathbf{F}_{\hat{G}}] = \mathrm{ENV}[\mathbf{E}^{\#}, \mathbf{F}_{G}]$ and $G^{(1)} - \mathrm{IND}$ is $G^{(1)}$-compact, the theorem of Harrington, Kechris and Simpson in 3.13 implies that $\mathrm{ENV}[\mathbf{E}^{\#}, \mathbf{F}_{\hat{G}}] < G^{(1)} - \mathrm{IND}$. The class of $G^{(1)}$ second order relations on $\mathfrak{A}$ coincides with the class of the second order inductive relations on $\mathfrak{A}$ by Theorems 5C.1 and 5C.2 of Moschovakis [EIAS] and hence $G^{(1)}$ has the prewell-ordering property. But then the main theorem of Aanderaa (1974) implies that $G^{(1)} - \mathrm{IND} < (\breve{G})^{(1)} - \mathrm{IND}$.

As a final remark, notice also the difference in "compactness" between the open game and the closed game quantifier. The Spector class $\mathrm{ENV}[\mathbf{E}^{\#}, \mathbf{F}_{\hat{G}}]$ has only "the minimum amount of compactness", since it is $\Sigma^0_2(\mathfrak{A})$-compact, but not $\Pi^0_2(\mathfrak{A})$-compact. On the other hand, the Spector class $\mathrm{ENV}[\mathbf{E}^{\#}, \mathbf{F}_{\hat{\breve{G}}}]$ is $(\breve{G})^{(1)}$-compact and hence it is $\Pi^0_n(\mathfrak{A})$-compact, for any $n \in \omega$.

References

Aanderaa, S.

[1974] Inductive definitions and their closure ordinals, in: *Generalized Recursion Theory*, edited by J. E. Fenstad and P. G. Hinman (North-Holland, Amsterdam), pp. 207–220.

ACZEL, P.

[1970] Representability in some systems of second order arithmetic, *Israel J. Math.*, **8**, 309–328.

[1972] Stage comparison theorems and game playing with inductive definitions, unpublished notes.

[1975] Quantifiers, games and inductive definitions, in: *Third Scandinavian Logic Symposium*, edited by S. Kanger (North-Holland, Amsterdam), pp. 1–14.

BARWISE, J.

[1975] *Admissible Sets and Structures* (Springer, Berlin).

[1978] Monotone quantifiers and admissible sets, in: *Generalized Recursion Theory II*, edited by J. E. Fenstad, R. O. Gandy and G. E. Sacks (North-Holland, Amsterdam), pp. 1–38.

ENDERTON, H. B.

[1967] An infinitistic rule of proof, *J. Symbolic Logic*, **32**, 447–451.

GRILLIOT, T. J.

[1971] Inductive definitions and computability, *Trans. Am. Math. Soc.*, **158**, 309–317.

HARRINGTON, L. A., and A. S. KECHRIS

[1975] On characterizing Spector classes, *J. Symbolic Logic*, **40**, 19–24.

HARRINGTON, L. A., A. S. KECHRIS and S. G. SIMPSON

[1973] 1-envelopes of type 2 objects, *Notices Am. Math. Soc.*, **20**, A-587.

HARRINGTON, L. A., and Y. N. MOSCHOVAKIS

[1974] On positive induction versus nonmonotone induction, mimeographed notes.

KECHRIS A. S., and Y. N. MOSCHOVAKIS

[1977] Recursion in higher types, in: *Handbook of Mathematical Logic*, edited by J. Barwise (North-Holland, Amsterdam), pp. 681–737.

KLEENE, S. C.

[1959] Recursive functionals and quantifiers of finite type I, *Trans. Am. Math. Soc.*, **91**, 1–52.

KOLAITIS, PH. G.

[1978a] Recursion in a quantifier vs. elementary induction, *J. Symbolic Logic*, **44**, 235–259.

[1978b] On recursion in **E** and semi-Spector classes, in: *Cabal Seminar 76–77, Proceedings of the Caltech–UCLA Logic Seminar, 1976–1977*, edited by A. S. Kechris and Y. N. Moschovakis, *Lecture Notes in Mathematics 689* (Springer-Verlag, Berlin), pp. 209–243.

MOSCHOVAKIS, Y. N.

[1967] Hyperanalytic predicates, *Trans. Am. Math. Soc.* **129**, 249–282.

[EIAS] *Elementary Induction an Abstract Structures* (North-Holland, Amsterdam).

[1974a] Structural characterizations of classes of relations, in: *Generalized Recursion Theory*, edited by J. E. Fenstad and P. G. Hinman (North-Holland, Amsterdam), pp. 53–79.

[1974b] On nonmontone inductive definability, *Fund. Math.*, **32**, 39–83.

[1976] On the basic notions in the theory of induction, in: *Logic, Foundations of Mathematics and Computability Theory, Proceedings of the Fifth Congress for Logic, Methodology and Philosophy of Science*, edited by Butts and Hintikka (Reidel), pp. 207–236.

RICHTER, W.

[1971] Recursively Mahlo ordinals and inductive definitions, in: *Logic Colloquium '69*, edited by R. O. Gandy and C. E. M. Yates (North-Holland, Amsterdam), pp. 273–288.

RICHTER, W., and P. ACZEL

[1974] Inductive definitions and reflecting properties of admissible ordinals, in: *Generalized Recursion Theory*, edited by J. E. Fenstad and P. G. Hinman (North-Holland, Amsterdam), pp. 301–384.

This page intentionally left blank

J. Barwise, H. J. Keisler and K. Kunen, eds., *The Kleene Symposium*
©North-Holland Publishing Company (1980) 391–414

A Proof-theoretic Approach to Non-standard Analysis with Emphasis on Distinguishing between Constructive and Non-constructive Results*

Shih-Chao Liu

Academia Sinica, Taipei, Taiwan

Dedicated to Professor S.C. Kleene on the occasion of his 70th birthday

Abstract: In this paper a constant, namely an operation without variables, is defined in the Zermelo–Fraenkel set theory ZF. This constant, which we denote by ∞, is such that (1) $\infty \in \omega \subset R$ and (2) the extension ZF* obtained from ZF by adding the axioms $0 < \infty$, $S(0) < \infty$, $S(S(0)) < \infty, \ldots$ is consistent relative to ZF, where **R** is the Archimedean real field defined as usual in ZF, and $S(x)$ is the successor operation $x \cup \{x\}$. Working within ZF* we develop a considerable part of non-standard analysis with ∞ as an infinity in ω and $1/\infty$ as an infinitesimal in **R**.

Notions such as constructive number and constructive operation (function) are defined in a new way. Some examples of non-constructive numbers and functions are given. On the other hand, a series of metatheorems are proved, which state sufficient conditions for numbers and functions so obtained to be constructive. Some relations among such notions as non-standard analysis, non-constructive number, undecidable sentence, intuitionism and so forth are pointed out and discussed with a view to shedding light on the nature of mathematics.

1. Introduction and preliminaries

In this paper development of a theory of non-standard analysis consists in establishing certain theorems and metatheorems of an axiomatic set theory ZF* which is an extension of the familiar Zermelo–Fraenkel theory ZF obtained by adding some new axioms. This approach to non-standard analysis is quite different from that of Keisler (1976) and of others who, following Robinson (1966), use model theory as their main tool. A constant (operation without variables) which we define in ZF and denote by ∞, becomes an infinite number in ω by virtue of the axioms of ZF*. The existence of such an infinite number ∞ in ω is the main feature of non-standard analysis.

Since the logic of ZF is the first-order classical logic and the law of excluded middle is allowed, we lay particular emphasis on the distinction

*This work is supported in part by the National Science Council of the Republic of China.

between constructive and non-constructive results. We establish many metatheorems (concerning real numbers) that can tell how much constructive information a ZF* theorem under consideration can contain in terms of concepts such as constructive natural number, constructive real number, constructive real operation (function) and so forth. These concepts will be explained in Section 1 below. Constructive numbers are useful for actual counting or measurement; but theorems containing no constructive information at all are not useful in such ways. Some examples of non-constructive natural numbers are given and hence examples of constant functions which are not constructive follow (see metatheorems 4.2–4.6 and Example 4.7).

Our attitude of preferring constructive numbers to non-constructive ones is similar to that of the intuitionists but do not hold strictly to the intuitionistic view in this paper, because our object-theory ZF* is based on a classical first-order logic and hence the laws of excluded middle and double negation can be used freely to get non-constructive elements. However, non-constructive elements are not always undesirable. Sometimes they may prove helpful even from the point of view of constructive mathematics. Indeed, the above-mentioned infinite number ∞ in ω is defined by using the law of excluded middle and is itself non-constructive in character. As can be seen in our development in this paper (especially Section 7) the introduction of ∞ together with the infinitesimal $1/\infty$, as real numbers supplies very expedient tools for dealing with integration and differentiation and thereby enables us to obtain a large number of constructive real operations. Thus non-standard analysis is not only of theoretical interest but serves a practical purpose.

Since our treatment of non-standard analysis is related to such concepts in logic and foundations as formalism, intuitionism, constructive numbers, finitary method, classical logic, undecidable sentences, the law of excluded middle and so forth, we will insert some philosophical remarks in this paper where they seem appropriate. We hope that these remarks will help to clarify or make more precise our idea of the nature of mathematics.

We start with ZF, and throughout this paper we assume

1.1. ZF *is consistent*

Constants, and operation in general, are very useful means to construct abbreviations for well-formed formulas in ZF. For example, suppose that $f(x), g(x_1,\ldots,x_n,y)$ are two well-formed formulas such that

$$\mathrm{ZF}\vdash\forall x_1,\ldots,x_n \mathrm{E}!y\, g(x_1,\ldots,x_n,y).$$

It is then agreed that the formula g defines an operation in ZF, which may be suitably denoted by say $G(x_1,\ldots,x_n)$, and that $f(G(x_1,\ldots,x_n))$ stands for the well-formed formula $\mathsf{E}y(g(x_1,\ldots,x_n,y)\&f(y))$. We assume that the reader has acquired enough knowledge about the rules to handle operations from the literature, say KARP (1967) and LIU (1974).

1.2. *Notations for some useful constants and operations*

As usual, ω is the least infinite ordinal; 0 is the empty set; $S(x)$ is the successor of x defined by $S(x)=x\cup\{x\}$; for any intuitive natural number n, $\ulcorner n \urcorner$ denotes the constant defined recursively by $\ulcorner 0\urcorner=0, \ulcorner 1\urcorner=s(\ulcorner 0\urcorner)$, $\ulcorner n+1\urcorner=S(\ulcorner n\urcorner)$. Informally speaking, although an intuitive natural number n is itself not an element in the formal set theory **ZF**, yet it has a counterpart, namely $\ulcorner n\urcorner$ in **ZF**.

2. The real field

Let the following constants, operations and formulas: **R** (the real field), **Q** (the rational field), **C** (the positive and negative integers), $x+y, x\cdot y, -x, 1/x, |x|, x<y$ (the ordering formula) be defined as usual so that they satisfy the following conditions:

2.1. $\mathsf{ZF}\vdash\omega\subset\mathbf{C}\subset\mathbf{Q}\subset\mathbf{R}$.

2.2. $\mathsf{ZF}\vdash\mathbf{R}$ is a complete ordered field with respect to $+$, $\cdot$, and $<$.

2.3. $\mathsf{ZF}\vdash(\forall x,y\in\mathbf{R})[x+(-x)=0\&(x\neq 0\rightarrow x\cdot(1/x)=\ulcorner 1\urcorner)\&(x>0\rightarrow x=|x|)\&(x<0\rightarrow -x=|x|)]$.

2.4. $\mathsf{ZF}\vdash x\in\mathbf{R}\rightarrow\mathsf{E}y(y$ is a nonempty proper subset of $\mathbf{Q}$ and x is the least upper bound of y in the sense of $<)$.

2.5. $\mathsf{ZF}\vdash x\in\mathbf{Q}\leftrightarrow\mathsf{E}y,z(y\in\mathbf{C}\&z\in\mathbf{C}\&z\neq 0\&x=y\cdot(1/z))$.

2.6. $\mathsf{ZF}\vdash x\in\mathbf{C}\leftrightarrow\mathsf{E}y(y\in\omega\&(x=y\vee x=(-y)))$.

2.7. $\mathsf{ZF}\vdash 0\in\omega\&[x\in\omega\&y\in\omega\rightarrow S(x)\in\omega\&(S(x)=S(y)\rightarrow x=y)\&(S(x)\neq 0)\&(x<y\leftrightarrow x\in y)]\&(0\in y\&\forall t(t\in y\rightarrow S(t)\in y)\&y\subset\omega\rightarrow y=\omega)$.

2.8. $\mathsf{ZF}\vdash x\in\omega\&y\in\omega\rightarrow(x+0=x\&x+S(y)=S(x+y)\&x\cdot 0=0\&x\cdot S(y)=x\cdot y+x)$.

2.9. $\mathsf{ZF}\vdash\mathbf{R}$ is Archimedean, i.e. $(\forall x\in\mathbf{R})(\mathsf{E}y\in\omega)[x<y]$.

3. The Constant ∞

By a well-known method we can find a primitive recursive function $p(m,n)$ such that for any two intuitive natural numbers m, n, $p(m,n)=0$ iff m is a Gödel number of a well-formed formula $A(x)$ and n is a Gödel number of a proof in ZF of a well-formed formula to which $A(\ulcorner m \urcorner)$ is reducible. Again we can find an operation $P(x,y)$ definable in ZF such that for any m, n, r, either $\mathrm{ZF}\vdash P(\ulcorner m\urcorner,\ulcorner n\urcorner)=\ulcorner r\urcorner$ or $\mathrm{ZF}\vdash P(\ulcorner m\urcorner,\ulcorner n\urcorner)\neq\ulcorner r\urcorner$ and furthermore we have (assume 1.1) $\mathrm{ZF}\vdash P(\ulcorner m\urcorner,\ulcorner n\urcorner)=\ulcorner r\urcorner$ iff $p(m,n)=r$. Under the assumption of 1.1 that ZF is consistent we can show (again by a well-known method) that

3.1. Metatheorem. *Let $A(x)$ denote a well-formed formula to which $\forall y(y\in\omega\rightarrow P(x,y)\neq 0)$ is reducible, and let q be a Gödel number of $A(x)$. Then $\mathrm{ZF}\nvdash A(\ulcorner q\urcorner)$, i.e. $A(\ulcorner q\urcorner)$ is not provable in* ZF *but $\mathrm{ZF}\vdash P(\ulcorner q\urcorner,\ulcorner i\urcorner)\neq 0$ for any intuitive i.*

3.2. Metadefinition. Let $B(y)$ stand for

$$y\in\omega\&\left(P(\ulcorner q\urcorner,y)=0\vee A(\ulcorner q\urcorner)\right).$$

Since the law of excluded middle holds in ZF we can easily show that $\mathrm{ZF}\vdash \mathrm{E}y\, B(y)$ and hence $\mathrm{ZF}\vdash \mathrm{E}!y(B(y)\&\forall z(z\in y\rightarrow\neg B(z)))$. Now we define ∞ as the constant satisfying

$$\mathrm{ZF}\vdash B(\infty)\&\forall z(z\in\infty\rightarrow\neg B(z)).$$

3.3. Lemma. $\mathrm{ZF}\vdash\infty\in\omega\&(P(\ulcorner q\urcorner,\infty)=0\vee A(\ulcorner q\urcorner))$.

3.4. Metatheorem. $\mathrm{ZF}\nvdash\infty=\ulcorner i\urcorner$ *for any intuitive i.*

Proof. Suppose $\mathrm{ZF}\vdash\infty=\ulcorner i\urcorner$. Then by Lemma 3.3, we get $\mathrm{ZF}\vdash P(\ulcorner q\urcorner,\ulcorner i\urcorner)=0\vee A(\ulcorner q\urcorner)$. By Metatheorem 3.1, it follows that $\mathrm{ZF}\vdash A(\ulcorner q\urcorner)$. This contradicts metatheorem 3.1. Hence metatheorem 3.4 is proved.

3.5. Metatheorem. $\mathrm{ZF}\nvdash\infty\in\ulcorner i\urcorner$, *for any intuitive i.*

Proof. Suppose $\mathrm{ZF}\vdash\infty\in\ulcorner i\urcorner$. Then by induction on i, we have $\mathrm{ZF}\vdash\infty=0\vee\infty=\ulcorner 1\urcorner\vee\cdots\vee\infty=\ulcorner i-1\urcorner$. By Lemma 3.3 we have $\mathrm{ZF}\vdash\infty=\ulcorner j\urcorner\rightarrow P(\ulcorner q\urcorner,\ulcorner j\urcorner)=0\vee A(\ulcorner q\urcorner)$. Since by Metatheorem 3.1, $\mathrm{ZF}\vdash P(\ulcorner q\urcorner,\ulcorner j\urcorner)\neq 0$, we further get $\mathrm{ZF}\vdash\infty=\ulcorner j\urcorner\rightarrow A(\ulcorner q\urcorner)$ for $j=0,\ldots,i-1$. We are led to $\mathrm{ZF}\vdash A(\ulcorner q\urcorner)$ which contradicts 3.1. Hence Metatheorem 3.5 is proved.

3.6 Metatheorem. $\mathsf{ZF} \nvdash \infty < \ulcorner i \urcorner$ *for any intuitive i.*

Proof. For, otherwise $\mathsf{ZF} \vdash \infty \in \ulcorner i \urcorner \cup \{\ulcorner i \urcorner\} = \ulcorner i+1 \urcorner$ which contradicts Metatheorem 3.5.

4. Constructive notions

4.1. Metadefinitions. Let ZF^* denote any consistent extension of ZF. A constant a is a *constructive natural number* in ZF^* if for some n, $\mathsf{ZF}^* \vdash a = \ulcorner n \urcorner$; a is a constructive integer if for some n, either $\mathsf{ZF}^* \vdash a = \ulcorner n \urcorner$ or $\mathsf{ZF}^* \vdash a = -\ulcorner n \urcorner$; a is a *constructive rational* if for some constructive integers b and c, $\mathsf{ZF}^* \vdash a = b \cdot (1/c) \& c \neq 0$; a is a *constructive real* if for any n, there is some constructive rational b such that $\mathsf{ZF}^* \vdash |a-b| < 1/\ulcorner n \urcorner$; an operation $F(x_1, \ldots, x_n)$ is a *constructive real operation* if for any n-tuple of constructive reals $a_1, \ldots, a_n, F(a_1, \ldots, a_n)$ is a constructive real.

The following five metatheorems (4.2–4.6) are stated under the assumption that ZF^* is just ZF.

4.2. Metatheorem. ∞ *is not a constructive natural number in* ZF.

Proof. By definition 4.1 and by Metatheorem 3.4.

4.3. Metatheorem. ∞ *is not a constructive integer in* ZF.

Proof. Suppose ∞ is a constructive integer. Then for some n, either $\mathsf{ZF} \vdash \infty = \ulcorner n \urcorner$ or $\mathsf{ZF} \vdash \infty = -\ulcorner n \urcorner$. Because of Metatheorem 4.2, the former case is excluded; hence $\mathsf{ZF} \vdash \infty = -\ulcorner n \urcorner$. We have $\mathsf{ZF} \vdash \infty \in \omega$ by Lemma 3.3 and have $\mathsf{ZF} \vdash \ulcorner n \urcorner \in \infty$. Then $\mathsf{ZF} \vdash (0 \in \infty \vee 0 = \infty) \& (0 \in \ulcorner n \urcorner \vee 0 = \ulcorner n \urcorner)$, and by Theorem 2.7, $\mathsf{ZF} \vdash 0 \leqslant -\ulcorner n \urcorner \& \ulcorner n \urcorner \leqslant 0 \& 0 \leqslant \ulcorner n \urcorner \& \ulcorner n \urcorner = 0 \& -\ulcorner n \urcorner = 0$. It follows $\mathsf{ZF} \vdash 0 = \infty$. This contradicts Metatheorem 4.2. Hence Metatheorem 4.3 is proved.

4.4 Metatheorem. ∞ *is not a constructive rational.*

Proof. Suppose ∞ is a constructive rational. Then there exist constructive integers a, b such that $\mathsf{ZF} \vdash \infty = a/b \& b \neq 0$. And there exist two intuitive natural numbers m, n such that ($\mathsf{ZF} \vdash a = \ulcorner m \urcorner$ or $\mathsf{ZF} \vdash a = -\ulcorner m \urcorner$) as well as ($\mathsf{ZF} \vdash b = \ulcorner n \urcorner$ or $\mathsf{ZF} \vdash b = -\ulcorner n \urcorner$). Thus we have only four possible cases: (1) $\mathsf{ZF} \vdash \infty = \ulcorner m \urcorner / \ulcorner n \urcorner$, (2) $\mathsf{ZF} \vdash \infty = -\ulcorner m \urcorner / -\ulcorner n \urcorner$, (3) $\mathsf{ZF} \vdash \infty = -\ulcorner m \urcorner / \ulcorner n \urcorner$, (4) $\mathsf{ZF} \vdash \infty = \ulcorner m \urcorner / -\ulcorner n \urcorner$. In any of the cases (1), (2), (3) and (4) we have $\mathsf{ZF} \vdash \infty \leqslant \ulcorner m \urcorner$. Evidently it contradicts Metatheorem 3.6. Hence ∞ is not a constructive rational.

4.5. Metatheorem. ∞ *is not a constructive real in* ZF.

Proof. Suppose ∞ is a constructive real. Then for some constructive integers a, b, $\mathsf{ZF}\vdash|\infty-a/b|<1/\ulcorner 1\urcorner\,\&\,b\neq 0$ and for some intuitive natural numbers m, n, we have (either $\mathsf{ZF}\vdash a=\ulcorner m\urcorner$ or $\mathsf{ZF}\vdash a=-\ulcorner m\urcorner$) as well as (either $\mathsf{ZF}\vdash b=\ulcorner n\urcorner$ or $\mathsf{ZF}\vdash b=-\ulcorner n\urcorner$). In any case we have $\mathsf{ZF}\vdash\infty\leqslant\ulcorner m+1\urcorner$ which contradicts Metatheorem 3.6. This proves Metatheorem 4.5.

4.6. Metatheorem. *The operation* $G(x)$ *satisfying* $\mathsf{ZF}\vdash\forall x(G(x)=\infty)$ *is not a constructive real operation.*

Proof. By Metatheorem 4.5.

4.7. Example. Let $H(x)$ denote the formula $(S\,\&\,x=0)\vee(\neg S\,\&\,x=\ulcorner 1\urcorner)$ where S is an undecidable sentence obtained by using Rosser's form (1936) of Gödel's theorem. Then we have $\mathsf{ZF}\vdash\mathrm{E}!x\,H(x)$. Let E denote the constant definable in ZF by $H(x)$, so that

$$\mathsf{ZF}\vdash H(E)\,\&\,(E=0\vee E=\ulcorner 1\urcorner)\,\&\,E<\ulcorner 2\urcorner\,\&\,E\in\omega.$$

We claim that for any intuitive natural number i the assumption that $\mathsf{ZF}\vdash E=\ulcorner i\urcorner$ leads to a contradiction. For suppose that $\mathsf{ZF}\vdash E=\ulcorner i\urcorner$. In case $i\neq 0$ and $i\neq 1$ we have $\mathsf{ZF}\vdash(E=0\vee E=\ulcorner 1\urcorner)\,\&\,E\neq 0\,\&\,E\neq\ulcorner 1\urcorner$; a contradiction. In case $i=0$ we have $\mathsf{ZF}\vdash H(0)$, $\mathsf{ZF}\vdash(S\,\&\,0=0)\vee(\neg S\,\&\,0=\ulcorner 1\urcorner)$, $\mathsf{ZF}\vdash S$, which contradicts the assumption S is undecidable. Similarly, in case $i=1$ we have $\mathsf{ZF}\vdash\neg S$; again a contradiction. Hence, by Definition 4.1, E is not a constructive natural number. Using the above facts we can easily show that there exists no constructive rational $\ulcorner n\urcorner/\ulcorner m\urcorner$ such that $\mathsf{ZF}\vdash|E-\ulcorner n\urcorner/\ulcorner m\urcorner|<1/\ulcorner 3\urcorner$. Hence, by Definition 4.1, E is not a constructive real. Then like $G(x)$ of Metatheorem 4.6 the constant operation $G'(x)$ satisfying $\mathsf{ZF}\vdash\forall x(G'(x)=E$ is not a constructive real operation. We observe that $G'(x)$ is different from $G(x)$ in that the constant value of $G(x)$ is an infinite number ∞, while that of $G'(x)$ is a finite number E. Of course, $G(x)$ and $G'(x)$ are only simple cases of non-constructive real operations. We hope that in a future paper we will be able to construct more complicated and interesting examples.

5. The extension ZF*

From now on we let ZF* be the extension of ZF obtained by adding the axioms $0<\infty,\ulcorner 1\urcorner<\infty,\ldots,\ulcorner i\urcorner<\infty,\ldots$ where i runs through all intuitive natural numbers.

Remark. Intuitively, these axioms cannot all be true, i.e. ZF* has no standard models. But formally ZF* is consistent. This is why the theory is called "non-standard". We further note that if we consider ∞ as an individual symbol in the language instead of a notation for a definable set in ZF, then the theory ZF* plus $\infty \in \omega$ can be easily shown to be consistent. For the main points of such an argument, the reader may see KLEENE (1976, p.762). However, since in that case the new theory has a language different from that of the original theory ZF, it can not serve all the purposes of this paper.

5.1. Metatheorem. ZF* *is consistent.*

Proof. Let $\psi_1, \ldots, \psi_n, \ulcorner i \urcorner_1 < \infty, \ldots, \ulcorner i \urcorner_m < \infty$ be any finite set of axioms of ZF* where $\psi_1, \ldots, \psi_n$ are axioms of ZF. We claim that this set of axioms does not lead to any contradiction. For, if

$$\mathrm{ZF} \vdash \psi_1 \& \cdots \& \psi_n \& \ulcorner i_1 \urcorner < \infty \& \cdots \& \ulcorner i \urcorner_m < \infty \rightarrow 0 = \ulcorner 1 \urcorner,$$

then we can choose a k such that $\mathrm{ZF} \vdash \ulcorner i \urcorner_j < \ulcorner k \urcorner$ for $j = 1, \ldots, m$ and hence $\mathrm{ZF} \vdash \ulcorner k \urcorner < \infty \rightarrow 0 = \ulcorner 1 \urcorner$. We then have $\mathrm{ZF} \vdash \ulcorner k \urcorner \nless \infty \& \infty \lessdot \ulcorner k \urcorner$ which contradicts Metatheorem 3.6. Hence the above arbitrary set of axioms of ZF* is consistent and therefore ZF* is itself consistent.

5.2. Metatheorem. $\mathrm{ZF}^* \vdash \ulcorner i \urcorner < \infty$ *for any intuitive i.*

Proof. These sentences are axioms of ZF*.

5.3. Metatheorem. ∞ *is not a constructive natural number in* ZF*.

Proof. By Metatheorem 5.2, Theorem 2.2 and Metatheorem 5.1.

5.4. Metatheorem. ∞ *is not a constructive integer in* ZF*.

Proof. Similarly as in the proof of Metatheorem 4.3 but here we use Metatheorem 5.3 instead of Metatheorem 4.2.

5.5. Metatheorem. ∞ *is not a constructive rational in* ZF*.

Proof. Use the easy metatheorem that for any positive constructive rational $\ulcorner m \urcorner / \ulcorner n \urcorner$, $\mathrm{ZF}^* \vdash \ulcorner m \urcorner / \ulcorner n \urcorner \lessdot \ulcorner m \urcorner < \infty$.

5.6. Metatheorem. ∞ *is not a constructive real in* ZF*.

Proof. Use the fact that for any constructive rational $c, \mathsf{ZF}^* \vdash c+1/\ulcorner 1 \urcorner < \infty$.

5.7. Metatheorem. *The constant operation* $G(x)$ *satisfying* $\mathsf{ZF}^* \vdash G(x)=\infty$ *is not a constructive real operation in* ZF^*.

5.8. Discussion. The constant ∞ definable in ZF is an element of the set ω that satisfies the five Peano axioms and so it is natural for us to call ∞ a natural number. However, unlike the constructive natural numbers 0, $\ulcorner 1 \urcorner$, $\ulcorner 2 \urcorner$,... in ω the number ∞ is not useful in actual counting or measurement; indeed it is a natural number only in name. The constant ∞ is then an example of a non-constructive natural number whose existence is provable by resorting to the law of excluded middle. Long before such an example was known the intuitionists began to suspect the validity of the law of excluded middle. In their endeavor to make sure that whenever the existence of a number is proved the proof used enables one to construct that number the intuitionists have proposed not to use the law of excluded middle, the law of double negation and other related classical principles as logical rules, with the following result: in a formal number theory, say the theory N in KLEENE (1962), a sentence of the form $\mathsf{E}x\, G(x)$ is provable iff $G(\mathbf{x})$ is provable for some constructive numeral $\mathbf{x}$ and a sentence of the form $A \vee B$ is provable iff A is provable or B is provable and so on. Thus for any undecidable A (which can be constructed by using ROSSER's form (1936) of Gödel's Theorem), even "$A \vee \neg A$" has such a plausible form it is not admitted as a theorem by the intuitionists because neither A nor $\neg A$ is provable even classically. After making the above observations about the prohibition of using some classical logical principles as a means of avoiding non-constructive number we would like to suggest that the intuitionists should take care of other occasions in which non-constructive numbers can arise. For example, when one extends the mathematical theory ZF to a new theory ZF^{**} by adding the single axiom

$$\mathsf{E}!y\big(y \in \omega \,\&\, P(\ulcorner q \urcorner, y)=0 \,\&\, (\forall z \in y)\big[P(\ulcorner q \urcorner, z) \neq 0\big]\big)$$

the constant definable by this axiom is a non-constructive number in ZF^{**}. The extension ZF^{**} is simply consistent but not ω-consistent. In such cases of making a consistent extension for a mathematical theory it also seems appropriate for the intuitionists to lay down some rules to preclude non-constructive numbers.

As another remark let us consider the following formula

$$\forall x\, \mathsf{E}!y\big(x=y \,\&\, (\mathsf{E}z \in \mathbf{R})\big[z \in \omega \,\&\, (P(y,z)=0 \vee A(y))\big]\big)$$

where $A(y)$ denotes $\forall t(t \in \omega \rightarrow P(y,t) \neq 0)$. This formula is not only provable in ZF but also has the property that for any constructive real a there is a unique constructive real b satisfying

$$\mathsf{ZF}\vdash a=b\,\&\,(\mathsf{E}z\in\mathbf{R})[z\in\omega\,\&\,(P(b,z)=0\vee A(b))].$$

Thus the formula can be used to define a constructive real operation $G(x)$ such that if x is a, then $G(x)$ is b. Since we are interested in constructive mathematics we are happy to accept and study this type of theorem. However, the intuitionists may object to taking the formula as a theorem because it involves non-constructive information; when $a=b=\ulcorner q\urcorner$ it is impossible to have any constructive real c satisfying

$$\mathsf{ZF}\vdash c\in\omega\,\&\,(P(\ulcorner q\urcorner,c)=0\vee A(\ulcorner q\urcorner)).$$

To prove this statement let us suppose that such a constructive real c exists. Then by Metatheorem 6.9 of the next section, $\mathsf{ZF}\vdash c<\ulcorner n\urcorner$ for some intuitive n. Using $\mathsf{ZF}\vdash c\in\omega$ and $\mathsf{ZF}\vdash\ulcorner n\urcorner\in\omega$ we further get $\mathsf{ZF}\vdash c\in\ulcorner n\urcorner$ by Theorem 2.7. By induction on the intuitive number n it follows that

$$\mathsf{ZF}\vdash c=0\vee c=\ulcorner 1\urcorner\vee\cdots\vee c=\ulcorner n-1\urcorner.$$

On the other hand, from our assumption

$$\mathsf{ZF}\vdash c\in\omega\,\&\,(P(\ulcorner q\urcorner,c)=0\vee A(\ulcorner q\urcorner))$$

it follows that

$$\mathsf{ZF}\vdash c=\ulcorner j\urcorner\rightarrow(P(\ulcorner q\urcorner,\ulcorner j\urcorner)=0\vee A(\ulcorner q\urcorner))\quad \textit{for } j=0,1,\ldots.$$

By the second part of Metatheorem 3.1, namely $\mathsf{ZF}\vdash P(\ulcorner q\urcorner,\ulcorner j\urcorner)\neq 0$, this result implies $\mathsf{ZF}\vdash c=\ulcorner j\urcorner\rightarrow A(\ulcorner q\urcorner)$. We conclude that $\mathsf{ZF}\vdash A(\ulcorner q\urcorner)$; a contradiction by Metatheorem 3.1.

More examples of operations of similar type will be provided later in Section 7, when integration and differentiation will be involved. The operations themselves are all constructive real operations but they are defined by using classical theorems which involve non-constructive numbers. We will resume our discussion of such examples in Discussion 7.8.

6. Non-standard analysis

In this section we will introduce the concepts of infinity and infinitesimal and the rules governing them. Some other fundamental concepts in analysis will also be introduced and studied.

6.1. Metadefinitions. A constant c definable in ZF^* is an *infinity* if for any intuitive natural number n, $\mathsf{ZF}^*\vdash\ulcorner n\urcorner<c\,\&\,c\in\mathbf{R}$.

6.2. Metatheorem. *The constant* ∞ *defined in Metadefinition* 3.2 *is an infinity.*

Proof. By Metatheorem 5.2.

6.3. Metatheorem. *If a, b are infinities, and if c is a positive constructive real, then* $a+b$, $a\cdot b$, $a+c$ *and* a/c *are infinities. Furthermore if* $\mathsf{ZF}^*\vdash 1/\ulcorner n \urcorner < c$ *for some n, then* $a\cdot c$ *is also an infinity.*

6.4. Metadefinition. A constant c is an *infinitesimal* if for any intuitive natural number n, $\mathsf{ZF}^*\vdash |c| < 1/\ulcorner n \urcorner \,\&\, c \in \mathbf{R}$. Let us denote $1/\infty$ by δ.

6.5. Metatheorem. *The constant* $1/\infty$, *namely* δ, *is an infinitesimal.*

Proof. Since $\mathsf{ZF}\vdash \infty \in \omega$, hence by Theorem 2.5, we have $\mathsf{ZF}^*\vdash 1/\infty \in Q \subset \mathbf{R} \,\&\, \delta \in \mathbf{R}$. Then Metatheorem 6.5 follows from the fact $\mathsf{ZF}\vdash\mathbf{R}$ is an ordered field with respect to $+$, $\cdot$ and $<$. Specifically,

$$\mathsf{ZF}^*\vdash \ulcorner n \urcorner < \infty \rightarrow 0 < \delta < 1/\ulcorner n \urcorner$$

for each n.

6.6. Metatheorem. *If c is an infinity, then* $1/c$ *is an infinitesimal. If c is an infinitesimal such that* $\mathsf{ZF}^*\vdash c > 0$, *then* $1/c$ *is an infinity.*

6.7. Metatheorem. *If a, b are infinitesimals and if c is a constructive real, then* $a+b$, $a\cdot b$, $a\cdot c$ *are infinitesimals; if furthermore* $\mathsf{ZF}^*\vdash 1/\ulcorner m \urcorner < c$ *for some m, then* a/c *is also an infinitesimal.*

6.8. Metatheorem. *If a, b are constructive reals and if c is a constructive real which is not an infinitesimal, then* $a+b$, $a\cdot b$, $-a$ *and* $1/c$ *are all constructive reals.*

6.9. Metatheorem. *For any constructive real c, there is always an intuitive natural number n such that* $\mathsf{ZF}^*\vdash c < \ulcorner n \urcorner$.

From the literature, in particular KARP (1967), we can find metatheorems such as the following: Given two operations $H(x_1,\ldots,x_n)$ and $G(y,x,x_1,\ldots,x_n)$, there is an operation $F(x,x_1,\ldots,x_n)$ satisfying

$$\mathsf{ZF}^*\vdash F(0,x_1,\ldots,x_n) = H(x_1,\ldots,x_n)$$
$$\&\,(i \in \omega \rightarrow F(i+1,x_1,\ldots,x_n) = G(F(i,x_1,\ldots,x_n),i,x_1,\ldots,x_n)).$$

Using the metatheorems we can define the following operations.

6.10. Definition. Given any operation $G(x,x_1,\ldots,x_n)$, then we let

$$\sum_{j<i} G(j,x_1,\ldots,x_n)$$

and

$$\prod G(j,x_1,\ldots,x_n)$$

be operations satisfying

$$\mathbf{ZF}^*\vdash \sum_{j<0} G(j,x_1,\ldots,x_n)=0\,\&\, \prod_{j<0} G(j,x_1,\ldots,x_n)$$

$$=\ulcorner 1\urcorner \,\&\, \Bigl(i\in\omega\rightarrow \sum_{j<i+1} G(j,x_1,\ldots,x_n)$$

$$=\sum_{j<i} G(j,x_1,\ldots,x_n)+G(i,x_1,\ldots,x_n)\Bigr)\,\&$$

$$\Bigl(i\in\omega\rightarrow \prod_{j<i+1} G(j,x_1,\ldots,x_n)$$

$$=\prod_{j<i} G(j,x_1,\ldots,x_n)\cdot G(i,x_1,\ldots,x_n)\Bigr).$$

In particular, when $G(i,a)$ is a we let $\Pi_{j<i}G(j,a)$ be denoted by a^i and when $G(i)$ is $i+1$, let $\Pi_{j<i}G(j)$ be denoted by $i!$.

6.11. Definition. Let $\Sigma_{j\leqslant a}G(j,x)$ denote $\Sigma_{j<a+1}G(j,x)$; let $\Sigma_{a\leqslant j<b}G(j,x)$ denote $\Sigma_{j<b-a}G(j+a,x)$. Similarly for Π.

In the following most theorems and metatheorems are simply stated without proof, because the technique for their proof is routine.

6.12. Metatheorem. *For any constructive natural number* a, $\Sigma_{j<a}(x^j/j!)$ *is a constructive real operation of the variable* x.

6.13. Definition. Given an interval $[a,b]$ and an operation $G(j,x)$, then by the *sequence of functions of* $G(j,x)$ we mean the set

$$\{\langle j,\{\langle x,G(j,x)\rangle : x\in[a,b]\}\rangle : j\in\omega\}$$

and by the *family of sequences of* $G(j,x)$ we mean the set

$$\{\langle x,\{\langle j,G(j,x)\rangle : j\in\omega\}\rangle : x\in[a,b]\}.$$

Given any constant c such that $\mathbf{ZF}^*\vdash c\in\omega$, we say $G(c,x)$ is *constructively uniformly continuous in* $[a,b]$ if for any n, there is some m such that

$$\mathbf{ZF}^*\vdash a\leqslant x\leqslant y\leqslant b\,\&\,|x-y|<1/\ulcorner m\urcorner\rightarrow|G(c,x)-G(c,y)|<1/\ulcorner n\urcorner.$$

We say that the sequence of functions of $G(j,x)$ is *constructively uniformly continuous in* $[a,b]$ *independent of* j, if for any n there is some m such that

$$\mathbf{ZF}^*\vdash j\in\omega\,\&\,a\leqslant x\leqslant y\leqslant b\,\&\,|x-y|<1/\ulcorner m\urcorner\rightarrow|G(j,x)-G(j,y)|<1/\ulcorner n\urcorner.$$

Given a constant x^* such that $\mathsf{ZF}^* \vdash x^* \in \mathbf{R}$, we say the sequence $G(j,x^*)$ is *constructively convergent* if for any n, there is some m such that

$$\mathsf{ZF}^* \vdash \ulcorner m \urcorner < i \,\&\, \ulcorner m \urcorner < j \to |G(i,x^*) - G(j,x^*)| < 1/\ulcorner n \urcorner.$$

We say that the family of sequences of $G(j,x)$ is constructively uniformly convergent if for any n, there is some m such that

$$\mathsf{ZF}^* \vdash a \leqslant x \leqslant b \,\&\, \ulcorner m \urcorner < i \,\&\, \ulcorner m \urcorner j \to |G(i,x) - G(j,x)| < 1/\ulcorner n \urcorner.$$

6.14. Metatheorem. *Given any interval $[a,b]$ where a, b are constructive reals, then the family of sequences of $\Sigma_{j<i}(x^j/j!)$ is constructively uniformly convergent in $[a,b]$.*

6.15. Metatheorem. *Given an interval $[a,b]$ where a, b are constructive reals and given any constructive natural number i^*, then the function $\Sigma_{j<i^*}(x^j/j!)$ of x is constructively uniformly continuous in $[a,b]$.*

6.16. Metatheorem. *Given any interval $[a,b]$ if the family of sequences $G(j,x)$ is constructively uniformly convergent and if for any constructive natural number i^*, the function $G(i^*,x)$ is constructively uniformly continuous in $[a,b]$, then the sequence of functions of $G(j,x)$ is constructively uniformly continuous independent of j.*

6.17. Definition. Let e^x denote the operation $\Sigma_{j<\infty}(x^j/j!)$.

6.18. Metatheorem. *For any interval $[a,b]$, a, b being constructive reals, $\Sigma_{j<i}(x^j/j!)$ is constructively uniformly continuous independent of i. In particular, the function e^x, namely $\Sigma_{j<\infty}(x^j/j!)$ is constructively uniformly continuous in $[a,b]$.*

Proof. By Metatheorems 6.14, 6.15 and 6.16.

6.19. Metatheorem. *For any interval $[a,b]$ where a,b are constructive reals, to any n there corresponds a m such that*

$$\mathsf{ZF}^* \vdash x \in [a,b] \,\&\, \ulcorner m \urcorner < i \to \Big|e^x - \sum_{j<i} (x^j/j!)\Big| < 1/\ulcorner n \urcorner.$$

In particular, for any constant c such that $\mathsf{ZF}^* \vdash c \in [a,b]$, we have

$$\mathsf{ZF}^* \vdash \ulcorner m \urcorner < i \,\&\, i \in \omega \to \Big|e^c - \sum_{j<i} (c^j/j!)\Big| < 1/\ulcorner n \urcorner.$$

We call e^c the constructive limit of the sequence

$$\left\{\left\langle i, \sum_{j<i} (c^j/j!)\right\rangle : i \in \omega\right\}.$$

Proof. By the fact of Metatheorem 6.14 that the family of sequences of $\Sigma_{j<i}(x^j/j!)$ is constructively uniformly convergent for the interval $[a,b]$ and the fact of Metatheorem 5.2 that for any intuitive m, $\mathsf{ZF}^*\vdash \ulcorner m \urcorner < \infty$.)

6.20. Metatheorem. *e^x is a constructive real operation of x.*

Proof. Let c be any constructive real. Then $\mathsf{ZF}^*\vdash c\in[c-\ulcorner 1\urcorner, c+\ulcorner 1\urcorner]$ where $c-\ulcorner 1\urcorner$ and $c+\ulcorner 1\urcorner$ are both constructive reals. By Metatheorem 6.19, to any n, there corresponds an m such that

$$\mathsf{ZF}^*\vdash \Big|e^c - \sum_{j<\ulcorner m\urcorner} (c^j/j!)\Big| < 1/\ulcorner n.2\urcorner .$$

Since by Metatheorem 6.12, $\Sigma_{j<\ulcorner m\urcorner}(c^j/j!)$ is a constructive real, hence by definition, the operation e^x is a constructive real operation.

6.21. Example. Let $F(x)$ be an operation such that

$$\mathsf{ZF}^*\vdash \big(x\in[0,1/\infty]\to F(x)=\infty\big)\,\&\,\big(x\in[1/\infty,\ulcorner 1\urcorner]\to F(x)=1/x\big)$$
$$\&\,\big(x\notin[0,\ulcorner 1\urcorner]\to F(x)=0\big).$$

It can be seen that in the interval $[0,\ulcorner 1\urcorner]$ the function $F(x)$ is uniformly continuous in the usual sense, but it is not constructively uniformly continuous. The behavior of $F(x)$ is illustrated by Fig. 1.

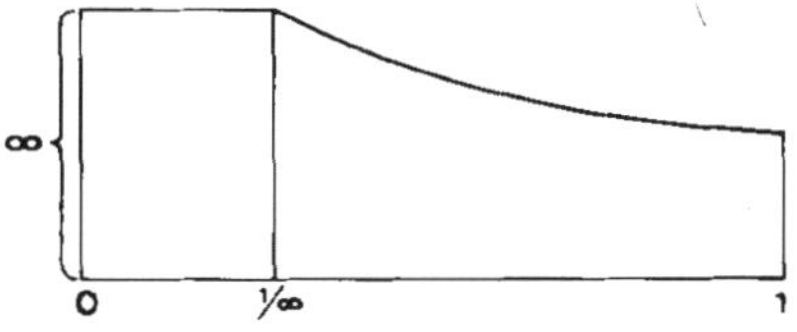

Figure 1. Behavior of the operation $F(x)$ of Example 6.21.

6.22. Example. Let $f(x)$ be an operation such that

$$\mathsf{ZF}^*\vdash \big(x\in[0,1/(\ulcorner 2\urcorner\cdot\infty+\ulcorner 1\urcorner)]\to F(x)=0\big)$$
$$\&\ \forall i\big[\, i\in\infty\ \&\big(1/(\ulcorner 2\urcorner\cdot i+\ulcorner 2\urcorner)\big)\leqslant x\leqslant\big(1/(\ulcorner 2\urcorner\cdot i+\ulcorner 1\urcorner)\big)\to F(x)$$
$$=\big(\big(1/(\ulcorner 2\urcorner\cdot i+\ulcorner 1\urcorner)\big)-x\big)/\big(\big(1/(\ulcorner 2\urcorner\cdot i+\ulcorner 1\urcorner)\big)-\big(1/(\ulcorner 2\urcorner\cdot i+\ulcorner 2\urcorner)\big)\big)\big]$$
$$\&\ \forall i\big[\, i\in\infty\,\&\big(1/(\ulcorner 2\urcorner\cdot i+\ulcorner 3\urcorner)\big)\leqslant x\leqslant\big(1/(\ulcorner 2\urcorner\cdot i+\ulcorner 2\urcorner)\big)\to F(x)$$
$$=\big(x-\big(1/(\ulcorner 2\urcorner\cdot i+\ulcorner 3\urcorner)\big)\big)/\big(\big(1/(\ulcorner 2\urcorner\cdot i+\ulcorner 2\urcorner)\big)-\big(1/(\ulcorner 2\urcorner\cdot i+\ulcorner 3\urcorner)\big)\big)\big]$$
$$\&\ \big(x\notin[0,\ulcorner 1\urcorner]\to F(x)=0\big).$$

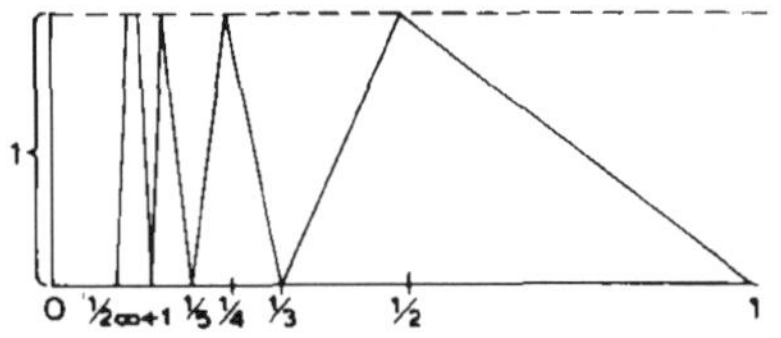

Figure 2. Behavior of the operation $F(x)$ of Example 6.22.

The function $F(x)$ is uniformly continuous in $[0, \ulcorner 1 \urcorner]$ but is not constructively uniformly continuous in $[0, \ulcorner 1 \urcorner]$. The behavior of $F(x)$ is illustrated by Fig. 2.

7. Integration and differentiation

7.1. Definition. Let x, y and ν be three variables. Then (ν is a *partition over the interval* $[x,y]$) stands for

$$\big(x \in \mathbf{R} \,\&\, y \in \mathbf{R} \,\&\, \nu \text{ is fcn} \,\&\, \mathrm{dom}(\nu) \in \omega \,\&\, \ulcorner 1 \urcorner \in \mathrm{dom}(\nu) \,\&$$

$$\mathrm{range}(\nu) \subset \mathbf{R} \,\&\, \nu(0) = x \,\& \quad \nu\big(\mathrm{dom}(\nu) - \ulcorner 1 \urcorner\big) = y \,\&$$

$$\forall t\big(t \in \mathrm{dom}(\nu) - \ulcorner 1 \urcorner \rightarrow \nu(t) < \nu\big(t + \ulcorner 1 \urcorner\big)\big)\big).$$

We also let (z is *a mesh of the partition* ν) stand for (ν is a partition and $\forall t(t \in \mathrm{dom}(\nu) - \ulcorner 1 \urcorner \rightarrow \nu(t + \ulcorner 1 \urcorner) - \nu(t) < z)$).

7.2. Metatheorem. *Let* $G(x)$ *(or* $G(i,x)$*) be a real operation, i.e.* $G(x)$ *satisfies* $\mathsf{ZF}^* \vdash \forall x(x \in \mathbf{R} \rightarrow G(x) \in \mathbf{R})$. *Let* $[a,b]$ *be an interval where* a, b *are constructive reals. Now suppose that* $G(x)$ *is constructively uniformly continuous in* $[a,b]$. *Then for any* n, *there is some* m *such that*

$$\mathsf{ZF}^* \vdash (\nu \textit{ is a partition of } [a,b]. \textit{ and } 1/\ulcorner m \urcorner \textit{ is a mesh of } \nu) \,\&$$
$$(\mu \textit{ is a partition of } [a,b] \textit{ and } 1/\ulcorner m \urcorner \textit{ is a mesh of } \mu) \rightarrow$$

$$\Bigg| \sum_{t < \mathrm{dom}(\nu) - \ulcorner 1 \urcorner} \big(\nu(t + \ulcorner 1 \urcorner) - \nu(t)\big) \cdot G\big(\nu(t + \ulcorner 1 \urcorner)\big)$$

$$- \sum_{t < \mathrm{dom}(\mu) - \ulcorner 1 \urcorner} \big(\mu(t + \ulcorner 1 \urcorner) - \mu(t)\big) \cdot G\big(\mu(t + \ulcorner 1 \urcorner)\big) \Bigg| < 1/\ulcorner n \urcorner.$$

Proof. We first have $b - a$ as a constructive real by Metatheorem 6.8. Then by Metatheorem 6.9, there is some p such that $\mathsf{ZF}^* \vdash b - a < \ulcorner p \urcorner$. Since

$G(x)$ is a constructively uniformly continuous real operation, there is some m such that

$$\mathsf{ZF}^*\vdash a \leqslant x \leqslant y \leqslant b \,\&\, |x-y| < 1/{}^{\ulcorner}m^{\urcorner} \rightarrow |G(x)-G(y)| < 1/{}^{\ulcorner}p{\cdot}n^{\urcorner}.$$

To complete our proof of Metatheorem 7.2 it suffices to demonstrate the following (here we understand $t+1$ stand for $t+{}^{\ulcorner}1^{\urcorner}$):

$$\mathsf{ZF}^*\vdash(\nu \textit{ and } \mu \textit{ are partitions of } [a,b] \textit{ with } 1/{}^{\ulcorner}m^{\urcorner} \textit{ as their mesh})\,\&$$
$$i+1\in\mathrm{dom}(\nu) \;\&\; j\in\mathrm{dom}(\mu) \;\&\; \nu(i)\leqslant\mu(j)<\nu(i+1)\leqslant\mu(j+1)$$
$$\rightarrow$$
$$\Big|\sum_{t<i+1}(\nu(t+1)-\nu(t))\cdot G(\nu(t+1)) - \sum_{t<j}(\mu(t+1)-\mu(t))\cdot G(\mu(t+1))$$
$$-(\nu(i+1)-\mu(j))\cdot G(\nu(i+1))\Big| \leqslant (\mu(j)-\mu(0))/(b-a)\cdot{}^{\ulcorner}n^{\urcorner}.$$

We prove this implication by induction on i. Assume the antecedent of this implication and further that for some i^*, j^*,

$$\nu(i^*)\leqslant\mu(j^*)<\nu(i^*+1)\leqslant\nu(i)\leqslant\mu(j^*+1)\leqslant\mu(j)<\nu(i+1)\leqslant\mu(j+1).$$

Since $i^*<i$, hence by the induction hypothesis, we have

$$\text{(7.2.1)}\qquad \Big|\sum_{t<i^*+1}(\nu(t+1)-\nu(t))\cdot G(\nu(t+1)) - \sum_{j<j^*}(\mu(t+1)$$
$$-\mu(t))\cdot G(\mu(t+1)) - (\nu(i^*+1)-\mu(j^*))\cdot G(\nu(i^*+1))\Big|$$
$$\leqslant (\mu(j^*)-\mu(0))/(b-a)\,{}^{\ulcorner}n^{\urcorner}.$$

We have the following equalities and inequalities:

$$\Big|\sum_{t<i+1}(\nu(t+1)-\nu(t))\cdot G(\nu(t+1)) - \sum_{t<j}(\mu(t+1)-\mu(t))\cdot G(\mu(t+1))$$
$$-(\nu(i+1)-\mu(j))\cdot G(\nu(i+1))\Big| =$$
$$=\Big|\sum_{t<i^*+1}(\nu(t+1)-\nu(t))\cdot G(\nu(t+1)) - \sum_{t<j^*}(\mu(t+1)-\mu(t))\cdot G(\mu(t+1))$$
$$-(\nu(i^*+1)-\mu(j^*))\cdot G(\nu(i^*+1))$$
$$+\big[(\nu(i^*+1)-\mu(j^*))\cdot G(\nu(i^*+1))$$
$$+\sum_{i^*+1\leqslant t<i}(\nu(t+1)-\nu(t))\cdot G(\nu(t+1)) + (\mu(j^*+1)-\nu(i))\cdot G(\nu(i+1))$$

$$+ \sum_{j^*+1<t<j} (\mu(t+1)-\mu(t))\cdot G(\nu(i+1)) + (\nu(i+1)-\mu(j))\cdot G(\nu(i+1))]$$

$$-[(\nu(i^*+1)-\mu(j^*))\cdot G(\mu(j^*+1))$$

$$+ \sum_{i^*+1<t<i} (\nu(t+1)-\nu(t))\cdot G(\mu(j^*+1)) + (\mu(j^*+1)-\nu(i))\cdot G(\mu(j^*+1))$$

$$+ \sum_{j^*+1<t<j} (\mu(t+1)-\mu(t))\cdot G(\mu(t+1)) + (\nu(i+1)-\mu(j))\cdot G(\nu(i+1))]\Big|$$

$$< (\mu(j^*)-\mu(0))/(b-a)\cdot \ulcorner n \urcorner + \quad \text{(by 7.2.1)}$$

$$+\Big|(\nu(i^*+1)-\mu(j^*))\cdot(G(\nu(i^*+1))-G(\mu(j^*+1)))$$

$$+ \sum_{i^*+1<t<i} (\nu(t+1)-\nu(t))\cdot(G(\nu(t+1))-G(\mu(j^*+1)))$$

$$+(\mu(j^*+1)-\nu(i))\cdot(G(\nu(i+1))-G(\mu(j^*+1)))$$

$$+ \sum_{j^*+1<t<j} (\mu(t+1)-\mu(t))$$

$$\cdot(G(\nu(i+1))-G(\mu(t+1)))+(\nu(i+1)-\mu(j))\cdot 0\Big|$$

$$< (\mu(j^*)-\mu(0))/(b-a)\cdot \ulcorner n \urcorner$$

$$+\Big[(\nu(i^*+1)-\mu(j^*)) + \sum_{i^*+1<t<i} (\nu(t+1)-\nu(t)) + (\mu(j^*+1)-\nu(i))$$

$$+ \sum_{j^*+1<t<j} (\mu(t+1)-\mu(t))\Big]\cdot(1/\ulcorner p\cdot n \urcorner)$$

$$= (\mu(j)-\mu(0))/(b-a)\cdot \ulcorner n \urcorner.$$

This completes the induction step; then Metatheorem 7.2 follows.

7.3. Discussion. In Metatheorem 7.2 we have not assumed that $G(x)$ is a constructive real operation. In fact, the non-constructive real operation $G(x)$ satisfying $\mathsf{ZF}^*\vdash \forall x(G(x)=\infty)$ is constructively uniformly continuous in $[a,b]$ and hence for this $G(x)$, Theorem 7.2 holds. On the other hand, Theorem 7.2 does not hold if $b-a$ is not a constructive real.

7.4. Definition. By Theorem 2.9, we have $\mathsf{ZF}\vdash\mathbf{R}$ is Archimedean, namely,

$$\mathsf{ZF}\vdash x\in\mathbf{R}\rightarrow \mathsf{E}y(y\in\omega \,\&\, x<y).$$

Thus we have

$$\mathsf{ZF}^*\vdash x\in\mathbf{R}\rightarrow \mathsf{E}y(y\in\omega \,\&\, x\cdot\infty \leqslant y+\ulcorner 1 \urcorner).$$

Let

$$D(x) \overset{\text{Df}}{==} \{y \in \omega : x \in \mathbf{R} \to x \leqslant (1/\infty)\cdot(y + \ulcorner 1 \urcorner)\}.$$

Then $\mathsf{ZF}^* \vdash \varnothing \neq D(x) \subset \omega$. Hence

$$\mathsf{ZF}^* \vdash \mathrm{E}!y(y \in D(x) \,\&\, \forall z(z \in y \to z \notin D(x))).$$

This theorem defines an operation which we denote by $N(x)$ so that $\mathsf{ZF}^* \vdash N(x) \in \omega \,\&\, N(x) \in D(x)$. From now on we use δ to denote $1/\infty$ as in Metadefinition 6.4, and have the following.

7.5. Theorem. $\mathsf{ZF}^* \vdash x \in \mathbf{R} \,\&\, 0 < x \to \delta \cdot N(x) < x \,\&\, x \leqslant \delta \cdot (N(x) + \ulcorner 1 \urcorner)$.

7.6. Definition. Let $[a,b]$ be an interval such that

$$\mathsf{ZF}^* \vdash a \in \mathbf{R} \,\&\, b \in \mathbf{R} \,\&\, b - a > 0.$$

Let $G(x)$ (or $G(i,x)$) be a real operation. We define

$$\int_a^b G(x)\mathrm{d}x = \Bigg(\sum_{j < N(b-a)} \delta \cdot G\big(a + \delta \cdot (j + \ulcorner 1 \urcorner)\big)\Bigg) + (b - a - \delta \cdot N(b-a)) \cdot G(b).$$

7.7. Metatheorem. *If $[a,b]$ is an interval where a, b are constructive reals with $\mathsf{ZF}^* \vdash a < b$, and if $G(x)$ (or $G(i,x)$) is a constructive real operation which is constructively uniformly continuous in $[a,b]$, then $\int_a^b G(x)\mathrm{d}x$ is a constructive real.*

Proof. Let n be any intuitive natural number. Then by Metatheorem 7.2, there is some m such that

$$\mathsf{ZF}^* \vdash \big(\nu \textit{ and } \mu \textit{ are partitions of the interval } [a,b] \textit{ with } 1/\ulcorner m \urcorner \textit{ as their mesh}\big)$$
$$\to \Bigg| \sum_{t < \mathrm{dom}(\nu) - 1} (\nu(t+1) - \nu(t)) \cdot G(\nu(t+1)) - \sum_{t < \mathrm{dom}(\mu) - 1} (\mu(t+1) - \mu(t)) \cdot G(\mu(t+1)) \Bigg| < 1/\ulcorner n \urcorner.$$

Since $b - a$ and $\ulcorner m \urcorner$ are constructive reals, we can use Theorem 6.9 to find an intuitive natural number r such that

$$\mathsf{ZF}^* \vdash (b-a) \cdot \ulcorner m \urcorner < \ulcorner r \urcorner \,\&\, (b-a)/\ulcorner r \urcorner < 1/\ulcorner m \urcorner.$$

Let us denote $(b-a)/\ulcorner r \urcorner$ by d. Then applying Metatheorem 7.2 we can show that

$$\mathsf{ZF}^*\vdash\Big|\sum_{j<N(b-a)}\delta\cdot G(a+\delta\cdot(j+1))+(b-a-\delta\cdot N(b-a))\cdot G(b)$$

$$-\sum_{j<\ulcorner r\urcorner}d\cdot G(a+d\cdot(j+1))\Big|<1/\ulcorner n\urcorner,$$

i.e.

$$\Big|\int_a^b G(x)\,\mathrm{d}x-\sum_{j<\ulcorner r\urcorner}d\cdot G(a+d\cdot(j+1))\Big|<1/\ulcorner n\urcorner.$$

We note that in showing the above inequality we should use the fact that

$$\mathsf{ZF}^*\vdash(b-a)-\delta\cdot N(b-a)<\delta<1/\ulcorner m\urcorner\,\&\,d<1/\ulcorner m\urcorner.$$

Since a, b, d, $\ulcorner r\urcorner$ are all constructive reals and since $G(x)$ is assumed to be a constructive real operation, hence by induction on the intuitive natural number r, we can show that $\sum_{j<\ulcorner r\urcorner}d\cdot G(a+d\cdot(j+1))$ is a constructive real. Let us denote $\sum_{j<\ulcorner r\urcorner}d\cdot G(a+d\cdot(j+1))$ by c. We have already established that

$$\mathsf{ZF}^*\vdash\Big|\int_a^b G(x)\mathrm{d}x-c\Big|<1/\ulcorner n\urcorner.$$

Hence $\int_a^b G(x)\mathrm{d}x$ is a constructive real. Metatheorem 7.7 is thus proved.

7.8. Discussion. Let us consider the expression $\int_z^y G(x)\mathrm{d}x$ where y, z are variables. We can extend the definition of 7.6 by putting $a=z$, $b=y$,

$$\int_z^y G(x)\mathrm{d}x=-\int_y^z G(x)\mathrm{d}x,\qquad \int_y^y G(x)\mathrm{d}x=0.$$

Then we have

$$\mathsf{ZF}^*\vdash y\in\mathbf{R}\,\&\,z\in\mathbf{R}\to\int_z^y G(x)\mathrm{d}x\in\mathbf{R}.$$

And $\int_z^y G(x)\mathrm{d}x$ becomes a real operation of the two variables y, z. By the Theorem 7.7, it is also a constructive real operation provided that $G(x)$ is constructively uniformly continuous in every interval $[a,b]$ for constructive reals a, b. We should note that in defining the integral $\int_z^y G(x)\mathrm{d}x$ we have introduced an operation $N(x)$ satisfying

$$\mathsf{ZF}^*\vdash x\in\mathbf{R}\,\&\,x>0\to N(x)<x\cdot\infty\,\&\,x\cdot\infty\leqslant N(x)+\ulcorner 1\urcorner.$$

$N(x)$ is by no means a constructive real operation. Thus we have here an example $\int_z^y G(x)\mathrm{d}x$ of constructive real operation which is defined by using some non-constructive element, namely $N(x)$. However, our definition of the integral $\int_z^y G(x)\mathrm{d}x$ has the advantage that it is in terms of plain summation without resorting to the elusive concept of limit.

7.9. Metatheorem. *Let* $[a,b]$ *be an interval where* a,b *are constants such that* $\mathsf{ZF}^*\vdash a\in\mathbf{R}\,\&\,b\in\mathbf{R}\,\&\,a<b$. *Let* $G(i,x)$ *be a real operation satisfying*

(1) *for the interval* $[a,b]$ *the family of sequences of* $G(i,x)$ *is constructively uniformly convergent as defined in Definition* 6.13; *and*

(2) *for any constructive natural number* r^*, *the operation* $G(r^*,x)$ *is constructively uniformly continuous in* $[a,b]$ *as defined in Definition* 6.13. *Then we have*

(3) *the sequence of functions of* $G(i,x)$ *are all constructively uniformly continuous in* $[a,b]$ *independent of* i;

(4) *if* $b-a$ *is a constructive real, then the sequence of integrals* $\int_a^b G(i,x)\mathrm{d}x$ *is constructively convergent; in particular, for any* n, *there is some* m *such that*

$$\mathsf{ZF}^*\vdash i\in\omega\,\&\,\ulcorner m\urcorner\leqslant i\rightarrow\left|\int_a^b G(\infty,x)\mathrm{d}x-\int_a^b G(i,x)\mathrm{d}x\right|$$
$$<1/\ulcorner n\urcorner.$$

(5) *if further* a, b *are constructive reals and* $G(i,x)$ *is a constructive real operation, then for any constructive natural number* i^*, $\int_a^b G(i^*,x)\mathrm{d}x$ *is a constructive real; consequently the integral* $\int_a^b G(\infty,x)\mathrm{d}x$ *is also a constructive real.*

Proof. Applying Metatheorem 6.16, we get (3) immediately from (1) and (2). To prove (4) we let n be any intuitive natural number. Since $b-a$ is supposed to be a constructive real, hence there is some p such that $\mathsf{ZF}^*\vdash b-a<\ulcorner p\urcorner$. By (1), there is some m such that

$$\mathsf{ZF}^*\vdash x\in[a,b]\,\&\,i,j\in\omega\,\&\,\ulcorner m\urcorner<i\,\&\,\ulcorner m\urcorner<j\rightarrow$$
$$|G(i,x)-G(j,x)|<1/\ulcorner p\cdot n\urcorner.$$

Thus we have

$$\mathsf{ZF}^*\vdash i,j\in\omega\,\&\,\ulcorner m\urcorner<i\,\&\,\ulcorner m\urcorner<j\rightarrow\forall x(x\in[a\cdot b]\rightarrow$$
$$|G(i,x)-G(j,x)|<1/\ulcorner p\cdot n\urcorner);$$

and

$$ZF^* \vdash i,j \in \omega \& \ulcorner m \urcorner < i \& \ulcorner m \urcorner < j$$

$$\to \left| \int_a^b G(i,x)\mathrm{d}x - \int_a^b G(j,x)\mathrm{d}x \right|$$

$$= \Bigg| \sum_{k < N(b-a)} \delta \cdot G(i, a+\delta\cdot(k+1)) + (b-a-\delta\cdot N(b-a))\cdot G(i,b)$$

$$- \sum_{k < N(b-a)} \delta G(j, a+\delta\cdot(k+1)) - (b-a-\delta\cdot N(b-a))\cdot G(j,b) \Bigg|$$

$$= \Bigg| \sum_{k < N(b-a)} \delta\cdot(G(i,a+\delta\cdot(k+1)) - G(j,a+\delta\cdot(k+1))) +$$

$$(b-a-\delta\cdot N(b-a))(G(i,b) - G(j,b)) \Bigg|$$

$$\leqslant \sum_{k<N(b-a)} \delta \cdot 1/\ulcorner p\cdot n \urcorner + (b-a-\delta\cdot N(b-a))\cdot 1/\ulcorner p\cdot n \urcorner$$

$$= (b-a)/\ulcorner p\cdot n \urcorner < 1/\ulcorner n \urcorner.$$

Hence the sequence of integrals $\int_a^b G(i,x)\mathrm{d}x$ is constructively convergent. This is the first part of (4); the second part follows immediately.

To prove (5) we let n be any intuitive natural number. By (4), there is some m such that

$$\left| \int_a^b G(\infty,x)\mathrm{d}x - \int_a^b G(\ulcorner m \urcorner,x)\mathrm{d}x \right| 1/\ulcorner n \urcorner$$

Since a, b, $\ulcorner m \urcorner$ are all constructive reals and $G(i,x)$ is assumed to be a constructive real operation, and since by (2), $G(\ulcorner m \urcorner, x)$ is constructively uniformly continuous in [a,b] hence by Metatheorem 7.7, $\int_a^b G(\ulcorner m \urcorner,x)\mathrm{d}x$ is a constructive real. Therefore, $\int_b^a G(\infty,x)\mathrm{d}x$ is a constructive real. This proves Metatheorem 7.9.

7.10. Definition. Let $G(x)$ be any real operation and c be a constant such that $ZF^* \vdash c \in \mathbf{R}$. Then we define

$$\mathrm{d}G(c)/\mathrm{d}x = (G(c+\delta) - G(c))/\delta$$

and call $\mathrm{d}G(c)/\mathrm{d}x$ the derivative of $G(x)$ at $x=c$. The derivative $\mathrm{d}G(c)/\mathrm{d}x$ is said to be *constructively smooth* if for some constructive reals a, b,

$$ZF^* \vdash a \leqslant c-\delta < c < c+\delta \leqslant b$$

and $\mathrm{d}G(x)/\mathrm{d}x$ is constructively uniformly smooth in $[a,b]$ in the sense explained as follows. Let $[a,b]$ be an interval with $\mathsf{ZF}^*\vdash a,b\in\mathbf{R}$. We say that $\mathrm{d}G(x)/\mathrm{d}x$ is *constructively uniformly smooth* in $[a,b]$ if for any n, there is some m such that

$$\mathsf{ZF}^*\vdash x,y,z\in[a,b]\,\&\,x\neq y\,\&\,x\neq z\,\&\,|x-y|<1/\ulcorner m\urcorner\,\&\,|x-z|<1/\ulcorner m\urcorner\rightarrow |(G(x)-G(y))/(x-y)-(G(x)-G(z))/(x-z)|<1/\ulcorner n\urcorner.$$

7.11. Metatheorem. *If $G(x)$ is a constructive real operation, if c is a constructive real and if $\mathrm{d}G(c)/\mathrm{d}x$ is constructively smooth, then $\mathrm{d}G(c)/\mathrm{d}x$ is a constructive real.*

7.12. Metatheorem. *If $G(x)$ is a constructive real operation and if $\mathrm{d}G(x)/\mathrm{d}x$ is constructively uniformly smooth in an interval $[a-\delta,b+\delta]$ where a, b are constructive reals, then the operation $\mathrm{d}G(x)/\mathrm{d}x$, i.e. $(G(x+\delta)-G(x))/\delta$ is a constructive real operation of x in $[a,b]$.*

7.13. Metatheorem. *Suppose $G(x)$ is a real operation and $\mathrm{d}G(x)/\mathrm{d}x$ is constructively uniformly smooth in an interval $[a-\delta,b+\delta]$ where a, b are constructive reals. Further suppose $\mathrm{d}G(x)/\mathrm{d}x$ is infinitesimal identically in $[a,b]$, i.e. for any n,*

$$\mathsf{ZF}^*\vdash a\leqslant x\leqslant b\rightarrow|(G(x)-G(x+\delta))/\delta|<1/\ulcorner n\urcorner.$$

Then $G(x)-G(a)$ is infinitesimal identically in $[a,b]$.

Sketch of an informal proof. Since $b-a$ is a constructive real, there is some m such that $|b-a|<\ulcorner m\urcorner$. Given any n, there is some p such that $|\mathrm{d}G(x)/\mathrm{d}x-(G(x)-G(z))/(x-z)|<1/\ulcorner 2\cdot m\cdot n\urcorner$ whenever $x,z\in[a,b]$ and $|x-z|<1/\ulcorner p\urcorner$. Let us denote $(x-a)/\ulcorner m\cdot p\urcorner$ by A. Then $A\leqslant(b-a)/\ulcorner m\cdot p\urcorner<1/\ulcorner p\urcorner$. Hence for $i<\ulcorner m\cdot p\urcorner$,

$$|\mathrm{d}G(a+i\cdot A)/\mathrm{d}x+(G(a+i\cdot A)-G(a+(i+1)\cdot A))/A| <1/\ulcorner 2\cdot m\cdot n\urcorner.$$

Since $\mathrm{d}G(a+i\cdot A)/\mathrm{d}x$ is infinitesimal, hence we have

$$|G(a+i\cdot A)-G(a+(i+1)\cdot A)|\leqslant A/\ulcorner m\cdot n\urcorner.$$

Then

$$|G(x)-G(a)|=|G(a+\ulcorner m\cdot p\urcorner\cdot A)-G(a)|\leqslant \sum_{i<\ulcorner m\cdot p\urcorner}|G(a+(i+1)\cdot A)-G(a+i\cdot A)|\leqslant\ulcorner m\cdot p\urcorner\cdot A/\ulcorner m\cdot n\urcorner$$
$$=(x-a)/\ulcorner m\cdot n\urcorner\leqslant(b-a)/\ulcorner m\cdot n\urcorner<1/\ulcorner n\urcorner.$$

Hence $G(x)-G(a)$ is infinitesimal.

7.14. Metatheorem. *Let* $[a,b]$ *be an interval where* a, b *are constructive reals and* $G(i,x)$ *is a constructive real operation satisfying*

(1) *for any* n, *there is some* m *such that*

$$\mathsf{ZF}^*\vdash x,y\in[a-\delta,b+\delta]\,\&\,i,j\in\omega\,\&\,\ulcorner m\urcorner<i\,\&\,\ulcorner m\urcorner<j\to|(G(i,x)-G(i,y))/(x-y)-(G(j,x)-G(j,y))/(x-y)|<1/\ulcorner n\urcorner;$$

(2) *for any constructive natural number* $i^*, \mathrm{d}G(i^*,x)/\mathrm{d}x$ *is constructively uniformly smooth in* $[a-\delta,b+\delta]$.

Then we have

(3) $\mathrm{d}G(i,x)/\mathrm{d}x$ *is constructively uniformly smooth in* $[a-\delta,b+\delta]$ *independent of* i, *i.e. for any* n, *there is some* m *such that*

$$\mathsf{ZF}^*\vdash i\in\omega\,\&\,a-\delta\leqslant x,y,z\leqslant b+\delta\,\&\,0<|x-y|<1/\ulcorner m\urcorner\,\&\,0<|x-z|<1/\ulcorner m\urcorner\to|(G(i,x)-G(i,y))/(x-y)-(G(i,x)-G(i,z))/(x-z)|<1/\ulcorner n\urcorner$$

(4) *for any constructive natural number* $\ulcorner m\urcorner$, *we have* $\mathrm{d}G(\ulcorner m\urcorner,x)/\mathrm{d}x$ *as a constructive real operation of* x *in* $[a,b]$

(5) *for any* n, *there is some* m *such that*

$$\mathsf{ZF}^*\vdash a\leqslant x\leqslant b\to|\mathrm{d}G(\infty,x)/\mathrm{d}x-\mathrm{d}G(\ulcorner m\urcorner,x)/\mathrm{d}x|<1/\ulcorner n\urcorner.$$

Hence $\mathrm{d}G(\infty,x)/\mathrm{d}x$ *is a constructive real operation of* x *in* $[a,b]$.

7.15. Metatheorem. *Suppose* $G(x)$ *is a constructive real operation constructively uniformly continuous in an interval* $[a,b+\delta]$, a, b *being constructive reals. Then* $\mathrm{d}(\int_a^x G(t)\mathrm{d}t)/\mathrm{d}x$ *is a constructive real operation and for any intuitive natural number* n,

$$\mathsf{ZF}^*\vdash a\leqslant x\leqslant b\to\left|\mathrm{d}\left(\int_a^x G(t)\mathrm{d}t\right)/\mathrm{d}x-G(x)\right|<1/\ulcorner n\urcorner.$$

8. Concluding remarks.

(1) Here we have two formal theories ZF and ZF* which are only two examples of theory that can formalize number theory, theory of analysis and set theory. These two examples show that we have considerable freedom in creating ideal objects in a theory without affecting its usefulness for actual counting or measurement if we have means by which to distinguish between what is constructive and what is not.

(2) However, whether an object in a theory is constructive in character or not can not be decided within the object theory itself but must use arguments in the metatheory.
(3) The number theory in a metatheory is the essential part of the metatheory because a metatheory is a study of concrete symbols and formulas and this can be replaced by a study of the Gödel numbers representing the formulas and symbols. In order to avoid worrying about the problem whether a number is constructive so that it can represent a concrete formula we had better to adopt an intuitionistic number theory in the metatheory. In such an intuitionistic number theory as developed, say by KLEENE in his (1952) every provable sentence can be guaranteed to be intuitionistically true.
(4) In this way we reduce the study of any formal mathematical theory to the study of an intuitionistic number theory which is not complete. Thus essentially the mathematical science progresses only by discovering new theorems in a formal intuitionistic number theory or by making suitable consistent extensions of this theory which can not be done by a computer.

Acknowledgment

The author wishes to express his thanks to Dr. Perry Smith for initiating him into the study of non-standard analysis. The author also thanks Dr. Smith for his criticism, corrections and comments made when the manuscript of this paper was prepared.

References

KARP, C.
[1967] A proof of the relative consistency of the continuum hypothesis, in: *Sets, Models and Recursion Theory*, edited by J. N. Crossley (North-Holland, Amsterdam), pp. 1–32.

KEISLER, H. J.
[1976] *Foundations of Infinitesimal Calculus*, (Prindle, Weber & Schmidt, Boston, MA).

KLEENE, S. C.
[1952] *Introduction to Metamathematics*, (North-Holland, Amsterdam).
[1962] Disjunction and existence under implication in elementary intuitionistic formalism, *J. Symbolic Logic*, **27**, 11–18.
[1976] The work of Kurt Gödel, *J. Symbolic Logic*, **41**, 761–778.

ROBINSON, A.
[1966] *Non-standard Analysis* (North-Holland, Amsterdam).

Liu, S. C.
[1975] A theory of first order logic with description, *Bull. Inst. Math., Academia Sinica*, **2**, 43–65.
Rosser, B.
[1936] Extensions of some theorems of Gödel and Church, *J. Symbolic Logic*, **1**, 87–91.

J. Barwise, H. J. Keisler and K. Kunen, eds., *The Kleene Symposium*
©North-Holland Publishing Company (1980) 415–421

Covering 2^ω with ω_1 Disjoint Closed Sets

Arnold W. Miller
Department of Mathematics, University of Wisconsin, Madison, WI 53706, U.S.A.

Dedicated to Professor S. C. Kleene on the occasion of his 70th birthday

Abstract: It is shown that 2^ω is the ω_1 union of meager sets does not imply 2^ω is the ω_1 union of disjoint non-empty closed sets and the latter does not imply CH.

In HAUSDORFF (1934) he showed that 2^ω is the ω_1 union of strictly increasing G_δ sets. It follows that 2^ω is the ω_1 union of disjoint non-empty $F_{\sigma\delta}$ sets. FREMLIN and SHELAH (1980) proved the following theorem.

Theorem 1. *The following are equivalent.*
(1) 2^ω *is the* ω_1 *union of strictly increasing* F_σ *sets.*
(2) 2^ω *is the* ω_1 *union of meager sets.*
(3) 2^ω *is the* ω_1 *union of disjoint non-empty* G_δ *sets.*

Proof.
(3)⇒(2) see FREMLIN and SHELAH (1980).
(2)⇒(1) Every meager set is contained in a meager F_σ set.
(1)⇒(3) Cover 2^ω with closed sets C_α for $\alpha<\omega_1$ so that no countable subcollection covers. Note that $C_\alpha - \bigcup\{C_\beta\colon \beta<\alpha\}$ are disjoint G_δ sets.

Theorem 2 (Luzin, see KURATOWSKI (1958a, p.348)). *Every* F_σ ($G_{\delta\sigma}$) *set in* 2^ω *can be written as the disjoint countable union of closed* (G_δ) *sets.*

Thus the only remaining case of disjoint ω_1 coverings of 2^ω by Borel sets is:

(C) 2^ω is the ω_1 union of non-empty disjoint closed sets.

Remark. By a theorem of Sierpinski (see Kuratowski (1958b, p.173)) the open unit interval cannot be written as the disjoint countable union of closed (in the closed unit interval) sets. Nevertheless (C) is equivalent to the same statement with 2^ω replaced by any uncountable Polish space.

Theorem 3. 2^ω *can be partitioned into* ω_1 *disjoint non-empty closed sets iff some uncountable Polish space can be iff all uncountable Polish spaces can be.*

Proof: If some uncountable Polish space can be partioned, then ω^ω can be, since every such space is the continuous image of ω^ω. Suppose $\omega^\omega = \bigcup\{C_\alpha: \alpha<\omega_1\}$ where the C_α are nonempty disjoint closed sets. By the proof of Lemma 7 we may assume each C_α is nowhere dense. It is easy to build $P \subseteq \omega^\omega$ compact perfect so that $\exists C_{\alpha_n}$ for $n<\omega$ such that each C_{α_n} is nowhere dense in P and $\bigcup\{C_{\alpha_n}: n<\omega\}$ is dense in P. P cannot be covered by countably many of the C_α's since then $P \cap \bigcup\{C_{\alpha_n}: n<\omega\}$ would be a dense meager (in P) G_δ set. Hence we conclude 2^ω can be partitioned. Next we show the unit interval $[0,1]$ can be partitioned. Assume $2^\omega = \bigcup\{C_\alpha: \alpha<\omega_1\}$ where the C_α are disjoint nowhere dense closed sets. By a back and forth argument it is not hard to show that for any two dense countable subsets of 2^ω there is a homeomorphism of 2^ω taking one to the other. Let E be $\{x\in 2^\omega: \exists n \forall m>n\ x(m)=1 \text{ or } \forall m>n\ x(m)=0\}$. We may assume that for every $\alpha<\omega_1$ $|C_\alpha \cap E| \leq 1$. Define the map F from 2^ω to $[0,1]$ by

$$F(x)=\sum\left\{\frac{x(n)}{2^{n+1}}: n<\omega\right\}.$$

Let $D_\alpha = F''C_\alpha$. Hence by lumping together the distinct pairs of D_α's which intersect we partition $[0,1]$. Now let X be any uncountable Polish space, we may assume X has no isolated points. Embed X into $[0,1]^\omega$, and if some projection of X contains an interval, then decompose that interval and pull the decomposition back to X. Hence we may assume X is zero dimensional. Thus either X contains a clopen set homeomorphic to 2^ω or it doesn't in which case X is homeomorphic to ω^ω and in either case we are done.

The following theorem was first proved by J. Baumgartner (unpublished) and rediscovered by the author and others.

Theorem 4. (C)$\not\Rightarrow$CH.

Proof. Let M be a model of $\neg$CH. Construct an ω_1 length c.c.c. Solovay and Tennenbaum (1971) extension. For $X \subseteq 2^\omega$ define the partial order $\mathbb{P}(X)$. Conditions are finite consistent sets of sentences of the form "$[s]\cap C_n=\varnothing$" or "$x\in C_n$" where $n<\omega, x\in X, s\in 2^{<\omega}$. Then $F=\bigcup\{C_n: n<\omega\}$ will be a meager (in fact measure zero) F_σ set covering X. (See Miller (1979) for similar arguments.) Iterate ω_1 times to get M_α for $\alpha \leq \omega_1$ so that $M_{\alpha+1}$ is gotten by forcing with $\mathbb{P}(2^\omega - \bigcup\{F^\beta: \beta<\alpha\})$ in M_α

creating the F_σ set F^α. An easy density argument shows that the F^α's are disjoint. By c.c.c. $M_{\omega_1} \models$ "$2^\omega = \bigcup\{F^\beta: \beta<\omega_1\}$." Note that in 2^ω any F_σ set is the countable union of disjoint closed sets. Since $F = \bigcup\{C_n: n<\omega\}$ implies

$$F = \bigcup_n C_n - \Big(\bigcup_{m<n} C_m\Big),$$

it is enough to see this for F_σ sets of the form $C \cap G$ where C is closed and G is open, but G is the disjoint union of countably many clopen sets.

Note that in the above model 2^ω is the ω_1 union of measure zero sets. Is this implied by (C)? The answer is no by the following theorem of STERN (1977), also discovered later but independently by K. Kunen.

Theorem 5. (C) *holds in any random real extension of a model of* CH.

Proof. Let $(\mathbb{B},\mu)$ be any measure algebra in the ground model M. Every element of 2^ω in $M^{\mathbb{B}}$ is random with respect to some Borel measure on 2^ω in M. (For any x such that $[\![x\in 2^\omega]\!]=1$ consider the Borel measure $\nu(B) = \mu[\![x\in B]\!]$.) Every Borel measure ν on 2^ω is regular (see ROYDEN (1968, p. 305)), so for any $E\subseteq 2^\omega$ Borel,

$$\nu(E) = \sup\{\nu(C): C\subseteq E \text{ and } C \text{ is closed}\}$$

and for any closed C,

$$\nu(C) = \inf\{\nu(D): C\subseteq D \text{ and } D \text{ is clopen}\}.$$

Since M models CH there are at most ω_1 Borel measures on 2^ω in M, so it is easy to construct disjoint F_σ sets F^α for $\alpha<\omega_1$ so that for every Borel measure ν in $M, \exists\alpha<\omega_1$ so that $\nu(\bigcup\{F^\beta: \beta<\alpha\})=1$.

Theorem 6. 2^ω *is the* ω_1 *union of meager sets does not imply* (C).

Proof. Any $C\subseteq 2^\omega$ closed is coded by a tree $T\subseteq 2^{<\omega}$ whose set of infinite branches

$$[T] = \{x\in 2^\omega: \forall n<\omega \quad x\restriction n\in T\}$$

is C. Perfect set forcing (SACKS, 1971) corresponds to forcing with perfect trees $T\subseteq 2^{<\omega}$ (perfects means $\forall s\in T$ there are incompatible extensions of s in T). $T\leqslant S$ iff $T\subseteq S$. Given $C_\alpha: \alpha<\omega_1$ disjoint non-empty closed subsets of 2^ω, $\mathbb{P}$ will be a suborder of perfect set forcing defined as follows:

$$T\in\mathbb{P} \text{ iff } T \text{ is perfect and for every } \alpha<\omega_1, C_\alpha \text{ is meager in } [T].$$

C meager in $[T]$ iff $\forall s\in T\,\exists t\supseteq s\ t\in T$ and $[T_t]\cap C=\varnothing$, where $T_t=\{r\in T: r\subseteq t \text{ or } t\subseteq r\}$. This modification is similar to that of Shelah.

Lemma 7. *$\mathbb{P}$ is not empty.*

Proof. For each $\alpha<\omega_1$ choose $x_\alpha\in C_\alpha$. Let $T=\{s\in 2^{<\omega}$: for uncountably many $\alpha, s\subseteq x_\alpha\}$. Then $T\in\mathbb{P}$.

Just as in perfect set forcing if G is $\mathbb{P}$-generic, then $x=\bigcup\bigcap G$ is an element of 2^ω and $G=\{T\in\mathbb{P}: x\in[T]\}$. Note that for any $\alpha<\omega_1$, $\Vdash$"$x\notin \bar{C}_\alpha$" is the closed set in the extension with the same code as C_α), because $\forall T\in\mathbb{P}\,\exists t\in T[T_t]\cap C_\alpha=\emptyset$, so $[T_t]\Vdash$"$x\notin\bar{C}_\alpha$."

Starting with M a model of CH an ω_2 iteration with countable support (as was done in LAVER (1976)) will be used to obtain a model N, where on each step some sequence of disjoint non-empty closed sets will be taken care of with the corresponding order $\mathbb{P}$. Provided sufficient care is taken, N will then model $\neg$(C). It will then suffice to show that $N\models$"$2^\omega=\bigcup\{C: C$ is closed nowhere dense and coded in $M\}$." For expository purposes we first show that the above statement holds when $N=M[G]$ for G $\mathbb{P}$-generic over M.

Lemma 8. *Let $T\in\mathbb{P}$ and $F\subseteq[T]$ finite.*

(a) If $T\Vdash$"$W_{i<N}\Theta_i$" *where $N<\omega$, then*

$$\exists S\leqslant T\, F\subseteq[S]\,\exists G\subseteq N\left[\text{card}\, G=\text{card}\, F \text{ and } S\Vdash\text{“}\underset{i\in G}{W}\Theta_i\text{”}\right].$$

(b) If $T\Vdash$"$\tau\in M$", then $\exists S\leqslant T\; F\subseteq[S]\;\exists G\in M$ countable and $S\Vdash$"$\tau\in G$".

Proof. Choose $n<\omega$ so that for every $x,y\in F$ $(x\neq y\Rightarrow x\restriction n\neq y\restriction n)$. For $x\in F$ let

$$R_x=\{t\in T\text{: }\exists m\geqslant n\; t=x\restriction m^\frown\langle 1-x(m)\rangle\}$$

and $R=\bigcup\{R_x\text{: } x\in F\}$. Choose $T'\leqslant T$ so that $R\subseteq T'$ and for all $s\in R$ $\exists m<N\; T'_s\Vdash$"Θ_m"(for (b): $\forall s\in R\,\exists x_s\in M\; T'_s\Vdash$"$\tau=x_s$" then let $S=T'$ and $G=\{x_s\text{: } s\in R\}$). Since $N<\omega$ $\forall x\in F\,\exists m_x<N\,\exists R'_x\subseteq R_x$ infinite so that for all $s\in R'_x$ $T'_s\Vdash$"Θ_{m_x}". Let $G=\{m_x\text{: } x\in F\}$ and $S=\bigcup\{T'_s: s\in\bigcup\{R'_x: x\in F\}\}$.

The stem of T is the unique $s\in T$ such that $T_s=T$ and $s^\frown\langle 0\rangle, s^\frown\langle 1\rangle\in T$. The nth level of T (Lev_n (T)) is defined by induction on $n<\omega$. Lev_0 $(T)=\{\text{stem of } T\}$.

$$\text{Lev}_{n+1}(T)=\{\text{stem of } T_{s^\frown\langle i\rangle}\text{: } s\in\text{Lev}_n(T) \text{ and } i=0,1\}.$$

For any $s\in T$ define x_s^T to be the lexicographical least element of $[T_s]$.

Definition. $T \leqslant^n S$ iff

(a) $T \leqslant S$ and $\mathrm{Lev}_n(T) = \mathrm{Lev}_n(S)$.

(b) $\forall t \in \mathrm{Lev}_n(S)\, x_t^S \in [T]$.

(c) $\forall t \in \mathrm{Lev}_n(S)$ if $x_t^S \in C_\alpha$ (α is necessarily unique if it exists, since the C_α are disjoint), then $\exists s \supseteq t\ s \in \mathrm{Lev}_{n+1}(T)$ such that $[T_s] \cap C_\alpha = \varnothing$.

Lemma 9. *If for each* $n<\omega$ $\ T^{n+1} \leqslant^n T^n$, *then* $\bigcap \{T_n\colon n<\omega\} = T \in \mathbb{P}$.

Proof. Since $\forall n \forall m [m \geqslant n \to \mathrm{Lev}_n(T^m) = \mathrm{Lev}_n(T)]$, T is perfect. Suppose for some $\alpha < \omega_1$ and $s \in T, [T_s] \subseteq C_\alpha$. Choose $n<\omega$ so that $s \subseteq t \in \mathrm{Lev}_n(T)$. By (b) $x_t^{T^n} \in [T]$, so $x_t^{T^n} \in C_\alpha$. But by (c) $\exists r \in \mathrm{Lev}_{n+1}(T^{n+1}) = \mathrm{Lev}_{n+1}(T)$ such that $[T_r^{n+1}] \cap C_\alpha = \varnothing$, contradiction.

Lemma 10. *Let* $T \in \mathbb{P}$ *and* $n<\omega$.

(a) *If* $T \Vdash$"$W_{i<N}\Theta_i$" *where* $N<\omega$, *then* $\exists S \leqslant^n T \exists G \subseteq N$ $\mathrm{card}\, G \leqslant 2^{n+1}$ *and* $S \Vdash$"$W_{i \in G}\Theta_i$".

(b) *If* $T \Vdash$"$\tau \subseteq M$ *is countable*", *then* $\exists S \leqslant^n T \exists G \in M$ *countable and* $S \Vdash$"$\tau \subseteq G$".

Proof. (a) Let $F = \{x_s^T\colon s \in \mathrm{Lev}_{n+1}(T)\}$. Applying Lemma 8(a) get $R \leqslant T$ with $F \subseteq [R], G \subseteq N, \mathrm{card}\, G \leqslant 2^{n+1}, R \Vdash$"$W_{i \in G}\Theta_i$". Since $F \subseteq [R]$ $\mathrm{Lev}_n(R) = \mathrm{Lev}_n(T)$. Let $D = \bigcup \{C_\alpha\colon F \cap C_\alpha \neq \varnothing\}$. Since this is a finite union D is nowhere dense in $[R]$. $\forall s \in \mathrm{Lev}_n(R)$ find $t_s \in R$ such that $t \supseteq s^\frown\langle 1\rangle$ and $[t] \cap D = \varnothing$. Let $S = \bigcup \{R_{s^\frown\langle 0\rangle}, R_{t_s} : s \in \mathrm{Lev}_n(R)\}$.

(b) Let $T_0 = T$. Using Lemma 8(b) and the argument above, build a sequence $T_{m+1} \leqslant^m T_m, G_m \in M$ countable for $m<\omega$ such that $T_m \Vdash$" the mth element of τ is in G_m." Then by Lemma 9 $S = \bigcap_{m<\omega} T_m \in \mathbb{P}$ and $S \Vdash$"$\tau \subseteq \bigcup_{m<\omega} G_m$". If in addition $\forall i<n$ $T_{i+1} \leqslant^n T_i$, then $S \leqslant^n T$.

Let $X = \{f \in \omega^\omega\colon \forall n\, f(n) < 2^{n^2}\}$. Suppose $T \Vdash$"$\tau \in X$", then using Lemma 10(a) build a sequence $T^{n+1} \leqslant^n T^n, T^0 = T, G^n \subseteq \omega$ with $\mathrm{card}\, G^n \leqslant 2^{n+1}$, and $T^{n+1} \Vdash$"$\tau(n) \in G_n$". Let $S = \bigcap_{n<\omega} T^n$, so $S \in \mathbb{P}$ by Lemma 9, and $S \Vdash$"$\forall n \tau(n) \in G_n$". But $C = \{f \in X\colon \forall n\ f(n) \in G_n\}$ is closed nowhere dense in X. Thus if G is $\mathbb{P}$-generic over M, then

$$M[G] \vDash \text{“}X = \cup \{C\colon C \text{ closed nowhere dense in } X \text{ and coded in } M\}\text{”}.$$

But X is homeomorphic to 2^ω, so

$$M[G] \Vdash \text{“}2^\omega = \cup \{C\colon C \text{ closed nowhere dense in } 2^\omega \text{ and coded in } M\}.\text{”}$$

We will do a Laver style iteration argument (LAVER, 1976). Assume for each $\alpha < \omega_2$ we have a partial order $\mathbb{P}_\alpha$ and a term $\langle C_\beta^\alpha\colon \beta<\omega_1\rangle$ so that $\Vdash_\alpha$"$\langle C_\beta^\alpha\colon \beta<\omega_1\rangle$ are disjoint nowhere dense closed subsets of 2^ω". Then for each $\alpha \leqslant \omega_2$ [$p \in \mathbb{P}_\alpha$ iff $\forall \beta < \alpha\ p \restriction_\beta \Vdash$"$p(\beta) \in \mathbb{P}(\langle C^\beta_\gamma\colon \gamma < \omega_1\rangle)$" and for all but countably many γ (called the support of p) $p(\gamma)$ is a canonical term for $2^{<\omega}$. Lemma 5 thru Lemma 10 of LAVER (1976) are proved in this case mutatis mutandis. (Change Lemma 6(i) to read: If $k < \omega$ and $p \Vdash$"$W_{j<k} \Theta_j$", then there is an $I \subseteq \{0, 1, \ldots, k-1\}$ with card $I \leqslant 2^{(n+1)i}$ and a p' such that $p' \leqslant^n_F p$ and $p' \Vdash$"$W_{j \in I} \Theta_j$." Also $\leqslant$ is reversed in LAVER (1976).)

In particular for any G $\mathbb{P}_{\omega_2}$-generic over M, $M[G] \models$"$\forall x \in \omega^\omega$ if $\forall n$ $x(n) < 2^{n^4}$, then $\exists g \in M \forall n$ card $g(n) \leqslant 2^{n^3}$ and $\forall n$ $x(n) \in g(n)$". Hence as above $M[g] \models$"2^ω is the ω_1 union of meager sets". Also there is a sequence $\langle W_\beta\colon \beta < \omega_2\rangle$ in M such that for each β, W_β is dense in $\mathbb{P}_\beta$ and card$(W_\beta / \equiv) \leqslant \aleph_1$. So by a bookkeeping argument we can insure that $M[G] \models$"For every sequence C_α: $\alpha < \omega_1$ of closed disjoint nowhere dense subsets of 2^ω, $\exists \beta < \omega_2 \langle C_\alpha\colon \alpha < \omega_1\rangle = \langle C^\beta_\alpha\colon \alpha < \omega_1\rangle$."

Remark. It easily follows from arguments similar to those above that no real in $M[G]$ is random over M, so $M[G] \models$"2^ω is the ω_1 union of measure zero sets" (see MILLER (1980)).

Tall remarks that Booth (1968, unpublished) proved that MA implies the closed unit interval is not the union of less than $|2^\omega|$ disjoint nonempty closed sets, and Weiss (1972, unpublished) rediscovered this and proved, for example, that MA implies no compact perfectly normal space is the union of κ many disjoint closed sets for any κ with $\omega < \kappa < |2^\omega|$.

References

FREMLIN, D. H., and S. SHELAH

[1980] On partitions of the real line, to appear.

HAUSFORFF, F.

[1934] Summen von $\aleph_1$ Mengen, *Fund. Math.*, **26**, 241–255.

KURATOWSKI, K.

[1958a] *Topology, Vol. 1* (Academic Press, New York).

[1958b] *Topology, Vol. 2* (Academic Press, New York).

LAVER, R.

[1976] On the consistency of Borel's conjecture, *Acta Math.*, **137**, 151–169.

MILLER, A.

[1979] On the length of Borel hierarchies, *Ann. Math. Logic*, **16**, 233–267.

[1980] Some properties of measure and category, to appear.

ROYDEN, H.
[1968] *Real Analysis* (Macmillan, New York).
SACKS, G.
[1971] Forcing with perfect closed sets, in: *Proc. Symp. in Pure Mathematics, Vol. 13, Part I* (Am. Math. Soc., Providence, RI).
STERN, J.
[1977] Partitions of the real line into F_σ or G_δ subsets, *C.R. Acad. Sci. Paris Sér. A*, **284** (16), 921–922.
SOLOVAY, R. and S. TENNENBAUM
[1971] Iterated Cohen extensions and Souslin's problem, *Ann. Math.*, **94**, 201–245.

This page intentionally left blank

J. Barwise, H. J. Keisler and K. Kunen, eds., *The Kleene Symposium*

Computational Characterization of Abelian Groups

*Peter M. Winkler**

Department of Mathematics, Emory University, Atlanta, GA 30322, U.S.A.

Dedicated to professor S.C. Kleene on the occasion of his 70th birthday

Abstract: A language-invariant characterization of abelian groups is obtained by showing that a group is abelian if and only if there is a finite machine which can compute any term in the group, in one pass of the variables.

In an algebra, i.e. a structure for a language with no relation symbols, a notion of "computation" may be obtained by considering successive applications of the algebra's operations upon variables. Thus, in a group, multiplying x_1 by x_2, inverting, and then multiplying by x_3 would constitute a computation. Each computation ends in a term (in this case, $(x_1x_2)^{-1}x_3$); computations ending in equivalent terms will be called equivalent.

While this notion is very primitive compared with, say, recursive algorithms, it does admit a classification in terms of space. This classification, which is analogous to computability by finite-state automaton, is used below to obtain a characterization of abelian groups which is language-invariant, i.e. independent of the operations used to present the group.

Specifically, an algebra A will be termed *one-pass* if there exists a finite number k, such that any computation in the language of A is equivalent in A to one which can be performed in the following manner: variables are input in subscript order, and it is never necessary to store more than k terms at one time. Thus, essentially, A is one-pass if a calculator could be designed which can perform any computation in A with just one pass of the data.

In WINKLER (1980) a more rigorous definition is given for this property, along with an algebraic equivalent; but the present definition will suffice for the application here. It turns out that the property of being one-pass behaves like a language-invariant form of commutativity; the one-pass semigroups, for example, include commutative and permutative semigroups. The theorem below tends to support the position that there is no good generalization for commutativity in a group.

*Partially supported by an Emory University summer research grant.

It will be convenient henceforth to use additive notation for groups, even when non-abelian. The following technical result will be useful.

Lemma. *Let G be an additive group with center Z, and let $\bar{a}=(a_0,a_1,\dots,a_k)\in G^{k+1}$. Suppose that for all $\bar{y}=(y_1,\dots,y_k)\in G^k$,*

$$a_0+y_1+a_1+y_2+\cdots+a_{k-1}+y_k+a_k=y_1+y_2+\cdots+y_k.$$

Then each $a_i\in Z$.

Proof. By induction on k, the $k=0$ case being vacuously true. From the above we have that for all $\bar{y}$,

$$(-(y_1+y_2+\cdots+y_{k-1})+a_0+y_1+a_1+\cdots+y_{k-1}+a_{k-1})+y_k$$
$$=y_k+(-a_k);$$

it follows that $a_k\in Z$ and for all $y_1,\dots,y_{k-1}$,

$$-(y_1+y_2+\cdots+y_{k-1})+a_0+y_1+a_1+\cdots+y_{k-1}+a_{k-1}$$
$$=-a_k.$$

But then

$$a_0+y_1+a_1+\cdots+a_{k-2}+y_{k-1}+(a_{k-1}+a_k)=y_1+\cdots+y_{k-1}.$$

Thus, by the induction hypothesis, each $a_i\in Z$ for $i\leqslant k-2$ and $a_{k-1}+a_k\in Z$; but $a_k\in Z$ and $(a_{k-1}+a_k)-a_k=a_{k-1}\in Z$, proving the lemma.

Theorem. *A group is abelian if and only if it is one-pass.*

Proof. If G is abelian, any computation will lead to a term which can be written in the form

$$m_1x_1+m_2x_2+\cdots+m_nx_n.$$

Any such computation can be done by computing successively x_1, $2x_1,\dots$, m_1x_1, $m_1x_1+x_2$, $m_1x_1+2x_2$, etc., additively inverting variables with negative coefficients. This can be accomplished with just two or three storage locations, so G is one-pass.

Conversely, suppose G is an additive one-pass group and fix the storage limit k. Let $\bar{x}$ stand for the $(k+1)$-tuple $(x_0,x_1,\dots,x_k)$ and consider the term

$$x_0+x_{k+1}+x_1+x_{k+2}+\cdots+x_{2k}+x_k.$$

In order to compute this term in one pass with only k storage registers, the $k+1$ variables $x_0,x_1,\dots,x_k$ must survive in storage as at most k terms

$$t_1(\bar{x}),t_2(\bar{x}),\dots,t_k(\bar{x})$$

at the point when the variable x_{k+1} is about to be input. It follows that an identity of the form

$$x_0 + x_{k+1} + x_1 + \cdots + x_{2k} + x_k = t(t_1(\bar{x}), \ldots, t_k(\bar{x}), x_{k+1}, \ldots, x_{2k})$$

must hold in G, where t and the t_i's are terms in the language of G. We claim this can be so only if G is abelian.

The terms t_i may be applied to $(k+1)$-tuples from any additive group, in particular the additive group $\mathbf{Q}$ of rational numbers. The map

$$(r_0, \ldots, r_k) \mapsto (t_1(r_0, \ldots, r_k), \ldots, t_k(r_0, \ldots, r_k))$$

is then a linear transformation from $\mathbf{Q}^{k+1}$ to $\mathbf{Q}^k$, regarded as vector spaces over $\mathbf{Q}$; and there must be a non-zero vector $(s_0, s_1, \ldots, s_k)$ in its kernel. By normalizing we may assume the s_i's are integers whose greatest common divisor is 1; we may then choose integers $n_0, \ldots, n_k$ so that

$$n_0 s_0 + n_1 s_1 + \cdots + n_k s_k = 1.$$

Let g be an arbitrary element of the group G and let $\bar{s}g$ stand for $(s_0 g, s_1 g, \ldots, s_k g)$. Since the subgroup generated by g is abelian, we have

$$t_i(\bar{s}g) = t_i(\bar{s})g = 0$$

for each i; hence for all $y_1, \ldots, y_k$,

$$\begin{aligned} s_0 g + y_1 + s_1 g + y_2 + \cdots + s_{k-1} g + y_k + s_k g \\ &= t(t_1(\bar{s}g), \ldots, t_k(\bar{s}g), y_1, \ldots, y_k) \\ &= t(0, \ldots, 0, y_1, \ldots, y_k) \\ &= t(t_1(0, \ldots, 0), \ldots, t_k(0, \ldots, 0), y_1, \ldots, y_k) \\ &= y_1 + y_2 + \cdots + y_k. \end{aligned}$$

By the lemma, each $s_i g$ is in the center of G, but then so is

$$\begin{aligned} c_0(s_0 g) + c_1(s_1 g) + \cdots + c_k(s_k g) &= (c_0 s_0 + \cdots + c_k s_k) g \\ &= 1g = g. \end{aligned}$$

We have proved that every element of G lies in the center, so G is abelian.

Reference

WINKLER, P.

[1980] Classification of algebraic structures by work space, *Algebra Universalis*, to appear.

This page intentionally left blank

5296473R00259

Printed in Germany
by Amazon Distribution
GmbH, Leipzig